AF725023

THE ROUTLEDGE HANDBOOK OF APPLIED CLIMATE CHANGE ETHICS

The *Routledge Handbook of Applied Climate Change Ethics* is a powerful reference source for the identification and exploration of the underlying ethical issues in climate change law and policy. Bridging theory with practice, it takes ethical engagement out of the classroom and into the halls of governance.

The *Handbook*'s 39 chapters – written by a diverse and inter-disciplinary team of experts from around the world – are case studies divided into five parts. Parts I–IV highlight the ethical issues that arise in climate change policy formation, from duties not to harm to duties to consider the views and voices of those who will be, or are being, harmed; from the role of human rights, justice, and democracy to how to identify and respond to disinformation and denialism. It also raises the ethics of various policy responses, such as cap-and-trade, carbon taxing, and geo-engineering. Part V offers a way forward, with strategies on how to expressly consider ethics in climate change policy formation, from negotiations to education, media, communication, and the power and potential of shaming.

The volume is essential reading for students, professors, and practitioners who wish to better engage with government and non-government organizations on climate policy, to better understand the practical application of the theory and philosophy of ethics, and to more strongly draft and defend ethical action in negotiating, drafting, and defending climate change law and policy.

Donald A. Brown is Scholar in Residence for Sustainability Ethics and Law at Widener University Commonwealth Law School in Harrisburg, USA. He has taught or lectured on climate change ethical/policy issues in 38 countries approximately 80 times. He also manages the award-winning website ethicsandclimate.org which contains over 200 articles and videos on climate ethics.

Kathryn Gwiazdon is Professor of Public International Law, Climate Change Law and Policy, and Human Rights Law at Northern Illinois University College of Law, USA. She is also the Executive Director of the Center for Environmental Ethics and Law, as well as Chair of the Ethics Specialist Group of the IUCN World Commission on Environmental Law.

Laura Westra is Professor Emeritus from the Department of Philosophy at the University of Windsor in Windsor, Canada. She is also Founder and Director of Global Ecological Integrity Group, a research group dedicated to pushing the boundaries of scholarly endeavor through inter- and trans-disciplinary engagement on matters affecting and governing the sustainability of life for both present and future generations.

ROUTLEDGE HANDBOOKS IN APPLIED ETHICS

Applied ethics is one of the largest and most diverse fields in philosophy and is closely related to many other disciplines across the humanities, sciences and social sciences. Routledge Handbooks in Applied Ethics are state-of-the-art surveys of important and emerging topics in applied ethics, providing accessible yet thorough assessments of key fields, themes, thinkers, and recent developments in research.

All chapters for each volume are specially commissioned, and written by leading scholars in the field. Carefully edited and organized, Routledge Handbooks in Applied Ethics provide indispensable reference tools for students and researchers seeking a comprehensive overview of new and exciting topics in applied ethics and related disciplines. They are also valuable teaching resources as accompaniments to textbooks, anthologies, and research-orientated publications.

Also available:

THE ROUTLEDGE HANDBOOK OF ETHICS AND PUBLIC POLICY
Edited by Annabelle Lever and Andrei Poama

THE ROUTLEDGE HANDBOOK OF ANIMAL ETHICS
Edited by Bob Fischer

THE ROUTLEDGE HANDBOOK OF FEMINIST BIOETHICS
Edited by Wendy A. Rogers, Jackie Leach Scully, Stacy M. Carter, Vikki Entwistle, and Catherine Mills

THE ROUTLEDGE HANDBOOK OF PHILOSOPHY OF PUBLIC HEALTH
Edited by Sridhar Venkatapuram and Alex Broadbent

THE ROUTLEDGE HANDBOOK OF THE ETHICS OF HUMAN ENHANCEMENT
Edited by Fabrice Jotterand and Marcello Ienca

THE ROUTLEDGE HANDBOOK OF APPLIED CLIMATE CHANGE ETHICS
Edited by Donald A. Brown, Kathryn Gwiazdon, and Laura Westra

For more information about this series, please visit: https://www.routledge.com/Routledge-Handbooks-in-Applied-Ethics/book-series/RHAE

THE ROUTLEDGE HANDBOOK OF APPLIED CLIMATE CHANGE ETHICS

Edited by
Donald A. Brown, Kathryn Gwiazdon, and Laura Westra

Routledge
Taylor & Francis Group
NEW YORK AND LONDON

Designed cover image: © Getty Images

First published 2024
by Routledge
605 Third Avenue, New York, NY 10158

and by Routledge
4 Park Square, Milton Park, Abingdon, Oxon, OX14 4RN

Routledge is an imprint of the Taylor & Francis Group, an informa business

© 2024 Taylor & Francis

The right of Donald A. Brown, Kathryn Gwiazdon, and Laura Westra to be identified as the authors of the editorial material, and of the authors for their individual chapters, has been asserted in accordance with sections 77 and 78 of the Copyright, Designs and Patents Act 1988.

All rights reserved. No part of this book may be reprinted or reproduced or utilised in any form or by any electronic, mechanical, or other means, now known or hereafter invented, including photocopying and recording, or in any information storage or retrieval system, without permission in writing from the publishers.

Trademark notice: Product or corporate names may be trademarks or registered trademarks, and are used only for identification and explanation without intent to infringe.

Library of Congress Cataloging-in-Publication Data
A catalog record for this title has been requested

ISBN: 978-0-367-48458-3 (hbk)
ISBN: 978-1-032-56364-0 (pbk)
ISBN: 978-1-003-03986-0 (ebk)

DOI: 10.4324/9781003039860

Typeset in Bembo
by codeMantra

To all those who make good trouble.

CONTENTS

CONTRIBUTORS

Hugh Breakey is Senior Research Fellow in Moral Philosophy at Griffith University. His research spans political theory, applied ethics, and legal studies, exploring the ethical challenges in diverse fields including peacekeeping, argument, climate change, sustainable tourism, private property, and international law.

Donald A. Brown is Scholar in Residence for Sustainability Ethics and Law, Widener University Commonwealth Law School, where he teaches International Environmental Law, Human Rights Law, and Climate Change Law. In November 2019, he was awarded by UNESCO in Paris the Avicenna Award for Excellence in Ethics and Science for his work on climate change ethics. Among other books, he is the author of *Navigating the Perfect Moral Storm: Climate Change Ethics in Light of a Thirty-Five Year Debate* (Routledge-Earthscan 2012).

Nicholas S. Bryner is Associate Professor of Law and holds the John P. Laborde Professorship in Energy Law at Louisiana State University's Paul M. Herbert Law Center. His research includes environmental law, public lands, climate change, and administrative law, with a focus on the United States and Latin America.

Peter D. Burdon is Associate Professor at the Adelaide Law School and a delegate for the National Tertiary Education Union. He is also Associate Dean (Learning and Teaching) for the Adelaide Law School. Peter is an expert in environmental theory and has written and edited books on Earth Jurisprudence and Earth Democracy. In 2017, he published *Hannah Arendt: Legal Theory and the Eichmann Trial* with Routledge.

Carl F. Cranor is Distinguished Professor of Philosophy and Faculty Member, Environmental Toxicology, at the University of California, Riverside. He has authored several books on the subject of law and ethical issues in regulating toxic substances, notably *Regulating Toxic Substances: A Philosophy of Science and Law* (Oxford University Press 1993), *Toxic Torts: Science, Law, and the Possibility of Justice* (Cambridge University Press, 2nd edition 2016), and *Tragic Failures: How and Why We Are Harmed by Toxic Chemicals* (Oxford University Press 2017).

Thiago de Araújo Mendes is a Postdoctoral Fellow in the Department of International Relations at PUC-Minas and former Secretary of Climate Change and Forests at the Ministry of Environment of Brazil. He is also the former UNFCCC Chair of CGE and Warsaw International

Mechanism on Loss & Damage. He has authored *From the Citizen's Constitution to REDD+: Domestic and International Policy Instruments Intersecting Climate Change and Forestry in Brazil from 1988 to 2018* (Revista Brasileira de Políticas Públicas e Internacionais 2019).

Cristiane Derani is Professor of International, Environmental, and Economic Law at Universidade Federal de Santa Catarina, Brazil, Fellow at the Center of Environment, Energy, and Natural Resources Governance at the University of Cambridge, Visiting Scholar at the Pantheon Sorbonne Paris 1, Expert of the United Nations Harmony with Nature program, Deputy Chair of the Ethics Specialist Group of the IUCN World Commission on Environmental Law, and Founding Member of the Ecological Law and Governance Association.

José Domingos Gonzalez Miguez is PhD candidate in International Relations at PUC-Minas and former Secretary of Climate Change and Forests at the Ministry of the Environment of Brazil. He is recognized for contributing substantially to the IPCC Nobel Prize and has authored "Equity, Responsibility and Climate Change," in *Ethics, Equity and International Negotiations on Climate Change* (Elsevier 2002).

Nigel Dower is Senior Lecturer Emeritus in Philosophy at the University of Aberdeen where he taught until 2004. Since then, he has pursued his interests in exploring ethics in a globalized world and has contributed to the voluntary sector (e.g., Aberdeen United Nations Association and the Food Ethics Council). His main research interests are in global ethics and the ethics and philosophy of development and environment. His publications include *World Ethics – the New Agenda* (Edinburgh University Press, 2nd edition 2007), *An Introduction to Global Citizenship* (Edinburgh University Press 2003), and *The Ethics of War and Peace* (Polity 2009).

Stephen M. Gardiner is Professor of Philosophy at the University of Washington, Seattle. His books include *A Perfect Moral Storm: The Ethical Tragedy of Climate Change* (Oxford University Press 2011), *Debating Climate Ethics* (Oxford University Press 2016), *The Ethics of "Geoengineering" the Global Climate* (Routledge 2020), and *Dialogues on Climate Justice* (Routledge 2022).

Geoffrey Garver teaches at McGill and Concordia Universities in Montreal and coordinates law and governance research for the Leadership for the Ecozoic program (http://www.l4ecozoic.org) of McGill University and the University of Vermont. His book *Ecological Law and the Planetary Crisis* was published in 2021 (Routledge).

Giada Giacomini is a Postdoctoral researcher at the Geneva Graduate Institute, Centre for International Environmental Studies (CH), and Max Weber Fellow at the European University Institute (Italy). She is working, respectively, on the accountability of international organizations and just transition litigation, and on environmental conservation and non-anthropocentric law. She recently published her monograph *Indigenous Peoples and Climate Justice: A Critical Analysis of International Human Rights Law and Governance* with Palgrave Macmillan (2022).

Maron Greenleaf is a sociocultural anthropologist, political ecologist, and legal scholar studying climate change, forests, and green economies. Currently, Maron is completing an ethnographic book manuscript on carbon offsets in the Brazilian Amazon and is starting new research on reforestation in post-industrial and post-agricultural landscapes. She also co-founded Dartmouth's Energy Justice Clinic, in which she works with community partners and Dartmouth students to understand and support socially just transitions to renewable energy.

Kathryn Gwiazdon is Executive Director of the Center for Environmental Ethics and Law, Chair of the Ethics Specialist Group of the IUCN World Commission on Environmental Law, and Professor of International Law, Climate Change Law and Policy, and Human Rights Law at Northern Illinois University College of Law. Recent publications include *The Law and Politics of Ecological Integrity* (Westra, Gwiazdon, Garver, eds., Springer 2024) and "A Crisis of Identity: the UN Sustainable Development Goals within an Unsustainable Law and Governance Framework" in *Ecological Limits of Development: Living with the Sustainable Development Goals* (Routledge 2021).

Johan Hattingh is Emeritus Professor of Philosophy at Stellenbosch University, South Africa, and former Dean of the Faculty of Arts and Social Sciences at the same university. In 2016, he was the President of the Ad Hoc Expert Group that drafted UNESCO's *Declaration of Ethical Principles in Relation to Climate Change* (2017).

Rainier A. Ibana teaches Ethics and Environmental Ethics at Ateneo de Manila University in the Philippines. He is Editor of *Budhi: A Journal of Ideas and Culture*. He served as UNESCO's primary educator of its MOOC course on "Climate Change Ethics: Lessons from the Global South" (FutureLearn 2016) and chaired UNESCO's COMEST Working Group on Environmental Ethics (2013–2018). The insights in the papers for this collection were initially developed in "The Ethical Implications of Applying the Laws of Thermodynamics and Biology's Ten Percent Law to Climate Change and Biodiversity" for the *International Social Science Journal* (2013).

Bruce Jennings is Senior Fellow at the Center for Humans and Nature and Adjunct Associate Professor of Health Policy in the Center for Biomedical Ethics and Society at Vanderbilt University. He has published widely on environmental ethics and political theory, bioethics, and public health. His most recent books are *Ecological Governance: Toward a New Social Contract with the Earth* (West Virginia University Press 2016); *Ethics,* Volume 4 in the *Encyclopedia of the Anthropocene,* 5 vols. (Elsevier 2017); and *Liberty and the Ecological Crisis: Freedom on a Finite Planet* (Routledge 2020).

Shi Jun is Professor at the Research Center for Green Development, Jinan University, China. He has published the following books in Chinese: *Chinese Value Position on International Climate Cooperation* (2016) and *Nature and Morality: An Ethical Reflection on Climate Change* (2014).

Kirk W. Junker is University Professor of Law, holds the Chair of US Law, is the Director of the Environmental Law Center, and is Head of the Examination Board for the International Master of Environmental Sciences at the University of Cologne, Germany. Recent publications include *Environmental Law Across Cultures: Comparisons for Legal Practice* (Routledge 2019), *US Law for Civil Lawyers* (CH Beck/Nomos/Hart 2021), and *Globalization, Environmental Law, and Sustainable Development in the Global South: Challenges for Implementation* (Routledge 2022).

Workineh Kelbessa is Professor of Philosophy at Addis Ababa University, Ethiopia. He is the author of *Traditional Oromo Attitudes towards the Environment: An Argument for Environmentally Sound Development* (2001) and *Indigenous and Modern Environmental Ethics: A Study of the Indigenous Oromo Environmental Ethic and Modern Issues of Environment and Development* (2011). His most recent publication is "African Worldviews, Biodiversity Conservation and Sustainable Development," 2022. His work focuses on environmental philosophy, African philosophy, comparative philosophy, development ethics, climate ethics, globalization, philosophy of love and sex, and Indigenous knowledge.

Bridget Lewis is Associate Professor in the School of Law at the Queensland University of Technology (Brisbane, Australia) where she researches environmental human rights. Her book *Environmental*

Human Rights and Climate Change: Current Status and Future Prospects (Springer 2018) critically examines the concept of a human right to a healthy environment and its relevance to the climate crisis. She is currently writing a monograph which explores the environmental rights of future generations.

Brendan Mackey is Professor and Director of the Griffith Climate Action Beacon, Griffith University, Queensland Australia. Brendan has published extensively in the fields of forest ecosystems, climate change, biodiversity, ethics, and sustainability. He was a coordinating lead author for the Intergovernmental Panel on Climate Change 6th Assessment Report 2022.

Michelle Maloney is an environmental lawyer and Earth-centered systems change advocate. She is Co-Founder and National Convenor of the Australian Earth Laws Alliance (AELA), Co-Director of the Indigenous led initiative Future Dreaming Australia, and Adjunct Senior Fellow at Griffith University. Michelle has more than 30 years' experience designing, managing, and evaluating systems change programs aimed at creating a regenerative, sustainable future.

Karla Matos is a Leadership on Environment and Development (LEAD) fellow, journalist and PhD candidate at the University of Lisbon, with a Master's degree in Social Ecology and more than 25 years of experience in environmental policy at the local, national, and international levels, including with the Ministry of Environment of Brazil as the Coordinator of their National Agenda 21 program. She currently serves on the Board of the Center for Environmental Ethics and Law (CEEL) and the Advisory Board of Pronatura Brazil.

Maria Silvia Muylaert de Araujo is a member of the IPCC/Working Group III to the 5th Assessment Report (AR5) of the Intergovernmental Panel on Climate Change (IPCC), Lead Author for Chapter 4 on "Sustainable Development and Equity," as well as a member of the IPCC/Working Group II to the 6th Assessment Report (AR6) of the IPCC, Lead Author for Chapter 18 on "Climate Resilient Development Pathways" and "Global to Regional Atlas." She is also a member of the Collaborative Program of the Ethical Dimensions of Climate Change coordinated by Rock Ethics Institute, Penn State University.

Alessandro Pelizzon is Associate Professor in the School of Law and Society at the University of the Sunshine Coast, a founder and current Executive Committee member of the Global Alliance for the Rights of Nature, and an expert of the UN Harmony with Nature program. His most recent publication is *Ecological Jurisprudence* (Edinburgh University Press, forthcoming).

Benjamin J. Richardson is Professor of Environmental Law at the University of Tasmania, Australia, and previously has worked in universities in Canada, England, the Netherlands, and New Zealand in advancing research and learning on corporate social responsibility, legal philosophy, Indigenous legal issues, and environmental governance. He is also active in the climate justice movement, including Extinction Rebellion.

Philip Robichaud is Assistant Professor of Ethics at the Vrije Universiteit Amsterdam. His areas of research specialization are moral responsibility, value theory, climate ethics, and the ethics of behavioral interventions such as nudges. He co-edited with Jan Willem Wieland the edited volume *Responsibility: The Epistemic Condition* and his journal publications have appeared in *Ethics*, *Philosophy and Phenomenological Research*, *The Monist*, and *Science and Engineering Ethics*.

Nicole Rogers is Professor of Climate Law at the Faculty of Law, Bond University, Queensland Australia. Nicole has published widely in the areas of wild law and interdisciplinary climate studies. From 2014 to 2017, she co-led the Wild Law Judgment project and is co-editor of *Law as if Earth Really Mattered: The Wild Law Judgment Project* (Routledge 2017). She is currently co-leading the Anthropocene Judgments project.

Azar Safari is Business and Human Rights Specialist and PhD researcher at the University of Amsterdam.

David Storey is Associate Professor of Practice in the Philosophy Department at Boston College, where he teaches courses on the history of philosophy, ethics, theology, technology, and climate change. He has authored *Naturalizing Heidegger: His Confrontation with Nietzsche, His Contributions to Environmental Philosophy* (SUNY 2015) and has published essays in *Environmental Ethics, Comparative and Continental Philosophy*, and *The New Atlantis*. In addition, he hosts 'Wisdom at Work: Philosophy Beyond the Ivory Tower,' a podcast that interviews philosophers working in the wild outside the academy.

Behnam Taebi is Professor of Energy and Climate Ethics and Scientific Director of the Safety and Security Institute at Delft University of Technology.

Prue Taylor is Deputy Director of the New Zealand Centre for Environmental Law and Senior Lecturer, School of Architecture and Planning, University of Auckland, New Zealand.

J. Timmons Roberts is Ittleson Professor of Environmental Studies and Sociology at Brown University and the Executive Director of the Climate Social Science Network. He has taught in and directed environmental science and policy programs at Brown, the College of William and Mary, and Tulane University. He recently served on the Board on Environmental Change and Society of the National Academy of Sciences and was awarded the Frederick Buttel Awards for distinguished scholarship from the American Sociological Association and the International Sociological Association.

Dina Lupin is Lecturer in the School of Law at the University of Southampton in the United Kingdom and a post-doctoral researcher in the Department of Philosophy at the University of Vienna. Her latest book, *Human Dignity and the Adjudication of Environmental Rights*, was published by Edward Elgar Publishing in 2020. She also serves as the Director of the Global Network for Human Rights and the Environment.

Ngozi Unuigbe is Professor at the University of Benin in Nigeria and an expert in International Environmental Law, Policy, and Ethics, with a research bias toward interrelated themes arising from traditional ecological knowledge's role in preserving Nature and addressing climate change and insecurity. She is a member of several organizations that focus on preserving Nature's rights and promoting ecological restoration such as United Nations Harmony with Nature, Global Ecological Integrity Group, Ecological Law and Governance Association, Earth Law Alliance, and the Center for Environmental Ethics and Law.

Steve Vanderheiden teaches political theory and environmental politics at the University of Colorado at Boulder. He is author of *Atmospheric Justice: A Political Theory of Climate Change* (Oxford University Press 2008), which won the Howard and Margret Sprout Award for the best book on international environmental politics, as well as *Environmental and Climate Justice* (Oxford University Press 2016).

Mirian Vilela is Executive Director of the Earth Charter International Secretariat and Earth Charter Center for Education for Sustainable Development at the University for Peace. She is also Coordinator and UNESCO Chair on Education for Sustainable Development with the Earth Charter.

ACKNOWLEDGMENTS

Donald A. Brown:

I have been lucky and honored to work with colleagues from around the world, many of whom I first met through annual meetings of the Global Environmental Integrity Group (GEIG). This group, founded by Laura Westra, uniquely held annual meetings comprising world class ethicists, scientists, economists, and lawyers. This is how I first met my co-editor, Kathryn Gwiazdon. I cannot think of a better experience for working on international environmental governance.

Kathryn Gwiazdon:

I would like to thank all of the authors and readers who see the power and potential of ethical arguments to move climate action, to Ron Engel for bringing me into this world of global ethics, and to Donald A. Brown for his inspiring and passionate commitment to use ethics to move hearts and minds. I would also like to thank Jordan, Evelyn, and Calvin for their love.

INTRODUCTION

Donald A. Brown and Kathryn Gwiazdon

Climate change is a multi-faceted crisis that humanity is failing to address. It is a crisis of existence: it threatens life on Earth – humans, nature, and the foundations of life. It is a crisis of governance: the scale and scope of its harms at this point in history is the result of choices by decision-makers, by actions and inaction in domestic and international law and policy. And it is a crisis of ethics: its harms, and the responses to it, are unequal and unjust. This *Handbook* seeks to address and confront these crises by looking at the too-often ignored aspect of climate change that could lead to powerful changes in law and policy: ethics. What are the ethics of its harms, the ethics of individual, state, and international response and responsibility, and the ethics of those who seek to prevent a response? And how can identifying the ethical issues, and responding with ethical arguments, inform and strengthen our current response – at all governance levels, with private and public actors?

Ethics is the inquiry into right and wrong behavior and when responsibility attaches. As such, it is the foundation of justice itself. What behavior harms others, what behavior helps or heals others, and who is responsible for the harm and the healing? It is not just some abstract philosophical debate for the ivory towers of academia, it is a *living* act, witnessed in our actions and inactions every day. And the value of this *living* act – of ethics alive and ethics applied – is even more important when considering the actions of leaders who make choices that affect others, who have a responsibility for their citizens and our interdependent, interconnected lives with one another and with the land upon which we depend. Do these leaders advance the common good? Do they honor their oaths to the rule of law and to their constituents as democratic representatives? Do they protect and advance human rights, truth, and justice? Or do they not? Do they harm? Do they hold themselves accountable for their harm? Do they deny, delay, or misinform? Applied climate change ethics addresses this all. It is the inquiry into individual, collective, private, and public action on highlighting and addressing harmful behavior as related to climate change. It looks at how civil society, government (and governments), and private actors choose to act to prevent, address, and remedy harms from climate change.

Climate change is a governance and ethics issue precisely because it is a justice issue. It is a civilization challenging problem that threatens life on Earth, but those most responsible for causing the problem are not those who are feeling its harms or feeling its harms most severely. In addition, those made most vulnerable to climate change, as well as those fighting alongside them, have little power in petitioning their governments, or the international community, to act more strongly or more urgently to prevent further harm or redress existing or historic harms. And even if processes

DOI: 10.4324/9781003039860-1

are in place – through legislation or the courts – the end result remains the same: civil society disempowerment, private industry empowerment, and state and global inaction.

The state is the powerholder when governing domestic pollution regulation and negotiating international agreements that could address climate change, and so, due to the institutional arrangements of state and global governance, it is their individual and collective action that ultimately determines what national and international action on climate change is advanced – or obstructed – and therefore what harm is being done to their nation and to their citizens, and by their nation and to other nations and citizens. Fundamental tenets of human rights and social contract theory advance that the purpose of a state is to protect its people from existential threats, represent majority will in law and governance, and advance institutions of justice, with proper checks and balances to prevent tendencies toward authoritarianism, oligarchy, and corruption; yet, with climate change, states are failing at every aspect of this most basic understanding of their purpose and responsibility. They deny the seriousness of climate change. They deny its existence. They deny their responsibility for the harms, or their responsibility for the harms that they are causing to others. Or they say or write beautiful words on crisis and responsibility, and then do little to bind themselves or other nations to act in any meaningful way.

Yes, states *technically* have the power to act on climate change and states are *technically* the powerholders in national and international law and governance; so, why are they not acting on a known harm? This begs the question: what if something or someone else is disempowering them? State action is being molded, manipulated, and even determined by non-state actors. We are witnessing an intentional disempowerment of democracy – of public institutions – by private actors: the very same private actors who are responsible for causing the harms that climate change action would remedy; indeed, the harms that are the responsibility of the state to remedy. Therefore, we are not only witnessing a governance crisis due to unequal power in civil society representation and even among states in global negotiations; we are witnessing a governance crisis in the intentional disempowerment of democracy *as a methodology to obstruct climate action*. This crisis that threatens life on Earth also threatens the most powerful industry on Earth: the fossil fuel industry, and the unfettered free market ideology that supports it. The fossil fuel industry has vast global resources available to it to resist anything that threatens its economic power, including democracy itself. It is an industry responsible for the economic vitality of many regions in the world, and it is an industry responsible for national and global *insecurity*. Nations and people have been made dependent upon their products, and through targeted, well-funded campaigns aimed at weakening democratic governance, nations and people continue to be dependent upon it. In order to secure their future profits, they make our collective future insecure. If the international community cannot find a way to cooperate on global carbon emissions, mass human extinction will occur. This is the science; this is the truth. Yet this truth – and the law and policy that should follow this truth – is being intentionally destabilized by those who benefit from destabilization. For humanity, our collective life is at stake; for the fossil fuel industry, and those who profit from it, their personal wealth is at stake. And they are winning.

They have entrenched themselves in academic institutions, hidden themselves in civil society front groups, and have succeeded in undermining legislative and judicial checks on the power of private money in public decision-making. They have offered and proffered, and manipulated, a language of science and economics and applied it to issues of justice and ethics, and not only have lawyers and judges and politicians accepted this language, but so have climate change advocates. Indeed, much of the public debate about climate change has been focused on issues created by those who oppose government policies on climate change. For example, they claim that governments should not act because of 'scientific uncertainty' about climate change or unjustifiable 'costs' to civil society or particular entities. This instrumental reasoning approach intentionally conceals the ethical dimensions of their arguments by making what appear to be 'value-neutral' factual claims. As a result, citizens and others engaged in climate policy issues are kept busy responding to factual assertions about science and economics instead of making more powerful claims about the ethics and justice issues involved, such as the duty not to harm or the accountability to change or remedy that harmful behavior. This is even more unfortunate because the international community has already agreed on certain ethical principles which undermine the instrumental argument. For instance, the UN Framework Convention on Climate Change requires that nations consider equity (i.e., fairness and justice) when setting their Nationally Determined Contributions. It also requires that nations apply the precautionary principle to climate action, which means that governments cannot use uncertainty to justify inaction on climate change.

Enlightenment philosophers were conscious of the fact that claims about what governments should do to achieve the common good could usually be reduced to a syllogism which contains factual premises and normative conclusions about what justice requires. Because of this, they believed that in order for citizens to fulfill their appropriate role in achieving the common good, they should be trained in disciplines that allow them to think critically about factual premises, and they should be trained in ethics to think critically about normative or justice conclusions. However, because of the dominance of technical issues of science and economics in climate policy controversies, practitioners rarely have the training or language needed to spot and evaluate the numerous ethical issues that arise in climate policy formation debates.

Ultimately, that is what this *Handbook* seeks to address. It aims to help citizens, practitioners, and politicians spot and evaluate ethical issues that arise in climate change policy formation, as well as provide a more powerful language to more effectively counter arguments based on instrumental rationality and self-interest. We believe that failing to account for ethics and justice in climate change law and governance is why climate change law and governance is failing. Those who obstruct climate change action disempower democracy because of its roots in ethics and justice, in human and collective rights, and in its requisite inquiry and dialogue into what is right and wrong when living together in one nation or on one planet. Injustice motivates; it has motivated the largest movements in human history, whether against racism, predation, or exploitation. And injustice is at the core of climate change harms. It is what creates them, and it is what continues them. We hope that this *Handbook* empowers advocates through the power of ethics and justice.

PART I

Ethical Issues That Arise in Governments' Greenhouse Gas Emissions Reduction Policy Formations

1

DUTIES OF NATION-STATES TO PREVENT ACTIVITIES WITHIN THEIR JURISDICTION THAT HARM PEOPLE AND NATIONS OUTSIDE THEIR BOUNDARIES

The Meaning of the No Harm Principle as Applied to International Relations

Nigel Dower

Introduction

This chapter examines the moral arguments for the 'no harm' principle in relation to climate change and what the implications are for the governments of nation-states and for their citizens, whether citizens are acting as individuals or through organizations. As well as identifying the various positive moral arguments that ought to have force with governments – both elected representatives and their civil servants – and with their citizens, it also explicitly identifies and answers two arguments that are often assumed by representatives of countries or their citizens why they do not have to do too much: first, in summary, that countries can also legitimately pursue their economic interests, partly because that is what their democratic electorates want; and second, that, as political realists argue, one's country comes first and the only real concern and responsibility of governments are to promote national interests. The argument will say that these are deeply flawed and inadequate ways of understanding our moral responsibilities.

It focuses in particular on the moral implications of the fact that in 1992 the United Nations Framework Convention On Climate Change (UNFCCC), negotiated at the Rio earth summit in 1992 and signed by virtually all countries, included the no harm principle (United Nations 1992: Preamble), and the fact that the Paris Agreement of 2015 provided targets to which countries were expected to make reductions commensurate with their economic circumstances in their GHG emissions, but broadly with the aim of holding the increase in the global average temperature to well below 2°C above pre-industrial levels and to make efforts to limit the temperature increase to 1.5°C above pre-industrial levels (Paris Agreement 2015). To accomplish this, every country must set a GHG emissions reduction target that represents the government's fair share of a remaining carbon budget that must constrain global carbon emissions to achieve the Paris Agreement's warming limit goals. Because delay in achieving a reduction target consumes some of the remaining carbon budget, reduction targets must identify both the amount of reduction and the date by which the reduction will be achieved.

DOI: 10.4324/9781003039860-3

The Preamble to the UNFCCC of 1992 says,

> Recalling also that States have, in accordance with the Charter of the United Nations and the principles of international law, the sovereign right to exploit their own resources pursuant to their own environmental and developmental policies, and the responsibility to ensure that activities within their jurisdiction or control *do not cause damage to the environment of other States or of areas beyond the limits of national jurisdiction.*
>
> *(emphasis added)*

This reflected the no harm principle that has been a widely recognized principle of customary international law whereby a State is duty-bound to prevent, reduce, and control the risk of environmental harm to other states (see also Voigt 2008). The UNFCCC principle is stronger than the customary international law principle because it talks of 'not causing' which is equivalent to 'preventing' and does not include merely reducing harm.

This is not a minor point, however, because it touches on a significant issue which we will have to consider more carefully later. There is after all a kind of tension between a thorough going no harm principle and what states are committed to doing under the Paris Agreement, which is to reduce their carbon impacts by certain amounts by certain dates. If it is acknowledged, as almost everyone would recognize now, that the levels of carbon emissions at current levels are contributing to harms now and will continue into the future until the target is fully met, it follows from this that no nation-state (particularly no economically advanced nation-state) can completely avoid causing levels of harm. So in endeavoring to act in the spirit of the no harm principle (UNFCCC 1992: Preamble) and acting earnestly to reduce carbon emissions according to certain targets and by certain dates (Paris Agreement 2015), it follows that, if we all collectively achieve this by progressively moving toward the basis of a brighter largely post-carbon future, then the various stages in that progression will have caused significant environmental harms, albeit gradually diminishing, and we have to accept that certain levels of environmental harm are acceptable. The challenge is that most countries governments and their citizens make it too easy for themselves in regard to what levels of environmental harm are in the meantime acceptable. The argument of this chapter is that the levels of acceptable harm should be understood as very low, and, if we understand the moral arguments correctly, then we will see why it is absolutely necessary to see them as very low. Complacency is not an option, as we are already on cusp of triggering catastrophe.

No Harm Principle as a Fundamental Basis of Morality

The principle that we should not do harm to other people is something that is at the core of almost any moral code or set of moral values (one has to add 'almost' because there may be very rare cases of very strange moral outlooks or indeed people without any moral outlook at all). People's views of what morality involves diverge considerably both as to the content of their values and principles and as to the source of justification (e.g., religious faith, reason, sentiment, or convention). But the idea that we should not harm one another is almost universally recognized. This is not surprising since without such a principle underlying moral rules about telling the truth, promise keeping, respecting property, and so on, society would not really be possible. This is not to say that the presumption not to harm may not be overruled by other moral considerations on occasion – it is not absolute – and historically the scope of morality has often been limited to members of one's own community or nation-state – so that not causing harm does not apply to people living outside one's community. But in the modern world, there has generally been a shift to recognizing that such a fundamental principle of morality should apply to all people everywhere and indeed not just to those living now but those who will be living in the future and may be harmed by the collective actions of the present (many would likewise extend the principle to include non-human life, for instance if it is accepted that we ought not to cause harm to higher animals including those whom we manage on farms).

To give an example: there may be considerable disagreement about whether we have general obligations to help others, e.g., those in poverty overseas, and if so to what extent, but there is less likely to be disagreement over the claim we should not deliberately harm others or do actions which we can reasonably foresee will harm others, other things being equal. That is why for instance in some ways the duty to prefer, where there is a choice, fairly traded goods over non-fairly or unfairly traded goods – that is, to prefer actions that avoid rather than contribute to, as the case may be, harms that farmers may endure elsewhere – is more significant than a duty to give aid. In the former case, we are avoiding or reducing actually contributing to a harm; in the latter case, we are merely failing to prevent or mitigate a harm caused by others or natural circumstances.

The example of fair trade serves also to illustrate a more general point. The basic moral principle of not harming one another (either intentionally or unintentionally but with an awareness of the real possibility) of course applies to many areas of our lives not merely to climate impacts. It is not merely that, as implied above, telling lies, promise breaking, stealing property, and so on cause harm. There are so many other general ways in which we conduct our lives – our economic choices, the way we choose to interact with other people, the attitudes we express in general conversation – in which the possibilities of harm are ever with us. Likewise in international relations, the idea of states not harming one another in customary international law covers many other areas, whereas the UNFCCC statement is specifically about environmental harm.

No Harm Principle as a Principle of Morality Applied to Climate Change

There are a number of aspects of the no harm principle which are relevant to climate change; explained here as 'beyond the local,' 'beyond the individual's direct agency or intention,' 'beyond the agency of individual persons,' and 'beyond and including reciprocity and agreement.' First, beyond the local. As already implied, climate harms prohibited by the no harm rule include those in other parts of the world or the whole world, as well as locally. Some effects in other parts of the world require the intermediate actions of many others, for instance, in complex economic chains linking the treatment of farmers in another country to the food one purchases. Other effects may require merely the laws of the natural world to operate, when for instance I actually cause carbon emissions myself and these emissions then go into the atmosphere and contribute to the spread of higher concentrations of carbon anywhere. This is so because CO^2 emissions mix to raise atmospheric concentrations around the world to relatively uniform levels. Others may be mixed cases – for instance, if carbon is emitted in a factory in another part of the world producing a gadget which I eventually buy and use, then I have been part of the causal chain resulting in higher carbon emissions elsewhere. We can call this 'harm at a distance' – distance in space and also in time since the harms caused by carbon emissions will be experienced by people in the future. The phrase harm at a distance is not meant to signify only at a distance (though that sometimes happens): it simply conveys the idea that the harms we cause are not merely local. They are often diffuse and are such that all our collective carbon emissions are pooled together to create a common and deteriorating atmosphere for us all to live in or possibly lead to 'tipping points' when negative change may occur much more rapidly or dramatically. Though some of the effects are already being experienced in all parts of the world and will in any case get much worse, the effects on the climate and on the economies of many poorer countries are already much worse (though their climate emissions are much lower), and this raises additionally specific issues concerning climate injustice and the need for compensatory action. One contemporary aspect of this is the commitment at COP26 in Glasgow in 2021 to the 'loss and damage' principle; although there is still much to be done in respect to implementing this mechanism, the idea that the developed countries should contribute to a fund to reduce the harms in poorer countries represented by the loss and damage caused by past and continuing carbon emissions is premised on the recognition that failing to do so continues to contribute to harm (this is not pursued here; see Brown [2020] on this aspect and for a full survey of the ethical and factual issues in climate change generally).

When considering beyond the individual's direct agency or intention, we need to make explicit certain elements of doing harm that already have been indicated in the above discussion. First, clearly harms may well be because of the cumulative/collective impacts of many myriads of actions performed by individuals. Second, the harm done by an individual is not merely or even primarily what he or she actually does – for instance the emitting of carbon by driving a petrol-fueled car – it is as much if not more to do with what is done by other agents often in other parts of the world, for instance building the car, which enables a person to have or do certain things. Our environmental or carbon impacts or footprints are often largely a consequence of what is done by others. If 'footprint' is defined more precisely to be the carbon emissions caused by one's behavior, then we need to factor in, in terms of one's moral responsibility, the 'carbon footprint of products,' i.e., the carbon emissions caused in the sourcing, manufacture, or transportation of goods (food, clothing, electronics etc.) which occur *before* one comes to own or use it. It varies from case to case: typically, the use of a car will involve more emissions than its production etc., but in other cases what we buy and own – the stuff we accumulate – may be rarely used or, if used, involve little carbon emissions, so involve less emissions than the pre-purchase production/transportation costs. This equally applies to a country where its collective carbon impact/footprint may have as much to do with what is produced in other countries as what it produces.

But we are responsible for them nonetheless if these negative side effects are reasonably foreseeable. And it is clear that from the beginning of the 1990s onward the evidence for the negative carbon side-effects of our actions was there to see. One of the most interesting aspects of environmental consciousness and environmental ethics has been the turning upside down of a traditional assumption in ethics – that we are much more responsible for what we deliberately or intentionally do than what we merely foresee will follow or is likely to follow. If what we do intentionally has significantly negative side-effects or contributes toward these, we cannot just dismiss these on the grounds that what we are aiming at is based on other arguments for doing so, such as, in the case of governments, stimulating economic growth. The traditional argument in the ethics of warfare that it is wrong to target civilians but alright to target enemy soldiers, even if one knows that many civilians will suffer and die as a side-effect was similarly flawed.

This however leads to a related point concerning the ethical significance of actions which may involve some harm but are directed to a goal of overall reducing harm. Here the harm is intended (not a side-effect) but as a means toward an outcome in which similar or other harms are reduced. An environmental campaigner like Al Gore may have traveled many miles by plane conveying the truth of *An Inconvenient Truth* which certainly contributed in itself to harm, but the goal was to persuade large numbers of people to reduce the harm they caused. This is relevant to the later discussion of what levels and kinds of harms can be regarded as morally acceptable. Gore's responsibility for carbon emissions attributable to his activities can be justified only if he becomes responsible for processes which effectively remove carbon from the atmosphere at levels significantly greater than amounts produced by his carbon-producing behavior. Yet such 'offsetting' behaviors raise a host of technical, factual, and ethical questions which are beyond the scope of this chapter, and these would need to be considered before offsetting efforts can be claimed to justify the GHG emissions activities. There is certainly no simple ethical answer to this kind of 'means-end' issue: much depends on the details of particular cases.

When considering beyond the agency of individual persons, the ethical force of the no harm principle applies not just to individual human agents (sometimes called natural persons), but equally to organizations and bodies, such as nation-states, business companies ranging from small firms to vast transnational companies, and civil society organizations such as churches or charities (sometimes called artificial persons, especially if recognized in domestic or international law). There are really two points here: first, these entities are capable of moral agency and are thus susceptible to moral assessment; and second, whatever differences of moral character they may have because of the kinds of agents they are, the no harm principle needs to apply to them all. For instance, some may argue that nation-states' governments or leaders, because of their 'trustee' or quasi-contractual role in relationship to the interests of their citizens, need to prioritize these interests over above the interests of others elsewhere in the world. Others may argue that the *raison d'etre* of business companies is to maximize profits, not to promote the public good. Churches may be primarily concerned with the well-being of their members and the spread of their faith. Charities and other civil society organizations exist to promote a particular goal specified in their governing documents. Even if we accept this kind of analysis of the *raison d'etre* of such bodies (and it is argued to be inadequate in many ways), this does not alter the fundamental fact that none of these bodies should engage in activities which cause harm to other natural or artificial persons anywhere in the world – or at least there must be a really good reason for so doing that does not simply fall back on the *raison d'etre* of the organization. The no harm principle in the 1992 UNFCCC places the responsibility on nation states. In fact, the no harm principle places a responsibility on nation states "to ensure that activities within their jurisdiction or control do not cause damage to the environment of other States or of areas beyond the limits of national jurisdiction" (UN 1992: Preamble). The no harm principle in the UNFCCC is not understood to create additional ethical obligations for entities other than nations; yet this chapter will explain that individuals have ethical obligations to not harm others under general ethical considerations.

And finally, with beyond and including reciprocity and agreement, it is implicit in the above discussion of the no harm principle that the idea that if an agent accepts the no harm principle, then he or she sees it as applicable to their behavior in certain areas of activity and acts accordingly, irrespective of whether other people accept the same moral position in regard to these areas of activity, or, even if they do, fail to act accordingly (because they are moral back-sliders or do not take it as seriously). Thus, in the early years of environmental consciousness about climate change, many individuals were seriously changing their lifestyles and reducing their contribution to environmental harms, even though they knew that most people were not accepting the need to do this or acting in the same way. And they did this of course very often because they were advocating change and wished to 'walk the talk.' We need to note that almost all human agents do accept the no harm principle of morality – without this society could hardly function, as noted before. But it is in the changing awareness of what kinds of actions constitute harm that inevitably there will be some ahead of others, acting as it were as vanguards for the new ethical challenge.

Once it becomes generally recognized either within a country or between countries that certain courses of action are necessary for everyone, with each person or country playing his or her or its part, then the moral force of sticking to the no harm principle becomes very strong. And this is the situation we are in now with regard to the need massively to reduce our carbon emissions. The very fact that there is widespread cooperation and what may be called reciprocal action is significant: but when it is supplemented with the fact that there are laws within countries or agreements and treaties between countries to act on a certain principle, in this case the no harm principle, then the case for full observance of it is very strong.

International Agreement

Let us next consider the moral force of international agreements. Clearly, we now have an international legal framework within which nations have agreed to play their parts to achieve certain internationally agreed warming limit goals. This obligation will require nations to reduce their greenhouse gas emissions by target amounts by certain dates. The two key documents of relevance here are the UNFCCC (1992) and Paris Agreement (2015). Whilst there can be disagreements about the nature/extent of the legal force of international agreements of various kinds (how soft or hard they are) – this chapter is not concerned with the legal aspects of international law – any such legal agreements carry with them moral force, which combined with the general thrust of the moral arguments already given form a powerful moral basis for that commitment.

It should however be noted that the relationship between morality and law is complex, and also that there may be normative conflicts within the sphere of morality and within the sphere of law. 'Legally required' does not entail 'morally required' (all things considered), nor vice versa. In the sphere of morality, the no harm principle cannot be an absolute because it may clash with other moral rules and even with itself where any alternative course of action causes some harm. Even if the international no harm principle is thought to generate an overriding legal obligation, it does not follow that in all circumstances morally speaking following it is the right thing to do. Even within the sphere of international law there may be circumstances in which there are clashes: the obligation to follow the no harm principle may come, for instance, into conflict with the right of a country to defend itself against attack which certainly involved significant harm, environmental and human. Nevertheless, the general trust of the current argument is that these moral considerations will in fact help to strengthen the resolve of governments to follow the no harm principle. Even in our current situation as governments move toward the 2050 goals (as discussed next), the obligation must be to minimize harm even if it cannot be avoided altogether.

The 'harms meantime' issue: the expectation is that if countries, with the cooperation and support of their citizens, do what is essential to make the reductions to achieve the targets, then global temperature could be stabilized somewhere between 1.5°C and 2°C above pre-industrial temperatures. But, as noted earlier, we have to recognize that an awful lot of harm in terms of deteriorating environmental conditions for very large numbers of people, apart from what is already bound to occur because of previous emissions, will continue until atmospheric GHG concentrations are stabilized (and indeed will continue after that but not getting worse, and possibly being ameliorated by subsequent restorative eco-systemic change).

No doubt it has been accepted by most people involved in forming these policies that it would not be reasonable to expect states with the support of their citizens to achieve this goal any quicker. So, one response might be that these short-term harms are reasonable because in a sense it is not possible to avoid them. This way of sidestepping the question of justification (since the question of justification would not arise) is implausible since there are so many possibilities for political choice. Another response might be that the harms are reasonable and justifiable because to attempt any greater reductions in carbon emissions at greater speed would have too great costs in terms of social and economic consequences. A third response is that what specifically a nation should do under law is determined by the principles of 'equity and common but differentiated responsibilities.' All nations are required to reduce their GHG emissions to levels which represent their fair, i.e., 'equitable' share of total acceptable global emissions, often referred to as a carbon budget, that must constrain the world to achieve the Paris Agreement's warming limit goals of as close as possible to 1.5°C but no more than 2.0°C. One of other of the latter two responses may seem sensible, but we have to acknowledge the consequence: some continuing harm until the targets are met is justified. Another way of putting it is to say that what needs to happen for a commitment to the Paris targets (or any future target of reaching net zero) to be achieved falls short of what would need to happen if adhering to the no harm principle was fully achieved now.

But ethically could and should a nation-state do more? Could a nation with the support of its citizens achieve a net zero-carbon emissions target needed to achieve the warming limit goal sooner than 2050? If it did, then less harm would be done by that country in the coming years. Although this looks like a technical issue (and is partly), it is partly an ethical issue to do with what people are willing to do partly for ethical reasons. That is, it would be dependent on the willingness of citizens to make more radical changes in their lifestyles sooner, in what they advocate to others and in what they signal to their governments they want them to do. That would be extremely unlikely to occur on a large enough scale, so if governments are really faced with pushing for change consistent with meeting certain social and economic objectives, the key question that needs to be faced is: what are those objectives? If the objectives are to maintain economic activity in such a way as to maintain vital social objectives, then it seems reasonable for countries to do this so long as they are able to achieve the agreed targets. If on the other hand the economic objectives are to maintain steady economic growth, and this goal is seen as being on a par with or even on certain occasions trumping the commitment to achieve the targets, then the targets will not be achieved, because the no harm principle in the background is not seen as an overriding moral principle within which other activities need to be framed. Of course, a country is legally responsible under the no harm principle for damages done outside their boundaries that are caused by emissions in excess of their obligations under the Paris Agreement, whether or not a country refuses to comply with their obligations due to their choice to pursue economic growth goals. The moral argument above merely reinforces the obligation to so frame, shape and possibly to some extent limit economic activity.

What is really needed is a reframing of the crucial function of economic activity. There are many theories being advanced (like ideas of the well-being economy, happiness economics, the circular economy), but one model that is particularly persuasive is advanced by Kate Raworth in *Doughnut Economics*, in which the overarching purpose of economic activity is not economic growth as such, but to sustain the conditions in which human beings can achieve social and personal well-being within the outer ring of ecological limits and within the inner ring below which humans should not fall into deprivation (Raworth 2017). The reality is that if a commitment to economic growth (as opposed to economic stability and lack of decline) is seen as being as important as if not more important than commitment to the targets, the targets will not be achieved. Only by nesting economic and other objectives within the controlling commitment to reaching the targets will the targets be reached. Again, this moral argument merely reinforces the argument based on the legal obligations countries have had to reduce their GHG emissions to levels that would prevent dangerous anthropocentric interference with the climate system since 1997 and should therefore be responsible for their failure to do so. Further consideration of the 'economic interests' will be returned to in a later section.

The Implications of Ethical Consistency

Over many years, there has been international condemnation of countries like Brazil which happen to have within their borders tropical rainforests and engage in chopping them down for economic gain – for instance for timber, or for clearing land for ranching or cash crops. There are various reasons for this condemnation – such as the fact that tropical rainforests are a major part of the lungs of the planet with the capacity to absorb carbon dioxide or are the home of rich biodiversity. Commonly enough, such countries will reply by saying that they have the right to use their own natural resources without interference. But this does not mean they can do whatever they want with such resources. The first part of the section of the UNFCCC quoted above actually reasserts the sovereign right to exploit their own resources, but then frames this immediately with the no harm principle. One way of emphasizing the idea that countries should not simply do what they will with the tropical rainforest is to see them, like the oceans, as the 'common heritage of mankind.'

What should interest us about this is that if that international condemnation is justified (as it surely is on environmental grounds generally and on no harm grounds in particular), then we must ask what are the implications of that condemnation? Surely in ethical consistency, it means that a country must be committed to ensuring that its own use of resources does not cause environmental problems either. If countries, particularly economically powerful countries, acquire on world markets vast amounts of oil over which they then have control, the same question arises considering the use of such resources (and more radically whether such resources, given that such resources would be likely to be used in a way that does cause unnecessary harm, should be acquired in the first place, certainly on the scale that they are). This line of argument of course merely gives support to the more general arguments already considered, but because almost all people will have a naturally strong reaction of moral condemnation of such activity in the rain forest, it should help to make vivid the troubling consequences of our more general carbon habits.

Resistance to the Force of the No Harm Argument

We now look at some of the arguments that are used by governments and people for not taking action or not taking action with sufficient urgency – with the implication, which may or may not be fully recognized, that the targets will not be achieved. This implication may not be recognized or rather accepted if the climate emergency is not accepted or it is thought its gravity is exaggerated and that all that is needed are changes over a longer time scale: with these factual issues we are not concerned in this chapter (though they are vitally important to address). The implication may also be resisted with the thought that whilst it is generally essential that countries and their citizens take urgent action to achieve the targets, these targets will be achieved or largely achieved, whether or not a country or individual plays their part or does so fully in pursuing them. There is a factual issue here in regard to countries (especially large countries) failing to play their part (i.e., it needs to be argued that any failure especially by larger countries to play their part will significantly affect the global outcome), but the main problem with this line of thought is that there is no moral justification for moral backsliding (that is, not complying but reaping the benefits of others complying). The moral principle of universalizability applies here: if something ought to be done by others, then one should do it oneself. This is a big issue in ethics which there is no space to pursue here. What we need to address here are arguments which are intended to justify a failure to act. First, there is the argument that promoting economic interests is a legitimate concern of governments, partly because it is what their citizens want or vote for. Second, there is the argument that each country is justified in prioritizing its own national interests even at the expense of other countries and general global impacts.

First, economic interests trump following the no harm principle. It has been argued above that, whilst economic concerns are valid, they need to be framed within an underlying commitment to environmental responsibility and not seen as of equal but competing value. The danger with the economics interests argument is that whilst it might accept that the no ham principle has some force, it denies that it has full overriding force. Much better is an approach already indicated in which economic activity should be framed and constrained by ecological limits; that is, choices of economic policy should not violate the no harm principle. Even the once much heralded 'triple bottom line' approach is inadequate in this respect (not demanding enough) as it is in danger of being used to compromise the priority of the no harm principle in this way. Here, some of the sources of resistance to this nesting approach need to be unpacked.

To some extent, this arises from typical assumptions about the nature of the state drawn from political theory. Whilst states may belong to some kind of society of states in which there are certain international norms, their sovereign role is to promote national interests, to meet the needs of their citizens, and in a democracy to follow the will of the people. On this basis, the national interest which clearly includes the national economic interest has some kind of priority over considerations to do with the interests of other countries and their people elsewhere. An analogy may be drawn: a society functions better if parents attend particularly to the interests of their own children rather than trying to promote the interests of everyone's children. Whilst there is some truth in this analysis, it remains the case as noted at the beginning that this does not entitle states to pursue policies which, however indirectly, cause known harm others in the world (any more than a parent's primary concern with his or her children justifies him or her in causing harm to other children) – and arguably making economic growth a primary target it in danger of doing just this – causing known harm elsewhere. Even though each county has agreed to comply with the no harm principle and it is already an international legal obligation, this is either ignored, paid lip service to, or treated as not overriding other considerations and hence not 'trumping' those considerations. One such consideration is the democratic argument already mentioned earlier.

Clearly the democratic argument complicates the issue but what this shows is that the moral arguments advocated here need to be accepted by citizens as well, so that they support if not advocate significant change and bear witness to it in their own lifestyles (by 'walking the talk'). The argument in any case is not merely about changing moral priorities in individuals, but in changing priorities and understandings of how individuals conceive of their own well-being. There needs to be a public conversation about how far and in what ways economic well-being and material 'stuff' are the key to well-being in general – perhaps, as Jason Heckel has argued, "less is more" (Heckel 2020).

In any case, as John Stuart Mill pointed out, what the majority wants in a democracy cannot determine everything, as it could, for instance, lead to intolerant behavior toward minorities (Mill 1859). His point about 'the tyranny of the majority' has equal application here. If it is the case that certain policies would have seriously detrimental effects on the rest of the world and thus for instance violate the rights of human beings elsewhere, then the fact that a majority wants the government to follow these policies does not establish that it should.

Next, we will look at the general prioritization of national interests. First, there is the argument that each country is justified in prioritizing its own national interests even at the expense of other countries' interests and the general 'global good.' Here the no harm principle is seen as having really no force or limited force. Whilst virtually no one would nowadays directly state this line of thought, it is in my opinion something that lingers in the background from thinking from an earlier era and is worth identifying explicitly. This argument stems from two traditional theories of international relations.

The first approach is skeptical realism or international skepticism, which simply denies that states are bound by moral considerations in their dealings with other states and either dismisses the claim that all human beings are equal or regards it as irrelevant to the conduct of states. The idea, put forward robustly by Thomas Hobbes, was that states as sovereign existed in anarchical world in which there was no common power to 'hold them all in awe' and since moral rules require somebody with sufficient power to enforce them in order to be valid, moral rules do not apply to international relations (see Bull 1985). Either universal morality is denied on the grounds that ethics is relative to particular societies or, even if it is granted, it is seen as irrelevant in a world where states merely struggle for advantage and power in competition with each other. From a realist point of view a nation-state's interest in such interdependence, institutions, and laws is purely prudential (i.e., self-interested) rather than ethical, and hence its commitment to international cooperation and international law is unstable because it is dependent on the contingencies of interest. So, on this view, even though a country has signed up to the no harm principle, its commitment to it is weak and certainly not to something that is overriding. Even on that basis though, most countries' governments would recognize that it simply was not in their long-term interests to see the global environment become degraded and the climate changed for the worse. States sometimes behave as if this realist analysis was still valid, but the model makes very little sense in the modern world with extensive international interdependence, international institutions, and international laws.

Second, there is the argument sometimes called internationalism or the 'morality of states' approach (or 'Grotianism,' after the thought of Grotius), which has been the dominant model of international relations since the Peace of Westphalia of 1648. This had traditionally been understood as saying that although states are governed by a moral framework in their relationships to each other in a 'society of states'; it is a framework in which it is legitimate for nation-states to put their own interests first, so long as they are bound by certain minimal rules of international cooperation. Maintenance of the international order was primary, along with the sovereignty of states, with international rules becoming accepted that govern in particular the conduct of war and the conduct of trade which required that agreements need to be honored (*pacta sunt servanda*). Although the idea of not harming other states was accepted, it was understood as having limited scope, and the idea that states might be obligated to support other states for instance with their development or the tackling of poverty was not part of the original picture (see Dower 2007 for a survey of positions in international relations theory).

Neither of these theories are the least bit relevant to our modern predicament, whatever limited validity they may have had in earlier centuries. The realist theory has become outdated as a theoretical approach, although in the absence of a world authority to enforce international laws, the temptation to act in contravention to them, particularly for powerful states, of course exists. The internationalist approach has essentially evolved given the evolution of international law and institutions and is in fact fully supportive of the idea the states ought to act on the no harm principle. The no harm principle along with many other international laws to do with the environment, health, trade, and many other aspects of modern life give the lie to a minimalist interpretation of internationalism as relevant to the modern world.

Underlying modern internationalism or a 21st century (and late 20th century) version of the morality of states can be called a 'global ethics' perspective (i.e., all people matter equally), which is also advocated by those who accept cosmopolitanism. Cosmopolitanism had historically been opposed to internationalism (and more starkly realism), but nowadays, although the theoretical basis of these theories is still somewhat different, the practical picture of what is required in the world is much the same, namely that we take seriously the extent of our international obligations, both moral and legal, as established in formal law and informal practice/custom. One influential way of characterizing the modern approach to internationalism is the theory of 'international citizenship' as expressed for instance by Linklater (1992). On this view states themselves are citizens of the society of states with citizen-like responsibilities, including for instance the promotion of human rights in the world. Whilst politicians, civil servants and diplomats are more likely to favor a version of internationalism or the 'morality of states' approach, a global (i.e., cosmopolitan) ethic is actually what underlies the general approach. Cosmopolitanism is sometimes resisted because of its political aspects, which in the hands of some thinkers leads to idea of world federalism or world government. But the ethical core of cosmopolitanism is accepted by most people (even if they simply call it a global ethic). At its core is the idea that all human beings have equal moral status, and that we all have responsibilities certainly not to harm or undermine the well-being of others and to some extent to positively promote it. The advantage of an explicit recognition of this cosmopolitan basis for international cooperation is that it makes it clearer that the task of tackling climate change rests with individuals as well as governments. As individual 'citizens of the world' or 'global citizens,' we need to accept our responsibility not just to follow the rules and regulations the government imposes on us to change our carbon habits, but to accept the need to make adjustments voluntarily and ahead of the game, and to use our energies to advocate the changes we see as needed in the world, if we are to preserve it in a remotely satisfactory state for future generations.

So, given a robust no harm principle based on the cosmopolitan premise that harm anywhere in the world is equally bad, it has been shown that national policies, including climate policies, that are based on national interests, and especially national economic interests, are deeply inadequate.

Conclusion

Since this chapter is meant to provide arguments and considerations that will help strengthen the moral resolve of both citizens and governments to play their part in achieving the internationally agreed targets and to do more rather than less in this regard so that the resultant leveling off of temperature is more likely to be nearer 1.5°C than 2.0°C increase, it may be helpful to summarize the key points. First, general aspects of the no harm principle include:

- The no harm principle is an almost universally accepted basic principle of morality and is equally applicable to individual human persons and to artificial persons such as companies or nation-states;
- It has global reach and extends to the future;
- Its significance applies to human acts seen as contributions to collective impacts;
- It often involves the agency of others that enables us to do and have what we do and have;
- It lies beyond what we intend and includes what is reasonably foreseen or foreseeable; and
- It applies to our behavior even if others do not do the same, but its moral force is reinforced by reciprocity of behavior and agreements to play parts in collective goals.

Second, concerning citizens, and understanding that as a general moral principle, the no harm principle covers many areas of our lives and cannot be fully realized:

- If a government sets a general target, as has happened in the case of carbon reduction, individuals need to play their part, which may not be merely following any directives given;
- Given the urgency of the need for environmental change, an individual may for moral reasons to go well beyond compliance with general expectations, but engage in advocacy through civil society;
- As global citizens, citizens accept in principle the full extent of global obligation; and
- Insofar as a serious commitment to the no harm principle requires major changes in lifestyles, there needs to be a serious re-evaluation of what are the key elements of the good life.

And lastly, concerning nation-states:

- Signing up to the two international agreements discussed does not merely create legal obligations, but also moral commitments to play their part effectively;
- The no harm principle of ordinary morality also makes the case for action, as does the quasi-moral force of customary international law;
- Although progressive movement toward meeting targets, e.g., in 2050, implies continuing harm meantime, the justification of such harm needs to be carefully monitored to ensure that this is in the spirit of the no harm agreement and does not allow a weakening of resolve because of economic concerns;
- The legitimate concern of governments to pursue economic goals needs to be firmly framed, i.e., limited within the no harm commitment and not seen as valid principle of equal standing alongside it; and
- The democratic argument (i.e., 'what the electorate want') should not be used to justify economic goals at the expense of the no harm principle, and a public conversation needs to be encouraged about the limits to democratic justification and about what the proper bases of the good life are.

References

Brown, D. (2020) *Seven Features of Climate Change that Citizens and the Media Need to Understand to Critically Evaluate a Government's Response to This Existential Threat and the Arguments of Opponents of Climate Policies*, https://ethicsandclimate.org (accessed 8 January 2021).

Bull, H. (1985) *The Anarchical Society – A Study of Order in World Politics*, London: Macmillan.

Dower, N. (2007) *World Ethics – The New Agenda*, 2nd ed., Edinburgh: Edinburgh University Press.

Hickel J. (2020) *Less Is More: How Degrowth Will Save The World*, London: William Heinemann.

Linklater, A. (1992) "What Is a Good International Citizen?," in P. Keal (ed.), *Ethics and Foreign Policy*, St Leonards: Allen and Unwin.

Mill, J.S. (1859) "On Liberty," in M. Philp & F. Rosen (eds.), (2015) *On Liberty, Utilitarianism and Other Essays*, 2nd ed., Oxford: Oxford World's Classics.

Raworth, K. (2017) *Doughnut Economics: Seven Ways to Think Like a 21st-Century Economist*, London: Random House.

United Nations (1992) *UNFCCC*, https://unfccc.int/resource/docs/convkp/conveng.pdf (accessed 27 October 2020).

United Nations (2015) *Paris Agreement*, https://unfccc.int/process-and-meetings/the-paris-agreement/the-paris-agreement (accessed 27 October 2020).

Voigt, C. (2008) "State Responsibility for Climate Change Damages," *Nordic Journal of International Law*, 77: 1–22.

2

DUTIES TO NOT HARM ECOLOGICAL SYSTEMS, PLANTS, AND ANIMALS

Michelle Maloney

Introduction

On August 10, 2021, the words 'duty of care' were partially spray-painted in large red letters across the front of Australia's Parliament House. It was remarkable because it was the first time Parliament House has been spray-painted, let alone with a legal phrase. It was only 'partially' spray painted because the activists managed to write 'duty of' in eight-foot red letters, before security guards stopped them, they were arrested for trespass, and their efforts were splashed across the nation's news.

The fact that the words chosen for this act of spray-painted vandalism of a Parliament House in Australia were 'duty of care' is significant. The action was taken after the release of the IPCC Report and in direct reference to an important case being heard in Australia: *Sharma v Minister for the Environment* [2021] FCA 560; 391 ALR 1. The action was driven by the prospect of terrifying anthropogenic climate change and the equally terrifying fact that the Australian Federal government at the time not only refused to act to address climate change but was actively blocking all efforts by its citizenry to make them act.

The *Sharma* case in Australia exemplifies the debate raging in many countries around the world: people know we all have a duty of care to protect present and future generations from runaway climate change, but many governments – often due to pressure from and collaboration with fossil fuel industries and other vested interests – deny having a duty of care or simply refuse to respond to climate change urgently enough. And if governments are denying that they have a duty of care to their own people – where does this position them regarding their duties towards the non-human world?

This chapter will explore what it means to have a duty of care, and a duty not to cause harm, to people and other members of the Earth community, and what this means within the climate changing world we live in. It will first briefly define what duties and obligations are and contrast them with the Western concept of rights. It will then introduce the wider context of climate litigation that is escalating around the world, in which citizens in many countries are demanding that their governments respond to climate change. The *Sharma* case in Australia will be used as an example of what citizens are arguing for with respect to duties (and what courts are finding) and contrast it with what some governments are still denying.

The chapter will then contrast the failures of current Western law and governance with respect to climate change, with two alternative ways of considering duties and obligations to the human

DOI: 10.4324/9781003039860-4

and non-human world. It will describe some of the core ideas within Earth jurisprudence, and then outline some of the most fundamental rules of law within Indigenous legal systems, by highlighting the Law of Obligation that exists within the legal systems of the First Nations Peoples of Australia.

The argument here is straightforward: most modern nation states – in domestic and international law – have failed to take duties seriously at all and have certainly failed to respond to their duties of care to people, plants, and animals with respect to climate change. If modern nation states were able to incorporate into their legal systems, even a small part of the key elements of Earth jurisprudence, and the ancient wisdom of First Laws – i.e., the deeply, *deeply* held understanding that we are all obligated to care for the living world and care for each other – then our legal systems and governance responses to climate change would look very different. Indeed, if modern nation states had a Law of Obligation to their citizens and all life on Earth, we would have acted when the science first indicated what greenhouse gases were doing to Mother Earth, and we wouldn't be facing the current horrors – and the dystopian future scenarios – that we are facing today.

Duties and the Role of Climate Litigation

In order to reflect on Western society's duties (or absence thereof) to care for life, including with respect to the impacts of climate change, it is first useful to define duties, obligations, and rights. A simple dictionary definition of 'duty' is: "obligatory tasks, conduct, service, or functions that arise from one's position" (as in life or in a group) (Merriam Webster Dictionary 2022). This refers to a positive act, something which must be done.

A simple definition of 'obligation' is: "Something (such as a formal contract, a promise, or the demands of conscience or custom) that obligates one to a course of action" (Merriam Webster Dictionary 2022). The reference to 'demands of conscience or custom' is especially important in this context – obligations can be much more than a legal requirement, something enforced by the legal system of the day; it refers to something deeper, something created by a society to support social and legal norms.

Rights can be described as "those things that one is morally or legally entitled to do or have" (Merriam Webster Dictionary), and rights are typically defined in relationship with duties – but not always. In Western law and governance, the concept of 'rights' is dense with theories, definitions, analyses, and critiques – while the world of 'duties' is often the neglected conceptual cousin. 'Duty' is often a foundation within legislation but is rarely the 'poster child' of litigation, debate, or new social movements. For example, we have the 'Rights of Nature' movement – but we don't have the 'Duty of Care for Nature' movement. That is what makes climate litigation around the world so important – it is not just about competing 'rights' – it highlights and articulates our duties and obligations.

Climate change litigation has arisen in more than 25 countries around the world, typically as a response from civil society groups trying to force action on climate change. Litigation has been directed at public and private companies, federal governments, city administrations, and insurance companies. The litigation has two broad categories: (1) public law actions against governments and public authorities, raising human rights and constitutional and administrative law arguments; and (2) private law actions based on areas of law such as tort, fraud, planning, and company law.

Globally, the cumulative number of climate change-related cases has more than doubled since 2015. Just over 800 cases were filed between 1986 and 2014, while over 1,500 cases have been brought in the last six years (2015–2021) (Highham & Setzer 2021). Overall, corporations are the most common claimants or plaintiffs in these cases, with governments being the most common defendants.

In Australia, there are 17 climate change litigation cases listed in the Federal Court at the time of writing. One of the most recent climate change litigation cases, the *Sharma* case, has raised the profile of our duties to each other and to the living world. The *Sharma* case was filed on September 8, 2020, in the Federal Court of Australia and was heard by Justice Mordecai 'Mordy' Bromberg. The case targeted the proposal to extend the Vickery open-cut coal mine in New South Wales, a State of Australia, which would increase the mine's coal production by about 25 percent and emissions by 100 million tonnes of greenhouse gases over its lifetime. The coal is destined for Australia's export markets.

The case was taken by eight children, led by Anjali Sharma (with the assistance of 86-year-old litigation guardian Sister Brigid Arthur). The lead claimant was a 16-year-old school student at the time the *Sharma* case began. She had reportedly long suffered from climate anxiety about the earth's potentially dire, climate changed future, and stated:

> A safe climate and an increase in fossil fuel production are incompatible. Given the impacts we've already experienced, and what the future holds if we don't act now, approving new coal mines will simply further harm young Australians. We are arguing that the Minister of the Environment has a duty to protect us all. This case aims to make sure that duty is fulfilled.
>
> *(Fairhurst 2021)*

The applicants argued that in deciding whether to approve the mine extension project under the Environment Planning and Biodiversity Conservation Act 1999 (Commonwealth [EPBC Act]), the Environment Minister has a duty to protect young people from the impacts of climate change; specifically, that the Minister owes them a 'duty of care.'

The basis for the applicants' claim that a duty of care exists was the law of negligence, which incorporates long-established principles relating to the foreseeability of possible future harm, and the relationship between the person whose actions cause or contribute to the harm and the person who may be harmed as a result of those actions. Applicants sought an interlocutory injunction to restrain the Environment Minister from exercising her power under the EPBC Act. They also sought a permanent *quia timet* injunction to restrain an apprehended breach of the duty of care. Much to the surprise of the Australian legal profession and civil society activists alike, when the decision was handed down on May 20, 2021, the Court held that the Environment Minister does indeed have a duty of care, a duty to take reasonable care not to cause the applicant children, or any Australian children, personal injury when exercising her power under s.130 and s.133 of the EPBC Act, to approve or not approve the extension project (*Sharma*: 513).

While the Court dismissed applications for injunctions and the Minister was not prevented from approving the extension project, the declaration by a court that the Government has a duty of care to its children was seen as a hugely important decision. David Barden, one of the lawyers representing the applicants, stated that the *Sharma* case, "is the first time in the world that a duty of care has been recognised, especially in a common law country" (Morton 2021). In other commentary, it was noted that:

> If supported by other courts, the recognition of the duty to take reasonable care could constrain the ability of both government and private entities to undertake projects that may lead to an appreciable risk of harm arising from the contribution of such projects to net carbon emissions.
>
> *(Dunlevie et al. 2021)*

The day immediately following the judgment, the Minister announced that the Federal government would appeal the decision. The Appeal was to be heard in the Full Federal Court,

What gained the most public attention was that by appealing the decision from the *Sharma* Case, the Environment Minister – an elected representative of the Australian people, speaking on behalf of the elected government of the day – was seen as denying that the government owed any duty of care to its young people. This appeal by the government stimulated anger within the Australian community and has coalesced action by groups such as Extinction Rebellion and youth climate action groups, including the aforementioned act of protest vandalism. It has also raised the consciousness of a nation toward what 'duty of care' actually means.

Despite the optimism among climate activist communities and environmental and human rights lawyers about the *Sharma* case, on 15 March 2022, the Full Federal Court overturned the decision in its unanimous judgment in *Minister for the Environment v Sharma* [2022] FCAFC 35. The appeal overturned the primary judge's finding that the Commonwealth Minister for the Environment (Minister) owed a novel duty of care at common law to Australian children who might suffer potential harm from the climate change implications of exercising her powers under the EPBC.

This decision has apparently ended the debate about whether a duty of care exists legally within the EPBC Act but has not stopped demands from the Australian public that their governments owe a duty of care to its people, regarding climate change.

Alternative Legal Approaches to a Duty of Care

While the rise of climate change litigation is important and potentially ground-breaking, it must also be remembered that this litigation is emerging because of the inherent failure of current Western law and governance to protect people and protect the living world from harm. Traditional environmental law has been criticized as being embedded in industrial society's pro-growth governance culture that simply legalizes severe environmental harm, rather than effectively protecting the Earth community (Campbell & Linzey 2009: 22). It has also been criticized for being incapable of calculating or managing the cumulative impacts of human activities and ignoring the reality of ecological limits (Guth 2008).

The past two decades have seen a growing movement of Earth-centered law and governance, which is challenging and creating alternatives to traditional environmental law. In this part of the chapter, the theory of Earth jurisprudence will be outlined as a touchstone for what Earth-centered governance sees as our duties towards the living world. This will be followed by a brief and insufficient (given the depth of the concept), introduction to the Aboriginal legal concept of the 'Law of Obligation.' Both Earth jurisprudence and the Law of Obligation throw into sharp relief how poorly Western law and governance currently handle our duties not to harm the living world and can provide a guide and inspiration to transforming our legal and governance systems.

Earth Jurisprudence

One of the foundational theories of Earth-centered law and governance in the 21st century, is Earth jurisprudence. The term 'Earth jurisprudence' was coined by deep ecologist, 'geologian' and Earth scholar, Thomas Berry (1914–2009).[1] In his 1999 book *The Great Work: Our Way Into the Future*, he proposed that the challenge for humanity is to understand the underlying, systemic reasons for the ecological crisis and to transform our relationship with the natural world from one of destruction to one of mutually beneficial support (Berry 1999: 7). He suggests that acting ethically and living within Earth's natural capacities requires that we look to a new jurisprudence, a new way of governing ourselves for the challenges and possibilities of the 21st century and beyond, so as to protect the integrity of Earth systems (Berry 1999: 161).

Earth jurisprudence is now seen by many Earth jurists as a foundation for the emerging field of ecological law, and one of the core foundations of 'Earth laws.' Earth laws is a framework that includes the theory of Earth jurisprudence, the legal issues and approaches included within the emerging field of 'ecological law' (https://elgaworld.org/), the developing law of ecocide, rights-based legal approaches including Rights of Nature and 'legal personhood for nature,' and other legal concepts being created within, and which directly challenge, Western traditions of law.

In Earth jurisprudence, Berry proposed that the primary cause of the ecological crisis is anthropocentrism, a belief by people in the Western industrialized world that we are somehow separate from, and more important than, the rest of the natural world (Berry 1999: 182). He argued that this anthropocentric worldview underpins all the governance structures of contemporary industrial society – economics, education, religion, law – and has fostered the belief that the natural world is merely a collection of objects for human use (Berry 1999: 4). Berry suggested that the time had come for an ecocentric, or Earth-centered, foundation for human societies. He stated that the mere existence of life and the laws of the emerging Universe, of Earth's functions, are the highest laws and human-made laws need to be in alignment with them.

Earth jurisprudence then, is an emerging theory of law and governance that requires a radical rethinking of humanity's place in the world, to acknowledge the history and origins of the Universe as a guide and inspiration to humanity, and to see our place as one of many interconnected members of the Earth community (Berry & Swimme 1992). Berry and the broader Earth jurisprudence movement acknowledge the inspiration and guidance that Indigenous cultures and Indigenous law can provide to industrialized societies and the development of Earth jurisprudence.

Many of the key elements of Earth jurisprudence and eco-centrism have long been debated in Western environmental philosophy and human ecology, and eco-centrism in the law has been explored by many writers (Stone 1972; Nash 1989; Bosselmann & Taylor 1994). As already noted, in *The Great Work* Thomas Berry suggests that industrial society's laws, religious traditions, educational institutions and economic systems need to reflect and protect the bio-physical realities of the Earth community. Earth jurisprudence advocates for a legal system with a number of critical elements.

First, Earth jurisprudence acknowledges that the Universe is the primary lawgiver. In contrast to the current Western legal system that sees human laws as the highest authority for human society (and implicitly, for all other life forms and ecological systems), Earth jurisprudence sees the laws of the Universe, the 'Great Jurisprudence' or 'Great Law' as providing the fundamental parameters of the Earth Community, including human societies (Burdon 2011). The concept of 'the Great Law' is seen as problematic even by many Western scholars. It is argued that it is very difficult to clarify what the Great Law is, or how to specify how people can live within those 'laws' (Burdon 2011; Pelizzon & Ricketts 2015). Nonetheless, the idea that nature or the Universe is the primary lawgiver sees Earth jurisprudence as explicitly advocating for an idea of human societies living within the rules or limits of the natural world. In other words, it places human relationships with the non-human world as primary.

A second foundational concept in Earth jurisprudence is the view that the Earth is an interconnected community and that there is a relationship-based existence between humanity and the rest of the Earth community. This contrasts with the current Western legal view that creates legal relationships among people, and between people and corporations (through constructs like property law), but commodifies and exploits all other aspects of the natural world (Cullinan 2011; Graham 2011). By framing the natural world as a community, Earth jurisprudence imposes greater constraints on humanity's actions than our current legal system does. By claiming that "the primary concern of the human community must be the preservation of the comprehensive community," Berry argued for a human world that works to ensure that all members of the Earth community can thrive and continue their evolutionary journey (Berry 1999: 580). Earth jurisprudence calls on people to see themselves as part of the living world, and supports the argument that human beings, through their relationships with each other and the non-human world, have an obligation and duty, to enable all life to persist, thrive, and evolve.

A third concept is the rights of nature, and the interconnected concept of the duties and responsibilities of humanity to protect and speak up for such rights. Berry argued that "nature's rights should be the central issue in any … discussion of the legal context of our society" (Berry 1999: 80). He proposed that the Earth community and all the beings that constitute it have rights, including the right to exist, to habitat, or a place to be and to participate in the evolution of the Earth community (Berry 2002: 214). This view directly challenges the core ideas within Western and other industrialized societies that see nature as merely the property of human beings. It also contrasts with the dominant approach in the Western legal system, which privileges and grants rights only to humans and selected human constructs, such as corporations or government entities. Granting rights to nature is a radical rethinking of the role of our anthropocentric legal system, yet the idea appears to be taking hold in many jurisdictions.

As noted by Iorns (2018), in many of the examples that now exist where nature (and in particular, rivers) have been given legal personality or other rights, it has actually been in an effort to uphold human responsibility to better protect them from degradation. She examines case studies from the USA, India, and Columbia and sees courts often interpreting the rights of nature in ways that better strengthen state and community capacity to enact their duty of care for the environment. For example, with respect to the Atrato River in Columbia:

> Whatever the practical realities of upholding rights in the Atrato region, the framing of the state of the River as a matter of human responsibility and guardianship, particularly with appointing people to act as legal guardians, is a big shift in perspective as well as law. With the accompanying focus on bio-cultural rights, together they lay the foundation for a shift in mind-set and practice.
>
> *(Iorns 2018)*

Indeed, as pointed out by Iorns (2018), and advocated for by Thomas Berry, shifting the perspective of people so that they see themselves in deep relationship with, and holding duties and responsibilities towards, the living world is a critically important way for humanity to reframe our culture and strengthen our governance and law to protect nature.

The Law of Obligation – One of the Foundations of Ancient First Laws

This chapter now turns to a short introduction to the remarkable governance system of one of the most ancient societies on earth: the Aboriginal or First Nations Peoples of the continent now known as Australia. Their deeply held, Earth-centered culture created one of the most long-lasting and stable societies in human experience, and one of the most important foundational ideas within their culture was a 'Law of Obligation'.

Aboriginal law is very different to – indeed, sometimes completely incomparable with – Western notions of law. As stated by eminent First Nations Law Professor Irene Watson:

> First Nations Peoples' law is of the beginning: of the first songs, sung by the ancestors. When the first steps were walked across the ruwe, country was sung into creation. Law conceived as a way of living is difficult to write about and cannot simply be described or easily translated into a foreign language that is empty of the ideas that our law ways carry. Our law was not written in the way in which the West conceives of writing. Law was painted in ceremonial design and symbols were marked on boundary markers, identifying traditional owners and their ngaitjis. *The differences between Nunga and non-Aboriginal legal systems are so extensive that*

> *there is no basis upon which comparison can be drawn* … (Aboriginal) law is the essential basis of social conduct: respect, reciprocity and caring for country to name a few. These ethical principles convey the essential nature of the law.
>
> *(Watson 2015: 22, emphasis added)*

Non-Indigenous people often know about Aboriginal peoples' connection to land ('Country') through a concept known as 'the Dreaming.' This English term – 'the Dreaming' or 'Dreamtime' – was created by Westerners studying Aboriginal people in Australia in the late 1800s.[2] Using this term to encapsulate the complexity of Aboriginal law, religion, and culture is simplistic and insufficient, but the concept entered the English language through the work of English-speaking anthropologists and has 'stuck.' 'The Dreaming' is often described as the time of creation, but as Nicholls (2014) explains:

> …(It) isn't something that has been consigned to the past but is a lived daily reality … it incorporates creation and other land-based narratives, social process including kinship regulations, morality, and ethics. This complex concept informs people's economic, cognitive, affective, and spiritual lives.

Nicholls (2014) goes on to explain that there is, in fact, no universal term to represent the complex beliefs and legal systems of the nations and peoples in the continent now known as Australia. Before 1788, there were more than 500 autonomous nations across the continent, with approximately 250 separate Aboriginal languages and between 600 and 800 dialects, and each of these nations had their own legal system,

> The Warlpiri people of the Tanami Desert describe their complex of religious beliefs as the Jukurrpa. … The Kija people of the East Kimberley use the term Ngarrankarni (sometimes spelled Ngarrarngkarni); while the Ngarinyin people (previously spelled Ungarinjin, inter alia) people speak of the Ungud (or Wungud). "Dreaming" is called Manguny in Martu Wangka, a Western Desert language spoken in the Pilbara region of Western Australia; and some North-East Arnhem Landers refer to the same core concept as Wongar – to name but a handful.
>
> *(Nicholls 2014)*

While acknowledging the diversity and autonomy of the hundreds of nations across the continent, Aboriginal political scientist and philosopher, Mary Graham, also notes that certain concepts can be described as 'pan-Aboriginal', as they are foundational and present across all different nations. Two such foundational concepts are the Relationist Ethos, and the Law of Obligation.

> Aboriginal relationality – traditionally the foundation of the Law – is an elaborate, complex and refined system of social, moral, spiritual and community obligations that provided an ordered universe for people. Within the context of this system, relationality embraced uncertainty and imprecision, consented to being driven by feeling, accepted and made room for conflict while regarding invasion and warlikeness as not only invalid but highly inefficient, resolved the contradiction between power and authority, provided coherence about the notion of a life, as Deleuze put it (citing Deleuze 2001: 27), and, finally, assumed and lived *as if* not only groups, but all people, are autonomous beings.
>
> The ethic that grew out of the Land/Human relationship could be called a custodial ethic. *This is an obligation to look after the Land that nurtures us, the ancient reciprocal relationship with*

> *nature, an ethic of looking after, stewardship, caring for, rather than a survivalist ethos with its rivalry and competition over resources and structural conflicts in hierarchies regarding power.* (Graham 2014 – emphasis added)
>
> In her forthcoming book *Future Law*, Graham further elaborates on the 'Law of Obligation,' as follows:
>
> The lives of Aboriginal people have been interwoven with specific others for millennia based on a custodial ethic handed down to descendants and consolidated in creative dramas (Dreaming Stories), rites and ceremonies, and language.
>
> The Law of Obligation is not like Western Law, not individually based or good faith based, not power wielding or hierarchical. There's no dominating or coercive sense of being *made* to carry out *duties*. Nor is there formally or officially any actual law called a Law of Obligation. Obligation is seen, understood, taught, and passed on generationally. It's more than a strong sense of responsibility, it is woven into the fabric of culture and has the highest intrinsic value. While we retain agency, the social, obligatory, custodial ethic of looking after family and Land, makes people want to fulfil their obligations. People grow up in a society where you're surrounded by others attending to obligations.
>
> The Aboriginal Law of Obligation is a regionally based system – so, for example, in Central Australia the term used is *Jukurrpa* meaning the Laws and protocols given to that Mob,[3] Walpiri, by their ancestors in that Place. Every other Place has their own language word for the same concept. But this concept or first principle, understood and adhered to by Mobs everywhere, is also an organizational principle, a moral code and rules for living and interacting with the environment. This conceptual foundation underpins the social and political ordering of Aboriginal society within regions across the whole continent.
>
> The Law of Obligation is an amalgamation of other social rules like reciprocity, looking after the Land and all its natural resources, sharing resources (like money and goods and housing/shelter), among family and relatives or the collective (the whole Mob), looking after family and relatives, looking after the wellbeing of, and being suitably deferential, to Elders, shared responsibility of relatives looking after each other's children, attending to the rules of kinship systems, playing your role in maintaining community harmony, being cooperative and collaborative with people within and outside one's own community, and attending the consensus – meetings and discussions regarding the wellbeing of, and general business concerns, of the Mob.
>
> *(Graham & Maloney 2023, forthcoming)*

This section then, has shown that in theories like Earth jurisprudence, and in the philosophy and culture of Aboriginal societies, there is a call to recognize the obligations and duties we have to be good custodians and to look after nature and look after each other.

Conclusions

It is quite striking to move from the brief overview of the *Sharma* case – where we see a modern nation state denying that it has a duty of care for its people, let alone turning its attention to non-human members of society – to the emergence within Western thinking of Earth jurisprudence and Earth-centered law and governance, and then into the ancient depths of the Aboriginal 'Relationist Ethos' and 'Law of Obligation.' The Aboriginal 'Law of Obligation' seems to be the antithesis of everything that has brought us to the brink of climate collapse: it is a deeply held understanding that we are in relationship with all things, and that we are obligated to care for all members of the living world and each other.

If modern nation states were able to incorporate even just a small part of the ancient wisdom of First Laws – and the deeply, *deeply* held understanding that we are all obligated to care for country (land, plants, animals) and care for each other – then our legal system and governance responses to climate change would look very different. And we would have acted on our obligations to care for life, when the first information about climate change was being explored and explained in the 1980s and 1990s. Now, our challenge in the Western and industrialized nations is to rapidly learn from and emulate core elements of Indigenous law, such as the 'Law of Obligation,' so that we can transform our governance systems to do everything that needs to be done, to care for the living world and restore our climate system and biosphere.

Notes

1 Berry often described himself as a "geologian" as he studied the Earth rather than theology. See Cullinan, C 2011, *Wild Law: A Manifesto for Earth Justice*, Green Books, United Kingdom, p. 12.

2 For an overview of how the term 'the Dreaming' entered the English language, see Nicholls, C 2014, "Dreamtime' and 'The Dreaming' – An Introduction', *The Conversation*, 23 January 2014, accessed 29 September 2021, https://theconversation.com/dreamtime-and-the-dreaming-an-introduction-20833.

3 'Mob' is a slang word often used in Australia, to refer to a collective of people. It can be used in various ways, depending on the context, and might refer to your Nation, language group, tribe, clan, or family.

References

Berry, T. (1999) *The Great Work: Our Way Into the Future*, New York: Crown.

Berry, T. (2002) "Rights of the Earth: We Need a New Legal Framework Which Recognizes the Rights of All Living Beings," *Resurgence*, September/October 2002.

Berry, T. & Swimme, B. (1992) *The Universe Story: From the Primordial Flaring Forth to the Ecozoic Era-A Celebration of the Unfolding Cosmos*, San Francisco, CA: Harper.

Bosselmann, K. & Taylor, P. (1994) "Governing the Global Commons: The Ecocentric Approach to International Environmental Law," in M. Prieur & S. Doumbe-Bille (eds.), *Droit de l'Environnement et Developpement Durable*, Presses Universitaires de Limoges.

Burdon, P. (2011) "The Great Jurisprudence," in P. Burdon (ed.), *Exploring Wild Law: The Philosophy of Earth Jurisprudence*, Adelaide: Wakefield Press.

Campbell, A. & Linzey, T. (2009) *Be the Change: How to Get What You Want in Your Community*, Kaysville, UT: Gibbs Smith.

Cullinan, C. (2011) *Wild Law: A Manifesto for Earth Justice*, Vermont: Chelsea Green Publishing.

Deleuze, G. (2001) *Pure Immanence: Essays on a Life*, New York: Zone Books.

Dunlevie, E., Hagedorn, M. & Meeking, E. (2021) "Liability and Climate Change Litigation: The Landmark Decision of Sharma v Minister for the Environment," *Russell Kennedy Lawyers*, August 3, 2021, accessed September 28, 2021. https://www.russellkennedy.com.au/insights-events/insights/liability-and-climate-change-litigation-the-landmark-decision-of-sharma-v-minister-for-the-environment.

Environment Planning and Biodiversity Conservation Act – EPBC. (1999) https://www.legislation.gov.au/Details/C2016C00777.

Fairhurst, T. (2021) "Students Launch Climate Crisis Class Action against Minister for Environment Susan Ley," *The Standard*, March 9, 2021, accessed September 28, 2021. https://www.standard.net.au/story/7159429/a-duty-to-protect-us-all-students-launch-climate-crisis-class-action/.

Graham, M. (2014) "Aboriginal Notions of Relationality and Positionalism: A Reply to Weber," *Global Discourse*, vol. 4, no. 1, pp. 17 – 22. https://doi.org/10.1080/23269995.2014.895931.

Graham, M. & Maloney, M. (2023, forthcoming) *Future Law: Tracking the Relationist Ethos*. Future Dreaming Report.

Graham, N. (2011) "Lawscape: Property, Environment and Law," *Journal of Environmental Law*, vol. 23, no. 1, pp. 160 – 164. https://doi.org/10.1093/jel/eqq027.

Guth, J. (2008) "Law for the Ecological Age," *Vermont Journal of Environmental Law*, vol. 9, no. 3, pp. 431–512. https://doi.org/10.2307/vermjenvilaw.9.3.431.

Highham, C. & Setzer (2021) "Global Trends in Climate Litigation: 2021 Snapshot," *London School of Economics and Political Science*, July 2, 2021, accessed September 28, 2021. https://www.lse.ac.uk/granthaminstitute/publication/global-trends-in-climate-litigation-2021-snapshot/.

Iorns, C. (2018) "From Rights to Responsibilities using Legal Personhood and Guardianship for Rivers," in B. Martin, L. Te Aho & M. Humphries-Kill (2019) (eds.), *ResponsAbility: Law and Governance for Living Well with the Earth*, Abingdon: Routledge, pp. 216-239.

Merriam Webster Dictionary (Online) accessed 20 November 2022. https://www.merriam-webster.com/dictionary/.

Minister for the Environment v Sharma (2022) FCAFC 35. https://www.fedcourt.gov.au/services/access-to-files-and-transcripts/online-files/minister-for-the-environment-v-sharma.

Morton, A. (2021) "Australian Court Finds Government Has Duty to Protect Young People from Climate Crisis," *The Guardian*, May 27, 2021, accessed September 28, 2021. https://www.theguardian.com/australia-news/2021/may/27/australian-court-finds-government-has-duty-to-protect-young-people-from-climate-crisis.

Nash, R. (1989) *The Rights of Nature: A History of Environmental Ethics*, Madison: University of Wisconsin Press.

Nicholls, C. (2014) ''Dreamtime' and 'The Dreaming' – An Introduction', *The Conversation*, January 23, 2014, accessed September 29, 2021. https://theconversation.com/dreamtime-and-the-dreaming-an-introduction-20833.

Pelizzon, A. & Ricketts, A. (2015) 'Beyond Anthropocentrism and Back Again: From Ontological to Normative Anthropocentrism, *The Australasian Journal of Natural Resources Law and Policy*, vol. 18, no. 2, pp. 105–124. https://search.informit.org/doi/10.3316/informit.760974676134958.

Sharma by Her Litigation Representative Sister Marie Brigid Arthur v Minister for Environment (2021) FCA 560. https://www.fedcourt.gov.au/services/access-to-files-and-transcripts/online-files/minister-for-the-environment-v-sharma.

Stone, C. (1972) "Should Trees Have Standing? Towards Legal. Rights for Natural Objects," *Southern California Law Review*, vol. 45, pp. 450–501.

Watson, I. (2015) *Aboriginal Peoples, Colonialism and International Law: Raw Law*, Abingdon: Routledge.

3

DUTIES TO CONSIDER THE VIEWS OF THOSE WHO WILL BE HARMED BY CLIMATE CHANGE

Rainier A. Ibana

Introduction

Ecology is defined as "the relation of organisms or groups of organisms to their environment" (Odum 1971: 3). The environment, however, also affects the behavior of organisms. Since there are many organisms to consider in an environment, the mutual relationships between organisms and their environment also affect other organisms that dwell within, and even those that are outside, an environment. These effects and counter effects produce a network of relationships among organisms and environments in more complex ways than what can be imagined by the human mind.

Reflexivity

Although the human mind transcends the capacity of the brain, the former depends on the multiplicity of axons that connect the latter's neurons. These interconnections exemplify the systemic complexity of the relationships among living organisms and their environments. The brain, however, must not be identified with the mind because latter transcends its organic origins by its capacity for reflexivity.

Reflexive thinking enables human beings to deliberately interfere with the interactions among objects, organisms, and their environments. Humans can anticipate the consequences of their interventions through the power of scientific and technological innovations. Human activities in the natural world can be designed and modified to prevent harming oneself and others.

The 'precautionary principle' was formulated by UNESCO's World Commission for the Ethics of Science and Technology (COMEST) precisely as a means of addressing the potentially irreversible harmful effects of human activities on the environment, future generations, endangered species, and human rights (UNESCO 2005: 14). Because of the complexity of effects and counter-effects of human intervention, however, the consequences of human activities are not readily foreseen. The simple act of breathing, for example, takes in the oxygen produced by vegetation. The surrounding vegetation also benefits from the carbon dioxide emissions of human exhalation. The leaves of these plants fall on the earth and produce the necessary nutrients for the soil, fungi, and other micro-organisms that support other forms of life.

This exchange of oxygen, carbon dioxide, and the other elements produced by plants is a product of millions of years of cosmic evolution. With the emergence of humans, the trajectory of evolution has become conscious of itself. This self-consciousness makes it possible to

DOI: 10.4324/9781003039860-5

counter-act the processes of entropy or biological degeneration through the management of the environment in such a way that the latter can enhance life instead of leading it towards destruction.

To broaden the scope of human understanding about a potential damage or harm, the public must be able to participate in the processes of deliberation on proposed human interventions. Public deliberation is a necessary condition to anticipate, prevent, and mitigate the adverse effects of introducing new technologies on the environment. Transparency of discourses, therefore, is an important measure that will enable the consideration of the views of those who will be potentially harmed by human-induced activities on the environment. Climate change, after all, has been found to be anthropogenic (Mcrae 2022). Its social and historical constructs can therefore be transformed, in principle, by human will and intelligence.

Oikos

Environments do not only refer to the wilderness such as deserts, forests, and oceans. It can be designed and built also by human ingenuity to suit their wants and needs. The word 'ecology,' after all, was derived from the term *oikos*, the Greek word for home. Construction builders, such as subdivision developers, are required to allocate green spaces in their housing projects (P.D. 1216). Prospective house and lot buyers, in fact, take the environmental conditions into account in their appraisal of prospective dwelling places.

The notion of home, however, can be expanded to include not only the immediate surroundings. Homes are located within neighborhoods, towns, provinces, countries, and regions of the world. These locations can become either vulnerable or resilient to environmental hazards depending on how human dwellings are designed and constructed. *Feng Sui*, the ancient Chinese belief in how homes are to be constructed, considers the sunlight, airflow, and rainfall in designing buildings and other infrastructures. This belief system is not only applicable to those who belong to Asian cultural circles. It can be useful also for other peoples in other places on Earth. It can even be applied to outer space.

Humans, moreover, are not the only entities that must be considered in the construction and development of environments. Other organisms, and even what appear to be inorganic objects, also contribute to the survival of humans in the same manner that other beings depend on humans for their existence. Humans can contribute to the resilience of ecological niches by increasing the network of relationships among organisms. The environment becomes more robust when the relationships among organisms support the various organisms that dwell in it. The more diverse the organisms, the more stable becomes an ecological niche. The UN Convention on Biodiversity (UNCBD) states that

> 'Biological diversity' means the variability among living organisms from all sources including, *inter alia*, terrestrial, marine, and other aquatic ecosystems and the ecological complexes of which they are part; this includes diversity within species, between species, and of ecosystems.

The variety and diversity of relationships among organisms and the environment is what matters in the determination of the quality of biodiversity in an ecological system. The conjunction 'and' in this quotation from the UNCBD is very significant.

The relationships among organisms and an environment must be balanced if they are to sustain their viability. There must be enough organisms to support a food web that keeps an environment alive and healthy. Voracious consumers that are not threatened by predators will eventually reduce their chances of survival if they do not consider the continuity of the lifecycles of organisms that serve their resource base.

Market demand for palm oil, for example, encouraged investors to convert rainforests into palm tree plantations. The biological diversity of rainforests was then reduced to monocrops. This reduction deprived other living organisms such as orangutans and other endangered species from their sustenance and ecological niche. Orangutans themselves valiantly exerted efforts to protect their homes (Olesen 2011).[1] Their resistance, however, came to naught because of the technological advantage of humans in ravaging the forests.

When rainforests were burned to make way for mono-crop plantations, greenhouse gases were released to the atmosphere. These harmful emissions significantly contributed to increasing the earth's temperature.[2] Water vapors that formed from rain-bearing clouds in the atmosphere evaporated from the waters due to the warm temperatures at the Earth's surface. These vapors are eventually dumped as heavy rains that cause floods, soil erosion, food insecurity, and loss of property and human lives along the banks of overflowing rivers and lakes. Rainforests could have absorbed these precipitations and prevented the release of harmful emissions that pollute the ozone layer. Dense ozone layers prevent the refracted heat from the Earth's surface from being released from the atmosphere thereby warming the air like an airtight pressure cooker.

Climate Justice

Junk food, cosmetics, and other products make use of palm oil. Junk food bingers, beauticians, and other consumers of palm oil products, such as fried foods and cosmetics, can become complicit in the production of greenhouse gas emissions. Endangered species, like the orangutans, are bound to be exterminated if consumers continue to mindlessly support products that make use of palm oil. Human populations must become conscious of their consumption patterns if they are to contribute to the protection of the planet and its vulnerable populations.

One of the paradoxes of the ethics of climate justice is the fact that those who did not significantly contribute to the pollution of the atmosphere are the ones who are most adversely affected by its adverse consequences. Less developed and developing countries, therefore, can legitimately demand compensation for the damage inflicted on them by the excessive carbon dioxide emissions generated by the intensive industrialization of developed economies. Schemes to compensate for these injustices, such as 'debt for nature swap'[3] and loss and damage compensation mechanisms, have been proposed to address this global imbalance.

It must be noted that wars, such as the on-going war in Ukraine, also induce pollution, killing and maiming human beings, and destroying agricultural products. They also indirectly starve others who are not directly engaged in combat. The smoke generated by burning ammunitions contribute to global warming while depriving human and animal populations from their much-needed nutrition requirements and reducing their livelihood opportunities from trading and earning a living.

It is important to consider the views of those that will be harmed by adverse environmental conditions such as climate change because its adverse effects are detrimental to everyone, including those who benefit from the extraction of the planet's natural resources. Climate migrants, those who have lost their homelands due to harsh weather conditions, will flood the gates of fortunate populations living in temperate regions. Migration will increase the density of affluent populations and produce unintended consequences such as racial violence, competition for opportunities, and over-crowding.

Local Knowledge

Although climate change is a global phenomenon, its adverse consequences are felt at the local level. Local knowledge claims by those who are immediately affected by climate change can contribute to the development of adaptation mechanisms that can delay or reduce the adverse effects of climate change. Local populations are the ones who dwell and live with the daily effects of harsh environments. They have special forms of knowledge and possess substantive information that are not immediately accessible to mere observers and visitors. Local people know the terrain and understand the signs provided by animals and plants that live around their environments. Even the migration patterns of birds and other species provide information about changing weather conditions in other parts of the world that lie beyond their immediate surroundings.

Because of their local knowledge, those who have become familiar with their environments have learned to adapt to the adversities caused by the changes of the seasons and the climate. These mechanisms of adaptation must be recognized and respected when introducing technologies that unintentionally disrupt local lifestyles. The introduction of faucets in far-flung areas, for example, makes it convenient for those who fetch water from artesian wells. But alternative forms of socialization must be provided for those who gather around the wells as a means of meeting and socializing with their neighbors.

Bridging the Gap between Scientific and Local Knowledge

The gaps between scientific and local knowledge are not easily bridged by the mere installation of information technologies. The language of science is usually unfamiliar to the local dialects. Scientific terminologies must be translated into the language of everyday life. But it takes a while to find the appropriate terminologies that can be readily understood by local populations while disasters suddenly strike when people are not yet fully prepared to deal with them. The social sciences can contribute tremendously to the continuing translation of local terminologies to the language of the sciences and vice-versa.

When typhoon Haiyan hit the provinces of Leyte and Samar in the Philippines in 2013, weather forecasters warned of 'storm surges' that will be generated by its accompanying strong winds. *Daluyong*, the local term for storm surge, however, was not familiar with the Manila-based weathermen. Consequently, the local population was not able to properly prepare for the catastrophe (Santos 2013). Aside from the scientific study of the adverse consequences of climate change, the appropriate terminologies about weather disturbances must be made readily available to local populations who are immediately and adversely affected by them.

Aside from the problem of translation, the sensibilities of local people must likewise be considered. Surveys of alternative livelihoods that can be explored and harnessed from the terrain of affected populations must be conducted. In explaining to foragers that cutting trees and burning wood to produce charcoal will cause more typhoons, for example, consideration must be made on how the affected parties will fuel their homes and prepare their food. Alternative livelihoods and affordable energy resources must be offered to them if the transformation of their traditional practices can be redirected effectively towards resource conservation. These alternative proposals will prevent the local populations' stark resistance and prevent confusion among them.

Sustainable ways of planting, cultivating, and harvesting trees can be introduced and explained to the stakeholders. Such explanations can help clarify how sustainable development initiatives can reduce carbon emissions, and eventually benefit the locality. Convectors from sunlight and solar power panels exemplify how heat and light can be generated to empower homes and kitchens at the local level. These local forms of adaptation to energy consumption reduce the cost of transporting energy and divert the saved energy resources for domestic consumption.

Special attention must be given to organisms whose expressions of distress are not articulated in terms of human language. Signs of environmental degradation can be detected in bleached coral reefs, marine animals that ingested microplastics, fish kills from red tides, mottled barks of trees due to acid rain, skin rashes on animals from new parasites that proliferate due to warmer weather conditions, etc. These cases require scientific methods of hypothesis formation, and the delicate processes of verification, falsification, and interpretation. Science can tremendously serve local knowledge claims.

Intermediate Technologies (Schumacher 1974)

Addressing the scientific problems generated by climate change at the local level needs intermediate technologies that will bridge the gap between the universality of scientific theories and the circumstances of the localities. Creative thinking will be required from engineers and entrepreneurs in the adaptation and invention of new products. Examples of these creative adaptations abound in the developing world. Solar lamps, made of plastic bottles infused with chlorinated water have been used to light up dark spaces by inserting them on roofs that allow the sunlight to refract their rays inside the shacks and homes of those who cannot afford to pay for electric bills (Laylin 2022). African night crawlers and worms that accelerate the conversion of decaying matter to become high-grade fertilizers have reduced household wastes while increasing the yield of agricultural produce (Richards 2022).

In Vietnam, rice fields that seem to have been rendered useless by salt-water intrusions were converted into shrimp farms (Vu 2021). This shift in livelihood programs has increased the income of farmers who have become fisherfolk instead. Ram pumps, generated by the flow of water itself, helped upland farmers who do not have access to electricity and power generators (Jocags 2017).

The ingenuous creativity of the marginalized sectors of society in coping with the odds presented by climate change depends on their creative capacity to make do with the material conditions that are available from their surroundings. But support mechanisms must be in place to scale up their inventions for the benefit of others who are similarly situated. The exchange of ideas among the disadvantaged sectors of society can go a long way in responding to the problems generated by climate change.

An important lesson that can be learned from these intermediate technologies is the deliberate effort of grassroots workers to align their efforts with, and recycle from, what is available in their natural surroundings. Their attempts to make use of what is already available from nature instead of extracting new resources and exporting them to other places can contribute to the overall conservation of the Earth's energy.

Fuel consumption and its accompanying costs are reduced, and local production is enriched, by reusing and recycling the material resources that are already extracted from the ground. These adaptive technologies also contribute to the mitigation of the causes of climate change. They justify the claim of local communities that they can also become legitimate partners, and not mere beneficiaries, of collective efforts to reduce the adverse effects of climate change.

Prudential and Aesthetic Judgments

The different contexts and problems faced by local communities and human individuals are as varied as each situation they are in. Aside from knowledge of the terrain, environmental adaptation requires the coordination of multiple factors such as the history of the soil and its previous vegetation, water resources, wind direction, and the other animals that depend on them.

Permaculture, the science of rearranging nature's resources so that they can feed on each other, is premised on the sustainable development of support mechanisms generated by the various inhabitants of an environment. Organic wastes are composted to fertilize the soil while water systems are allocated in lower elevations where aquatic plants and animals can be grown and cultivated. These arrangements are a result of observing the dynamic relationships of the organisms in an environment. Their survival and flourishing are made possible through human interventions that follow the contours of the natural world.

Locating organisms in their proper places where they can flourish and support one another can be compared to artistic ways of rearranging the material world. Symbiotic relationships are organized and relocated to create feedback loops that increase the productive capacity of organisms that dwell therein. Not unlike classical works of art wherein propriety and the consideration for the material stuff that constitutes the artwork guides the development and emergence of new creations, permaculture likewise takes the role of each element in an environment in the propagation and sustainability of the whole ecological niche.

This requires a kind of judgment that takes the different parts of a dynamic whole into account. This holistic perspective requires a long-term and distant point of view wherein the different parts of the whole contribute to the survival of the other parts that constitute the whole environment in complementary and supportive ways. The survival of one part can affect the survival of other parts and the well-being of the whole niche depends on the well-being of each of its constituent parts.

Ecological niches, moreover, are not to be taken in isolation from the broader context that it finds. The productivity of farmlands depend on the healthy condition of the mountains that surround it, the drainage system that allows water to flow into itself, and the seasons that suit the cycle of planting, cultivating, and harvesting various types of products. *The Precautionary Principle* considers the multiple variables that must be considered in the performance of measures that prevent harming vulnerable subjects. COMEST compared this principle to handling court cases:

> Just as in legal court cases, each case will be somewhat different, having its facts, uncertainties, circumstances, decision-makers, and the element of judgment cannot be eliminated.
> *(COMEST 2005: 15)*

Nevertheless, jurisprudence still dictates that when rendered uncertain about the norms that will be followed to pronounce judgment on cases that are encountered for the first time, it is best to follow the precedents that govern cases that are similarly situated. When it comes to ethical judgments where there are no precedents because of the unique circumstances that humans usually find themselves in, the prudential virtues of individuals and communities can serve as guideposts for what is considered right or wrong under the prevailing circumstances.

Simple Lifestyles

The cultivation of 'simple lifestyles so that others may simply live' has become the slogan of those who declutter their surroundings to expand their dwelling spaces. The garbage for some people could become treasures for others. Recycling and reusing materials could go a long way to extend the lifecycle of products that have been extracted from nature.

One of the obscene paradoxes of the contemporary world is the attempt to reduce obesity in well developed economies while underdeveloped nations harbor malnourished and stunted children. Improving the quality of life requires the moderation of consumption patterns on the part of those who have more to spare for others while providing the appropriate nutrients for depraved populations. Reducing the scarcity of food among the poor and improving quality of dietary requirements of the privileged sectors of society can improve the well-being of the human population in general. The emphasis here is on well-being rather than ostentatious consumption of the world's resources. This redistribution is not only applicable to food and nutrition, but in terms of clothes and dwelling places also. One need not wear imported designer clothes when local alternatives are available. The latter will generate employment among local populations while reducing transportation costs due to importation. Houses need not be framed by metals when bamboo grass that grows abundantly and replenishes easily has proven to contain more tinsel strength when cured properly.

Social Capital

This holistic approach to the challenges posed by climate change implies that the cooperation of all sectors of society, in their various capacities, must contribute to the solution of this global problem. The most adversely affected parties by the tragic consequences of climate change are not able to fully contribute to its mitigation. The climate crisis has added another burden to those who have been impoverished already by the prevailing inequalities brought about by the unequal distribution of power, lack of material resources, and impoverished cultural opportunities.

The lack of material conditions to support the interests of vulnerable populations can be compensated by their cooperation with others. Cooperation increases the power of individuals not only from the collective aggregation of their contributions. The aggregation itself produces a distinctive identity that exceeds the summation of its parts. Considering the interests of those who are affected by climate change, therefore, improves the quality of responses to the problems caused by climate change.

The qualitative improvement of the ideas produced by societies that deliberate on their actions before executing them is not owned by their members. Better arguments and deeper insights emerge out of the exchange of opinions among the participants of the discourse. By responding to objections and comments by other participants, intellectual blind spots are identified and reconsidered. The process of deliberation functions like the feedback loops of production systems that improve and increase the overall energy of the system itself.

Social capital, the degree of cooperation that can be expected from others to make societies function effectively, is a legitimate resource for the improvement of the quality of life. Studies have shown that those who can rely on others to support their economic endeavors also have increased chances of success in their enterprises (Kenton 2022).

So-called 'capital' need not merely depend on the accumulation of financial resources alone or be evaluated in terms of the surplus value of the wages of laborers, or the speculative value of natural resources from the environment. Productivity can be enhanced also by the cooperative and collective effort of the whole community. In the case of climate change and global warming, this cooperation requires the joint efforts of humanity.

The Moral and Intellectual Solidarity of Humankind (COMEST 2013: 15)

Human beings cannot abdicate their responsibility for activities that cause global warming and their adverse effects on the environment and other living beings. The intricate relationships between organisms and the environment serve as a paradigm to understand the needed social cooperation by all human beings to address the problem of climate change. Global problems, such as climate change, moreover, cannot be solved by individual nations or corporations alone.

Global warming cannot be comprehensively addressed by merely reducing and minding consumption patterns. When large human populations are starving and depend on their natural environments for sustenance, asking them to exercise frugality could mean life and death in relation to their present needs. It is incumbent for affluent individuals and developed countries to assist those who are in dire need, especially when disasters strike, and it is beyond the latter's means to cope with catastrophic calamities. The advanced development of most industrialized countries, after all, was fueled by energy sources that induced global warming and adversely placed vulnerable populations at a disadvantage.

Cooperation, moreover, can happen on many fronts: from a peaceful resolution of conflicts to avoid the atrocities of war that release carbon emissions from burning fossil fuels and weapons of mass destruction, to the humble means of adaptation by the poor and marginalized sectors of society to the adverse effects of climate change. These burdens of resolving the problems of climate change mitigation and adaptation, must be shared by everyone. Only by means of cooperation can we prevent further damage to the planet and the other the living organisms that cohabit our shared living spaces.

Conclusion

Considering the views of those who will be harmed by climate change requires allowing the vulnerable sectors of society to participate in public and global discourses. The disadvantaged sectors are legitimate partners, and not mere beneficiaries, of the solutions that address global problems, especially because they are the ones who are adversely affected by natural calamities caused by global warming. They must be compensated for their loss of lives and livelihood so that they can build back better homes and move on with their lives following the requirements of human rights and the enhancement of their dignity as human beings.

Because humans depend on natural resources to achieve a decent life, it is important to protect the natural resource base that supports life by increasing productivity and properly distributing the fruits of production according to the principles of equity and justice. Increased productivity and equal distribution of resources require the support of science and technologies that are not readily accessible to local communities. Prudential judgments on the viability of local initiatives and the adaptability of new technologies to address the problems posed by climate change must be exercised to avert unintended adverse consequences that are not immediately anticipated by well-intentioned agents who introduce new ways of dealing with the natural world.

The social capital of vulnerable populations can be enhanced by cooperating with them in the organization and enhancement of livelihood programs that conform with the contours of the natural world. Permaculture and the enhancement of biodiversity can serve as models for the kind of cooperation that is needed in the social world. Moreover, since humans are imbued with intellectual capabilities that transcend the limitations imposed by nature, the intellectual solidarity of humankind must be exercised by sharing best practices that have succeeded in addressing the problems generated by climate change.

Notes

1 "The Production of Palm Oil," available at https://theproductionofpalmoil.weebly.com/palm-oil-versus-orangutans.html (Accessed on October 28, 2022).

2 See, Union of Concerned Scientists, "Palm Oil and Global Warming," available at https://www.ucsusa.org/sites/default/files/legacy/assets/documents/global_warming/palm-oil-and-global-warming.pdf (Accessed July 21, 2023).

3 UNDP, "Tools," *Sustainable Finance Hub*. Available at https://sdgfinance.undp.org/sdg-tools/debt-nature-swaps (Accessed on October 25, 2022).

References

Jocags. (2017) *What Is a Ram Pump and How It Works?* Gilman Rampump & Enterprises. Available at https://gilmanrampump.com/howrampumpworks/. Accessed on October 25, 2022.

Kenton, W. (2022) *Social Capital: What Is Social Capital? Definition, Types, and Examples,* Investopedia. Available at https://www.investopedia.com/terms/s/socialcapital.asp. Accessed on October 27, 2022.

Laylin, T. (2022) *MIT Students Install 10,000 Revolutionary Solar Bottle Lamps in Manila Slums*, Inhabitat. Available at https://inhabitat.com/mit-students-install-10000-revolutionary-solar-bottle-lamps-in-manila-slums/. Accessed on December 19, 2022.

Mcrae, M. (2022) "What Is Anthropogenic Global Warming?" *Science Alert*. https://www.sciencealert.com/anthropogenic-global-warming. Accessed on December 19, 2022.

Odum, E. (1971) *Fundamentals of Ecology*, Philadelphia, PA: W.B. Saunders Company.

Olesen (2011) "Palm Oil Versus Orangutans," *The Production of Palm Oil*. Available at https://theproductionofpalmoil.weebly.com/palm-oil-versus-orangutans.html. Accessed on October 28, 2022.

Philippine Presidential Decree 1216 (PD 1216) (1977) Available at https://housepinoy.com/real-estate-philippines/defining-%E2%80%9Copen-space%E2%80%9D-residential-subdivision-and-amending-sec-31-pd-957-pd-1216. Accessed on October 25, 2022.

Richards, S. (2022) "African Night Crawlers: All you Need to Know," *Worms Kill Waste*. Available at https://wormskillwaste.com/african-nightcrawlers/. Accessed on December 19, 2022.

Santos, L. A. (2013) "Storm Surge: Lost in Translation and Interpretation," *Devex*, November 15, 2013. Available at https://www.devex.com/news/storm-surge-lost-in-translation-and-interpretation-82311. Accessed on October 24, 2022.

Schumacher, E. F. (1974) *Small Is Beautiful: Economics as if People Mattered*, London: Abacus Edition.

United Nations Convention on Biological Diversity (UNCBD) (1992) Available at https://www.cbd.int/convention/text/. Accessed on December 19, 2022.

Vu, K. (2021) "Faced with Climate Challenges, Vietnamese Farmers Switch to Shrimp," *Reuters*. Available at https://www.reuters.com/business/environment/faced-with-climate-challenges-vietnamese-rice-farmers-switch-shrimp-2021-05-06/. Accessed on December 19, 2022.

World Commission on the Ethics of Scientific Knowledge and Technology (COMEST) (2005) *The Precautionary Principle*, Paris: UNESCO.

World Commission on the Ethics of Scientific Knowledge and Technology (COMEST) (2013) *Background for a Framework of Ethical Principles and Responsibilities for Climate Change Adaptation*, Paris: UNESCO.

4
THE ETHICS OF CLIMATE TARGETS
Extinction Rebellion v. The Climate Establishment

Stephen M. Gardiner

Introduction

The foundational international treaty on climate change, the United Nations' Framework Convention on Climate Change (UNFCCC), states as its ultimate objective the avoidance of "dangerous interference with the climate system" (United Nations 1992). 'Dangerous' is a normative term, and the aim of avoiding danger emerges from the ethical ambition to protect current and future generations. Yet, when specifying what is required to meet this ambition, major actors typically avoid normative language and instead propose descriptive targets. For example, the Paris Agreement specifies the aim of staying well below 2°C and pursuing efforts to limit warming to 1.5°C. Similarly, the activist organization Extinction Rebellion (XR) has demanded that governments "act now to … reduce greenhouse gas emissions to net zero by 2025" (Extinction Rebellion 2020/2022).

This chapter argues that disputes about climate targets can be understood only by taking seriously the ethical foundations of climate policy, and the centrality of concern for justice. Unfortunately, too heavy an emphasis on 'the numbers' obscures what is at stake and what really needs to be done. The chapter illustrates this through looking at the discrepancy between XR's demand and more mainstream proposals to meet the Paris goals. It then argues that the position of the climate establishment is ethically vulnerable, and that more radical proposals such as XR's should be taken seriously. The chapter concludes by briefly raising questions about the kinds of ethical principles that ought to guide climate targets.

A Troublesome Picture

Some of the biggest problems of our times are obscured by a naïve but persistent picture of how public policy works (or should work). According to this picture, appropriate social action comes in three basic steps:

(Step 1) Scientists determine the goals that need to be met;
(Step 2) Policy experts (e.g., scientists, economists, politicians, diplomats, engineers, etc.) determine how to meet those goals; then
(Step 3) Society as a whole (e.g., governments, firms, civil society groups) implements those solutions.

 DOI: 10.4324/9781003039860-6

One specific instantiation of the picture involves climate change:

> (Climate step 1) Science tells us that the goal of international climate policy should be to avoid exceeding 2°C in global temperature rise, and to pursue 1.5°C;
> (Climate step 2) The mainstream policy community tells us that this means achieving net zero emissions by 2050; therefore,
> (Climate step 3) The international community should move swiftly to implement the changes necessary to achieve net zero by 2050.

This basic framing of the climate problem seems implicit in much public discussion. Many states, NGOs and activist organizations emphasize numerical climate targets such as 2°C, 1.5°C, and net zero by 2050.[1] In doing so, they typically cite scientific reports. Similarly, prominent climate advocates such as Greta Thunberg often speak of the need to "listen to the scientists" or "follow the science" when they push for enhanced climate action (e.g., Ortiz 2019). At first glance, then, it is natural to think of climate targets as primarily a scientific and policy issue, in a way that reflects the three-step model. Notably, this appears to leave little or no room for ethics. The implicit assumption appears to be that, even if the ultimate motivation for addressing climate change is ethical, once that background motivation is in place, we can swiftly leave ethics behind and focus on the numbers.

This chapter will argue that the first impression is mistaken. In fact, the central issues surrounding climate targets are ethical, rather than scientific. One sign that the naïve picture is misleading emerges from digging just a little deeper into the scientific reports themselves. For instance, the most authoritative source in mainstream climate science – the United Nations' Intergovernmental Panel on Climate Change (IPCC) – clearly rejects it. First, and most generally, the IPCC makes clear that value judgments are required "at almost every turn in decision making about climate change," and that this is partly the domain of ethics, broadly construed, including moral philosophy (IPCC 2014: 37).[2] Second, the IPCC singles out determining what constitutes 'dangerous anthropogenic interference with the climate system' as its leading example of where value judgments are needed (IPCC 2014: 37). Third, and of special relevance here, the IPCC offers establishing mitigation targets as its main illustration of where value judgments are central: "For example, setting a target for mitigation involves judging the value of losses to people's well-being in the future, and comparing it with the value of benefits enjoyed now" (IPCC 2014: 37). Clearly, then, the IPCC rejects the naïve picture and regards ethics as essential to climate targets.

A second sign that the naïve picture is misleading is that the most prominent mainstream targets are contested by many leading scientists and activist groups, who push for more aggressive action. One notable example is XR, a grass-roots activist organization that describes itself as "a decentralized, international and politically non-partisan movement using non-violent direct action and civil disobedience to persuade governments to act justly on the Climate and Ecological Emergency" (Extinction Rebellion 2020/2022), or sometimes 2025 for rich countries and 2030 for poor countries (e.g., Read 2021). XR has very few high-level goals, but one is to aim for net zero by 2025 (Extinction Rebellion 2020/2022). This is in sharp contrast to the mainstream goal of net zero by 2050.

At first glance, this difference between mainstream policy and XR might be assumed to turn on an empirical dispute about what is possible. If so, some might say that there ought to be a presumption that the dispute should be resolved in favor of the mainstream view. After all, the mainstream goal emanates from the policy community, which has considerable data and in-depth analysis on its side; by contrast XR is a loosely formed activist organization and does not appear to offer a detailed defense of its alternative goal, of getting to zero much more quickly. Consequently, it is natural to think that, if society should 'follow the science,' it ought to support the 2050 target and forget about XR.

Unfortunately, this dismissal of XR is much too complacent. For one thing, the impression the dismissal creates of the dispute – as between hard science and amateur conjecture – is faulty. For another thing, when that picture is rejected, it becomes clear that ethics, not science, is at the center of the issue. As a result, foregrounding ethics has important implications for thinking about targets. One implication is that the case against XR's target is not nearly so strong as it initially appears. A second implication is that the case in favor of the more moderate mainstream is much weaker that it seems. A third implication is that the weakness of the mainstream position suggests that, even if one thought that XR goes too far in its proposal – and that is not a claim being made here – some alternative radical proposals may still be warranted and indeed much more appropriate than the mainstream position. This is of enormous practical relevance. If the mainstream is much too timid from an ethical point of view, then a more aggressive approach to climate policy is required.

What's Possible?

Many climate goals are expressed using the language of 'what's possible.' For example, the Paris Agreement puts forward the aim of reaching a global peak in emissions "as soon *as possible*," and includes the idea that "each Party's successive nationally determined contribution will … reflect its *highest possible ambition*" (UNFCCC 2015: Article 4, emphases added).

On the surface, appeals to 'what's possible' sound like straightforward claims about *physical* feasibility. When we say that something is not physically feasible, we mean that the constraints imposed by physical laws and processes entail that it cannot be achieved. One example would be the inability of a normal human being to leap over the Empire State Building in one bound; another would be the impossibility of powering the current infrastructure of the USA with a single AAA battery.

Is it physically impossible to get to net zero by 2025? Looked at narrowly, the answer is clearly 'no' (Gardiner 2011). Human actors (including states, corporations, local communities, and so on) could simply switch off everything that uses fossil fuels, cease other emissions-causing activities, and then refuse to start them again (ever). This is perfectly *physically* possible. It does not require violating any laws of nature. It does not even require new technologies. It just requires stopping. This could be done tomorrow. Humanity could simply call an end to the fossil fuel era at 9 a.m., Eastern Standard Time. Moreover, there is no need for policy experts to tell us that. It is a basic physical fact about the world.

We can conclude that instant decarbonization in 2025 is perfectly physically possible, and so too is complete decarbonization in 2025, or more generally within a five-year period of whenever you are reading this. Wherever the real dispute between the mainstream and XR is, it is not there. So, where is it? The answer lies in ethics. If there is a good reason to resist sudden decarbonization, it probably rests on an ethical claim: the impacts would be so severe as to make the cure worse than, or at least comparable to, the disease because a very quick cessation of carbon emissions would bring on a humanitarian disaster. This ethical claim rests on an implicit assumption: that a sudden ban of fossil fuels (for instance) would result in a dramatic loss of global infrastructure, affecting many of the most vital functions of societies, including the provision of food, medical care, and employment. So, for example, farms would not be able to run their tractors, hospitals would lose electricity, mobility would be severely impaired, and so on. In a world that is heavily dependent on fossil fuels for energy, and heavily dependent on energy for almost everything else, a very quick cessation of emissions would have dire consequences for human welfare (and likely for the welfare of many nonhumans too).

The humanitarian argument clearly raises an important consideration, and one that speaks strongly at least against instant cessation of emissions now (tomorrow morning, circa April 2021, as I write this). Perhaps it also speaks against cessation in 2025, or even within a five-year window. Still, both claims would require further investigation. We return to that in a moment. For current purposes, however, the central point is narrower: the heart of the question of targets at stake here is ethical, not scientific. It concerns how best to protect current and future people (and nonhuman nature) against both the threats posed by climate change and the threats posed by sudden cessation.

XR takes one view. It believes that the threats of climate change are so severe that drastic action and a swift transition away from fossil fuels are urgently needed. The mainstream policy community takes the view that a more gradual transition is sufficient. Implicitly, this suggests a deeper disagreement about the magnitude of each risk: how much climate risk is acceptable; how much transition risk is acceptable. Nevertheless, the notions of acceptability at stake will be importantly ethical: they will concern (among other things) how much protection for current and future generations is warranted, what the ethical responsibilities of those making decisions about the risks are, and especially what justice demands.

Is XR in trouble here? Is it obviously unethical? Specifically, does XR's advocacy of a sudden, dramatic transition put XR at a disadvantage relative to the mainstream, given the humanitarian argument? Perhaps surprisingly, this becomes less clear when one looks at what the mainstream view itself requires in terms of rapid climate action, and at the context of the overall climate risk.

To begin with, it is worth emphasizing that mainstream approaches to climate action are themselves very demanding. One might even say that they are radical in their own right. For example, a special report from the IPCC in 2018 suggested that both the 2°C and 1.5°C targets could still be achieved, but the conditions were stringent. For 1.5°C, the IPCC stated that global net emissions must decline by about 45 percent from 2010 levels by 2030, and to net zero by around 2050. This, they added, "would require *rapid and far-reaching transitions* in energy, land, urban and infrastructure (including transport and buildings), and industrial systems ... *unprecedented in terms of scale*, ... [with] *deep emissions reductions in all sectors*" (IPCC 2018, emphases added). While the prospects for meeting the 2°C target were more optimistic, doing so still required a 25 percent cut by 2030 and net zero by 2070. Notably, meeting either temperature target would involve a very substantial effort, mandating a fast transition in global energy infrastructure.

Given this, it is not obvious that the mainstream view has significant advantages over XR in being less radical. For one thing, while it is conceivable that cuts on the order of 50 percent in less than a decade provide greater protection against humanitarian disaster during the transition than a complete halt in five years, that claim is largely a conjecture at this point. For another thing, greater protection does not necessarily mean adequate protection. Perhaps humanitarian disaster could come either way. Since both social transformations – XR's and the mainstream's – are radical, there are humanitarian risks for both approaches, and presumably many other ethical risks too. Third, and more importantly, protecting against such risks is likely to be a central task for *any* ethical transition. XR and the mainstream both face challenges here – but then, so too would any serious proposal for targets. Among other things, this implies that background assumptions about what kinds of social mechanisms and norms should be involved in a rapid energy transition need to be laid bare. Again, this seems to require discussion of ethics (see the next section).

Let us turn now to overall climate risk. There are profound risks on the other side as well – of going too slowly. XR is motivated to push for a quicker transition than the mainstream in large part because of the threat of severe -indeed potentially catastrophic – impacts on humans, their civilizations, and the nonhuman world. Importantly, while the international community has settled on 2.0°C and 1.5°C as the crucial targets for avoiding dangerous climate change, reaching much higher temperatures is clearly possible. For instance, for intermediate emissions scenarios, the IPCC projects 2.1°C–3.5°C for 2100 and 2.3°C–4.6°C for 2200; and for a very high emission scenario, 3.3°C–5.7°C by 2100 and 6.6°C–14.1°C by 2200 (IPCC 2021). Notably, the upper bounds of these projections are an order of 3.5°C–5.7°C by 2100 and 4.6°C–14.1°C by 2200. To put that in perspective, a rise in average global temperature of even around 5°C would be of similar magnitude to an ice age shift, albeit in the warmer direction. Moreover, the kinds of climate shift projected by the IPCC would also happen much more quickly than an ice age transition – over a few decades or a century or two, rather than thousands, or tens of thousands of years. Crucially, there is good reason to believe that humanity would struggle to cope with such fast, high-magnitude changes in the global environment. For example, some believe that global agriculture may begin to collapse at around 4°C, which may be reached later this century. The stakes then are very high.

Given this, XR is surely correct that the ongoing political inertia on climate action is very concerning. Since the Rio Earth summit of 1992, when the UNFCCC was negotiated, global emissions have risen dramatically (by 40–60 percent, depending on which gases one includes), and the average global temperature is up by more than 1.2°C already over preindustrial baselines. Conventional approaches to climate policy appear to be failing, and arguably even in disarray (e.g., Gardiner 2022).

It is for such reasons that XR is advocating for such a rapid transition, on the order of five years. Their view appears to be that the nature and magnitude of the threat to the future of humans and nonhumans is severe, that mainstream attempts to protect against this threat are failing dramatically, and that radical measures are needed to overcome this failure and neutralize the threat.[3] These are, in part, ethical judgments. They include implicit judgments about intolerable risk, justifiable sacrifices, and the responsibilities of the current generation. While the precise content of these judgments is not explicit, it seems clear that claims of their general shape are neither inexplicable nor obviously unreasonable given the basic science. There are strong grounds for thinking that the current generation of decision-makers in society, like that before it, is failing in its responsibilities to the future. In short, there is a case to answer. Rival approaches that would resist XR should take on that burden of proof. Yet, as we shall see below, this is a burden that it is not easy for the mainstream to meet.

Baselines

Why might one question the mainstream climate targets, and their adequacy for protecting current and future generations? One problem is the baselines against which the mainstream projections are being made, which appear to show an unethical lack of ambition.[4] In what follows, the focus will be on three main issues: probability, climate sensitivity, and negative emissions.

The first baseline issue concerns *probability*. The Paris Agreement calls for holding global temperature rise below 2°C and pursuing efforts to limiting it to 1.5°C. Many analysts interpret this as requiring at least a 50 percent chance of staying below each threshold, but some model has higher probabilities, such as 66 percent, 90 percent, or even 95 percent. As one might expect, asking for a higher likelihood pushes up the temperature estimate. For example, just before the Paris meeting, Climate Action Tracker projected that the 2015 commitments would provide a 50 percent chance of limiting global warming to 2.7°C and a 66 percent chance of limiting it to 3.0°C, based on a mid-range climate sensitivity (Climate Action Tracker 2015).

This raises an important question: which probability should global climate policy be aiming at? The core objective of the UNFCCC, and so of international climate policy, is to protect current and future generations from dangerous anthropogenic interference with the climate system.

So, which likelihoods of holding temperatures below 2°C or at 1.5°C would count as offering a reasonable level of protection for the future? Like 'dangerous,' 'reasonable level of protection' is a normative, and ultimately ethical, notion, raising important questions of justice and responsibility.

The first problem is *procedural*. It appears that neither the modelers, nor, more importantly, the parties to the Paris Agreement are called on to answer the ethical questions. Specifically, there is no attempt to justify the choice of probabilities on ethical grounds. This procedural problem is quickly stated but no less profound for that, and I will return to it below.

The second problem is *substantive*. Offhand, providing the future with only a 50–50 chance of avoiding dangerous climate change seems irresponsible, and shamefully so. It is difficult to imagine any serious person saying that we had successfully or meaningfully protected our children and grandchildren based on such odds. 50–50 is after all the equivalent of a coin-toss. Even aiming for a 66 percent probability of success accepts a one in three chance of failure. This hardly seems a robust commitment to protecting the future from a severe threat. Recall that climate change threatens severe and potentially catastrophic outcomes for billions of people, and perhaps even an existential threat to humankind. Arguably, in such circumstances, aiming for higher probabilities, such as 90 or 95 percent, would be more appropriate, and more in keeping with an appropriate standard of care for the future. Yet these would, of course, be much more demanding objectives.[5]

Similar concerns arise with the second baseline issue, the modeling of *climate sensitivity*. There is uncertainty about precisely how sensitive the climate is to extra emissions, and the mainstream approach seems unduly complacent in the face of this uncertainty. The first issue is that while conventional models typically employ a mid-range climate sensitivity, choosing a high sensitivity would make an enormous difference. For instance, some models suggest that a high sensitivity implies that the carbon budget for 1.5°C had already been breached by 2021, while others have projected that a high sensitivity would result in a 95 percent probability of a temperature rise of 5.2°C by 2100 based on the Paris NDCs. Given such projections, the ethical concern should be clear: it seems unreasonably complacent to focus solely on the mid-range climate sensitivity, and ignore the risks associated with a higher sensitivity. To do so jeopardizes the future in a way that is difficult to justify. At the very least, the mainstream approach requires further argument in order to be ethically defensible.

The second issue is that mainstream climate targets are largely based on models that assume that future climate change will be linear, and consequently do not take the prospect of nonlinear or even abrupt climate responses seriously enough. It is well known that positive feedback mechanisms (e.g., ocean circulation, rapid permafrost melt) may accelerate climate change, and that there are likely to be 'tipping points' in the climate system. Although we do not know exactly when and where such nonlinear behavior will be triggered, it is likely to result in serious and possibly catastrophic negative impacts. According to leading experts, such climate surprises "require an emergency response" (Lenton et al. 2019). However, conventional mitigation policies do not focus on them, and so arguably do not afford sufficient protection against dangerous climate change. Again, the mainstream appears unreasonably complacent.

To illustrate the ethical concerns more directly, consider two related points. First, given the risk of tipping points and high-sensitivity, and the kinds of impacts they ultimately imply, it is important to question the way in which mainstream projections are normally presented. For one thing, to say, as people often do, that existing Paris commitments 'put the world on track' for 2.7°C seems seriously misleading when this means *only* a 50–66 percent chance under a mid-range climate sensitivity. For another thing, it seems problematic to emphasize the chance of *success* without at the same time emphasizing the implications of *failure*. For instance, instead of 'on track for 2.7°C' based on a mid-range sensitivity, why not say instead that the Paris pledges 'threaten the future with 5.2°C,' given that this is the projection for a high climate sensitivity under 95 percent probability?

The second point is the importance of questioning the mainstream goal of achieving net zero as late as 2050. Again, this is based on a mid-range climate sensitivity and appears to ignore the risks of a high sensitivity. In doing so, the mainstream view appears indifferent to the considerable risks associated with a high sensitivity. One might say that it neglects the compelling thought that protecting the current generation and the future requires a significant safety margin against such risks. To which one might add that anything short of that violates any reasonable standard of care.

Consider an analogy. Imagine you are a parent being told that your child is 'on track' for graduating from high school. Initially, this seems good news. However, suppose you then find out that the school means by 'being on track' only that the probability of graduation is 50–66 percent. Surely, you would be alarmed; probably, you would also feel seriously mislead, and perhaps even betrayed. Moreover, suppose that it then emerges that the 50–66 percent probability is actually for the *average* child, but the school knows that the rate of failure is much higher for higher-risk students, and has no firm idea of whether your child is higher risk. As a parent, it seems likely that you would now be outraged by the misleading nature (indeed flippancy) of the 'on track' remark. Chances are, you would go on to sound the alarm about both the state of the school and its messaging. Presumably, you would demand that the system take stronger action, including by providing better safety margins and a higher standard of care.

If so, when it comes to the climate case, you would be more like XR than the mainstream. For these kinds of concerns are a key reason why XR calls for a much shorter time horizon for net zero emissions. As the philosopher and XR activist Rupert Read puts it:

> XR calls for biodiversity-destruction and climate-deadly carbon emissions to be stopped by 2025 in rich countries, by 2030 worldwide. Only an eye-wateringly demanding date along those lines – impossible (within the bounds of politics-as-usual), and realistic (about what we actually need) – gives us some margin for error, some room for manoeuvre. If the massive uncertainties about the progress of the climate crisis don't turn out in our favour.
>
> *(Read 2021)*

A third baseline issue is equally worrying. Surprisingly, mainstream projections rely on the assumption that over this century there will be huge contributions to mitigation from *negative emissions technologies*, that is technologies that remove carbon dioxide from the atmosphere. They also assume that such technologies will be available on a large scale as soon as 2030. The role assigned to such technologies is critical. According to an influential study published in *Science*, "it is impossible for the Paris commitments to reach 1.5°C with a 50% chance without significant negative emissions," and "even for 2°C, very few scenarios have explored mitigation without negative emissions" (Anderson and Peters 2016: 182).

Unfortunately, society is not currently equipped with such technologies. For this reason, the authors argue that the mainstream projections involve "*an unjust and high-stakes gamble*" (Anderson and Peters 2016: 183, emphasis added). Specifically, "the scenarios *assume* that the large-scale rollout of negative-emission technologies is technically, economically, and socially viable," yet the main technology being relied on to deliver the negative emissions, Bioenergy with Carbon Capture and Storage (BECCS), "remains a *highly speculative technology*" (what is sometimes called 'a technological imaginary') (Anderson and Peters 2016: 182, emphasis added). Moreover, if BECCS and other negative emissions technologies "fail to deliver at the scale enshrined in many IAMs, their failure will be felt most by low-emitting communities that are geographically and financially vulnerable to a rapidly changing climate." When the authors call this "*an unjust and high-stakes gamble*," their point seems to be that it is ethically irresponsible to base climate policy on such questionable assumptions, especially since the fate of billions of people is at stake (Anderson and Peters 2016: 183, emphasis added).[6]

In summary, the mainstream view relies on assumptions about baselines that look ethically contentious and tend to make the conventional targets seem unduly weak and poorly justified. In particular, the targets do not look like robust attempts to deliver on the pledge of protecting current and future generations against dangerous anthropogenic interference.

In choosing mid-range probabilities, mid-level climate sensitivity, and great optimism about negative emissions technologies, the mainstream seems to place less importance on protecting the future than one might reasonably expect, or which can be ethically defended. One might say that the establishment picture 'stacks the deck' against the future. In light of this, more aggressive targets start to look more reasonable. Even if XR goes too far – and again, I am not saying that they do go too far – they do seem to have a point. Moreover, their more aggressive target seems to be raising the right kinds of questions given the gravity of the situation.

Principles

How might the recognition that climate targets are importantly ethical help us to improve the conversation around such targets? A sensible first suggestion is that we should demand more transparency about the ethical assumptions underlying specific targets. Specifically, those putting forward targets should be explicit about how they arrive at them, what the ethical underpinnings are, and how they would justify their approach compared to rival attempts to establish targets. They should also take care to highlight these features of their targets, making them highly visible to the public, rather than (say) hidden in technical details.[7]

A second sensible suggestion is that proposals for targets should specifically address key values, especially when there are tradeoffs being made and where questions of justice are central. One approach would be to offer guiding principles for formulating targets. In particular, we could develop a set of relevant principles of moral salience that help to identify pressing moral considerations that should be taken into account. This is especially important when some are at risk of being neglected, as is likely in a morally-challenging context like climate change (see Gardiner 2011).

What kinds of principles seem promising? One way to proceed would be to begin with principles employed in other contexts. Elsewhere, I have argued for a set of principles for governing geoengineering that I call 'the Tollgate Principles.' Arguably, these might be reasonably extended, with suitable modifications, to global climate governance and policy more generally. So, it is worth asking what the approach might look like when applied specifically to global climate targets. Although a deep analysis and development will have to wait for another time, a brief discussion may be useful for the purposes of illustration.

Suppose the Tollgate Principles for geoengineering were adapted to the context of climate targets so as to deliver a set of corresponding 'Tollgate Target Principles.' These principles can be divided into three broad categories. The first category covers what might be loosely described as procedural issues. Principles in this category emphasize the general character of the enterprise of setting targets, and the implications for how that enterprise should be approached:

- First Tollgate Target Principle (Framing): Global climate targets should be set by or on behalf of the global, intergenerational and ecological public, in light of their interests and other ethically-relevant norms;
- Second Tollgate Target Principle (Authorization): Global climate targets should be set by bodies acting on behalf of (e.g., representing) the global, intergenerational, and ecological public, with appropriate authority and in accordance with suitably strong ethical norms (e.g., justice, political legitimacy);

- Third Tollgate Target Principle (Consultation): Decisions about global climate targets should be made only after proper notification of, and consultation with, those materially affected and their appropriate representatives, and after due consideration of their self-declared interests and values; and
- Fourth Tollgate Target Principle (Trust): The setting of global climate targets should be organized so as to facilitate reliability, trust and accountability across nations, generations, and species.
- Fifth Tollgate Target Principle (Ethical Accountability): Robust governance systems (including of authority, legitimacy, justification, and management) are needed to implement appropriate targets.

The first two Tollgate Principles emphasize the foundational role played by the global, intergenerational and ecological public, and the need of the targets to reflect their concerns, interests, and authority, in keeping with strong ethical norms. The third and fourth Tollgate Principles highlight the necessity for consultation and trust in pursuit of these aims. The fifth principle contends that robust governance structures will be needed.

Notably, taken together, these five principles suggest a mixed verdict on current approaches to setting targets. On the one hand, they seem to offer some support for the role of the UNFCCC, for agreements like Paris, and for a body like the IPCC, all of whom have been influential in setting the conventional targets of 1.5°C and 2.0°C, and the mainstream ways of modeling those targets. For instance, such mechanisms seem preferable to relying solely on unilateral national-level approaches to climate policy, especially ones that concentrate on annual emissions and shy away from questions of global targets and the longer-term future.

On the other hand, there are reasons to be critical. Consider, for instance, doubts about whether the conventional targets reflect the concerns of a global, intergenerational public, and are appropriately authorized by such a public. For example, it is widely thought (first) that international climate agreements are driven primarily by the richer, more powerful nations and emphasize their interests at the expense of the more vulnerable (Dimitrov et al. 2019); and (second) that the commitment of powerful parties to the 1.5°C threshold is 'merely aspirational,' even though many more vulnerable countries view it as necessary for their very survival. Such problems suggest violations of the first and second Tollgate Target Principles. Similarly, in light of such points, it is difficult to argue that conventional targets respect the third and fourth principles, dealing with consultation and trust. Moreover, arguably, the general weakness of the response so far seems to reflect a neglect of the fifth principle (cf. Gardiner 2011, 2014).

Related issues arise with the second category of Tollgate Principles, which concern technical feasibility and predictability:

- Sixth Tollgate Target Principle (Technical Availability): For a global climate target to be policy-relevant, ethically defensible forms of it must be technically feasible on the relevant timeframe; and
- Seventh Tollgate Target Principle (Predictability): For a global climate target to be policy-relevant, the emergence of ethically defensible forms of implementation must be reasonably predictable in the relevant timeframe and in relation to the threat being addressed.

These principles appear to be violated by the reliance on unproven technologies, such as BECCS, particularly on such a large-scale. They also seem in tension with the lack of specific plans to implement climate policies that meet the mainstream targets on the timescales needed.

The third set of target principles focuses on the content of climate policies:

- Eighth Tollgate Target Principle (Protection): Global climate targets should focus on protecting basic ethical interests and concerns (e.g., human rights, capabilities, fundamental ecological values), and be socially and ecologically preferable to other available climate targets or approaches;
- Ninth Tollgate Target Principle (Respecting General Ethical Norms): Global climate targets should respect general ethical norms that are well-founded and salient to global environmental policy (e.g., autonomy, justice); and
- Tenth Tollgate Target Principle (Respecting Ecological Norms): Global climate targets should respect well-founded ecological norms, including norms of environmental ethics and governance (e.g., sustainability, precaution, respect for nature, ecological accommodation).

Taken together, these principles require attention to a range of important norms that play a central role in the focus of climate targets and their implementation. Notably, conventional targets appear to be opaque with respect to many of the aims, such as protecting human rights, ensuring sustainability, and combatting injustice. They also appear to be in conflict with other aims of the Tollgate Principles, such as reasonable precaution (as noted above).

In conclusion, the illustration suggests that the principle-based approach is useful. The Tollgate Target Principles provide a reasonable diagnosis of many defects of current climate targets, and some guidelines for what is needed to overcome those defects.

In light of this, it is reasonable to ask whether we can push further on ethical principles in particular, and especially those likely to be included under the Ninth Principle, respecting general ethical norms. So, again by way of illustration, the paper will close with a few remarks about how that project might move forward.

To begin with, one central concern is protecting the least well-off and most vulnerable. An obvious approach is to prioritize the interests of the global poor, and in particular to protect what Henry Shue calls 'subsistence emissions.' Specifically, Shue tells us: "those living in desperate poverty ought not to be required to restrain their emissions, thereby remaining in poverty, in order that those living in luxury should not have to restrain their emissions" (Shue 1993). Among other things, this suggests the need to introduce what might be labeled the 'Luxuries First Principle' (LFP): emissions required only for luxuries should be subject to much more stringent restrictions than those required for subsistence.

The LFP seems worthy of serious discussion. First, it has immediate application. Note, for example:

> The combined emissions of the *richest one per cent* of the global population account for *more than twice the poorest 50 per cent*. The elite will need to reduce their footprint by a factor of at least 30 to stay in line with the Paris Agreement targets.
>
> *(UNEP 2020, emphasis added)*

Second, the LFP seems highly relevant to climate targets (e.g., Capstick et al. 2020). In particular, if the primary worry about going to zero quickly (as XR proposes) is the risk of humanitarian catastrophe, then it seems that most luxury emissions will not be protected. In light of this, we might propose that principles like XR's 'net zero by 2025' might be justified in a modified form.

In particular, one might advocate for 'net zero *luxury* emissions in five years' rather than 'net zero in five.' Notably, other things being equal, 'net zero luxury emissions in five' does not raise the same kind of humanitarian threat as 'net zero in five.'[8]

The LFP appears to be a serious way of pushing forward the central concern of protecting the most vulnerable. Some, however, may not be satisfied with it, and want to push even harder. Darrel Moellendorf, for example, proposes that we define dangerous anthropogenic interference with the atmospheric system – or dangerous climate change as he prefers to call it – in terms of poverty. Given this, he proposes his Anti-Poverty Principle (APP):

> policies and institutions should not impose any costs of climate change or climate change policy (such as mitigation and adaptation) on the global poor, of the present or future generations, when those costs make the prospects for poverty eradication worse than they would be absent them, if there are alternative policies that would prevent the poor from assuming those costs.
>
> *(Moellendorf 2014: 22)*

It is tempting to think that Moellendorf is simply echoing Shue. However, on the face of it, his approach goes much further. Moellendorf states that poverty defines what counts as dangerous, and that minimizing poverty should be the overarching objective of global climate policy. Given this, he claims that any policy that extends global poverty should be avoided, where this includes policies that would combat climate change, such as strict emissions cuts. This claim has the implications for targets. Notably, in his 2014 book, Moellendorf appears to accept the conventional 2°C target but to reject 1.5°C on the grounds that the costs might be very high and so damaging to the global poor (Moellendorf 2014: 24). However, in more recent work, Moellendorf argues that 1.5°C is defensible and should be the main target (Moellendorf 2022, chapter 1).

One reason to think that Moellendorf was initially too quick to reject the 1.5°C target is the risk of epistemic injustice. It is notable that the strongest advocates for 1.5°C have been the most vulnerable countries (Lenferna 2018). It is also relevant that advocates for 1.5°C have sometimes questioned the values of those who support 2°C instead. Importantly, the dispute about 1.5°C is therefore as much about ethics as it is about science. It is about who deserves protection and about whose voices should be listened to when protection is discussed. For example, in Copenhagen in 2009, Lumumba Stanislaus Di-Aping of the Sudan, the Head of the G-77 negotiating group of developing countries, went so far as to claim that promoting 2°C amounts to *facilitating genocide*:

> [The draft text] asks Africa to sign a suicide pact, an incineration pact in order to maintain the economic dominance of a few countries. It is a solution based on values, the very same values in our opinion that funneled six million people in Europe into furnaces.
>
> *(BBC 2009)*

Moellendorf's overall approach faces more general problems. One is that his particular anti-poverty principle seems myopic. Climate change affects more than the global poor. Threats of widespread suffering and death arise to many other populations, such as to those living in areas affected by wildfires and hurricanes. Preventing dangerous anthropogenic interference with the climate system ought to involve protecting those populations as well as the global poor. It should also include other objectives, such as preserving cultures, protecting wild animals, and so on.

A second issue is that Moellendorf's anti-poverty principle seems too uncompromising. We do not normally believe that poverty trumps all other moral values. Nor do we believe that any reduction in the risk of poverty (no matter how small) would trump any reduction in other risks (however large). For instance, most people would probably endorse prolonging poverty for one day if it were necessary to save millions of lives or to protect massive areas of the natural world.

A third problem is that Moellendorf's anti-poverty principle appears to pay too little attention to context. It is invoked as a constraint on climate policy. But climate change is not the only contributor to poverty and probably not (or not yet) the most dominant. Therefore, a more attractive anti-poverty principle would apply to the global system as such, not just to climate: it would be a *genuinely global anti-poverty principle*, not a climate-specific anti-poverty principle. Moreover, here there are reasons for optimism. Suppose we had a genuine global commitment to eliminate poverty. If so, reducing poverty might be more readily achievable through means other than climate policy, so that in practice climate action need not be sharply constrained by it. For example, perhaps the best approach would be to apply an anti-poverty principle primarily to trade policy, rather than to climate targets such as 1.5°C.

The debate about how to understand poverty constraints is ongoing and important. There are, however, other kinds of principle lurking in debates about targets. One maintains that there are further limits on the kind of sacrifice that should be expected, and climate policies must respect those limits.

The most prominent versions of this approach assume that the limits are one's of self-interest, and in particular national self-interest. So, it is sometimes claimed that climate policy should not demand sacrifices of national self-interest, since otherwise they become politically-infeasible (e.g., Weisbach in Gardiner & Weisbach 2016). I have argued extensively against such views elsewhere and will not rehearse those arguments again here (e.g., Gardiner in Gardiner & Weisbach 2016, Gardiner 2021).

More interesting limits are those that suggest what we might call a principle of generational self-defense. For instance, some worry that proposals like 'zero by 2025' make unreasonable demands of the current generation or groups within it. This kind of concern appears to underlie much of the rhetoric in other areas of the climate debate as well.

Now, on the one hand, it is worth recognizing that some claims of self-defense may be reasonable in the climate context. So, for example, given the history of environmental injustice and environmental racism, there seems to be a considerable threat that many of the costs of a climate transition (and especially the most severe) would in practice be made to fall on vulnerable communities and groups. In context, this may be catastrophic for such groups, or perhaps we should say exacerbate already extreme situations (see Whyte 2019). Thus, one might conclude that any reasonable climate policy should involve protections against environmental injustice. This claim has recently gained prominence in the USA, as the Biden administration has pledged to put environmental justice "at the heart of everything we do" (Rozsa 2021).

On the other hand, other appeals to self-defense do not pass muster. Consider, for example, the claim that no group should make any sacrifice of its perceived self-interest, or that luxury emissions are untouchable. What then might limit reasonable claims to generational self-defense? The traditional literature provides some guidance here, but there are also reasons to doubt the adequacy of that literature. While this is not the place to pursue that topic, it is worth noting that the limits to permissible self-defense may be sharp (e.g., Gardiner 2013, 2016; Fruh and Hedahl 2019). To illustrate, consider the guidance coming from just war theory. One consideration is the principle of proportionality: for example, one cannot legitimately claim that one is entitled to kill someone in self-defense simply because they have momentarily stepped on one's toe. Similarly, one might argue, one cannot rightfully insist on maintaining a very high-consumption, luxury culture when doing so poses a profound threat to billions of people.

That said, it is not the purpose of this paper to settle on a comprehensive account of the ethical principles needed to guide climate targets. Instead, the main point is that such principles are necessary. Without ethical guidance, climate targets become unmoored, and risk being seen as merely arbitrary. Moreover, it seems highly likely that disputes about ethics, principles, and what they imply are already implicit in current policy disagreements. For example, it is a reasonable conjecture that XR believes that its target of zero in five years does not violate a reasonable principle of generational self-defense, is not ruled out by a responsible approach to addressing global poverty, and even perhaps is compatible with a luxury-first principle. By contrast, the mainstream view of net zero by 2050 appears to be much more indulgent of the current generation. For instance, it appears implicitly to give a high priority to shorter-term, narrowly-economic conceptions of national self-interest, to maintaining provisions of luxuries, and perhaps to accepting a greater long-term impact on the world's poor and on future generations. These are contrasting visions of the basic situation and the underlying ethics. Future discussions of climate policy should take this seriously rather than hiding behind 'the numbers.' Anything less suggests at least a woeful lack of transparency, and perhaps also an unethical abdication of responsibility.

Notes

1 In 1990, at the Rio Earth Summit, the nations of the world agreed to avoid "dangerous anthropogenic interference" with the climate system. By 2015, with the Paris agreement, this had come to be interpreted as requiring holding the increase in global temperature to "well below" 2°C, while "pursuing efforts" to limit it to 1.5°C [article 2.1(a)].

2 More generally, the IPCC states "Something is dangerous if it leads to a significant risk of considerable harm ... Judging whether human interference in the climate system is dangerous therefore divides into two tasks. One is to estimate the risk in material terms: what the material consequences of human interference might be and how likely they are. The other is to set a value on the risk: to judge how harmful it will be. The first is a task for natural science, but the second is not [Section 3.1]." It goes on to say: "Judgements of value (valuations) are called for, ... the task falls to decision makers... Social sciences and humanities can contribute ... disciplines, including ethics (moral philosophy), decision theory, risk analysis, and the normative branch of economics, investigate, analyze, and clarify values themselves'. In addition, the IPCC says: "There is a further element to decision making. People and countries have rights and owe duties towards each other. These are matters of justice, equity, or fairness. They fall within the subject matter of moral and political philosophy, jurisprudence, and economics."

3 They say: "We are in the midst of a climate and ecological breakdown. ... our world is in crisis and life itself is under threat ... We are in the midst of a mass extinction of our own making and are governments are not doing enough to protect their citizens, our resources, our biodiversity, our planet, and our future" (Extinction Rebellion 2022).

4 Drawn from Gardiner (2023).

5 Thus, we might ask, are the 50 and 66 percent probability baselines overly generous to the current generation? Is this a case of intergenerational buck-passing? And does the general silence and complacency on this issue suggest moral corruption?

6 Sadly, the fact that much mainstream discussion of Paris conveniently obscures this issue looks like a prime case of moral corruption, of a deep distortion of the ways in which we talk about climate policy that helps to disguise what is really going on, ethically-speaking (see Gardiner 2011, 2022, 2023).

7 Brown et al. (2018) make similar recommendations for other, more downstream issues of implementing targets.

8 At least, this is so in cases where we can assume that net zero luxury emissions does not itself threaten subsistence, as it might if demand for luxury goods fueled subsistence for many. Of course, this might be addressed in other ways – such as through social guarantees of subsistence, at least in the period of transition (e.g., basic income, which may be desirable for other reasons).

References

Anderson, K., & Peters, G. (2016) "The Trouble with Negative Emissions," *Science* 354(6309): 182–183.

BBC. (2009) *Copenhagen Reaction in Quotes*, December 19.

Brown, D.A., Breakey, H., Burdon, P., Mackey, B., & Taylor, P. (2018) "A Four-Step Process for Formulating and Evaluating Legal Commitments Under the Paris Agreement," *Carbon & Climate Law Review* 12(2): 98–109.

Capstick, S., Khosla, R., & Wang, S., et al. (2020) "Bridging the Gap – The Role of Equitable Low-Carbon Lifestyles," In *Emissions Gap Report 2020*. UNEP. Available at: https://wedocs.unep.org/20.500.11822/34432

Climate Action Tracker (2015). Available at: https://climateactiontracker.org.

Dimitrov, R., Hovi, J., Sprinz, D., Sælen, H., & Underdal, A. (2019) "Institutional and Environmental Effectiveness: Will the Paris Agreement work?," *WIREs: Climate Change* 10(4): 1–12.

Extinction Rebellion (2020/2022) Website [accessed April 2020 and May 2022].

Fruh, K. & Hedahl, M. (2019) "Climate Change Is Unjust War: Geoengineering and the Rising Tides of War," *The Southern Journal of Philosophy* 57(3): 378–401.

Gardiner, S.M. (2011) *A Perfect Moral Storm: The Ethical Tragedy of Climate Change*, Oxford: Oxford University Press.

———. (2013) "The Desperation Argument for Geoengineering," *Political Science & Politics* 46(1): 28–33.

———. (2014) "A Call for a Global Constitutional Convention Focused on Future Generations," *Ethics and International Affairs* 28(3): 299–315.

——— (2016) "In Defense of Climate Ethics," in S.M. Gardiner & D.A. Weisbach (eds) *Debating Climate Ethics*, Oxford: Oxford University Press.

——— (2017) "Climate Ethics in a Dark and Dangerous Time," *Ethics* 127: 430–465.

——— (2021) "*Debating Climate Ethics* Revisited," *Ethics, Policy & Environment* 24(2): 89–111.

——— (2022) "Is the Paris Climate Agreement Another Dangerous Illusion?" in *The Norwegian Academy of Science and Letters Yearbook 2021* (pp. 157–174), Novus forlag: Oslo.

——— (2023) "Climate Targets and Moral Corruption," *American Philosophical Association Public Philosophy Blog*. Available at https://blog.apaonline.org/2023/04/10/climate-targets-and-moral-corruption/.

Gardiner, S.M., & Fragnière, A. (2018) "The Tollgate Principles for the Governance of Geoengineering," *Ethics, Policy & Environment* 21(2): 143–174.

Gardiner, S.M., & Weisbach, D.A. (2016) *Debating Climate Ethics*, Oxford: Oxford University Press.

IPCC. (2014) *Climate Change 2014: The Physical Science Basis. Working Group I Contribution to the Sixth Assessment Report of the Intergovernmental Panel on Climate Change*, Cambridge: Cambridge University Press.

IPCC. (2018) "Summary for Policymakers," *Global Warming of 1.5°C*. Available at: https://www.ipcc.ch/site/assets/uploads/sites/2/2019/05/SR15_SPM_version_report_LR.pdf

IPCC. (2021) *Climate Change 2021: The Physical Science Basis. Working Group I Contribution to the Sixth Assessment Report of the Intergovernmental Panel on Climate Change*, Cambridge: Cambridge University Press.

Lenferna, A.L. (2018) *Equitably Ending the Fossil Fuel Era: Climate Justice, Capital and the Carbon Budget*. Doctoral dissertation, University of Washington.

Lenton, T., Rockström, J., Gaffney, O., Rahmstorf, S., Richardson, K., Steffen, W., & Schellnhuber, H.J. (2019) "Climate Tipping Points – Too Risky to Bet Against," *Nature*. Available at: https://www.nature.com/articles/d41586-019-03595-0

Moellendorf, D. (2014) *The Moral Challenge of Dangerous Climate Change*, Cambridge: Cambridge University Press.

———. (2022) *Mobilizing Hope: Climate Change and Global Poverty*. Oxford: Oxford University Press.

Ortiz, E. (2019) "Teen Climate Activist Great Thunberg Tells Congress: 'Unite Behind the Science," *NBC News*, September 18.

Read, R. (2021) *Parents for a Future*, Norwich: UEA Publishing Project.

Rozsa, M. (2021) "Joe Biden Outlines Ambitious Climate Change Agenda," *Salon*, January 28.

Shue, H. (1993) "Subsistence Emissions and Luxury Emissions," *Law & Policy* 15(1): 39–60.

United Nations. (1992) *United Nations Framework Convention on Climate Change*, Nairobi: United Nations.

United Nations Environment Programme. (2020) *Emissions Gap Report 2020*, Nairobi: United Nations.

United Nations Framework Convention on Climate Change [UNFCCC]. (2015) *Adoption of the Paris Agreement*, United Nations.

Whyte, K.P. (2019) "Indigeneity in Geoengineering Discourses: Some Considerations," *Ethics, Policy & Environment* 21(3): 289–307.

5

THE ROLE OF HUMAN RIGHTS IN SETTING A NATIONAL GREENHOUSE GAS EMISSIONS TARGET

Bridget Lewis and Donald A. Brown

Introduction

This chapter argues that those who see climate change as a civilization challenging moral and ethical problem will find that many practical lessons can be learned from human rights law and its philosophical foundations that should help achieve a greater response to climate change, consistent with national, regional, and individual ethical and moral obligations. These lessons will include: (a) substantive conclusions about obligations that follow when specific human rights are violated; (b) procedural lessons about increasing compliance with rights obligations that can be seen by examining almost 65 years of continuing development of human rights law at the international, regional, and national scale; and (c) specific ideas about how to get nations to take their ethical and equity obligations seriously in international climate change negotiations. The chapter will end with recognition of some challenges to a human rights approach to climate change, yet with an explanation why, despite these challenges, greater use of human rights obligations should be made to find a solution to climate change.

Moral and Legal Rights

Individual human rights are widely acknowledged to be derived from the idea that all human beings should be treated with dignity and respect. Furthermore, if one assumes that each and every individual person is entitled to respect and dignity, the human rights which have been widely acknowledged by most of the nations of the world are simply deductions from the obligation to treat every human being with respect.

The *Universal Declaration of Human Rights* (UDHR) adopted by the international community in 1948 begins with the following:

- "Whereas recognition of the inherent dignity and of the equal and inalienable rights of all members of the human family is the foundation of freedom, justice, and peace in the world"; and
- "Whereas it is essential, if man is not to be compelled to have recourse, as a last resort, to rebellion against tyranny and oppression, that human rights should be protected by the rule of law" (UDHR 1948: Preamble).

And so human rights are understood to apply to all people. Furthermore, governments have a duty to protect all citizens' rights by law. This duty is understood to impose a responsibility for nations

DOI: 10.4324/9781003039860-7

to adopt laws to protect human rights. If climate change is violating human rights, therefore, nations have duties to pass laws to prevent climate harms. The duty to do this does not depend upon prior law. The duty precedes legislative action.

That human rights exist prior to legislative action that creates them has been grounded in several philosophical ethical traditions including, for instance, natural law theories, the deontological theories of Kant and others, some utilitarian or consequentialists theories such as the theory of John Steward Mill, and theories of justice derived from fair process including John Rawls' theory of justice as fair process (Moeckli et al. 2017). And so human rights are often claimed to be self-evident truths discoverable by reason. Therefore, there is no need for legislative action to claim that governments have a duty to take action to prevent threats to life, food security, and human health from climate change.

Human Rights Impacts of Climate Change

In 2009, Simon Caney published a seminal paper on the relationship between climate change and human rights (Caney 2009). The paper explains that climate change violates many human rights including three of the most fundamental and least controversial rights: (1) right to life, (2) right to health, and (3) right to subsistence. Climate change violates the right to life because a changing climate will and is killing people through more intense storms, floods, droughts, and heat waves. Climate change will violate the right to health by increasing the number of people suffering from disease, death, and injury from heatwaves, floods, storms, fires, and droughts; increases in the range of malaria and dengue fever, and the burden of diarrhoeal diseases; and cardio-respiratory morbidity associated with ground level ozone. Climate change will violate the right to subsistence by increasing: (a) droughts which will undermine food security, (b) water shortages, (c) sea level rise which will put some agricultural areas under water, and (d) flooding which will lead to crop failure.

Caney explains that other human rights are affected by climate change but an understanding that climate change violates these three rights puts the claim that climate change violates human rights on the most uncontroversial grounds. The three rights to life, health, and subsistence are widely recognized as moral entitlements, but are also enshrined in domestic, regional, and international human rights law and therefore legally protected in most jurisdictions. The right to life, for instance, is guaranteed in article 6 of the *International Covenant on Civil and Political Rights* (ICCPR 1966) and is also found in the various regional human rights treaties. The right to health is guaranteed in article 12 of the *International Covenant on Economic, Social and Cultural Rights* (ICESCR 1966) while the right to subsistence is reflected in both the right to an adequate standard of living (ICESCR: Article 11) and the right to self-determination (ICESCR and ICCPR: Common Article 1). These and other rights are all recognized in the UDHR. In addition, many of the human rights that have now been widely acknowledged as binding obligations of states to their citizens have been recognized for hundreds of years including in documents such as the Magna Carta in Great Britain in 1215, the English Bill of Rights in 1689, the French Declaration on the Rights of Man in 1789, the US Bill of Rights in 1791, and the United Nations Charter in 1945.

And so, most of the human rights violated by climate change have been acknowledged to be binding on governments for a very long time. Because of this, a strong argument can be made that once climate change science had concluded that human activities causing climate change were threatening human lives, health, and food security, nations and others causing climate change had a duty to take action to reduce the threat of climate change despite the absence of a treaty or other consensually arrived at international agreement that would require them to reduce greenhouse gas (GHG) emissions. In fact, most nations of the world are members of at least one if not multiple regional or international human rights agreements, and the *Paris Agreement* on climate change recognizes the need for states to adhere to these existing human rights obligations in addressing climate change (Paris Agreement 2015: Preamble; Rajamani 2018). This provides a solid foundation for arguing that climate change must be addressed as a human rights imperative.

Human rights obligations are understood to include the duties to respect human rights, to protect citizens from human rights violations, and to fulfill human rights enjoyment. The duty to respect human rights obliges governments to refrain from taking any action that would interfere with the enjoyment of human rights. The duty to protect requires governments to protect citizens not only from the acts of the government that would deprive citizens of human rights but also to protect citizens from the climate change causing activities of entities within the nation's jurisdiction. And so, under a human rights approach to climate change, nations have a duty to take action to prevent high emitting entities within their jurisdiction. The duty to fulfill means that nations must enact laws that are necessary to assure that citizens will enjoy human rights. Therefore, governments have duties to pass laws on climate change that will assure that all citizens enjoy the free and full exercise of their rights.

Legal Enforcement of Human Rights

While climate change violates human rights in a moral sense in that it interferes with individuals' ability to enjoy their fundamental entitlements to life, health, and subsistence, whether climate change amounts to a *legal* violation of rights as guaranteed under international and domestic law is a somewhat more difficult proposition.

According to human rights law all rights have: (a) a right-holder, usually a citizen, that is the party who has the right, (b) an object, or what the right is, and (c) an addressee, the party who must make the right available (Moeckli et al. 2017). For human rights violated by climate change, the right-holder is all citizens whose rights are being violated, the addressee is the nation or other government entity, courts, or legislatures, and the object is the rights to life, health, and sustainable food among other human rights infringed by climate change. Applying this framework to establish a legal violation in the context of climate change has proved challenging.

In one of the first cases to attempt to litigate climate change as a human rights issue, the Inter-American Commission on Human Rights declined an application by Inuit peoples arguing that the United States' failure to address climate change constituted a violation of their human rights under the *American Declaration on the Rights and Duties of Man* (1948). Despite being able to point to widespread interferences with the Inuit peoples' livelihoods and culture, the Commission responded that it could not determine that these interferences constituted a violation of the rights protected in the declaration (Inuit Petition to the Inter-American Commission on Human Rights 2005; Osofsky 2007; Peel and Osofsky 2018).

The indirect, cumulative, trans-boundary, and long-term nature of climate impacts pose a challenge for claimants attempting to show a violation of human rights law. Traditionally, international human rights law has positioned states as the principal duty-bearers, with obligations owed to any person within their territory or under their jurisdiction. Climate change raises challenging questions about how to account for the global and long-term effects of states' emissions, and how to prove that a state or group of states have 'caused' the interference with human rights, when all states' historic and ongoing emissions contribute to the problem of global warming. Further, human rights law has typically been applied in a reactive manner, responding to actual interference of individuals' rights. It is less capable of addressing predicted harms, especially where there is uncertainty about the precise nature of those impacts (Knox 2009: 163; Office of the High Commissioner for Human Rights 2009).

However, since the Inuit peoples' pioneering attempt to use human rights law to address climate change, there have been some shifts in legal jurisprudence, and advances in our understanding of climate change have helped to overcome some of the challenges relating to causation and apportionment of responsibility (Marjanac and Patton 2018). Most famously, in 2019, the Supreme Court of the Netherlands upheld a ruling in the *Urgenda* case that the Netherlands' failure to implement tougher regulations to cut GHG emissions amounted to a violation of the state's obligations, including obligations under the *European Convention on Human Rights* (*State of the Netherlands v Urgenda Foundation* 2019).[1]

Recent cases before the Committee on the Rights of the Child (*Sacchi et al v Argentina et al.* 2019) and the European Court of Human Rights (*Cláudia Duarte Agostinho and others v Portugal and 32 Other States* 2020) offer an important opportunity for international bodies to revisit the issues that have in the past been obstacles to human rights-based climate litigation. These cases, both pursued by children and young people, argue that all states have an obligation to prevent the human rights harms caused by climate change by doing their 'fair share' to cut global GHG emissions. Significantly, they seek to enforce this obligation against foreign states, arguing that the foreseeability of climate impacts effectively extends states' jurisdiction to include all those affected by their intentional and ongoing emission of GHGs.

In 2021, the *Sacchi* complaint was ruled to be inadmissible by the Committee on the Rights of the Child because the Applicants had not exhausted their local remedies. However, it still represents an important landmark in the evolution of a human rights-based approach to climate change. The *Duarte Agostinho* case is still on foot and has been fast-tracked and referred to the Grand Chamber of the European Court of Human Rights. Regardless of the outcome in these cases, they have arguably already had an impact by drawing attention to the plight of those affected by climate change (particularly children and young people). Further, the language of human rights has a powerful moral force which may help to prompt stronger action, even in the absence of a legal victory (Bodansky 2014: 511).

Benefits of Human Rights in Addressing Climate Change

Returning to Caney, because climate change clearly violates human rights in a moral sense, certain things follow. These consequences for policy include:

- Because human rights are violated, costs to those causing climate change entailed by policies to reduce the threat of climate change are not relevant for policy. That is, if a person is violating human rights, he or she should desist, even if it is costly. The abolition of slavery was immensely costly to slave owners yet because basic human rights were violated costs to the slave owner of abolishing slavery were not relevant;
- If climate change is a human rights problem, compensation is due to those whose rights have been violated;
- If one has a right not to suffer a particular harm, then it is wrong to violate that right because one can pay compensation. It is for instance wrong to assault someone even if the person assaulted can be paid compensation for the harm;
- The human rights approach generates both duties for mitigation and adaptation, and imposes important standards for the implementation of mitigation and adaptation policies (Pedersen 2011; Lewis 2015);[2]
- Human rights apply to each and every human being as they are based on the idea that all human beings are born free and entitled to certain rights;
- If the human rights of the most vulnerable are being violated, they need not bear the burdens of mitigating the threats; and
- Human rights usually take priority over other human values such as efficiency and promoting happiness.

The universality of human rights noted above offers other benefits as well. In particular, because human rights are enjoyed by all people, a response to climate change which is grounded in human rights may help to address the inter- and intra-generational injustices inherent in climate change. These injustices flow from the fact that developed states have historically contributed most to the problem of climate change, but people in developing states are enduring the worst environmental and economic impacts (Caney 2006). Similarly, current and past generations have generated the emissions which have led to global warming, but climate change will significantly impact current children and future generations (Brown Weiss 2008). This intergenerational injustice is compounded by the fact that children and future generations also have no say in policy decisions about how to manage climate risk. Recognizing that all people have equal rights might reinforce the idea that developed states have a responsibility to address climate change.

A further point to note is that states' obligations to address the human rights impacts of climate change do not supplant their other obligations to respect, protect and fulfill human rights. The need to fulfill economic, social, and cultural rights in particular has often been pointed to as a justification for not taking stronger action to cut emissions. For the least-developed states, there are justice arguments here as well, since industrialized states were able to achieve economic growth through the use of fossil fuels, yet developing states are being asked to limit the use of fossil fuels in their economic development strategies. Advancements in renewable energy and arrangements for finance and technology transfer between developed and developing states help to alleviate this tension. Further, there is increased recognition of states' obligations of international cooperation within human rights frameworks, which not only recognize the shared responsibility for addressing climate change but may also help to overcome some of the jurisdictional limitations on enforcement identified above (Knox 2016). However, the reality is that developing states have many demands on their limited resources and some prioritization of human rights obligations is inevitable.

The most recent attempts at human rights-based climate litigation are argued mindful of the reality that existing atmospheric concentrations of carbon dioxide equivalent (CO_2e) are already creating serious interferences with human rights. This can be seen in recent bushfires, cyclones and floods across the world, depletion of Arctic sea-ice, rising sea levels in the South Pacific, and desertification of agricultural land. Given that a proportion of emissions will remain in the atmosphere for millennia, we are already locked into some level of warming, and every additional ton of CO_2e makes this problem worse. There is growing evidence that we have already reached, or are perilously close to, a number of planetary boundaries and tipping points that could have catastrophic consequences.

This reality removes the excuse that some states have used that the human rights implications of climate change are 'future' problems that they cannot yet be held accountable for (Murphy 2020). The young petitioners currently before the European Court of Human Rights have pointed to recent bushfires in Portugal to demonstrate that they have already experienced violations of their human rights. It also highlights the fact that every state's emissions, no matter how small, contribute to global warming and the human rights violations that it creates. States can no longer rely on the argument that their own emissions do not cause global warming because they are only a 'drop in the ocean.'

The reality that climate change is already occurring, and the gravity of the human rights impacts that are already being experienced, has shifted the focus away from mitigation somewhat and towards adaptation and loss and damage. While this is a sad reflection of the international community's failure to prevent the harmful impacts of climate change, it may have the effect of strengthening human rights as both a language and a legal tool for prompting states to act. The obligation to protect citizens from climate harms through adequate adaptation measures may prove easier to enforce than an obligation to reduce emissions. At the same time, the prospect of hefty compensation claims may spur governments to take both adaptation and mitigation action as a matter of urgency.

The fact that CO_2e emitted to date will continue to interfere with human rights for centuries to come raises a difficult question about the extent to which we owe obligations to protect the human rights of future generations. Human rights law has generally not been applied to future generations, and there are limited avenues for future generations' interests to be formally included in decision-making processes or legal actions. There has been some progress in recent years, mostly based on constitutional rights which include notions of intergenerational equity (*Ashgar Leghari v Federation of Pakistan* 2015; *Demanda Generaciones Futuras v Minambiente* 2018). At present, the best hope may be that successful human rights claims will provide a knock-on benefit for future generations through stronger climate action (Knox 2020). Even without these legal victories, however, an understanding of climate change based on human rights emphasizes the inherent rights of all people, regardless of when they are born, and has potential to contribute to intergenerational climate justice.

Notes

1 A number of other cases have confirmed that states are under an obligation to address the impacts of climate change on their citizens' human rights. See, e.g. *Ashgar Leghari v Federation of Pakistan* (2015); *Demanda Generaciones Futuras v Minambiente* (2018).

2 For example, adaptation that is achieved through relocation of communities must be consistent with their rights, particularly the right to free, prior and informed consent. Similarly, mitigation through changes to land use (e.g. relocation of land from agriculture to biofuel production) should avoid creating other human rights implications (e.g. impacts on food security).

References

American Declaration on the Rights and Duties of Man. (1948) Organization of American States.

Ashgar Leghari v Federation of Pakistan (2015) Lahore High Court, Case No 25501/2015.

Bodansky, D. (2014) "Climate Change and Human Rights: Unpacking the Issues," *Georgia Journal of International & Comparative Law* 38(3): 511

Brown Weiss, E. (2008) "Climate Change, Intergenerational Equity, and International Law," *Vermont Journal of Environmental Law* 9(3): 615

Caney, S. (2006) "Cosmopolitan Justice, Rights and Global Climate Change," *Canadian Journal of Law and Jurisprudence* 19(2): 255–278

Caney, S. (2009) "Climate Change, Human Rights and Moral Thresholds," in S. Humphreys (ed.) *Human Rights and Climate Change*, Cambridge: Cambridge University Press.

Cláudia Duarte Agostinho and others v Portugal and 32 Other States. (2008) European Court of Human Rights, App No 29371/20.

Demanda Generaciones Futuras v Minambiente. (2018) Supreme Court of Colombia, Case No 11001-22-03-000-2018-00319-01.

International Covenant on Civil and Political Rights. (1966) 999 UNTS 171, opened for signature 16 December 1966, entered into force 23 March 1976.

International Covenant on Economic, Social and Cultural Rights. (1966) 993 UNTS 3, opened for signature 16 December 1966, entered into force 3 January 1976.

Inuit Petition to the Inter-American Commission On Human Rights To Oppose Climate Change Caused By The United States Of America. (2005) Inter-American Commission on Human Rights.

Knox, J. (2009) "Climate Change and Human Rights Law," *Virginia Journal of International Law* 50: 163.

Knox, J. (2016) *Report of the Special Rapporteur on the Issue of Human Rights Obligations Relating to the Enjoyment of a Safe, Clean, Healthy and Sustainable Environment.* Human Rights Council, 1 February 2016.

Knox, J.H. (2020) "Constructing the Human Right to a Healthy Environment," *Annual Review of Law and Social Science* 16(1): 79–95.

Lewis, B. (2015) "Balancing Human Rights in Climate Policies," in O. Quirico and M. Boumghar (eds.) *Climate Change and Human Rights* (pp. 39–52), Abingdon: Routledge.

Marjanac, S. and Patton, L. (2018) "Extreme Weather Event Attribution Science and Climate Change Litigation: An Essential Step in the Causal Chain?" *Journal of Energy & Natural Resources Law* 36(3): 265–298.
Moeckli, D. et al. (eds.) (2017) *International Human Rights Law* (3rd ed). Oxford: Oxford University Press.
Murphy, K. (2020) "Australia Asks UN to Dismiss Torres Strait Islanders' Claim Climate Change Affects Their Human Rights", *The Guardian*, 14 August 2020. http://www.theguardian.com/australia-news/2020/aug/14/australia-asks-un-to-dismiss-torres-strait-islanders-claim-climate-change-affects-their-human-rights
Office of the High Commissioner for Human Rights. (2009) *Report of the Office of the United Nations High Commissioner for Human Rights on the Relationship Between Climate Change and Human Rights*, 15 January 2009.
Osofsky, H.M. (2007) "The Inuit Petition as a Bridge?: Beyond Dialectics of Climate Change and Indigenous Peoples' Rights," *American Indian Law Review* 31: 675, at 688, 692–697.
Paris Agreement. (2015) UNTS vol. 3156, opened for signature 12 December 2015, entered into force 4 November 2016.
Pedersen, O. (2011) "The Janus-Head of Human Rights and Climate Change: Adaptation and Mitigation", *Nordic Journal of International Law* 80(4): 403–423.
Peel, J. and Osofsky, H.M. (2018) "A Rights Turn in Climate Change Litigation?" *Transnational Environmental Law* 7(1): 37–67.
Rajamani, L. (2018) "Human Rights in the Climate Change Regime: From Rio to Paris and Beyond," in J.H. Knox and R. Pejan (eds.) *The Human Right to a Healthy Environment,* Cambridge: Cambridge University Press.
Sacchi et al. v Argentina, Brazil, France, Germany and Turkey. (2019) Committee on the Rights of the Child, Comm Nos 104-108/2019.
State of the Netherlands v Urgenda Foundation. (2019) Supreme Court of the Netherlands, Case No 19/00135.
Universal Declaration of Human Rights, United Nations, 1948, art 21.3.

6

APPROPRIATE APPLICATION OF EQUITY

The Case Study of Brazil's National Determined Contributions

Maria Silvia Muylaert de Araujo

Introduction

The majority of the climate policy literature refers to equity as applied to mitigation and finance. This does not exhaust all aspects of equity in acting on climate change policy and issues related to sustainable development. Another important application of equity is climate adaptation. The equity-based climate policy framework should link mitigation, adaptation, and their respective costs. There are important equity interlinkages between mitigation and adaptation climate policies that need to be more deeply discussed. The climate policy context has a clear governmental trajectory of decisions with important global commitments, such as the 1997 Kyoto Protocol and the 2015 Paris Agreement. The topic of equity, responsibility, and related concepts has been at the center of the climate debate since the very first beginning (UNFCCC 1992: Articles 2 and 3), involving mitigation and adaptation topics, although with more emphasis in the mitigation issue, with a crescent new focus on the importance of the adaptation in the last few years.

Equity and fairness are important for considering pathways that limit warming to 1.5°C in a way that is livable for every person and species. They recognize the uneven development status between richer and poorer nations, the uneven distribution of climate impacts (including on future generations) and the uneven capacity of different nations and people to respond to climate risks. This is particularly true for those who are highly vulnerable to climate change, such as indigenous communities in the Arctic, people whose livelihoods depend on agriculture or coastal and marine ecosystems, and inhabitants of small island developing states. The poorest people will continue to experience climate change through the loss of income and livelihood opportunities, hunger, adverse health effects and displacement (IPCC SR 1.5 Chapter 5: 479). It is important to look at the equity interlinkages of both mitigation and adaptation climate policies to help the climate policy formation of policymakers and citizens. How has the Brazilian National Determined Contribution (NDC) as a whole implicitly and explicitly considered the topic of equity and related concepts for both mitigation and adaptation issues?

Equity and Related Concepts: The 2015 Brazil INDC and 2020 Brazil 'First' NDC

The 2015 Brazilian Intended NDC (INDC) and the updated 2020 'First' NDC use the concept of equity mainly linked to historical responsibilities. The historical responsibility deals with mitigation, not with adaptation. The goal is to establish a scientific, robust, quantitative, fair, and

DOI: 10.4324/9781003039860-8

objective indicator to prevent dangerous anthropogenic interference with the climate system. This indicator intends to offer an irrefutable precise relation between GHG emissions and temperature increase in numbers, linking the moment of GHG emissions to the moment of the effect. This historical responsibility proposal of the 2015 INDC, not adopted internationally, considers several ethical perspectives, and political discussions between developed and developing countries:

> Most of the current concentration of greenhouse gases in the atmosphere is a result of emissions since the industrial revolution (the post-1750 period). Current generations are bearing the costs of past interference with the global climate system, resulting from human activities and consequent greenhouse gas emissions, primarily by developed countries, during the last two centuries. Similarly, current human activities around the world will affect the climate system over the next centuries. In order to build a fair and equitable global response to climate change, it is therefore of central importance to link cause (net anthropogenic greenhouse gas emissions) and effect (temperature increase and global climate change) …
>
> *(5)*

> The specific and relative role of each actor's emissions to global climate change can be determined using the global mean surface temperature as an indicator. Each individual actor's contribution to temperature increase should take into consideration differences in terms of starting points, approaches, economic structures, resource bases, the need to maintain sustainable economic growth, available technologies, and other individual circumstances. Establishing the series, in all sectors, of anthropogenic greenhouse gas emissions by sources and removals by sinks allows the estimation of the relative share of total temperature increase attributable to an individual country. The relative responsibility of a given country in relation to the global mean surface temperature increase can be estimated with a high level of confidence. Hence, the marginal relative contribution to the global average surface temperature increase is a relevant measure to evaluate responsibility in the global effort to limit temperature increase to 2°C compared to pre-industrial levels.
>
> *(5–6)*

Whereas the 2020 NDC treats capacity as a technical issue linked to energy power generation capacity, the 2015 INDC mentions the concept of capacity mainly in the context of the National Adaptation Plan (NAP). The focus is on the government and specific sectors conditions to adopt climate adaptation measures

> The Government of Brazil gives particular attention to the poorest populations, in terms of improving their housing and living conditions, bolstering their capacity to withstand the effects of severe climate events …
>
> It should be further noted that Brazil seeks to enhance its national capacity in water security (National Water Security Plan) and conservation and sustainable use of biodiversity (National Strategic Plan for Protected Areas, as well as the implementation of the Forest Code, particularly concerning protected areas …
>
> The National Adaptation Plan will provide a basis for Brazil to strengthen the country's adaptation capacity, assess climate risks and manage vulnerabilities at the national, state, and municipal levels. Through the NAP, Brazil's vision for its adaptation undertakings is to integrate, where appropriate, vulnerabilities and climate risk management into public policies and strategies, as well as to enhance the coherence of national and local development strategies with adaptation measures.
>
> *(3)*

Cooperation among developing countries to enhance capacity to deal with the problem of climate change is another important topic in the 2015 INDC,

> Recognizing the complementary role of South-South cooperation, on the basis of solidarity and common sustainable development priorities, Brazil will undertake best efforts to enhance cooperation initiatives with other developing countries, particularly in the areas of: forest monitoring systems; biofuels capacity-building and technology transfer; low carbon and resilient agriculture; restoration and reforestation activities; management of protected areas; increased resilience through social inclusion and protection programmes; capacity building for national communications and other obligations under the Convention, in particular to Portuguese speaking countries.
>
> *(4)*

Table 6.1 shows a summary and comparison of the concepts of equity and related topics taken into account for mitigation and adaptation in the 2015 INDC and the 2020 NDC.

Table 6.1 Concepts of equity and related topics taken into account for mitigation and adaptation in the 2015 INDC and the 2020 NDC

Equity and related concepts	*2015 Brazil INDC for the Paris agreement*	*2020 Brazil First NDC for the Paris agreement*[1]
Equity	Only linked to mitigation, not to adaptation. Equity directly relates to historical responsibilities of GHG emissions.	Only linked to mitigation, not to adaptation. Equity directly relates to historical responsibilities of GHG emissions. This 'updated' NDC document includes the word moral although without deep reflection, "This understanding provides for the main moral foundation underpinning the principle of common but differentiated responsibilities, which is one of the central pillars of the international climate change regime, constituted by the UNFCCC, its Kyoto Protocol, and the Paris Agreement." (7)
Responsibility	Considers the principles of the UNFCCC, in particular, Article 3, the principle of common but differentiated responsibilities and respective capabilities that deal with mitigation and adaptation. The document also considers the historical responsibility in the context of mitigation.	Considers the principles of the UNFCCC, in particular, Article 3, the principle of common but differentiated responsibilities and respective capabilities that deal with mitigation and adaptation. The document also considers the historical responsibility in the context of mitigation. It also refers to the lower responsibility of the country when accounting for the anthropogenic GHG emissions per capita.
Equality	Not referred	Not referred

(Continued)

Equity and related concepts	*2015 Brazil INDC for the Paris agreement*	*2020 Brazil First NDC for the Paris agreement*[1]
Capacity	Capacity is mainly in the context of the National Adaptation Plan. The focus is on the government and specific sectors' conditions to adopt climate adaptation measures, such as capacity to withstand the effects of severe climate events; capacity in water security (National Water Security Plan); biofuels capacity building and technology transfer; and capacity-building for national communications.	Capacity as a technical issue is linked to energy-power generation capacity.
Sustainable development	Sustainable development is referred to in the long-term vision for a transition toward energy systems based on renewable sources and the decarbonization of the global economy by the end of the century, in the context of sustainable development. Brazilian contribution toward achieving the objective of the Convention, in the context of sustainable development but recognizing that emissions will grow to meet social and development needs (challenges regarding poverty eradication, education, public health, employment, housing, infrastructure, and energy access).	Refers to sustainable development in Brazil with several multilateral environmental treaties and a wide range of laws and public policies in the area of sustainable development. It mentions the importance of the Sustainable Development Mechanism (SDM) provided for under Article 6, Paragraph 4 of the Paris Agreement (2015).

Source: Author.

Conclusion

The Brazilian approach discussed here is not intended to replace the IPCC listed criteria in the 5th Assessment Report but is a permissible interpretation of the equity principles which IPCC concluded to be understood to comprise four key dimensions: responsibility, capacity, equality, and the right to sustainable development (IPCC 2014: 317). Both the 2015 INDC and the 2020 NDC present equity and related concepts seriously and treat the national climate proposals with almost the same perspectives. Theoretically, they are very well constructed. Nevertheless, there are important differences in the document of 2020 in practical terms that hide what should matter.

The numbers of the first document (2015) changed in the first updated document (2020) although the commitments in percentage are the same. This was because there was a new national GHG emissions and sequestration inventory launched in 2016. This new inventory used new methodology revisions after the 2015 INDC. The result is that the amount of the GHG emissions for the base year (2005) increased. Then, the reductions of GHG emissions became easier to acquire in 2025 and 2030 (37 percent and 43 percent respectively to 2005).

This kind of technicality presented by the Brazilian government in the 2020 NDC is matter of criticism by all respected institutions and professionals. The lack of compromise with the long-term commitment is the most important issue to discuss in terms of ethics. Unfortunately, equity and related concepts received secondary importance in the 2020 NDC as the long-term efforts to acquire climate change stabilization were jeopardized.

Note

1 The 2020 Brazil First NDC was updated again in March 2022 ("Brazil First NDC – second update"), and the word 'moral' is entirely missing from the document. In addition, the phrase "loss and damages" appeared and the horizon of Brazilian neutrality in GHG emissions changed from 2060 to 2050. Available at https://unfccc.int/sites/default/files/NDC/2022-06/Updated%20-%20First%20NDC%20-%20%20FINAL%20-%20PDF.pdf.

References

Brazil. (2015) *Federative Republic of Brazil Intended Nationally Determined Contribution Towards Achieving the Objective of the United Nations Framework Convention on Climate Change*. Available at http://www4.unfccc.int/submissions/INDC/ Published%20Documents/Brazil/1/BRAZIL%20iNDC%20english%20FINAL.pdf.Brazil. (2020) *Paris Agreement Brazil's Nationally Determined Contribution (NDC)*. (updated submission). Available at https://unfccc.int/sites/default/files/NDC/2022-06/Brazil%20First%20NDC%20%28Updated%20submission%29.pdf?download.

Intergovernmental Panel on Climate Change (IPCC). (2014) *5th Assessment Report, Contribution of Working Group III to the Fifth Assessment Report of the Intergovernmental Panel on Climate Change*. Cambridge University Press: Chapter 4.

IPCC. (2018) *Special Report (SR) 1.5, Chapter 5: Sustainable Development, Poverty Eradication and Reducing Inequalities*. Available at https://www.ipcc.ch/report/sr15/chapter-5-sustainable-development-poverty-eradication-and-reducing-inequalities/.

The Paris Agreement. (2015) Available at https://unfccc.int/sites/default/files/english_paris_agreement.pdf.

United Nations Framework Convention on Climate Change. (1992) Available at https://unfccc.int/resource/docs/convkp/conveng.pdf.

Further Reading

Brown, D., Breakey, H., Burdon, P., Mackey, B., and Taylor, P. (2018) "A four-step process for formulating and evaluating legal commitments under the Paris Agreement." *Carbon & Climate Law Review*, 12(2): 98–109.. https://doi.org/10.21552/cclr/2018/2/5

Brown, D., and Taylor, P. (2018) *National Climate Justice Research Project on Ethics and Justice in Formulating National Climate Change Policies. Lessons Learned from Analyses of 23 Countries' Consideration of Ethics and Justice in Formulating Climate Policies Derived from the First and Second Round Reports*. Available at https://nationalclimatejustice.org/summary-0f-findings/

Carvalho, N., Berrêdo Viana, D., Muylaert de Araújo, M.S., Lampreia J., Gomes, M.S.P., and Freitas, M.A.V. (2020) "How likely is Brazil to achieve its NDC commitments in the energy sector? A review on Brazilian low-carbon energy perspectives." *Renewable & Sustainable Energy Reviews*, 133: 110343.

IPCC Special Report on Climate Change, Desertification, Land Degradation, Sustainable Land Management, Food Security, and Greenhouse gas fluxes in Terrestrial Ecosystems. Summary for Policymakers. Final Draft for Government Review. Date of Draft: 29th April 2019.

Miguez, J., and Santhiago de Oliveria, A. (2011) "The importance of historical responsibility in the context of international regime on climate change." In Experts from BASIC Countries (Eds.), *Equitable Access to Sustainable Development: Contribution to the Body of Scientific Knowledge*. Expert Forum: Beijing, Brasilia, Cape Town, and New Delhi: 24–36.

Muylaert, M.S., Campos, C.P.L., and Rosa, L.P. (2007) "Sustainable development and equity: Greenhouse gases historical contribution by sectors." *Renewable & Sustainable Energy Reviews*, 11: 988–997.

Muylaert, M.S., Cohen, C., Rosa, L.P., and Pereira A.S. (2004) "Equity, responsibility and climate change." *Climate Research*, 28: 89–92.

Muylaert, M.S., Silva, C., and Campos, C.P.L. (2009) "Land use change sector contribution to the carbon historical emissions and the sustainability – Case study of the Brazilian Legal Amazon." *Renewable & Sustainable Energy Reviews*, 13: 696–702.

7

MITIGATION DUTIES OF POOR AND VULNERABLE COUNTRIES

Steve Vanderheiden

Introduction

Setting out the key objectives and normative constraints of the cooperative international scheme to respond to climate change that its terms and processes would found, the 1992 United Nations Framework Convention on Climate Change (UNFCCC) called upon signatory parties to "protect the climate system for the benefit of current and future generations, on the basis of equity and the common but differentiated responsibilities and respective capabilities" of state parties to the convention (Article 3, Principle 1). These key principles – of equity and the 'common but differentiated responsibilities and respective capabilities' (or CBDR+RC) of states – have over the decades since the convention generated considerable attention from climate ethicists as well as observers of international climate politics. Together, they provide guidance for how the burdens associated with climate change mitigation ought to be shared among and between states.

While these two burden-sharing principles rule out some arrangements that obviously fail to follow any plausible interpretation of either ideal, their application to international climate policy remains contested given the ambiguity in how each is understood as well as in how the two principles relate to each other. Following the UNFCCC text, which held that developed country parties to the convention should 'take the lead' in reducing their greenhouse gas (GHG) emissions, the 1997 Kyoto Protocol assigned GHG emissions caps to developed but not developing countries, assigning no mitigation duties to the latter and allowing them to increase their emissions without penalty. At first blush, this decision to temporarily exempt developing countries from binding mitigation commitments may appear to be consistent with the ethical principles declared in the climate treaty, since affluent developed countries (with their far larger per capita emissions) are primarily responsible for causing climate change, and as the result of a long historical pattern of highly inequitable access to carbon-based energy.

However, this decision to temporarily exempt developing countries from legally binding mitigation duties was immediately contested. It was challenged by the United States as unfair, with the U.S. Senate citing exemptions from mitigation duties for China and India as a core basis for its resolution promising to refuse ratification of the treaty, and the Bush administration later citing those exemptions in justification of its withdrawal from the treaty framework in 2001. The exemption of developing countries from binding mitigation duties remained controversial throughout the duration of the protocol, which saw its first commitment period expire in 2012 (the treaty was extended through a second commitment period until 2020, but did not go into force until the day before its expiration given intractable disagreements about its terms). Based largely upon

DOI: 10.4324/9781003039860-9

this decision at Kyoto to delay binding mitigation duties for developing countries until developed country parties had undertaken significant mitigation actions, in combination with the withdrawal of the United States from the treaty framework, some advocates for developing countries began to urge that the exemption be made permanent, and that the world's developed countries be held solely responsible for mitigating climate change.

It remains controversial whether the UNFCCC principles allow or justify a permanent exemption from all mitigation burdens for poor countries (the so-called 'Bali Firewall'), as some claim (Smith 2010). Likewise controversial is the question of whether and how countries like China might at some point "graduate" from non-Annex 1 to Annex 1 (developing to developed for purposes of the convention) status. Some, for example, have suggested that rapidly developing countries (with their rapidly developing per capita emissions rates) could at some stage of development be reclassified as Annex 1 and thus be assigned binding mitigation burdens (Ackerman et al. 2013), which China has steadfastly rejected. While the 2015 Paris Agreement discontinued the centrally assigned burden-sharing framework of the Kyoto protocol, replacing binding national mitigation targets with a 'pledge and review' system of 'nationally determined contributions' (NDCs) toward mitigation efforts, the controversy about whether or not poor and vulnerable countries should have some kind of mitigation duties remains. The same UNFCCC principles remain in force even without the centrally assigned mitigation burdens of the Kyoto framework, and those principles can nonetheless be used to evaluate various NDCs. The question of how much (if anything) the world's poor and vulnerable countries must do to contribute toward the world's collective mitigation goals remains one of the central ethical and political questions for international climate policy.

Insofar as this question about fair mitigation burdens for developing countries can be addressed through examination of the UNFCCC's guiding principles, the focus must be upon how equity or CBDR applies to such countries. Within climate ethics scholarship as well as through activist work on climate justice, various conceptions of these principles have been proposed and applied to this question. Given variation among plausible conceptions, there exists a range of defensible burden-sharing arrangements rather than a single best position on how to interpret and apply these principles from the climate convention. As this range is cast by the Intergovernmental Panel on Climate Change (IPCC) in its fifth assessment report:

> There is a basic set of shared ethical premises and precedents that apply to the climate problem that can facilitate impartial reasoning that can help put bounds on the plausible interpretations of 'equity' in the burden-sharing context. Even in the absence of a formal, globally agreed burden sharing framework, such principles are important in establishing expectations of what may be reasonably required of different actors [...] These principles are equality, historical responsibility, capacity, and right to sustainable development.
>
> *(IPCC 2014: 317–318)*

In considering duties of poor and vulnerable countries to contribute toward global mitigation efforts, this chapter relies primarily upon these ethical principles, which bound the equity and CBDR commitments of the UNFCCC, examining how they inform this question, but shall also consider several other ethical issues that are applicable to this question.

Understanding and Applying the Core Ethical Principles

In determining the extent and basis of climate change mitigation duties for poor and vulnerable countries, those four principles noted above offer an instructive starting point. The principle of "historical responsibility" is usually associated with the CBDR principle, which is widely viewed as requiring countries with historically greater per capita emissions to be assigned proportionately larger mitigation burdens (thus differentiating national responsibility). Equality is associated with the UNFCCC's equity principle and is most commonly applied to individual greenhouse emissions rights, manifesting in such claims as those for assignment of national emissions allowances on an equal per capita basis (Vanderheiden 2008). Capacity originates in the +RC addendum to the CBDR principle and refers to differential national abilities to reduce emissions, for which per capita national income is a common surrogate. Finally, the right to sustainable development is frequently asserted against overly restrictive national emissions limits for those countries in earlier development stages, at least insofar as such restrictions limit development opportunities. Like the other three principles, this right is reflected in the text of the climate convention, which treats such development as an imperative for developing countries that would allow them "better to address the problems of climate change" (Article 3, Principle 4).

All four principles in various ways could plausibly be seen as limiting the extent of mitigation burdens for poor countries, but none rules out the assignment of some mitigation duties to these countries (even if these come at no cost to them), either on its own or in combination with others. Since such countries have relatively low historical emissions, the "historical responsibility" understanding of the CBDR principle would assign them relatively lighter (if not necessarily zero) current or near-future mitigation burdens. Indeed, poor and vulnerable countries that have historically contributed little toward climate change may by the CBDR principle have a strong case that they should not be burdened at all by global mitigation efforts, but being unburdened by mitigation duties is not the same as having no such duties. Per capita emissions limits could be set at a level where such limits served as a benefit rather than a burden—for example, if poor countries could sell their unused emissions rights to rich countries that need to purchase them to comply with their own mitigation commitments.

Similarly, since the residents of these countries have used less of the planetary capacity that is to be equitably shared—termed 'emissions absorptive capacity' (Blomfield 2013) and conceived as the shared resource that allows emissions without environmental impact and thus forms the basis for rights or entitlements to emit GHGs—they are often held to be entitled to relatively greater per capita access to that capacity in the future. Hence, considerations of equity as well as differentiated responsibility weigh in favor of lesser (but not zero) mitigation duties for low-emitting developing countries. Poor countries typically also have significantly less capacity to reduce emissions, given their comparative capacity as indicated by per capita income (a commonly used surrogate). Again, these applications of ethical principles would recommend relatively light mitigation duties but would not justify an exemption from all such duties.

Finally, protection of the development rights of the global poor would likewise weigh against high mitigation burdens that would inhibit their opportunities for development, as strict limits on national emissions would otherwise tend to do. However, sustainable development rights would not justify a blanket exemption from any sort of national emissions limits, and indeed may even require such limits. Sustainable development rights may allow continued emissions growth for a period, since development (as it is standardly understood and measured) usually results in increases in per capita emissions (for example, from increased mobility or purchasing power), some allowance for emissions growth may be necessary to accommodate development imperatives. However, concern for mitigation targets necessitates that this development be pursued in a sustainable manner, relying upon low or zero carbon sources of energy rather than following the carbon-intensive development path taken by most countries of the global North. Often, such low or zero carbon energy sources are more expensive and/or require extensive capital investment. These costs in turn raise questions about how sustainable development is to be financed, and there are good ethical reasons for thinking that affluent countries ought to provide most or all of such development assistance. However, a primary reason for this international finance of sustainable development is to fulfill development rights while doing so sustainably, which requires developing countries to accept mitigation duties.

While no plausible interpretation of any of these four ethical principles allows for *unlimited* emissions growth within poor or vulnerable developing countries, which is what a full exemption from mitigation duties entails, plausible interpretations of the principles do allow for a range of potential burden-sharing arrangements. Those attaching more weight to historical emissions would assign lighter mitigation duties to late-developing and high-emitting countries like China (now the biggest overall carbon emitter) because its cumulative emissions over the past two centuries represent a smaller share of cumulative global emissions than does its current emissions relative to current world emissions. Those understanding differentiated national capacities for mitigation in terms of per capita income may view some poor country's mitigation responsibilities differently than would those understanding such capacity in terms of territorial resources for low-carbon energy development. Those (Duus-Otterström & Hjorthen 2019) measuring the carbon footprint of an export-driven economy like China's in terms of *production emissions* (i.e., those resulting from industrial production within its national borders) would assign it greater mitigation burdens that would those measuring this footprint in terms of *consumption emissions* (i.e., attributing that carbon footprint to the country in which the good or service was consumed).

These four principles rest upon different philosophical justifications or theories of justice for their normative force. Historical responsibility appeals to an account of corrective justice in which those responsible for causing harm incur some remedial burden to redress it. While such an account could also be used to justify compensation for those adversely affected by climate change, as with the 'loss and damage' mechanism that would finance relief efforts for victims of climate-related harm, it can also be applied to mitigation burdens. As Henry Shue writes about this way of thinking about the ethics of climate change (1999: 533),

> if whoever makes a mess receives the benefits and does not pay the costs, not only does he have no incentive to avoid making as many messes as he likes, but it is also unfair to whoever does pay the costs.

Because rich countries were able to pollute more during a time in which such pollution was unrestricted, Shue argues, fairness now requires that they undertake larger mitigation burdens as a matter of historical responsibility as well as to compensate for the climate-related harm they have caused.

One could plausibly invoke historical responsibility in defense of requiring affluent developed countries to pay the full costs associated with climate change remedies, since poor developing countries have only relatively low per capita emissions that arguably could not be further reduced without serious harm. Here, the idea is that countries cannot be held responsible for emissions that they cannot be expected to avoid, like those associated with bare subsistence. This involves denying that countries should be held responsible for their *subsistence emissions* (Shue 1993), assigning mitigation duties instead only on the basis of national *luxury emissions* (i.e., those generated by human activities that are not necessary for meeting basic needs, an analysis through which very poor countries would not be at all responsible for climate change). Countries whose past emissions are associated only with bare subsistence should not now be required to pay for climate change mitigation, according to this argument, since only through their luxury emissions can countries acquire such remedial liability. But this argument is about allocating the *costs* of mitigation, not which countries should participate in mitigation actions like restrictions on GHG emissions or protection of carbon sinks. This distinction between paying and acting often manifests in calls for affluent developed countries to finance mitigation actions within poor developing ones, as with the Clean Development Mechanism and forest programs under the UN's Collaborative Programme on Reducing Emissions from Deforestation and Forest Degradation in Developing Countries (UN-REDD).[1]

Although it also recommends that high-emitting developed countries bear the largest share of mitigation burdens, the equality principle rests upon a quite different ethical foundation as well as a different way of casting the core problem of climate ethics. Rather than viewing the core ethical problem of international climate cooperation as a *burden-sharing* problem, which is fundamentally about how much various states should do or pay for mitigation, those utilizing the equality principle typically view it as a *resource-sharing* problem, which asks how much of a shared resource link emissions absorptive capacity various states are entitled to use through their greenhouse emissions. Equality (along with a resource-sharing conceptual framework) features prominently in arguments calling for affluent countries to rapidly reduce their emissions while allowing poor developing countries to continue to increase theirs until both converge upon per capita emissions equality (Bows & Anderson 2008). Here again, the principle could justify an arrangement whereby all mandatory emissions reductions were assigned to developed countries but not one in which developing countries were entirely exempted from all emissions limits (including limits on future emissions), as is implied by the claim that such countries have no mitigation duties of any kind.

Capacity is often invoked as a secondary principle to responsibility, as in the "greenhouse development rights" framework proposed by Eco-Equity (Baer et al. 2009), but draws upon a distinct theoretical foundation. Whereas the responsibility draws upon retributive or corrective justice in assigning remedial burdens to responsible parties, capacity draws upon consequentialist moral reasoning. Countries that have greater capacity to reduce their emissions (whether from affluence or ready availability or low-carbon energy sources) are from this principle tasked with doing more to mitigate climate change because they can do more good (i.e., they can yield more carbon abatement with lower impact on resident populations), not because they are more responsible for causing the problem.

Another way of contrasting these two kinds of ethical theorizing compares the backward-looking (i.e., based primarily on actions undertaken in the past and their effect) focus of the responsibility principle compared to the primarily forward-looking (i.e., based on considerations about future actions and impacts) focus of capacity-based theories. Hybrid approached to climate justice like that proposed by Eco-Equity assign greater mitigation duties to affluent developed countries both because they are more responsible for having contributed more to climate change and because they have greater capacity to reduce those emissions (and can thus yield more environmental benefits at lower social cost). Such hybrids are sometimes challenged by questions about how to treat high-emitting but relatively undeveloped countries like Qatar, given that the two principles pull in opposite directions, but otherwise reinforce judgements about which countries should be assigned the greatest burdens, as both affluence and responsibility are frequently associated with carbon-intensive development paths. While having low capacity might initially appear to justify an exemption from all mitigation duties for poor (and thus low capacity, at least by one common measure) states, as before it would only require that mitigation projects within or requirements upon such states be financed by the international community. On the other hand, reliance upon capacity alone to inform international mitigation burden-sharing arrangements would be objectionable for failing to take account of differentiated responsibilities of various countries.

The right to sustainable development is not so much an ethical principle as it is an important countervailing interest that is designed to balance imperatives of sustainability with those of human development. As a right that is acknowledged in the UNFCCC and declared within several international human rights conventions, it is typically justified by the equality principle, which is violated by wide extant inequality among persons depending upon their nation of birth or residence, at least insofar as this results from different development stages. It likewise borrows from the capacity principle in allowing relatively undeveloped countries to prioritize development over mitigation (not to reject mitigation duties altogether, but to use relatively more available resources to promote development objectives, compared to societies at higher development stages). It also requires sustainable forms of development, which differ from (and are typically more costly than) the high-carbon development paths taken by most of the world's developed countries, which raises related questions about how such development is to be financed. Because it stresses that development must be sustainable, this principle also cannot be used to justify an outright exemption from national mitigation duties, which would allow for high-carbon development paths that would not be environmentally sustainable.

Climate ethicists disagree about which is the most plausible interpretation of these various principles, as well as about how to prioritize the various principles and to apply them to formulate for differentiating mitigation responsibilities, and it is not a task of this chapter to adjudicate among competing candidates for which position is best (academic ethicists must insist that their own position is best in order to be published). Rather, its task is to identify the positions and their supporting arguments or ethical justifications, but before turning to those positions and arguments, one final remark about this range of plausible interpretations of the ethical principles relevant to national mitigation burden-sharing arrangements is warranted. That is, the plurality of plausible interpretations of ethical principles still disallows some burden-sharing arrangements, including those demanding too much of poor and/or vulnerable developing countries (expecting them to bear equal costs or achieve equal percentage-based reductions from current emissions, for example) as well as those proposals that require too little (as discussed above) of them.

Arguments for and against Including Poor and Vulnerable Countries in Mitigation

Having surveyed the various ethical principles that have been declared in the climate convention and applied by scholars and activists to analysis of the fairness or equity of various burden-sharing arrangements, we now turn to another kind of argument for or against inclusion of poor or vulnerable countries within cooperative international mitigation efforts. Rather than appealing to principles of justice or fairness in mitigation burden-sharing (or resource-sharing), these arguments appeal to practical consequences resulting from inclusion of poor countries in cooperative international mitigation efforts or their exemptions from mitigation duties that are part and parcel of such efforts.

Arguments of this sort rely upon consequentialist ethical analysis, which defines right actions in terms of the relatively good consequences that they are expected to yield, relative to alternatives. Utilitarianism, which is the best-known variety of consequentialist ethics, defines the morally right action as the one that brings about the greatest good for the greatest number (or minimizes harm if some such harm is unavoidable). On this basis, one consequentialist reason to include low-emitting developing countries within a mitigation burden-sharing system would be to protect against carbon leakage, which occurs when carbon-intensive industry or activities migrate from one geographical location to another in order to evade carbon emission regulations that apply in the original location but not in the new one. Since carbon leakage can undermine the efficacy of GHG emissions controls where they exist and also deter the adoption or ambition of such controls, it can be expected to increase climate-related harm.

Carbon leakage occurs when polluting industries relocate to jurisdictions without carbon pricing or emissions controls rather than paying costs associated with those controls. Such costs can arise from carbon pricing schemes like carbon taxes, from market-based instruments like ca-and-trade systems, or from 'best-available-technology' regulations that require polluters to use only efficient systems designed to minimize carbon (among other) emissions from industrial activities. Insofar as costs associated with carbon pollution vary by jurisdiction, as when some countries have relatively strong mitigation policies in place while others have weak policies or no policies at all, the resulting regulatory disparities create an incentive for polluters to flee the former and relocate in the latter. Similar to a 'race-to-the-bottom' based on labor costs, where manufacturing industries fled relatively high-wage developed countries to relocate in relatively low-wage developing countries, carbon regulation creates another kind of race-to-the-bottom.

Many carbon-intensive industries have already fled countries or subnational jurisdictions like states or provinces that have introduced emissions controls, and many more have threatened to do so if the jurisdictions in which they reside adopt such controls. Countries or states stand to lose jobs and revenues while competing jurisdictions lowering or eliminating those costs stand to gain them, effectively aligning state imperatives for economic and revenue growth with those of allowing rather than controlling greenhouse pollution. Often, the mere threat to relocate to a location with lower 'regulatory burdens' can provide sufficient incentives for local, subnational, or national governments to eschew or minimize carbon regulation. The result is to deter and to exert downward pressure upon the ambition of existing national mitigation efforts along with undermining those efforts by merely relocating rather than reducing the production of harmful GHGs. Allowing any country a full regulatory exemption from mitigation duties would also allow and incentivize carbon leakage to that country.

In terms of its contribution to climate change, GHGs are indifferent to the geographic origin of their emission, causing the same insidious impact upon global climate whether emitted from an affluent developed country or a poor developing one. However, local air and water pollutants often accompany those industries and activities associated with carbon pollution, so migration of polluting industry to poor developing countries would exacerbate disparities between rich and poor in exposure to anthropogenic pollution-related health hazards, which raises additional environmental justice concerns. Carbon leakage would thus harm those residing in poor and vulnerable developing countries in two ways: they would be exposed to more climate-related harm as well as exposed to more local sources of harmful pollution. Since these adverse impacts are expected, the evaluation of any plan to exempt some countries from binding mitigation duties would have to take them into account in ethical evaluation.

Two objections might be raised against citing carbon leakage from affluent developed to poor developing countries as an ethical consideration against exempting the latter from binding mitigation duties. First, one might object that while carbon leakage may undermine mitigation efforts, it can also benefit residents of poor developing countries by encouraging development as industry flees climate regulations in the global North and relocates to the global South. The flip side of affluent developed countries losing jobs and revenues as polluting industry is offshored may be that poor developing countries potentially benefit by this relocation. While exacerbating climate change is bad for the global poor, so also is poverty and its associated impacts. Insofar as the consequences of policy alternatives determine their moral rightness or wrongness, why not also consider the potentially beneficial consequence of incentivizing industry relocation from the world's affluent countries to its poorer ones? Indeed, do the world's poor not a have a right to the kind of development that residents of affluent countries now enjoy, and would relocation of carbon-intensive industry not facilitate that kind of development?

Here, one might recall the recognition within the climate convention of a right to *sustainable* development, which would not necessarily be afforded by luring polluting industry to relocate to poor countries through the promise of exemption from environmental protections. Indeed, the right to sustainable development seeks to balance imperatives of development (e.g., poverty reduction, economic mobility, social inclusion) with those of ecological sustainability, but recognizes the two as intertwined. Industrial development that brought new jobs but also polluted local air and water and so imperiled resident populations would not be sustainable. It may be unfair to expect poor countries to forego the benefits of the same carbon-intensive forms of development that affluent developed countries enjoyed in the past, but there are better ways of addressing that unfairness than allowing more countries to develop unsustainably. For this reason, sustainable development is typically understood as requiring affluent countries to assist poor ones with development pathways that can provide the social and economic benefits without the local and global ecological costs. Indeed, participation in a global mitigation framework offers a means for financing sustainable development as it creates incentives for donor countries or private investors to finance projects that reduce emissions while also yielding social benefits.

Second, one might object that trade measures like carbon tariffs that are designed to offset the effects of the regulatory disparities that cause carbon leakage are often unfair to poor countries. In order to allay worries about impacts upon the competitiveness of carbon-intensive industry, supporters of cap-and-trade policies have proposed carbon tariffs that are designed to equalize the impact of domestic climate policies on foreign firms. If, for example, a domestic mitigation policy added $30 per ton of carbon to the cost of domestic manufacture of some good, these tariffs would charge a $30 per ton tariff on imported versions of the same good if it was manufactured in a country without carbon pricing. But carbon tariffs like this can amount to unfair (and perhaps illegal) forms of protectionism and are viewed by some critics as punitive toward firms located in developing countries (but which did not relocate there as the result of regulatory evasion). Here, the objection against characterizing regulatory disparities that allow for carbon leakage as wrong because of the climate consequences of that leakage casts measures designed to prevent that leakage as a medicine that is more harmful than the disease.

In recognition of the potential unfairness of both uninhibited offshoring of carbon pollution and punitive border measures designed to mitigate leakage as well as protecting domestic industry, Robyn Eckersley (2010) proposes a scheme of consumption-based carbon accounting and carbon tariffs designed around the climate convention's CBDR principle, arguing that such a policy could hold affluent consumers responsible for their emissions (regardless of whether carbon-embedded goods were produced domestically or abroad) which would assist in development imperatives rather than punishing poor countries for their underdevelopment. However, her proposal depends upon both producing and consuming countries participating in international mitigation efforts such that emissions from the production of goods or services that are exported back to affluent countries could be charged to those importing countries. A full exemption from any kind of mitigation duties would negate the benefit of consumption-based carbon accounting that Eckersley proposes. Opportunities to use border measures to assist poor developing countries to transition away from fossil fuels—a core aspect of their sustainable development—depends on their participation in cooperative mitigation efforts.

Here, we might return to perhaps the most compelling reason for requiring that poor and vulnerable developing countries participate in cooperative international mitigation efforts, which may allow them to increase their domestic emissions to enable development but nonetheless requires some limit on those increases. Creating a system that allowed and encouraged regulatory evasion such as manifests in the offshoring of greenhouse pollution could also be wrong in that it fails to hold responsible parties to account for their contributions to climate change, instead facilitating the shirking of that responsibility through leakage. Consider, for example, an industry involved in regulatory evasion in order to continue supplying consumers with carbon-intensive goods or services at the lowest possible cost. It relocates to a poor country with no emissions controls and then exports its carbon-embedded goods back to the affluent country whose environmental regulations it evaded. Whether one holds the producer or the consumer responsible for the continued production and consumption of the polluting good, the system that allowed the regulatory disparity to exist and for rich countries to offshore their greenhouse emissions would fail to hold either party responsible for their ongoing contributions to climate change. Indeed, it would reward parties for successfully evading such responsibility. Neither would pay the social costs of their polluting, imposing these instead upon the poor and vulnerable residents of the country to which the regulation-avoiding industry fled, both in their increased exposure to local pollutants but also from exacerbated climate change. Here, the prospect of carbon leakage weighs clearly against allowing such regulatory evasion, not so much because of its practical consequences but because of the unfairness it engenders and rewards.

Conclusions: Mitigation Duties for Poor and Vulnerable Countries

As we have now seen, neither the four ethical principles declared through the climate convention nor the consideration of its practical consequences would recommend that poor or developing countries be exempted from all mitigation duties or binding emissions limits. On the other hand, these considerations all recommend that affluent and high-emitting countries be assigned significantly greater mitigation duties, and perhaps that mitigation duties for poor countries provide resources and incentives for sustainable development rather than net burdens. Indeed, the climate convention's reference to 'common but differentiated responsibilities' implies that all countries share some 'common' mitigation responsibilities, even if these are differentiated among them on the basis of principles or practical considerations like those cited above. Differentiated national capacities for mitigation, like differentiated current or historical responsibility for greenhouse emissions, would similarly warrant *less* remedial responsibility for climate change, but would not exempt poor countries from having *any* such responsibilities.

Poor developing countries may not be required to reduce their emissions from current levels, or to bear the costs of mitigation actions undertaken within their borders, but the ethical principles enshrined within the climate convention still require that they participate in collective mitigation efforts, for example by limiting their emissions growth more than might otherwise occur or by participating (if primarily as recipients of development aid and as sites for projects) in sustainable development efforts financed by the international community. Concern for equality in per capita sink access or for sustainable development may well allow for considerable increases in per capita emissions for those residing in low-emitting developing countries, but limits on such allowances for emissions growth are still required. Since cooperative international mitigation efforts must simultaneously dramatically reduce ongoing anthropogenic contributions toward climate change while also addressing the unfairness of historical resource use patterns that have created the environmental problem, ethical considerations as well as those of efficacy in international mitigation efforts both require some common mitigation responsibilities, even if they also require that these be highly differentiated among countries for reasons noted above.

Note

1 https://www.un-redd.org/about/programme.

References

Ackerman, F., Shanda, X., Stern, N., Li, L., Eklund, K., Hallding, K., Gang, F., & Edenhofer, O. (2013) *The Economics of Climate Change in China: Towards a Low-Carbon Economy*. London: Earthscan.

Baer, P., Athanasiou, T., Kartha, S., & Kemp-Benedict, E. (2009) "Greenhouse Development Rights: A Proposal for a Fair Global Climate Treaty," *Ethics, Place and Environment* 12(3): 267–281.

Blomfield, M. (2013) "Global Common Resources and the Just Distribution of Emission Shares," *Journal of Political Philosophy* 21(3): 283–204.

Bows, A., & Anderson, K. (2008) "Contraction and Convergence: An Assessment of the CCOptions Model," *Climatic Change* 91: 275–290.

Duus-Otterström, G., & Hjorthen, F.D. (2019) "Consumption-Based Emissions Accounting: The Normative Debate," *Environmental Politics* 28(5): 866–885.

Eckersley, R. (2010) "The Politics of Carbon Leakage and the Fairness of Border Measures," *Ethics & International Affairs* 24(4): 367–393.

Intergovernmental Panel on Climate Change. (2014) *Climate Change 2014: Mitigation of Climate Change. Contribution of Working Group III to the Fifth Assessment Report of the Intergovernmental Panel on Climate Change* [Edenhofer, O., R. Pichs-Madruga, Y. Sokona, E. Farahani, S. Kadner, K. Seyboth, A. Adler, I. Baum, S. Brunner, P. Eickemeier, B. Kriemann, J. Savolainen, S. Schlömer, C. von Stechow, T. Zwickel and J.C. Minx (eds.)]. New York: Cambridge University Press.

Shue, H. (1993) "Subsistence Emissions and Luxury Emissions," *Law & Policy* 15(1): 39–60.

——— (1999) "Global Environment and International Inequality," *International Affairs* 75(1): 533.

Smith, C. (2010) "The Bali Firewall and Member States' Future Obligations within the Climate Change Regime," *Law, Environment and Development Journal* 6: 284.

Vanderheiden, S. (2008) *Atmospheric Justice: A Political Theory of Climate Change*. New York: Oxford University Press.

PART II

Ethical Issues That Arise from the Responsibility to Respond to Unavoidable Climate Change Harms

8

DEMOCRACY AND JUSTICE IN CLIMATE CHANGE ADAPTATION

Bruce Jennings

[Global warming] is bound to effect economic and political stability and to change our coastlines and our lives, but it could also be the making of a new world—one worth getting excited about all over again.

Lynall Watson, Heaven's Breath: A Natural History of the Wind (2019: 236)

[Humankind is causing change at a rate that is] beyond the adaptive capacities of biological evolution.. . [But we may still] countermand the curses we have laid upon the earth.

Tim Robinson, Listening to the Wind (2019: 411)

Introduction

Climate action is a grassroots social movement. Climate governance is a legal and regulatory authoritative exercise of power. They each have essentially two components. One is the mitigation of deleterious climate change, mainly by curtailing its causal roots in human activities that add carbon dioxide and other greenhouse gases to Earth's atmosphere. The other component is adaptation that involves changing human activity in ways designed to accommodate anticipated large scale and interlocking changes in bio-geophysical and ecological systems. This involves making changes in political and economic institutions and practices. It also requires widespread alterations in collective and individual behavior.

Both mitigation and adaptation measures will be crucially important in the 21st century, and both require the development of a complex apparatus of governance and planning that to this day remains underdeveloped so that its potential contribution to developing a more democratic and just political morality for ecological governance has gone largely unnoticed. Of the two, while it might seem pessimistic or even defeatist, adaptation is actually the more potentially radical and far-reaching. Mitigation has been conceptualized so that it fits under preexisting frameworks of environmental regulation (Kysar 2010). It has been likened to "pollution control on steroids" (Ruhl 2010). Global warming is directly linked to the concentration of carbon in the atmosphere and the oceans, so climate change mitigation efforts have readily quantifiable numeric goals (parts per million atmospheric concentrations, or global mean temperature limits). While still politically and ideological controversial, mitigation is not a regulatory anomaly. Yet precisely for that reason in American environmental law, at least, mitigation has been vulnerable to legal opposition and hurdles that have slowed it to a snail's pace (Ruhl & Salzman 2020).

DOI: 10.4324/9781003039860-11

For its part, adaptation cuts across all kinds of regulatory and jurisdictional categories and is much more challenging methodologically. It might be thought of as all hazard emergency planning on steroids (Jennings & Arras 2016). It involves questions of land use planning and property rights. It may have to draw upon technological innovation and deliberate ecosystemic management that will go against the grain of many tenets of conservation management. It must plan for differential impacts in various areas and for large-scale changes in biodiversity and human migration patterns, rendering the geographically bordered nation state a problematic unit of planning, governance, and social control.

One reason for the broad scope of climate change adaptation is well expressed by the following observations by Yeb Sano, lead negotiator from the Philippines at COP19:

> Disasters are never natural. They are the intersection of factors other than physical. They are the accumulation of the constant breach of economic, social, and environmental thresholds. Most of the time disasters are a result of inequity and the poorest people of the world are at greatest risk because of their vulnerability and decades of maldevelopment, which I must assert is connected to the kind of pursuit of economic growth that dominates the world; the same kind of pursuit of so-called economic growth and unsustainable consumption that has altered the climate system.
>
> *(Sano 2013)*

In recent years, as carbon levels, recorded temperatures, and severe weather events have proliferated – and their frequency has increased – more attention is being paid to adaptation activities. To provide a manageable focus to this very large subject, this chapter will concentrate on ethical and political ideals relevant to the process of deliberation and debate that lead to the formulation of adaptation plans. Discussion of related issues, such as the implementation and subsequent evaluation of such plans, will have to be set aside for another occasion. Adaptation planning brings climate science, ecology, and conservation biology into contact with public health and similarly oriented perspectives and movements in cognate fields such as prospective emergency planning, adaptive environmental management, and land-use planning. It draws orientation from social epidemiology, community-based participatory research, deliberative planning, and the building of 'learning communities' and 'learning organizations,' notions drawn from management science (Dewulf 2013; Forester 1999; Schön and Rein 1997).

In this chapter, adaptation planning will be viewed as a collective process that is closely tied to environmental law and governance, economic production and distribution, and cultural and social imaginaries or worldviews. Climate adaptation planning is one important venue in which fundamental ethical choices about economic and environmental policy will be explicitly subject to deliberation and dialogic decision-making. The necessity of climate change adaptation presses upon us the question of not only *how* adaptation is accomplished, but also what *should* be accomplished through successful adaptation. If adaptation planning is conceptualized and conducted narrowly, as a matter of technical expertise, then it is likely that only the 'how questions' will get attention. But if it is conceptualized more broadly as a civic practice, as a matter of inclusive participatory democracy, then the 'should questions' may receive concerned attention. If they do, then the equity and democratic legitimacy of the activity of adaptation – or the lack thereof – will be made explicit.

As research on the ethics of climate change develops, one frequent approach is to identify and embrace variously described conceptual or value frameworks made up of ethical principles or very general normative rules (Kass 2001). This chapter follows a relational rather than a principle-based approach. From the perspective of relational theorizing, ethical goals, such as social justice and democratic participation, do not function in adaptation planning as prescribing a particular course of action, end point, or outcome that is rationally justified. Instead of the end points, relational theorizing treats ethical goals more like compass points or headings to guide the adaptation and the ongoing social and natural symbiotic relationalities it promotes (Jennings 2019). Such compass points are considerations like community carrying capacity, ecosystemic and social-psychological resilience, public trust, and parity among diverse perspectives and interests – including the interests of nonhuman forms of life and ecosystems.

This approach to ethical and social theory is appropriate because the infrastructure of adaptation planning necessarily will be diverse and localized (Fresque-Baxter & Armitage 2012). And this, in turn, is determined by how a dynamic planetary transformation such as climate change manifests itself. Adaptation will be place-based because it must be calibrated to specific conditions of change. Sea level rise means one thing in Miami and another in Denver, but there will be some implications of it for adaptation planning in each city (e.g., coastal de-development and in-migration, respectively). Heat waves are different things in Phoenix, Tokyo, Chicago, Paris, and Oslo and their respective hinterlands. Nonetheless, adaptation planning strategies can be theorized, if not implemented, in a general way since local adaptation measures will be similar enough in a region that a larger scale pattern will become discernable over time (Smit et al. 2000).

In light of this, it is all the more important to grasp adaptation planning as a dynamic activity, a form of practice with external goals and inherent values. And the dialogic activity of planning, from which a guiding plan emerges on a given scale, requires the coordination of many groups, disciplines, and interests and draws on numerous bodies of knowledge and expertise. The role of ethics in formulating plans for climate change adaptation in particular communities, regions, or the nation as a whole is to define reasonably just, humane, and responsible parameters for action and decision-making. Theorists of adaptation have recognized the importance of normative (ethical and cultural) as well as explanatory (scientific) factors:

> Values translate into action because they frame how societies develop rules and institutions to govern risk, and to manage social change and the allocation of scarce resources. [The socially constructed nature of adaptation suggests] four meta-domains whose roles need to be explored in this social construction of adaptation limits: ethics (how and what we value), knowledge (how and what we know), risk (how and what we perceive), and culture (how and why we live) …
>
> *(Adger et al. 2009: 338)*

Moreover, one of the most instructive and practical contributions relational theorizing and ethical analysis can make is by highlighting the ways in which adaptation planning is not only a necessary activity but a potentially transformative one. This certainly pertains to the ways in which the physical and ecosystemic landscape will be transformed by adaptation measures, the adaptation of physical infrastructure, so to speak. But it also pertains with equal importance to the transformation of the ethical and political infrastructure of our societies. Currently, the explicit political morality of liberal democratic and late capitalist societies systematically fails to challenge the inequitable distribution of wealth and power and underestimates the change potential inherent in civic learning and revitalization. A new ethical and political infrastructure of adaptation – and its corresponding 'regime of inequality' – will focus our attention on questions of environmental and economic justice and questions of social citizenship and participatory democratic governance (Piketty 2020). Taking adaptation to a new geophysical and ecological world seriously, and actually beginning to do something about it, makes those questions unavoidable.

Mitigation and Adaptation

The terms mitigation and adaptation have been with us from the very beginning of the international climate change governance era. Historians of the global discussion concerning climate change have noted that in the 1990s,

> interest in adaptation was overwhelmed by concern about the need to reduce greenhouse gas emissions and stabilize atmospheric greenhouse gas concentrations. …Critics felt that belief in the potential value of adaptation would soften the resolve of governments to grasp the nettle of mitigation and thus play into the hands of the fossil fuels interests….
>
> *(Schipper & Burton 2009: 1, 7)*

Thus, for about two decades, mitigation of greenhouse gas levels in the atmosphere was the watchword, and adaptation was relatively neglected.

Mitigation works on the geophysical causes of climate change and their associated human behaviors. The Intergovernmental Panel on Climate Change (IPCC) said that mitigation strategies involve "an anthropogenic intervention to reduce ae sources or enhance the sinks of greenhouse gases" (Ruhl 2010: 366, note 1). Adaptation, by contrast, is focused on vulnerabilities and disruptions associated with climate changes that we have reason to believe are inevitable, although their precise magnitude and impact may still be causally affected by our continuing mitigation and precautionary preparations. The IPCC defines adaptation as, "Adjustment in natural or human systems in response to actual or expected climatic stimuli or their effects, which moderates harm or exploits beneficial opportunities" (Ruhl 2010: 366, note 2).

Climate governance encompasses both mitigation and adaptation, and it should coordinate them. However, coordination between the essential preventive and precautionary functions of mitigation and the ameliorating and responsive functions of adaptation has been managed poorly (Urwin & Jordan 2008). Today, there is a discernable pivot toward adaptation, but not because mitigation efforts have failed. Indeed, the single most important component of mitigation – transition from a carbon-based to a renewable fuel energy system – is well underway, although other forms of mitigation concerning agricultural practices, land-use planning, biodiversity loss, fresh water conservation, net-zero building design, and mass transportation have not made as much headway. We are also learning that financial incentives exist that promote rather than impede the transition to renewable energy technologies and systems (Bond 2018; Global Commission on Geopolitics 2019). Successful mitigation need not be based on governmental regulation alone but can arise from factors at work in the private sector as well (Vandenbergh & Gilligan 2017).

The problem is not that mitigation has failed, but that it has come too late. Its focus on the longer term was perhaps a drawback as well as an advantage. As the years, then decades plodded on, when it came to climate change, it seemed as if society still had time to turn things around. If we had the will, there would be a way. And we would get the will when fear got our attention. Well, either climate change was moving faster and with more stealth than we expected (it was), or we were more obtuse than we realized (we were).

What Kind of Activity Is Adaptation Planning?

On the threshold of what new age, what new land, do we stand? The new land currently forecast by the best science we have and by the most reasonable expectations and conjectures is neither just nor democratic. It is not one of equality, empowerment, or progressive human self-realization. Instead, it is a land of the gradual erosion of those conditions that have made our humanity possible – relative geophysical planetary stability, Holocene temperate climate, biodiversity, ecological resilience. Such changes represent geological business as usual, perhaps, if one takes the very long view; but they will be a profound devolution from the perspective of human history measured in decades and centuries (Lynas 2020).

Among those who take the need for adaptation seriously, there are two schools of thought about the kind of undertaking adaptation planning is and should be. One sees adaptation planning as a technical and expert service provided by experts and purchased by imperiled consumers, either individually or collectively. Its goal is protection and security. The other school of thought sees adaptation planning as a civic practice engaged in by citizens in pursuit of a common good of equity and resilience in their communities. These perspectives are contrasted further below and it is argued that understanding adaptation planning as a form of civic practice is the most ethically promising and practically sustainable path for future reflection and research in the field of applied climate ethics. Initially, however, we need to clarify the notion of 'civic practice.'

The notion of a practice is a term of art in philosophical and social scientific studies of purposeful human agency (MacIntyre 2007; Schön 1984; Schön & Rein 1997). It refers to activities and relationships that are expressive, systematic, and in many cases institutionalized. Practices can be engaged in for instrumental reasons, treated as a means to some other end. However, they are also pursued in order to realize inherent, intrinsic values, such as a kind of flourishing or virtue for an individual and a kind of shared condition of well-being and common good for a community. Medicine is a practice in this sense, as are many other professions. It can be done in order to receive money, and it can be done in order to relieve the suffering and serve the health of others. Finally, a practice can be either private or public, individualistically oriented and libertarian, or communally oriented and associational or 'civic.' A civic practice is one that pertains to and affects not only the rights and interests of private individuals, but also the values and obligations of the community as a whole. With these terms in place to fix ideas, my thesis is that climate change adaptation planning is a practice that has societal as well as individual significance. It is not only a means to external ends, such as population safety and efficient environmental management in the face of novel climatic conditions, but also an activity that inherently promotes a culture of interdependence and mutuality (Adger et al. 2013).

To be sure, climate adaptation is a manifestation of public authority acting on the basis of technical expertise, and adaptation planning is certainly a part of that. However, planning has a legislative dimension as well in the sense that it involves value-oriented, pluralistic, and deliberative decision-making (Purdy 2010). This can be done by members of an elected or appointed body, and it can also be performed in a more directly democratic and participatory way. This 'parliamentary' (place of talking) aspect of adaptation planning is crucial for the normative development of the plan and the culture within which it must function. The ethical acceptability of an emergency plan is a function both of the substantive content of its provisions (how effective and equitable they are) and of the process through which those provisions are discussed, formulated, argued about, and, ultimately, agreed to.

What assumptions are being made when climate change adaptation is treated as a technical service best left to experts? This is what I call the consumer model of adaptation planning (Jennings & Arras 2016: 9–12, 77–80). It assumes that scientific, social, and cultural knowledge of a specialized kind is sufficient for the task. It assumes that local and contextual knowledge of particular communities and places is not required. By contrast, what I call the civic model of adaptation planning understands the ability to plan, be prepared, and be resilient as acquired capabilities – effective functionings – of communities and individuals (these ideas will be re-visited when we consider various conceptions of justice below). From a civic perspective, climate change adaptation planning and response are forms of activity that ordinary citizens ought to engage in out of a sense of membership and solidarity. The recognition of solidarity perceives that everyone is a part of a community of common interest and common vulnerability. The recognition of membership perceives that we have a responsibility for others and for the health of our shared community as a whole. From a consumerist perspective, adaptation planning is fundamentally a specialized service that fearful and vulnerable individuals purchase (through their taxes) for their own protection and *submit to.* From a civic perspective, citizens *engage in* adaptation planning and its implementation. The consumerist perspective makes passive and privatizes those for whom the plan is created. The civic perspective draws out their agency and makes them public citizens.

Of course, this does not mean that there is no room for technical expertise or hierarchical authority in adaptation planning conceived as a civic practice. It does mean that some entrenched thinking about citizen engagement and democratic participation in planning processes and systems will have to give way to new paradigms and concepts. Often discussions of public engagement, community collaboration, and democratic deliberation have assumed that adversarial conflicts are inevitable between activist and attentive grassroots participants based on efforts to maintain and extend power and control over planning discussions and outcomes. However, that paradigm of hierarchical and self-interested competition is giving way to an emphasis on deliberative networks and collaborative structures that can mediate among groups with different levels of technical and scientific information and differing value priorities. Evidence is mounting that civic planning of this kind is a practically achievable aspiration (Collins & Ison 2009). The struggle to retain interest group power can recede from center stage in our politics as the struggle to achieve justice and serve the common good advances.

Successful climate change adaptation planning must tap a preexisting fund of civic responsibility, a sense of justice, and concern for others in need. But adaptation planning does more than simply presuppose these virtues, it provides an occasion to foster them. And it will do so if the making of the plan, as well as the health and safety benefits of it, is shared in a meaningful and participatory way. Fear and self-interest will no doubt be strongly in evidence as the threats of climate change become more apparent in the coming decades. But climate change adaptation planning can move communities beyond these motivations to a sense of common purpose and solidarity. In this way, communities become resilient ethically and socially as well as environmentally. Planning as a civic practice can be an important resource for meeting that objective. It is less like a commercial contract between seller and buyer (provider and client) and more like a social contract, an agreement to be entered into by all and that establishes commitments of responsibility for each (Adger et al. 2013). Preparedness for the coming challenge of climate and ecosystemic adaptation is something that citizens enact, not something that individuals purchase and consume (Adger 2003).

Some caveats are in order at this point. Stressing civic considerations in this conception of climate change adaptation does not mean that adaptation planning should wait until preexisting barriers to full inclusion are overcome, or until the extent of racism and poverty are reduced and greater social justice is achieved. Adaptation planning must cope with society as it is, if only because time is running out. And this emphasis on the possibly transformative effects that participation in planning should not be read to suggest that climate change adaptation will be the sole – or even the principal – instrument of social reform. Many other activities must converge on the problem of civic renewal and resilience. Nonetheless, in studying climate change adaptation, it is important to remain open to the conceptual argument and empirical evidence that various forms of adaptation planning can reinforce the renewal of our civic life and our liberal democratic values (Childress & Bernheim 2003).

Moreover, climate change adaptation is not only about protecting a human community; it is also ultimately about a sense of place, its land, nature, and folklore. It contains a forward-looking vision of how the natural and the built environment can benefit all its inhabitants in the future. At its best, climate change adaptation preserves the past, protects the present, and promotes a more secure, resilient future. It can make us safer. It can also make us better ethically and politically.

Climate change adaptation planning offers the opportunity to nurture and strengthen the dimension of civitas and the common good in our lives together. Here are some of the ways this works. The memory of recent disasters, or the prospect of impending ones, focuses the mind and seems to stretch the attention span as few other issues do. Participatory planning efforts can energize entire communities and be part of a more general effort to revitalize communities economically and replenish social capital. It can foster collaboration among the varied organizations that comprise civil society and the public sphere, which are the infrastructure of civic life and are integral to the ability to recover from disaster and dislocation (Adger 2003). If civic life and civil society have become depleted – so that some areas turn into civic deserts, akin to food deserts, where neither healthy communications nor healthy diets are readily available – then adaptation planners will have to find a way to build them back, because civil society and its social capital are every bit as essential as physical sea barriers and wetlands to control storm surge.

It is true that community participation does not always produce positive climate change mitigation or adaptation outcomes (Spence 2019). Siting decisions concerning wind power generating towers are one example where grassroots activism has generated delay or even blocked proposed projects. This has led some analysts to argue for less democratic participation and more top-down authoritarian governance (Jennings 2016: 171–196; Ruhl & Salzman 2020). Such cases tend to be examples of advocacy rather than deliberative participation. They are driven largely by the protection of vested group or corporate interests and not by a contextual assessment of the common good. Perhaps as the civic practice of community-based adaptation planning develops and matures, it will teach American political culture how to tell the difference and facilitate a shift from single-issue advocacy to deliberative judgment.

The trick is to allow adaptation planning, undertaken before disasters in the relative calm of normal times, to range widely and searchingly and to build up what one might call an ethos of adaptive duty. Adaptation from a civic perspective is not only about the relatively sudden disasters that cannot be ignored or the manifestations of global climate change in the hurricanes, floods, droughts, heat waves, megafires, pandemics, and biodiversity loss. It must also be brought to bear on the long, persistent social emergency of inequality, racism, violence, and structural injustice – the legacies of slavery and colonial exploitation. We gear up for blows that wound society. Why then should we ignore slow poisons that deform, deface, and sap our sense of community and the common good, from which it is much harder to recover? The new social infrastructure that climate change adaptation will design need not be itself unjust. On the contrary, one of the tests of ethically legitimate adaptation is that it takes advantage of the opportunities that climate change, paradoxically, may open up for greater social equality and justice. This will be the silver lining, perhaps, of the storms of our grandchildren (Hansen 2009).

Just Adaptation

To be sure, expansion of social justice may not be the first thing that people think about when they contemplate the future conditions proffered by climate change. But problems and challenges to justice and equity on a global scale have been widely debated and have achieved a special importance in the context of climate change partly because its dynamic impacts vary in severity in different areas of the world. In many parts of the world, there is a great disparity between the small amount of climate damage done by relatively low carbon emissions of generations of people living there and the severity of the losses climate change is visiting upon them (Paavola & Adger 2006). Over the last two centuries, the world's poor did not even partake equitably of the wealth generated by the energy extracted from fossil carbon.

This disparity between high and lower emission countries was formally addressed by the Warsaw International Mechanism for Loss and Damage associated with Climate Change Impacts

(WIM) in 2013. Since then, the concept of loss and damage has been expanded to include the adverse effects associated with other slow onset and extreme events that cannot be adapted to, such as the inundation of entire island nations due to sea level rise. It also includes both monetizable impacts and intangible losses and damages, often referred to as 'non-economic losses' (Calliari et al. 2020: 1). But economic loss and damage and non-economic losses – essentially losses of culture, agency, and individual and social capability – exist within nations and regions as well and must be addressed in adaptation plans and in their roll-out and implementation. Like matters of equity and distributive justice always do, answers to these questions entail redistribution of wealth and power and create winners and losers. In the case of adaptation this is especially true because, as Yeb Sano quoted earlier pointed out, the vulnerabilities that adaptation planning is designed to address often arise from and feed upon deep-seated, pervasive patterns of social injustice (Thomas & Twyman 2005). Hurricane Katrina in the New Orleans area in 2005 provided one of the most graphic American illustrations of this phenomenon. Although that natural disaster wreaked havoc upon rich and poor alike, the poor and marginalized, neglected for so long, bore the brunt of the catastrophe (Cooper & Block 2006; Hartman & Squires 2006).

How are questions of distributive justice and equity often defined and conceptualized? There are three main approaches that are influential in both theoretical and applied ethics. These are referred to as neo-utilitarian justice, fairness justice, and capability justice. Each is germane to climate change adaptation ethics, although they are based on different value priorities. Taken as compass headings rather than mandatory rules, the principles developed within each of these theories of justice lead to quite different normative evaluations of adaptation priorities and policies. Each of these theories have their own formulation of the condition that defines justice or rightness in the world. These involve well-being; respect for duty and the intrinsic value of persons; and flourishing under social conditions conducive to the actualization of inherent natural capabilities. Their respective right-making conditions are: (1) aggregate net benefit for a population; (2) fair and equitable treatment of each individual as a person deserving of rights and respect; and (3) social and structural conditions that support individual and collective actualization of a wide range of potential human capabilities (Persad 2019).

Neo-utilitarian or welfarist theories of justice hold that justice is the maximization of aggregate net health benefit in a population (Goodin 1995). Here, a slogan capturing this might be 'the greatest benefit for the greatest number.' Benefit is variously defined in works following this approach. It includes health and safety, happiness, utility, and well-being. This informs climate policy in several ways. Some utilitarian philosophers argue that non-human beings that are capable of experiencing pain and suffering should be included under the requirements of justice, but in planning circles, this perspective on justice is largely anthropocentric utilitarianism; the interests of other species may matter, but only human interests are counted. Within the human population, depending on how broadly the 'greatest number' is demarcated in space and time, aggregate net benefit justice might have a present time bias or a national, ethnic, or geographic bias, or it might equally facilitate international and/or intergenerational sharing of resources in the pattern it defines as just. However, this approach tries to make ethics as quantifiable and quantitative as possible, and it is not clear how to apply it in these broad senses, especially broadly over time because the preferences and experiences of future people who do not yet exist will be difficult to specify.

Fairness theories of justice hold that justice is a distribution of benefits not based on aggregate net benefit but on the patterns of fair, non-discriminatory treatment of individuals, which requires an equal distribution of basic social resources unless an unequal pattern would be beneficial to the least well-off persons in society (Rawls 2001). Here, a slogan capturing this might be 'respect other persons by treating them as ends with inherent worth, never as mere means to someone else's ends.' Fairness is based on the basic right of equality of opportunity to achieve species typical functioning (health) and to live one's life within the normal opportunity range available to individuals in a society. If inequality cannot be avoided (due to scarcity of resources, for example), justice defined as fairness should give priority to the needs and interests of the least well-off. While it stresses duties to respect the individuality and rights of persons rather than the maximizing of the interests of people in the aggregate, it is not clear how fairness theories of justice can accommodate the intergenerational questions of justice that climate ethics and adaptation ethics confront. Will adaptation of a new climatic world of greater scarcity, for example, require the curtailment of some rights of contemporary persons for the sake of the rights and needs of future persons?

Capability theories of justice hold that justice is not about the aggregate distribution of resources; it is about the patterns of the use of resources to achieve self-realization, relations of solidarity amid diversity, and the actualization of a wide repertoire of capabilities that produce individual and collective flourishing (Nussbaum 2011; Sen 2009). Here, a slogan capturing this might be 'to each a world conducive to the realization of unique potential, from each an openness to diversity and change.' Fairness theories emphasize the equal opportunity to receive resources. Capability theories stress the background social conditions that enable and empower each individual to use resources to promote their human flourishing and development. Human flourishing or the human good starts from inherent potentialities within the evolved being of all species, including humans, and is realized through social conditions that allow for the effective functioning of these potential capacities over the life span. Justice involves institutional arrangements and social supports that make the realization of capacities possible.

How do these approaches to justice bear on adaptation planning? Neo-utilitarian approaches to justice are the most commonly found in discussions of the ethics of both climate change planning and preparedness and public health emergency preparedness and response. This kind of planning normally talks about populations and aggregates rather than individuals, although individuals and groups with special vulnerabilities and needs are usually given special consideration in adaptation or response plans. The main problem, which has arisen in the discussion of seeing adaptation planning as a commodity or expert service or a civic practice, is that neo-utilitarian justice is flawed by an individualistic, aggregative notion of a population. Many would argue that an adequate conception of justice for climate change needs a more holistic notion of a community and the common good rather than aggregate net benefit distributed across a large number of individuals. Moreover, the neo-utilitarian view has a narrowly hedonistic notion of well-being or benefit. These generally are the criticisms that the capability theory of justice would level at neo-utilitarianism.

Fairness theories take issue with the aggregate net benefit notion in a different way. Their complaint is that net-benefit can be achieved with a highly inegalitarian distribution of scare resources. To illustrate this issue, the following case hypothetical is instructive:

> Consider, for example, a ward with 100 patients, 50 of whom require one pill and 50 of whom require two pills to recover. The patients are otherwise similar. The clinic has 50 pills and must decide how to distribute them. To achieve the best outcome, all 50 pills should be given to the patients who need only one to recover. However, to give each patient an equal chance to recover, entitlement to treatment should be awarded randomly. Seventeen fewer cures would result.
>
> *(Brock & Wikler 2006: 264)*

Which pattern of distribution should generally be favored in adaptation planning? In that context, the aggregate net benefit justice may not always pertain to life-or-death issues, but sometimes it will, and ethically serious stakes foreseeably will be involved, nonetheless. If justice means maximizing the wellbeing of the greatest number of people regardless of the equity of the pattern of resource distribution, then seemingly both well-being and fairness cannot both be served. How should we think about which is more important and more congruent with the other goals and with the culture and sensibility that should inform the adaptation planning process? Can preserving fairness and equity in distributional plans or policies (of which adaptation planning will have to define several) ethically be given priority over efficiency and total welfare?

Finally, the distinctiveness of capability theories of justice lies in the fact that they are not primarily theories of *distributive* justice at all; they are theories of *structural* justice. Capability theories assume that all human beings have a wide spectrum of potential capabilities absent some specific impairment, in other words, a normal species' typical potentiality range. This is a constitutive or ontological matter, not a distributive one. What justice pertains to are the factors that influence the realization of those potentials, the social determinants of actualized capabilities and functionings that people can develop. Developmental human self-realization and flourishing is the main focus for capability theories. Aggregated outcomes or fair, equitable sharing move from center stage, and the spotlight of justice falls on the range, diversity, and richness of the activities and lives individuals are enabled to live in a given time and place. From this structural point of view, the distribution of goods is not what drives the structure of society; rather, it is the structure of society that brings about the richness or paucity of human capabilities and functionings, without which even abundant resources will be of little good. Justice is not about what people can have, but who they can be.

Between Substance and Process

None of these three theories of justice yield univocal solutions to the questions of balancing happiness, dignity, or human self-realization and flourishing. Those who study applied ethics often treat processes of democratic deliberation as a crucial supplement to these substantive ethical theories (Daniels & Sabin 2002; Gutmann & Thompson 1996). In this combination of prescription and process, a number of possible tradeoffs might be plausibly justified by conflicting sets of values, so the task is to formulate fair rules for a process that will serve to legitimate a particular social choice. In this respect, two of the ethical dimensions of climate change adaptation planning – justice and democratic deliberation and participation – interconnect.

Adaptation planning should be inclusive, provide for parity of voice, and consist of reasonable (evidence-based) deliberation and debate concerning various adaptation provisions and ideas. Yet, the question remains what will motivate people to accept and follow an adaptation plan, when they do not agree with its provisions? And what will motivate compliance and cooperation even when the plan has provisions that will require personal sacrifices, such as financial or nonmonetizable loss, on their part? The answer to these questions is legitimacy. People will accept private sacrifice when they recognize doing so to be a public duty. People cooperate when they consider the adaptation process to have been respectful and just to all, and they regard the plan itself to be a legitimate expression of their own civic voice (despite their private disagreement) as well as the voice of the entire community (Jennings 2003).

Let us conclude by turning to some important aspects of climate change adaptation legitimacy. First, publicity or transparency in decision-making. Some believe that climate change adaptation involves the kind of "tragic choices" that are too disruptive be discussed openly (Calabresi & Bobbitt 1978). Those who view adaptation planning as a kind of civic practice maintain that when it comes to matters of behavior change, institutional change and economic redistribution, publicity and transparency are prerequisites of the plan's legitimacy, regardless of how much sacrifice the plan calls for. Adaptation governance directed by technical experts behind closed doors must be the exception rather than the rule in this domain and used only in the most sensitive cases involving law enforcement, national security, or as a temporary expedient to prevent panic and save lives. Otherwise, closed door planning invites an unaccountable decision procedure that can obscure conflicts of interest, corruption, and discrimination against minorities or the socially marginalized. Even absent abuses such as these, secrecy undermines public trust and hampers the successful implementation of the plan.

Second, an appeals process. Those who disagree with a certain value ordering or who believe they or others have been unfairly disadvantaged by an adaptation planning choice should be able to appeal the decision to responsible and responsive authorities. This will help ensure that adaptation provisions are impartially interpreted and applied. Moreover, the existence of an appeals process testifies to the belief that all persons have an inherent moral standing that should be recognized and respected appropriately. That includes a right to have their grievances aired and addressed. When conjoined to the publicity condition, the appeals requirement can provide society with a public record of criticisms bearing on allocation criteria and of official responses to them. That is important so that adaptation planning can learn from past mistakes, and it provides reassurance that builds confidence in further iterations of the planning process.

Lastly, let us consider claims based on evidence, or proposals based on relevant reasons. Some defenders of deliberative democratic procedures have proposed that limits be placed on the kinds of reasons that might legitimately be advanced in such public deliberations (Gutmann & Thompson 1996). The only reasons that should count in public allocation decisions are those that could be accepted as relevant by fair-minded people who are disposed to find mutually justifiable terms of cooperation. Part of the legitimacy of climate change adaptation stems from its association with scientific knowledge and objectivity. Although adaptation is an inherently value-laden activity as well, value disputes that are mediated and resolved under the aegis of verifiable factual knowledge have the strongest degree of public trust and support.

Conclusion

W. Neil Adger and colleagues argue that:

> ...An adaptable society is characterized by awareness of diverse values, appreciation and understanding of specific and variable vulnerabilities to impacts, and acceptance of some loss through change. The ability to adapt is determined in part by the availability of technology and the capacity for learning but fundamentally by the ethics of the treatment of vulnerable people and places within societal decision-making structures.
>
> *(Adger et al. 2009: 350)*

Caring about what is done to vulnerable people and places and justly recognizing their moral standing by publicly standing up and speaking out is called solidarity. The importance of adaptation planning to successful climate change adaptation lies partly in the inherent value of preparedness in the face of foreseeable change and other 'known unknowns.' But the capability to adapt and the capability of a society to be solidaristic are closely linked. Planning activities over several years in local communities also contribute to adaptation capability and to social solidarity by making explicit the underlying values shaping decisions and actions and by subjecting those values and activities to a learning process through deliberation and accountability. Participatory civic practices, such as adaptation planning, contribute significantly to civic learning, which is the foundation of sustainable solidarity and justice. When civic learning and scientific learning work together, legitimate and adaptive climate change preparedness planning and governance will be within our grasp.

References

Adger, W. N. (2003) "Social Capital, Collective Action, and Adaptation to Climate Change," *Economic Geography*, 79: 387–404.

Adger, W. N., Barnett, J., Brown, K., Marshall, N., and O'Brien, K. (2013) "Cultural Dimensions of Climate Change Impacts and Adaptation," *Nature Climate Change*, 3: 112–117.

Adger, W. N., Dessai, S., Goulden, M., Hulme, M., Lorenzoni, I., Nelson, D. R., Naess, L. O., Wolf, J., and Wreford, A. (2009) "Are There Social Limits to Adaptation to Climate Change?" *Climatic Change*, 93: 335–354.

Adger, W. N., Quinn, T., Lorenzoni, I., Murphy, C., and Sweeney, J. (2013) "Changing Social Contracts in Climate-change Adaptation," *Nature Climate Change*, 3: 330–333.

Bond, K. (2018) *2020 Vision: Why You Should See Peak Fossil Fuels Coming*, London: Carbon Tracker at https://www.carbontracker.org/reports/2020-vision-why-you-should-see-the-fossil-fuel-peak-coming/.

Brock, D. W., and Wikler, D.(2006) "Ethical Issues in Resource Allocation, Research, and New Product Development," in Dean T. Jamison, Joel G. Breman and Anthony R. Measham, et al. (eds.), *Disease Control Priorities in Developing Countries*, New York: Oxford University Press, 2nd edition, 259–270.

Calabresi, G., and Bobbitt, P. (1978) *Tragic Choices*, New York: W. W. Norton.

Calliari, E., Serdeczny, O., and Vanhala, L. (2020) "Making Sense of the Politics in the Climate Change Loss & Damage Debate," *Global Environmental Change*, 64: 1–10.

Childress, J. F., and Bernheim, R. G. (2003) "Beyond the Liberal and Communitarian Impasse: A Framework and Vision for Public Health," *Florida Law Review*, 55(5), 1191–1231.

Collins, K., and Ison, R. (2009) "Jumping off Arnstein's Ladder: Social Learning as a New Policy Paradigm for Climate Change Adaptation," *Environmental Policy and Governance*, 19: 358–373.

Cooper, C., and Block, R. (2006) *Disaster: Hurricane Katrina and the Failure of Homeland Security*, New York: Times Books.

Daniels, N., and Sabin, J. (2002) *Setting Limits Fairly*, Oxford: Oxford University Press.

Dewulf, A. (2013) "Contrasting Frames in Policy Debates on Climate Change Adaptation," *WIRE's Climate Change*, 4: 321–330.

Forester, J. (1999) *The Deliberative Practitioner: Encouraging Participatory Planning Processes*, Cambridge, MA: MIT Press.

Fresque-Baxter, J. A., and Armitage, D. (2012) "Place Identity and Climate Change Adaptation: A Synthesis and Framework for Understanding," *WIRE'S Climate Change*, 3: 251–266.

Global Commission on the Geopolitics of Energy Transformation. (2019) *A New World: The Geopolitics of the Energy Transformation*. Abu Dhabi: Irena at https://www.irena.org/publications/2019/Jan/A-New-World-The-Geopolitics-of-the-Energy-Transformation

Goodin, R. E. (1995) *Utilitarianism as a Public Philosophy*, New York: Cambridge University Press.

Gutmann, A., and Thompson, D. (1996) *Democracy and Disagreement*, Cambridge, MA: Harvard University Press.

Hansen, J. (2009) *Storms of My Grandchildren: The Truth about the Coming Climate Catastrophe and Our Last Chance to Save Humanity*, New York: Bloomsbury.

Hartman, C., and Squires, G. D. (2006) *There Is No Such Thing as a Natural Disaster: Race, Class, and Hurricane Katrina*, New York: Routledge.

Jennings, B. (2003) "On Authority and Justification in Public Health," *Florida Law Review*, 55: 1241–1256.

Jennings, B. (2016) *Ecological Governance: Toward a New Social Contract with the Earth*, Morgantown, WVA: West Virginia University Press.

Jennings, B. (2019) "Relational Ethics for Public Health: Interpreting Solidarity and Care," *Health Care Analysis: Journal of Health Philosophy and Policy*, 24(4): 4–12

Jennings, B., and Arras, J. (2016) "Ethical Aspects of Public Health Emergency Preparedness and Response," in B. Jennings, J. Arras, D. Barrett and B. Ellis (eds.), *Emergency Ethics: Public Health Preparedness and Response*, New York: Oxford University Press, 1–103.

Kass, N. E. (2001) "An Ethics Framework for Public Health," *American Journal of Public Health*, 91: 1776–1782.

Kysar, D. A. (2010) *Regulating from Nowhere: Environmental Law and the Search for Objectivity*, New Haven, CT: Yale University Press.

Lynas, M. (2020) *Our Final Warning: Six Degrees of Climate Emergency*, London: Fourth Estate.

MacIntyre, A. (2007) *After Virtue: A Study in Moral Theory* (3rd edition), South Bend, IN: University of Notre Dame Press.

Nussbaum, M. C. (2011) *Creating Capabilities: The Human Development Approach*, Cambridge, MA: Harvard University Press.

Paavola, J., and Adger, W. N. (2006) "Fair Adaptation to Climate Change," *Ecological Economics,* 56: 594–609.
Persad, G. (2019) "Justice and Public Health," in A. C. Mastroianni, J. P. Kahn and N. E. Kass (eds.), *The Oxford Handbook of Public Health Ethics,* New York: Oxford University Press, 33–46.
Piketty, T. (2020) *Capital and Ideology,* Cambridge, MA: Harvard University Press.
Purdy, J. (2010) "The Politics of Nature: Climate Change, Environmental Law, and Democracy," *Yale Law Journal,* 119: 1122–1209.
Rawls, J. (2001) *Justice as Fairness: A Restatement,* Cambridge, MA: Harvard University Press.
Robinson, T. (2019) *Listening to the Wind: The Connemara Trilogy, Part One,* Minneapolis, MN: Milkweed Editions.
Ruhl, J. B. (2010) "Climate Change Adaptation and the Structural Transformation of Environmental Law," *Environmental Law,* 40(2): 363–436.
Ruhl, J. B., and Salzman, J. (2020) "What Happens When the Green New Deal Meets the Old Green Laws?," *Vermont Law Review,* 44: 693–722.
Sano, Y. (2013) *Address to the UN Climate Change Convention (COP19) Warsaw,* November 11, 2013 https://www.climatechangenews.com/2013/11/11/its-time-to-stop-this-madness-philippines-plea-at-un-climate-talks/
Schipper, E. L. F., and Burton, I. (2009), "Understanding Adaptation: Origins, Concepts, Practice and Policy," in E. L. F. Schipper and I. Burton (eds.), *The Earthscan Reader on Adaptation to Climate Change,* New York: Earthscan.
Schön, D. A. (1984) *The Reflective Practitioner,* New York: Basic Books.
Schön, D. A., and Rein, M. (1997) *Frame Reflection: Toward the Resolution of Intractable Policy Controversies,* New York: Basic Books.
Sen, A. (2009) *The Idea of Justice,* Cambridge, MA: Harvard University Press.
Smit, B., Burton, I., Klein, R. J. T., and Wandel, J. (2000) "An Anatomy of Adaptation to Climate Change and Variability," *Climatic Change,* 45: 223–251.
Spence, D. B. (2019) "Regulation and the New Politics of (Energy) Market Entry," *Notre Dame Law Review,* 95(1): 327–396.
Thomas, D. S. G., and Twyman, C. (2005) "Equity and Justice in Climate Change Adaptation Amongst Natural-Resource-Dependent Societies," *Global Climate Change,* 15: 115–124.
Urwin, K., and Jordan, A. (2008) "Does Public Policy Support or Undermine Climate Change Adaptation? Exploring Policy Interplay Across Different Scales of Governance," *Global Environmental Change,* 18: 180–191.
Vandenbergh, M. P., and Gilligan, J. M. (2017) *Beyond Politics: The Private Governance Response to Climate Change,* New York: Cambridge University Press.
Watson, L. (1984 [2019]) *Heaven's Breath: A Natural History of the Wind,* New York: New York Review Book.

Further Reading

A Clear Discussion of the International Debate Over Responsibilities for Climate Change and International Compensations to Hard Hit Developing Countries.

Adger, W. N., Dessai, S., Goulden, M., Hulme, M., Lorenzoni, I., Nelson, D. R., Naess, L. O., Wolf, J., and Wreford, A. (2009) "Are There Social Limits to Adaptation to Climate Change?" *Climatic Change,* 93: 335–354.

Provides an Overview of the Components and Distinctiveness of Climate Change Adaptation and the Questions it Poses for Existing Environmental Law.

A Thorough Discussion of the Institutional and Cultural Factors Affecting the Success or Failure of Adaptation Planning in Particular Communities.

Calliari, E., Serdeczny, O., and Vanhala, L. (2020) "Making Sense of the Politics in the Climate Change Loss & Damage Debate," *Global Environmental Change,* 64: 1–10.

Provides an Overview of Thinking About Deliberative Public Participation in the Planning Field. Is Critical of a Power Based Competitive Model Popularized by Arnstein and Describes an Alternative Network Communications and Social Learning Model of the Relationships Between Elites and Citizen Participants in Planning Settings.

Collins, K., and Ison, R. (2009) "Jumping off Arnstein's Ladder: Social Learning as a New Policy Paradigm for Climate Change Adaptation," *Environmental Policy and Governance,* 19: 358–373.

Ruhl, J. B. (2010) "Climate Change Adaptation and the Structural Transformation of Environmental Law," *Environmental Law,* 40(2): 363–436.

9

SOCIALLY CONSTRUCTED EXPECTATIONS OF CERTAINTY IN CLIMATE SCIENCE

Kirk W. Junker

The Exigence

It is not a coincidence that the states that most benefit from technology are also the states with the greatest monetary wealth. Yet it is most often precisely in these states that both greenhouse gas (GHG) emissions are the highest and that scientific uncertainty is most often the excuse offered for failing to mitigate the high GHG emissions. For a scientist, to say that science is 'uncertain' is as banal as saying that water is 'wet.' Yet this simple excuse has traction with citizens, regulators, and legislators. This handbook contribution will trace the social construction of this trope – 'scientific uncertainty' – in an effort to get past it and leave it behind. This will enable fruitful discussion of ethical choices to mitigate climate disruption under the conditions of science as science understands them, including its inherent uncertainty. The misguided expectation of certainty in climate science serves to emphasize two fundamental points about the nature of knowledge itself: first, the expectation of certainty in science is misinformed and misguided, and second, there are informed and methodical paths for making ethical choices to mitigate climate disruption.

The Climate Change Convention Explicitly Excludes Certainty

Ethical decisions are not limited to formal expressions in order to be effective, but when they are formally expressed, they are expressed as civil society decisions and can be recorded in binding sources of law. When faced with the climate crisis, already in 1992, the states of the world answered the question as to how to proceed – not with caution, but with *pre*caution, and precaution was agreed as a legal principle. Article 3 of the United Nations Convention on Climate Change (UNFCCC) is a binding legal commitment for all of the 197 ratifying states, including the high emitters, to

> Take precautionary measures to anticipate, prevent or minimize the causes of climate change and mitigate its adverse effects. Where there are threats of serious or irreversible damage, lack of full scientific certainty should not be used as a reason for postponing such measures.
>
> *(1992)*

One might expect that with such a clear legal commitment, no further discussion would be necessary regarding the commitment of states to act. If a state does not want to be bound to the principle any longer, it would of course be free to leave the convention, but no state has

DOI: 10.4324/9781003039860-12

left the convention. So how can it be that so many years later, many states, foremost among them the high-emitters, are still trotting out this old excuse, which was precisely and bindingly rejected by all states in 1992?

To begin to answer that question, one is reminded that as a matter of public international law, a 'principle' is a primary source of binding law, on the same level as a treaty, according to Article 38 of the UN Statute for the International Court of Justice (1945). The precautionary principle is one such binding principle. Then, in the clearest example of a belt-and-braces legal act one can imagine, the principle is explicitly also agreed in the UNFCCC, making it doubly clear that all signatory states are bound by it. Thus, the only question is 'to what acts or forbearances are the states bound?'

To answer this question, there are four features of the precautionary principle to be highlighted. First, included in the legal principle of precaution is that any party wishing to act must first demonstrate that the planned actions will not cause harm. In law, this is known as 'shifting the burden of proof.' Regarding the climate crisis, that means the parties have agreed to shift the burden *from* the state or the citizens, who would otherwise need to prove after the fact that an action has caused harm, *to* the party that is proposing the action, before the action is attempted. There are practical results to this shift in the burden of proof. For example, rather than the public's tax money being spent on the state's energy and time to prosecute a harm to the public committed by a private party, the private party that wishes to act must spend its time, energy, and money to demonstrate that the proposed action, if carried out, will *not* harm the public's interests.

A second feature, although subtle, goes to questioning the position of the original burden of proof. Why was the burden initially with the public and the state such that it is regarded as a change – a 'shift' – to put the burden on the person potentially causing the harm? What does a society consider to be its base position or the base position of the individual when constructing legal relationships between the society and the individual? Why would any society assume the base position to be that an individual can do as he pleases until the society can prove he has harmed others? Even the utilitarianism of Bentham and Mill limits one's rights at the point where "another's nose begins" (Holmes, quoted in Caplan 2013). We have many different legal tools, including taxes, permits, and licenses to make clear that certain types of activities, left to the decision of the individual, would not be acceptable to the society, and in those situations, we do not wait until the society can prove harm. Must the society wait until after a battery, rape or murder has occurred to prohibit those acts? Of course not. So, in fact, we are well-accustomed to societies prohibiting harmful activity before it occurs and only in exceptional circumstances do we allow otherwise prohibited acts to take place, such as in cases of self-defense.

A third noteworthy feature of the precautionary principle is the element of time. As Article 3 of the UNFCCC makes explicit, some harms are so severely damaging or irreversible that the promise to reverse the harm, if the harm were to occur, is an insufficient reason to allow an individual to proceed. As climate science has clearly demonstrated, the act of loading GHG into the atmosphere is so severely damaging, and not simply reversible, that it demands the scrutiny of examination before the act, not afterward.

A fourth feature of the precautionary principle merely puts an obvious point into words – we must take action although certainty is lacking. This feature is neither the beginning point nor the most important point of the principle, but it requires the most scrutiny because it has been haunted by the grossest mischaracterizations of the practices of science. Notice that the language of Article 3 does not say 'in instances of scientific uncertainty,' which would suggest that outside those instances, science is certain. It says that "lack of full scientific certainty should not be used as a reason for postponing such measures" (UNFCCC 1992: Art. 3). In other words, the drafters recognized that this excuse was common currency, so they had to explicitly say that just because the nature of science always and already includes uncertainty, one cannot use that inherent aspect of science as a reason to say, 'let's wait.' Given that uncertainty is part of science, upon what would one be waiting?

Why Did We Dismiss Certainty? Because Science Is Never Certain

Before turning to how uncertainty has been socially and rhetorically constructed and specifically applied to the climate crisis, we should examine the putative role of certainty in science more generally. The debacle of the world's handling of climate disruption has a basis in the misunderstanding of the nature of science. Fossil fuel manipulators have convinced politicians, legislators, and others that science generally or normally produces certainty. Scientists tell us it does not.

Scientists from Anaximander to Epicurus to Heisenberg[1] included concepts of uncertainty and indeterminacy in their theories and practices. Indeterminacy means that one cannot determine all parameters at the same time in a given system (Heisenberg 1925; Healey 1981). "Uncertainty is not only inherent in predicting chaotic classical systems, it also appears to be basic to our descriptions of the fundamental laws of physics – as described in the measurement axiom of standard quantum mechanics" (Palmer & Hardaker 2011). Max Born demonstrated the integral nature of indeterminism in quantum mechanics. And Richard Feynman

> recognized that doubt, or uncertainty, is an inherent part of the scientific method, and therefore an inherent element to scientific prediction. By contrast, predictions that have little or no basis in science can often be recognized as such, by a complete failure to acknowledge the existence of uncertainty.
>
> *(Palmer & Hardaker 2011: 4681, in reference to Heisenberg's 1927 principle)*

More recently, physicist Carlo Rovelli, called one of the one hundred most influential thinkers in the world (Global Thinkers 2019), clearly states

> what I see as the deepest misunderstanding about science... is the idea that science is about certainty... Science is not about certainty. Science is about finding the most reliable way of thinking at the present level of knowledge. Science is extremely reliable; it's not certain. In fact, not only is it not certain, but it's the lack of certainty that grounds it. Scientific ideas are credible not because they are sure but because they're the ones that have survived all the possible past critiques, and they're the most credible because they were put on the table for everybody's criticism.
>
> *(Rovelli 2015: 75)*

Rovelli makes an important distinction when he notes that the strong part of science – the data – does not characterize the nature of science. Consistent with the name we give to the enterprise – 'science' or 'knowledge' – he maintains instead that knowledge is to understand "something about reality essentially by changing something in the conceptual structure we use to grasp reality" (Rovelli 2015).[2] Knowledge (*scientia*) is therefore understood as conceptual structures, not uninterpreted collections of data. According to Rovelli, "one can only speak of 'certainty' in the mathematics of data, not in knowledge (science)" (Rovelli 2015; see also Rovelli 2014).

> Even within classical physics, with its deterministic and precisely known laws, the evolution of many systems is nevertheless uncertain because these laws are chaotic – weather prediction is but one example. For other systems, such as the economy, we have no underlying 'laws,' and, additionally, the predictions can themselves influence the system about which these predictions are made. It is sometimes said that the difference between a weather forecast and an economic forecast is that the weather forecast cannot change the weather, whereas an economic forecast can change the economy. Matters are less clear cut these days, where a climate-change forecast can in principle change societal carbon footprints, and hence the climate.
>
> *(Palmer & Hardaker 2011: 4682)*

It is very rare that scientists confirm an absolute truth or certainty, since "[u]ncertainty is inevitable at the frontiers of knowledge" (Palmer & Hardaker 2011: 4684). Consequently, practicing science, using science, and talking about science cannot be said to depend upon certainty. Some researchers conclude that although communicating uncertainty outside the scientific arena is an important issue, it is also possible that "the notion of probabilistic prediction as too complex for the public to understand" (Palmer & Hardaker 2011: 4684). As evidence of this problem, one might note that while most citizens understand basic probabilities with relatively limited parameters, such as with sports gambling, they may still blame a meteorologist for poor work in his or her predictions if the weather does not happen as predicted, even if the meteorologist used the probabilities of meteorology precisely according to the science. The particular topic of weather forecasting is relevant for climate discussions:

> If in fact the public were more exposed to weather prediction as inherently probabilistic, perhaps there would be more acceptance of the simple fact that one's view about the risk of dangerous climate change would not be framed in the black and white terms of 'belief" and 'skepticism'.
> *(Palmer & Hardaker 2011: 4683)*

But even if we can overcome this hurdle, and all become cognizant of probability and uncertainty as concepts inherent to science, the problem remains of how one makes decisions *in the light of* uncertain scientific input (Palmer & Hardaker 2011: 4684). At its foundation, this problem is found within what some call 'traditional scientific practices.' Within science, the scientist is encouraged to limit the amount of conscious decision-making and choice that will be necessary. So an environmental researcher is often led to believe that as between false positives (type I error) and false negatives (type II error), choices to reduce false positive error should be made because it conforms to 'traditional scientific practices' that seek to reduce the chances of a false positive (Lemons & Junker 1996: 396, citing Simberloff 1987). Accordingly, by minimizing type I error, it is said that science decisions based on speculative thinking will be reduced (Lemons & Junker 1996: 396, citing Miller 1993), thus making clear that the value is to limit choice and decision-making. But even if a practice can narrow the need for choices to be made, that does not mean that certainty can be obtained. It just delays answering the question of how we make decisions in the light of uncertain scientific input. Alternatively, if the value of the exercise is to benefit common interest, then when analyzing human activities for which there is a reasonable and evidence-driven basis that they are dangerous, an ethical argument can be made that avoiding a false negative (type II error) with high levels of probability is in fact in the common interest.

The trope of 'scientific uncertainty' is liberally spread throughout civic discourse without even assigning it an agreed meaning. It seems that those who use the term 'uncertainty' while in a defensive mode, use it to imply that if one is uncertain, then one knows nothing, thereby producing a fictional dichotomy of choice – certain knowledge merits action, but anything less equals knowing nothing and therefore deserves no action. That attitude was explicitly dismissed by Article 3 of the UFCCC, but nevertheless still echoes around the chambers of decision makers. Ironically, the same persons who want to continue *status quo ante* in their daily carbon practices, and thus demand certainty from science before supporting change, are the same persons who are happy to take pills and undergo surgeries, drive cars across bridges and through tunnels,[3] and use every sort of magnetic field and electrical appliance, when all of these are released to the public without certainty that the pill will have only the intended effects, that the bridge will stand or that the magnetic waves of the appliance will not harm the body. Thus, it must be the psychological comfort and convenience of the *status quo ante*, not the science, that drives decisions.

In summary, if we interrogate what we mean by knowledge, we find a much broader set of possibilities than 'certainty' and 'nothing.' Uncertainty is an unavoidable aspect of any scientific endeavor, and climate science is no exception. The fifth assessment report of the Intergovernmental Panel on Climate Change (IPCC) of 2013 used the word 'uncertain' or its derivatives (e.g., 'uncertainty') more than 2200 times in the report of Working Group I alone – around 1.5 times per printed page. Uncertainty is found throughout the narrative of climate disruption. There is uncertainty about the physics, such as when we ask how much warming can we expect with a doubling of carbon dioxide from preindustrial levels? There is uncertainty about the economics, such as when we ask how much will it cost to mitigate and how much money do we save if we do not mitigate? Perhaps most importantly, there is even uncertainty about human values and choices, as when we ask how likely is it that humanity will act to control the risk from climate disruption (Lewandowsky et al. 2015)?

If Scientists Do Not Expect Certainty, Why Does Anyone Else Expect It?

As Carlo Rovelli has made clear, it is a slip in thinking to equate the attributes of data with science. Data must be interpreted or else it has no meaning. And it is in the interpretation of the data that uncertainty is omnipresent. The interpretations themselves are choices and decisions and include the use of values. The strength of the data encourages scientists and non-scientists alike to speak of 'value-free data,' and from there, it is a short slip to speak of 'value-free science.' If in fact data has a value-free relationship to the material world, the uncertainty of science would still demand that science must use something other than the data to interpret it, and there one sees the limit of the 'value free' assignation. Moreover, if in fact science *were* constituted by a value-free relationship to the material world, that would be precisely the reason that it has no decision-making power in the social world. Philosopher Leslie Stevenson characterizes the 'value-free' reputation of science by noting that science is supposed to investigate facts and not values, and that the discovery of truth is the only value that is recognized by scientists, thus leaving decisions that apply scientific results to society as a whole (Stevenson 1989).

Psychologists Mandel and Tetlock researched what they call the "myth of value-neutral scientific advertising" and found that scientists too, as humans, find it difficult to "separate the epistemic functions of their judgments (accuracy) from the social-economic functions (from career advancement to promoting moral-political causes that 'feel self-evidently right')" (Mandel & Tetlock 2016: 451). To further debunk the fiction of value-neutral scientific work, ethicist, and lawyer, Donald A. Brown distinguishes the dual roles for scientists – first, the "research role" wherein "strict and rigorous procedures are called for and no scientist should make claims when acting in this role that have not been demonstrated"; and second, the "public policy role," wherein science is expected to go beyond investigating and proving causation, to warning the public of significant potential risks, in advance of the proof being completely demonstrable before the harm occurs (Brown 2011).

Despite science itself not characterizing its practices as producing certainty, and despite states asserting in agreement via Article 3 of the UNFCCC that uncertainty in science is not an acceptable excuse for inaction, there is still a narrative of science as though it does produce certainty and because it does or can, we should wait for that to happen before we take action to mitigate anthropogenic climate disruption. How did this narrative (Junker 1998)[4] become constructed? There are two chief categories for the sources of the mischaracterization: the intentional mischaracterizations and the unintentional or socially constructed characterizations.

Intentional Misinformation

A first and most obvious reason that the 'wait for certainty' excuse gets so much traction is that there has been a considerable intentional effort to use it to misinform citizens (Brown 2013). The intentional misinformation regarding climate disruption has many aspects, including misinformation about climate science specifically, but also misinformation about science generally (Oreskes & Conway 2010; see also Oreskes 2015).[5] A simple, but effective tactic has been to insert the word 'debate' into climate discussions so often, that it becomes like the surname of the topic, even for climate activists. It might come as a surprise to persons in the United States, for example, that in the rest of the world, when the topic is discussed in English, the word 'debate' is not included. As UN Secretary General Antonio Gutiérrez has emphasized, it is a climate "crisis" (not a debate) (UN 2018). The major tactic of these groups "was and continues to be manufacturing uncertainty constantly asserting that the evidence is not sufficient to warrant regulatory action" (Dunlap & Jacques 2013). It is ironic that money interests try to point to uncertainty in science as a reason to ignore science and rather follow money interests instead, because when it comes to certainty and predictability, even physics insists it is not certain, and by comparison to physics, the uncertainty of economics makes it painfully lacking in its ability to predict anything. Yet these same money interests do not advocate against economic choices based upon its extreme uncertainty.

It has been well documented that fossil fuel economic interests fund statements from all sorts of organizations that are dressed up as science, or as neutral parties, or as researchers, or even as government spokespersons (Brown 2013; see also Boykoff & Boykoff 2004; see also Hansson 2017)). Unless a viewer, reader or listener digs to learn who is funding those statements, the costume labels that they wear can be persuasive. This is not the first time that industries, and so-called 'think tanks' such as the Heritage Foundation, Competitive Enterprise Institute, George C. Marshall Institute, and the Cato Institute, as well as other political and economic interest groups, using false experts, have methodically created and circulated misinformation on issues (Jacques et al. 2008). This strategy has been used in many countries, including the United States, in matters concerning vinyl chloride, lead paint, electricity generation, tobacco use, guns, the ozone layer, acid rain, aspirin, herbicides, and asbestos.[6]

But the creation and dissemination of that sort of intentional misinformation is only part of the explanation as to why misinformation and mischaracterizations are effective in persuading citizens. They are also effective because citizens are receptive to the misinformation and mischaracterizations. Citizens have been socialized to accept these fossil fuel sales talks, including the mischaracterization of science as being certain, before the bad seed is even strewn. Thus, although it is important to expose the mouthpieces that intentionally misinform, it is also important to address and correct the readiness of the audience to accept the misinformation.

Unintentional Social Conditioning

Social Conditioning May Begin with Schoolbooks

Over and against the overt and intentional mischaracterizations of certainty, there are assumptions that are not expressed, but are implied in ideologies, and may therefore prove to be the most recalcitrant and insidious of the mischaracterizations. Citizens are receptive to misinformation about uncertainty in climate science because we are already socialized to believe in certainty. The President of the (United States) National Academy of Sciences, Marcia McNutt, has made clear that science is "not a body of facts" but a "method for deciding whether what we choose to believe has a basis in the laws of nature or not" (Achenbach 2015).[7] McNutt emphasizes that students must be taught to think scientifically and often they are not taught this well. The systematic misunderstanding of the nature of science could begin with schoolbooks, whether they be in physics (Norton 1994), chemistry (Abd-El-Khalick et al. 2008), or biology (Chiappetta & Fillman 2007). And at higher levels of science education, when it comes to the method of science,

> Even many college students don't really understand what evidence is. The scientific method doesn't come naturally – but if you think about it, neither does democracy. For most of human history neither existed. We went around killing each other to get on a throne, praying to a rain god, and for better and much worse, doing things pretty much as our ancestors did.
>
> *(Achenbach 2015, quoting Andrew Shtulman of Occidental College)*

The very construction of the phrase 'scientific uncertainty' implies that there is also scientific certainty, and further, the use of the phrase often implies that because scientific certainty is possible, it is the norm and the goal and anything less – uncertainty – is incomplete and cannot – and as a matter for ethics, *should not* – be acted upon. But actions to mitigate climate disruption are matters of choice, not the product of a mathematical equation. How we make those choices can be by gut feelings (themselves a product of exposure to ideas that we cannot articulate), or social preferences. On behalf of nearly every citizen of the world, 197 governments have made one of those choices already – the precautionary principle – based upon cognitive support for ethical choices regarding harm,[8] risk of harm, risk of loss, and distribution of resources. The fiction that we can and should only act when science produces certainty has resulted in continued inaction, and the consequent unethical and destructive results of that inaction. Humans have the ability to make informed choices and must exercise that ability, not wait for mathematics to make it appear to be easy with the false promise of 100 percent likelihood that they have chosen correctly.

Thus, in education, one finds references to 'scientific uncertainty' that infer the mischaracterization that normal science is certain, such as when a chapter is titled "Risk Management and Scientific Uncertainty" in a popular book in the United States for environmental law students (Findley & Farber 2008). In that chapter, several court cases are cited in which the courts have indeed used the term 'scientific uncertainty,' but they have done so to emphasize the unattainability of certainty and even admonished the litigants not to expect certainty from science. From the chapter title alone, however a readership of students who are conditioned to expect certainty would infer that scientific uncertainty is the exceptional circumstance in which risk must be managed. Other court cases in that same chapter talk about probability, but not in the context of a goal of certainty. Instead, the courts quoted the probability of a particular harm using a statistical method.

While aggressive media, corporate, and political forces are often regarded as the primary cause of the public's climate confusion, a review of five decades of science education relating to climate in general, and climate disruption in particular, reveals that basic climate science has not been well addressed in national and state education standards or science education curricula in the United States, which is nearly always the largest GHG emitter *per capita*. Moreover, key misconceptions and misinformation about basic climate science are strongly held by students, teachers, and public audiences (McCaffrey & Buhr 2008).

Certainty Is a Feature of Religion, Not Science

Yuval Harari makes an eye-catching claim in his popular history of humankind: "The Scientific Revolution has not been a revolution of knowledge. It has above all been a revolution of ignorance" (2014: 279). He reminds us that until the Scientific Revolution, people did not believe in progress, but rather regress – they were falling away from a prior time of the golden age of certainty and perfection. He goes on to point out that the Scientific Revolution enabled knowledge to occur because it began from the base position that we do not know, and from there we can acquire knowledge and make progress, whereas previous ages began from the position that what we had already known in the past golden age has since been lost, and we have fallen from that age and fall further from its wisdom with the passing of time. Religions assert that laws are not the result of human caprice but are "ordained by an absolute and supreme authority" (Harari 2014: 234). That assertion consequently places at least some fundamental laws beyond challenge, thereby ensuring social stability and offering that some things are 'certain.'

By contrast,

> Scientific thinking is a constant reminder to us that we don't know the answers. In religious thinking, this is often unacceptable… Many religions, or some religions, or some ways of being religious, are based on the idea that there should be a truth that one can hold onto and not question. This way of thinking is naturally disturbed by a way of thinking based on continual revision, not just of theories but of the core ground of the way in which we think.
>
> *(Rovelli 2014: 223)*

Science can provide the description, but not the prescription for decisions. Sociologist of science Steve Fuller makes that point that once a scientific project is underway, it might seem to be very mathematically determined, but if we ask the question why did we choose this project or later, how do we use the results, science has nothing to say (Fuller 1988). Science is unable to set its own priorities. It is also incapable of determining what to with its discoveries.

> [F]or a purely scientific view point it is unclear what we should do with our increasing understanding of genetics. Should we use this knowledge to cure cancer, to create a race of genetically engineered supermen, or to engineer dairy cows with super-sized udders? It is obvious that a liberal government, a Communist government, a Nazi government and a capitalist business corporation would use the very same scientific discovery for completely different purposes, and there is no *scientific* reason to prefer one usage over others.
>
> *(Harari 2014: 305)*

To answer these questions and make decisions, one must have a method and tools to evaluate the matter. The cognitive approach to ethics enables systematic and methodical choices to be made. If an individual or society bases practices on authority, rather than on rational decision making, and one does not understand how science works, then one might easily embrace the method of science as being like the method of religion – rule by an authority that has the capability of being certain. But just as belief is not knowledge, religion is not science.

Legal Practice

When certainty is expected of science, it is not expected by scientists themselves, but rather by non-scientists. The mischaracterization is formalized in the practice of law. Law is possibly the strongest social institution in which facts are formally decided (Shapiro 1983).[9] Although one can look to law for concrete expressions of civil society's ethical choices to mitigate climate disruption, as we do when we look to the UNFCCC, one can also find seeds of the misguided expectation of certainty from science planted by legal practices. There is in fact some evidence that the notion of certainty itself comes to us early in the English language through law. The word 'certain' arrived with the Norman conquest of England and that Norman sense of usage can still be heard when English follows French word order, such as in the term 'a day certain,' rather than standard English word order, as in 'a certain day.' This sense of the word "certain" was already recorded in law by 1297, with the phrase "a certyn rente" (Simpson 1989: 1050). If by comparison, one looks today at the term 'certainty' as used in the law, problems abound. The problems began when physicians were asked to conclude causes of illness or death, and lawyers invented 'reasonable medical certainty' as the test phrase. However,

> The term 'reasonable medical certainty' has no scientific meaning. Its legal meaning is at best ambiguous, at worst misleading. It is not required by the U.S. Federal Rules of Evidence, nor any other evidence code. More importantly, the term ('scientific certainty') is problematic for a different reason – misleading the jury – and should be excluded under U.S. Federal Rule 403 for that reason alone.
>
> *(Gianelli 2010: 40–41)*

Although the majority of US federal and state courts no longer require the standard of scientific certainty for admitting evidence, the phrase has stuck in citizens' minds such that there is an expectation of it. According to physicist and lawyer, David H. Kaye, the problem is that the phrase "reasonable degree of scientific certainty" combines two words of concern – "scientific" and "certainty," but has no scientific meaning. He reiterates that the notion of a "reasonable degree of scientific certainty" that lawyers tended to use, was very likely invented by lawyers themselves. Scientists have no use for this phrase, outside of the courtroom. According to Kaye:

> A reasonable degree of scientific certainty' is not a defined concept in scientific disciplines or even in law. … It is legal mumbo jumbo derived from archaic cases in which lawyers discovered that if a medical doctor did not utter the incantation 'to a reasonable degree of medical certainty,' his testimony might be excluded because doctors were not supposed to talk about mere probabilities. Modern cases usually recognize that suitably explained information about less-than-certain possibilities can be helpful in various circumstances, but experts want to (or are induced to) incant on only 'medical certainty' but also 'clinical certainty,' 'psychological certainty,' 'psychiatric certainty,' 'engineering certainty,' 'architectural certainty,' 'ballistic certainty,' 'professional certainty,' and even 'forensic certainty' and 'legal certainty.'
>
> *(DOJ US NCFS 2016, citing Kaye 2010: 82)*

Not just individual researchers, but the institutions of the law have felt it necessary to redirect the citizen's understanding of certainty. Both the courts and the US Department of Justice have indicated that the law does not and should not entertain notions of certainty to test legal evidence. In the US Supreme Court case of *Daubert v. Merrell Dow Pharmaceuticals Inc.*, 509 U.S. 579 (1993), which redefined that which would be accepted as 'science' in US federal courts, the Court noted that "it would be unreasonable to conclude that the subject of scientific testimony must be 'known' to a certainty; arguably, there are no certainties in science" (1993). Elsewhere, judges have noted that

> Science itself may be highly uncertain and controversial with respect to many of the matters that come before the courts. Scientists often express considerable uncertainty about the dangers of a particular substance. And their views may differ about many related questions that courts may have to answer.
>
> *(National Research Center et al. 2011)*

In the US case of *Ethyl Corp. v. EPA*, 541 F. 2d 1 (1976), when discussing the certainty of environmental medicine, a federal appeals court noted that "awaiting certainty will often allow for only reactive, not preventive regulation" (1976). The court added that if the statutory intent is to prevent environmental catastrophe, preventive decisions, even if uncertain, cannot be labeled as illegitimate, as was suggested by the Ethyl Corporation in the case. And in the US Supreme Court case of *Industrial Union Department, AFL-CIO v. American Petroleum Institute*, 448 U.S. 607 (1980), in which the US Occupational Safety and Health Administration (OSHA) regulated carcinogens, the Court stated that OSHA was not bound by the burden of proof requirement "to support its finding that a significant risk exists with anything approaching scientific certainty."[10]

Lawyers tend to want to inject terms like 'scientific certainty' or 'reasonable scientific certainty' as though these labels would qualify the evidence as being a product of science itself. According to the US Department of Justice's (DOJ) National Commission on Forensic Science, in its report "Testimony Using the Term 'Reasonable Scientific Certainty,'" the term "reasonable degree of certainty" was first used in the context of scientific evidence in the examination of a witness in the case *Herbst v. Levy* (1935), when the witness was "asked whether he could determine with reasonable scientific certainty the cause of the capsizing of the boat" (DOJ US NCFS 2016). Then in 1969, a US federal appeals court linked the phrase to admissibility of evidence. The DOJ study makes clear that this phrase was not the mandate of the court in the case, but a question invented by a trial lawyer. As mentioned above, it seems that lawyers developed the idea for medical testimony when witnesses in civil cases were asked about the potential consequences of an injury or illness. The DOJ Commission explained that

> The legal community should recognize that medical professionals and other scientists do not routinely use 'to a reasonable scientific certainty' when expressing conclusions outside of the courts. Such terms have no scientific meaning and may mislead factfinders [jurors or judges] when deciding whether guilt has been proved beyond a reasonable doubt. Forensic service providers should not endorse or promote this use of this terminology.
>
> *(DOJ US NCFS 2016: 1)*

The DOJ Commission further emphasizes that the terminology of reasonable certainty was not necessary, and that the term "reasonable degree of scientific [or any other discipline] certainty" has no place in the judicial process" (DOJ US NCFS 2016: 5). The Commission recommends that "Rather than use 'reasonable… certainty' terminology, experts should make a statement about the examination itself, including an expression of the uncertainty in the measurement or in the data" (DOJ US NCFS 2016: 5). The Commission emphatically concluded that "the term 'reasonable degree of scientific [or any other discipline] certainty' has no place in the judicial process" (DOJ US NCFS 2016: 5). Thus, the courts and the Department of Justice made clear that legal decisions must be made in the light of uncertainty, and that is where the focus of climate decisions must be as well.

The Status Quo Ante and Laissez Faire Assumptions from Economics

Despite private litigation having invented unhelpful phrases about certainty, the law does at times also offer something useful in helping us to get past our misunderstanding of science and our expectations of decisions being dictated by mathematical certainty. For example, a foundation of legal procedure is that the 'moving party' (the plaintiff in a civil case or the state in a criminal case) carries the burden of proof, as was listed as one of the four features of the precautionary principle above. The defendant, civil or criminal, needs prove nothing, and so long as the moving party cannot meet its burden of proof, the defendant is without liability. So, it is not a balancing of arguments, but rather the burden to prove the facts of which someone complains, and if he or she cannot prove those facts, the other argument (if any) does not even need to be heard. If the plaintiff cannot set it up, the defendant need not knock it down. To build that foundation, we must make an assumption about what constitutes ordinary life and who therefore is the party asking for something extraordinary, that is, 'moving' the state to intervene and therefore must be the one to prove facts. *Laissez-faire* economic thinking begins with an unquestioned *a priori* sense of how the world 'ordinarily' is. But that state of socio-economic relations is a choice, and there are other choices. From that beginning, so the thinking goes, one has a right to keep things the same. If those status quo conditions harm others by polluting the air they breathe, it is the citizens or the state who would need to take an action – that is, be the moving party – to prove the facts of the pollution that violate the law, and thereby ask the state to stop the pollution that another person has caused. However, if by employing the precautionary principle, through imposition of a tool such as requiring an emissions permit for potentially harmful activities, then the emitter becomes the moving party, and therefore the emitter carries the burden of proof. Enshrining the precautionary principle in climate law and institutions is a choice to recognize the especially dangerous and irreversible harms to persons and property caused by GHG and to force the emitter to be recognized not as 'ordinary' or 'systems normal,' but as a moving party that wants to introduce something out of the ordinary and disrupt climatic conditions, which in turn can or will harm other persons. The permit is the legal tool through which institutionalization of the precautionary principle occurs and it is most effective precisely because it shifts the burden of proof to the emitter, thus changing the *status quo ante* from the assumption that GHG emissions are normal and that one has a right to emit them, therefore requiring harmed persons to take action and be the moving party against the GHG emitter and thereby carry the burden of proof.

Having witnessed more than thirty years of inaction and resistance to mitigating climate disruption, we know who intentionally put the uncertainty into society's mind and why they did it, but we must also consider why society did not see the error in its response. It could be that the high-emitting states regarded themselves as exceptions to the rules to which they had agreed. It could be that the high-emitting states had not thoroughly considered what it would mean to be legally-bound to the precautionary principle. It could simply be a case of buyer's regret, but if it is so, then these same states have the responsibility to their citizens and other states to communicate that they do not wish to subject themselves to international law in this regard. If so, then they must withdraw from the UNFCCC, wherein the binding precautionary principle is found. It could be that the high-emitting states, while professing to support the international legal order, intended from the beginning not to honor their commitments, due to social desirability bias, as discussed below, and thereby test international resolve to enforce the commitment. Whatever the reason, we are many years past the legally-binding commitment, and high-emitting states continue to emit, and to feel comfortable with the very weak and unsupportable excuse of waiting for scientific certainty. This demonstrates that a piece of the inaction puzzle is the way in which society has generally accepted the position that emitters are ordinary and normal, and by enjoying the base position of the *status quo ante*, are entitled to exercise their rights first, leaving harmed persons in the undesirable legal position of being the moving parties who must carry the burden to prove harm.

System Justification

Is the desire for certainty a psychological feature of humans? If so, citizens do not even need to be intentionally deceived about the nature of science – they may in fact project their own personal desires for certainty onto science. To wait for certainty is to wait for a socially-constructed fiction to suit human desires, as Samuel Beckett portrays so well in his theater play, *Waiting for Godot* (1956). Humans cope with the material world through social stabilization practices such as communicating with one another in a predictable and safe manner,[11] even when they run counter to observable and empirical evidence.

Political psychologist John T. Jost began his research to test whether it was true that poor people do not rebel against the unjust distribution of wealth because they believe they will someday also be rich. His empirical evidence demonstrated that this explanation is lacking. According to Jost, the answer instead lies in the theory of system justification.

> In social science, theories of relative deprivation, moral outrage, identity politics, and collective action all propose that people should object, perhaps even violently, to stark forms of inequality. And certainly, some do but many do not, so we also need to know why men – and women – do not rebel in the face of injustice.
>
> *(Jost 2017)*

The theory of 'system justification' helps to explain: (a) why so many people accept the social, economic, and political systems that affect them as more or less legitimate and justified, and (b) how, as a result of this process, the disadvantaged sometimes participate in their own subjugation. System justification is the tendency to defend, bolster, and justify aspects of the societal status quo, often at a non-conscious level of awareness. According to Jost, it works as a motivator because it addresses fundamental human needs to *reduce uncertainty*, threat, and social discord (Jost 2017, emphasis added).

As this thinking goes, if the system was already that way, it seems to justify that it continues that way. Thus, treating the emissions of GHG as the *status quo ante* makes change difficult. Jost and colleagues "have sought to understand system justification processes, their social and psychological antecedents, and their consequences for *important public policy issues such as climate change*, economic inequality, and voting behavior" (Jost 2017, emphasis added).

But although changing an apparently justified system may be difficult, it is not impossible. One needs only to look at reactions around the world to wars or to emergency measures in response to virus epidemics or pandemics for evidence. As these emergency measures become commonplace, they are talked about as the new 'normal.' If that is what a crisis response looks like, one must ask why climate disruption is not psychologically recognized as crisis. Despite evidence that far more people will have died from climate disruption than COVID-19 (UN Global Compact 2020), we treat the climate disruption as normal. With the climate crisis, there has not even been a 'wear a mask or stay at home' crisis response, let alone wartime sacrifices, because we allow that the GHG system is justified.

Has the 'Social Desirability Bias' Uncovered a Step Forward?

If and when citizens do understand the nature of science to include uncertainty, then we are faced with making choices. The desire to say that the data mathematics of economics, chemistry or physics will dictate citizens' choices is a failure to recognize the descriptive and therefore limited nature of the methods and practices of economics and science. From these descriptions, we need prescriptions. Informed citizens still must make responsible choices.

In social science, the object of study – a person – is a knowing object and is conscious of his or her social context. In a situation like political polling, these objects of study may therefore filter their response to interrogation through what political pollster Robert Cahaly calls a 'social desirability bias.' It sounds scientific. It simply means unethical choice. Cahaly caught the media's attention when, in 2016, contrary to other pollsters, he predicted that Donald Trump would be elected President of the United States. Cahaly explained his insight by noting that persons recognize dominant civil society values and may claim to hold those civil values when asked, but then take actions in different ways, including socially-undesirable ways, such as voting for a candidate who openly displays antisocial or socially unacceptable behavior. The deception serves to relieve a person who is making choices against civil society – in this case, the voter – from needing to defend the socially indefensible (Russonello & Lyall 2020; see also Knowles 2020). If the expressed dominant value in a civil society is to make the ethical choice to mitigate climate impacts, voters, consumers, and other actors may well say that they support climate mitigation, but that might only be due to their knowing that when asked in civil society, it is the 'socially desirable' answer, even if they act to the contrary. The fact that socially-desirable deceptions can play such a decisive role also however reveals a glimmer of hope. In looking at a socially desirable response, one can at least observe that the respondent recognizes a dominant social norm of civil society, even if he or she then privately chooses to act otherwise. It would seem, therefore, that the presence of the social desirability bias is in fact an indication of a step in the right direction – the quiet, but as yet unacted upon, recognition of the ethically just answer to the problem. With one step further, these same people just might act upon it.

Conclusion – Waiting for the Godot of Certainty

Misinformation is not only intentionally perpetrated by blunt instruments that deny the specific issues of climate disruption, or even generally deny science itself, but misinformation is also asserted to misinform listeners, watchers, and readers of the basic and foundational ways in which knowledge is produced. Descriptive enterprises such as science or economics are neither constructed nor practiced to provide prescription, and prescription is what societies and governments need in order to take present action to plan future practices that will alleviate the catastrophically harmful impacts of climate disruptions. There are of course fossil fuel money interests that intentionally deny climate disruption, undermine science, and destabilize knowledge through stating or implying that until science produces certainty, no action against their interests should be taken. And those money interests will continue to do so.

But at the same time, the world is becoming more enlightened as to how misinformation is perpetrated. There is still much work to be done to stop those intentional and destabilizing misinformation forces. More work is also needed however to understand and explain the social structures that result in unintentional misinformation and misdirected expectations of certainty, whether they be from educational shortcomings, religion, economics, law, system justification, or the practices of social desirability. What *is* certain is that the world's states have agreed to mitigate GHG and they have dismissed the excuse of waiting for scientific certainty. The decision has been made, and it is legally binding. All that remains is to implement it.

Acknowledgments

The author would like to thank two members of the University of Cologne's Environmental Law Center: Moritz Röhrs for his very helpful research and Mrinalini Shinde for her careful proofreading and editorial improvements of the text.

Notes

1 In this famous paper, physicist Heisenberg wrote of the *Unschärferelation* (relation of unsharpness) which is sometimes called the *Unbestimmtheitsrelation* (indeterminacy relation) and less often, the *Unschärfeprinzip*. (principle of unsharpness).
2 Rovelli says further that data is the stuff from which we can build knowledge – or disprove theories – but "the data alone is not equal to knowledge and is therefore not equal to science. … Science has nothing to do with the assembling of data and the ways of organizing the assembly of data. It has everything to do with the way we think, and with our mental vision of the world." See also, Rovelli (2014).
3 Civil engineers for instance have a different way of characterizing the quality of their scientific proposals and findings, based not on certainty, but on confidence intervals which are always values less than one hundred per cent.
4 A lasting treatment of narrative in law is provided in Robert Cover (1983–1984).
5 In Oreskes 2015, the author states, "that a reason for the cultural persistence of doubt is what we may label the 'fact of uncertainty'. Uncertainty is intrinsic to science, and this creates vulnerabilities that interested parties may, and commonly do, exploit, both by attempting to challenge the specific conclusions of technical experts and by implying that those conclusions threaten other social values."
6 Sometimes the intent goes beyond misinformation to intentional harm by encouraging addictions. We learned that the tobacco industry did study the harmful effects of smoking and did understand it was harmful. Then they made smoking even more addictive by adding more addictive elements into their

cigarettes. We have now also learned that the fossil fuel industry and their paid-for "think tank" research studies knew that their product would lead to climate change, and like the cigarette industry with smoking, the oil companies set out to intentionally cause even more fossil fuel addiction. "The fossil fuel age has to end: that's the message of today's [IPCC] report. To have any chance of avoiding the chaos, droughts and rising tides of 1.5 degrees or more of global warming, we must massively and speedily transform our society to kick our fossil fuel addiction See (Munić 2018). The effects of behavior in the USA are not limited to the USA. "No matter how poorly placed they find themselves now," writes Delhi journalist Snigdha Poonam, Indian youth "make up the world's largest ever cohort of like-minded young people, and they see absolutely no reason why the world shouldn't run by their rules." The effect, Poonam says, will be to "change our world in ways we can't yet imagine. ...This new generation is motivated: it combines the cultural values of the traditional Asian family with the life goals of the American teenager." See Jack, J. (2018), 'India has 600 million young people — and they're set to change our world,' *The Guardian, for a book review of Dreamers: How Young Indians Are Changing the World.* Available at https://www.theguardian.com/commentistree/2018/jan/13/india-600-million-young-people-world-cities-Internet.

7 Geophysicist Marcia McNutt, Fellow of the Royal Society, President of the National Academy of Sciences of the United States, past Editor-in-chief of *Science*, member of the National Academies of Sciences, Engineering, and Medicine advisory committee, where she chaired the climate intervention committee, former director of the United States Geological Survey, science adviser to the United States Secretary of the Interior, and professor of marine geophysics at the Stanford University.

8 For a discussion of the context of cognitivism in the ethics of science, see Junker, K.W. and Ott, K. (1998) "More Than Just Doing What's Right: Ethics in Science," in: *Science and the Public*, Milton Keynes.

9 Once introduced, its use in law expanded and multiplied. The language of the law was an influence on the development of Enlightenment science, as with the migration of the concept and word "fact." See also, Shapiro (2000).

10 In paragraph 1, the Court states: "The burden was on OSHA to show, on the basis of substantial evidence, that it is at least more likely than not that long-term exposure to 10 ppm of benzene presents a significant risk of material health impairment. Here, OSHA did not even attempt to carry such burden of proof. Imposing such a burden on OSHA will not strip it of its ability to regulate carcinogens, nor will it require it to wait for deaths to occur before taking any action. The requirement that a 'significant' risk be identified is not a mathematical straitjacket; OSHA is not required to support its finding that a significant risk exists with anything approaching scientific certainty; and the record in this case and OSHA's own rulings on other carcinogens indicate that there are a number of ways in which OSHA can make a rational judgment about the relative significance of the risks associated with exposure to a particular carcinogen."

11 Berger and Luckmann (1966).

References

'2019 Global Thinkers,' *Foreign Policy*, 1 December. Available at https://foreignpolicy.com/2019-global-thinkers/ (Accessed: March 3, 2021).

Abd-El-Khalick, F., Waters, M. and Le, A. (2008). 'Representations of Nature of Science in High School Chemistry Textbooks Over the Past Four Decades', *Journal of Research in Science Teaching*, 45(7), pp. 835–855.

Achenbach, J. (2015). 'Why Do Many Reasonable People Doubt Science?', *National Geographic*, March. Available at https://www.nationalgeographic.com/magazine/2015/03/science-doubters-climate-change-vaccinations-gmos/ (Accessed: 3 March 2021).

Beckett, S. (1956). *Waiting for Godot: A Tragicomedy in Two Acts*. London: Faber and Faber.

Berger, P. and Luckmann, T. (1966). *The Social Construction of Reality: A Treatise in the Sociology of Knowledge*. London: Penguin Books, p. 57.

Boykoff, M.T. and Boykoff, J.M. (2004). 'Balance as Bias: Global Warming and the US Prestige Press', *Global Environmental Change,* 14(2), pp. 125–136.

Brown, D.A. (2011). 'On Confusing Two Roles of Science and Their Relation to Ethics', *Ethics and Climate*. Available at https://ethicsandclimate.org/2011/06/02/on_cofusing_two_roles_of_science_and_thier_relation_to_ethics/ (Accessed: 4 March 2021).

Brown, D.A. (2013). *Climate Ethics: The Perfect Moral Storm*. Abingdon: Routledge, pp. 98–102.

Caplan, A. (2013). 'Liberty Has its Responsibilities: Holding Non-Vaccinators Liable for the Harm They Do', *Human Vaccines & Immunotherapeutics*, 9(12), pp. 2666–2667.

Charter of the United Nations and Statute of the International Court of Justice, June 26, 1945, 59 Stat. 1031 [the Charter], 1055 [ICJ Statute], T.S. No. 993 [I.C.J. Statute at 25], 3 Bevans 1153 [I.C.J. Statute at 1179].
Chiappetta, E.L. and Fillman, D.A. (2007). 'Analysis of Five High School Biology Textbooks Used in the United States for Inclusion of the Nature of Science', *International Journal of Science Education*, 29(15), pp. 1847–1868.
Daubert v. Merrell Dow Pharmaceuticals, Inc., 509 U.S. 579 (1993).
Department of Justice, US National Commission on Forensic Science (2016). *Testimony Using the Term "Reasonable Scientific Certainty.* Available at https://www.justice.gov/archives/ncfs/file/795336/download (Accessed: March 3, 2021), pp. 1–5, 4.
Dunlap, R.E. and Jacques, P.J. (2013). 'Climate Change Denial Books and Conservative Think Tanks: Exploring the Connection', *American Behavioral Scientist,* 57(6), pp. 699–731.
Ethyl Corp. v. Environmental Protection Agency, 541 F.2d 1 (D.C.Cir.1976).
Findley, R.W. and Farber, D.A. (2008). *Environmental Law in a Nutshell.* New York: Thomson West, 7th ed., pp. 165–192.
Fuller, S. (1988). *Social Epistemology, Part IV.* Bloomington, IN: Indiana University Press.
Gianelli, P. (2010). 'Reasonable Scientific Certainty: A Phrase in Search of a Meaning', *Criminal Justice,* 25(1), pp. 40–41.
Hansson, S.E. (2017). 'Science Denial as a form of Pseudoscience', *Studies in History and Philosophy of Science*, 63, pp. 39–47.
Harari, Y.N. (2014). *Sapiens: A Brief History of Humankind.* London: Penguin Random House, p. 279.
Healey, R.A. (1981). 'Indeterminism,' in Bynum, W. F., Browne, E. J. and Porter, R. (eds.) *Dictionary of the History of Science.* Princeton: Princeton University Press, pp. 202.
Heisenberg, W. (1925). 'Über Quantentheoretische Umdeutung Kinematischer und Mechanischer Beziehungen,' *Zeitschrift für Physik*, 33(1), pp. 879–893.
Herbst v. Levy, 279 Ill. App. 353, 358 (Ill. App. Ct. 1935).
Indus. Union Dept. v. Amer. Petroleum Inst., 448 U.S. 607 (1980)
Jacques, P.J., Dunlap, R.E. and Freeman, M. (2008). 'The Organisation of Denial: Conservative Think Tanks and Environmental Scepticism', *Environmental Politics,* 17(3), pp. 349–385.
Jost, J.T. (2017). 'A Theory of System Justification', *Science Brief,* June. Available at https://www.apa.org/science/about/psa/2017/06/system-justification.aspx (Accessed: 3 March 2021).
Junker, K.W. (1998). 'Law and Science Serving One Master … Narrative,' in Scanlon, E., Hill, R. and Junker, K.W. (eds.) *Communicating Science: Professional Contexts.* London: Routledge, pp. 249–269. A lasting treatment of narrative in law is provided in Cover, R.M. (1983–1984). 'The Supreme Court 1982 Term, Foreword: Nomos and Narrative,' *Harvard Law Review,* 97(4), pp. 4–74.
Kaye, D.H. (2010). *The Double Helix and the Law of Evidence.* Cambridge, MA: Harvard University Press, p. 82.
Knowles D. (2020). '"Shy" Trump Voters Will Power His Win, Says Pollster Who Called 2016 Race', *Yahoo! News*, 23 October. Available at https://news.yahoo.com/shy-trump-voters-will-power-his-win-says-pollster-who-called-2016-race-203448623.html (Accessed: 3 March 2021).
Lemons, J. and Junker, K.W. (1996). 'The Role of Science and Law in the Protection of National Park Resources', in Wright, R.G. (ed.) *National Parks and Protected Areas: Their Role in Environmental Protections.* London: Wiley-Blackwell, p. 397
Lewandowsky, S., Ballard, T. and Pancost, R.D. (2015). 'Uncertainty as Knowledge,' *Philosophical Transactions of the Royal Society,* 373(2055), p. 20140462.
Mandel, D.R. and Tetlock, P.E. (2016). 'Debunking the Myth of Value-Neutral Virginity: Toward Truth in Scientific Advertising', *Frontiers in Psychology*, 7, p. 451. Available at https://www.frontiersin.org/article/10.3389/fpsyg.2016.00451 (Accessed: 3 March 2021).
McCaffrey, M.S. and Buhr, S.M. (2008). 'Clarifying Climate Confusion: Addressing Systemic Holes, Cognitive Gaps, and Misconceptions Through Climate Literacy', *Physical Geography*, 29(6), pp. 512–528.
Miller, A. (1993). 'The Role of Analytical Science in Natural Resource Decision Making', *Environmental Management*, 17, pp. 563–574.
Munić, J. (2018). 'Energy Atlas: Facts and figures about renewables in Europe', *Heinrich Böll Foundation.* Available at https://www.boell.de/en/2018/04/24/energy-atlas-2018-figures-and-facts-about-renewables-europe (Accessed: 3 March 2021).
National Research Council et al. (2011). *Reference Manual on Scientific Evidence.* Washington, DC: National Academies Press, p. 3.
Norton, J.D. (1994). 'Science and Certainty', *Synthese,* 99, pp. 3–22.
Oreskes, N. (2015). 'The Fact of Uncertainty, the Uncertainty of Facts and the Cultural Resonance of Doubt,' *Philosophical Transactions of the Royal Society,* 373(2055), p. 20140455.

Oreskes, N. and Conway, E.M. (2010). *Merchants of Doubt: How a Handful of Scientists Obscured the Truth on Issues from Tobacco Smoke to Global Warming*. New York: Bloomsbury Press.

Palmer, T.N. and Hardaker, P.J (2011). 'Handling Uncertainty in Science. *Philosophical Transactions of the Royal Society*, 369(1956), pp. 4681–4684, 4682.

Rovelli, C. (2014). 'Science Is Not About Certainty,' in Brockmann, J. (ed.) *The Universe*. New York: Harper Perennial, pp. 214–228, 222.

Rovelli, C. (2015). *Anaximandre de Millet, ou la Naissance de la Science*. Paris: Dunod.

Russonello, G. and Lyall, S. (2020). 'The One Pollster in America Who Is Sure Trump Is Going to Win', *The New York Times*, 8 November. Available at https://www.nytimes.com/article/trafalgar-group-poll.html (Accessed: 5 June 2023).

Shapiro, S. (1983). *Probability and Certainty in Seventeenth-Century England: A Study of the Relationships Between National Science, Religion, History, Law, and Literature*. Princeton: Princeton University Press.

Shapiro, B. (2000). *A Culture of Fact. A Culture of Fact: England, 1550–1720*. Ithaca, NY: Cornell University Press.

Simberloff, D. (1987). 'Simplification, Danger and Ethics in Conservation Biology,' *Bulletin of the Ecological Society of America*, 68(2), pp. 156–157.

Simpson, J.A. (ed.) (1989). *Oxford English Dictionary*. Oxford: Oxford University Press, 2nd ed., Vol. II, pp. 1050.

Stevenson, L. (1989). 'Is Scientific Research Value-Neutral?', *Inquiry*, 32(2), pp. 213–222.

United Nations. (2018). *Secretary-General's Remarks on Climate Change*. 10 September. Available at https://www.un.org/sg/en/content/sg/statement/2018-09-10/secretary-generals-remarks-climate-change-delivered. (Accessed: 3 March 2021).

United Nations Framework Convention on Climate Change. (1992). Art. 3, May 9, 1992, S. Treaty Doc No. 102-38, 1771 U.N.T.S. 107.

United Nations Global Compact. (2020). *COVID-Related Hunger Could Kill More People Than the Virus*. Available at https://unglobalcompact.org/take-action/20th-anniversary-campaign/covid-related%20hunger-could-kill-more-people-than-the-virus (Accessed: 3 March 2021).

10

ETHICAL OBLIGATIONS TO DEVELOP AND SUPPORT A PRECAUTIONARY PRINCIPLE TO GUIDE NATIONAL RESPONSES TO CLIMATE CHANGE

Carl F. Cranor

Introduction

In creating social institutions and recommending subsidiary endeavors that must be guided within them, those decisions are at a minimum typically the outcome of various kinds of empirical evidence and social values/normative considerations. Three important areas of social life in which this is illustrated are the law, regulatory science, and implementation of science in support of institutional commitments, with my final focus on international global warming agreements. These three topics are considered in order to clarify some major components that guide the creation of institutions and guide how their decisions are to be implemented.

The ultimate target of this chapter is to indicate some ethical/normative issues in decisions concerning global warming, but the background is prepared by illustrating that some ethical/normative values are not unusual, and even common, in social and scientific decisions. Ethical values in science may seem to be an unusual, possibly even a controversial topic. However, once we understand a broader approach to institutional decisions, this will not be problematic. In science and science policy areas, both epistemic and non-epistemic values/normative judgments must be considered and invoked.

The chapter begins with some generic ideas from the law, an institutional endeavor in which ethical/normative values are obvious and essential. Next is a discussion of normative issues in toxicological science and the application of that science to regulatory toxicological decisions. Finally, ethical/normative issues that one might expect to find in research on global warming and in decisions about how to address unresolved issues in global warming are considered.

The Law

Various laws or areas of law facilitate our living together in the broadest terms. One area seeks to prevent the grosser forms of violence, theft and deception (criminal law), another provides redress for harms caused to another (the tort law), while others serve various social purposes, such as permitting us to buy, sell, and trade with one another in a rule guided manner (contract law).

An issue common to the criminal law, to torts, and to science concerns in the regulatory law is how institutions address uncertainty and how they resolve uncertainties that can arise when explicit rules do not settle the issues at stake. The law has long addressed this issue and has some

 DOI: 10.4324/9781003039860-13

principles to guide decision-making under uncertainty. Three generic legal devices are pertinent with all supported by normative considerations: legal burdens of proof, legal standards of proof, and legal presumptions.

Burdens of Proof: Consider legal burdens of proof in the criminal law and torts. The law like many areas of scientific research gives substantial pride of place to the existing status quo; this is exhibited in the legal status quo (LSQ) between parties when they engage in litigation. If Sam is alleged to have criminally assaulted someone, (a) the LSQ is that he is presumed to be not guilty until and unless, (b) he is proven guilty by the procedures of the criminal law. If Sally is alleged to have damaged Jane's garage, Sally is also presumed to not be liable until her damaging Jane's garage is established in accordance with the procedures of civil litigation.

Roughly speaking, the LSQ may or may not be generically considered a desirable state of affairs, but in the law, it may not be changed or disturbed unless and until appropriate legal processes have been followed and there is sufficient legal and factual support for modifying it. In short, legally a person is presumed 'innocent until proven guilty (or liable).' The law's commitment to the LSQ, like most of science's commitment to the scientific status quo, is conservative; whatever is or exists is accepted or is OK, until there is legally sufficient evidence to modify it. In the law, people have based their lives and conduct on legal relationships that exist and these should remain in place until there is a good legal reason for changing them.

However, *who* must present a factual and legal case for changing the LSQ and *how much* and *what kind* of factual and legal material must be marshaled to justify legal change depends upon the area of law. In the criminal law, the state prosecutor or the attorney general of a jurisdiction has the legal obligation or burden of proof to show that an accused person (Sam in the above example) is guilty of the crime in question (assault against another). In civil litigation, if Sally is alleged to have damaged Jane's garage, typically Jane, with the assistance of an attorney, not an agent acting for the government, has the legal obligation or burden of proof to establish that Sally damaged her garage. However, in the criminal law even though the prosecutor has the burden of proof, if the defendant raises a defense of insanity that at the time of presumably committing a crime, in some jurisdictions the defendant may have the burden of proof to provide sufficient evidence of insanity and in other jurisdictions the prosecutor may continue to have the burden of proof to show the defendant was sane.

The differences in the burden of proof for changing the LSQ have been justified by reference to normative or ethical considerations, depending upon what is at stake for the different parties. The burden of proof in civil litigation – who must legally justify changing the status quo – in different circumstances may be assigned (a) to the person who would change the LSQ, but also sometimes to other parties, such as (b) to one who has ready access to knowledge about the facts in question, (c) to handicap a disfavored contention (e.g., to one who would claim another committed fraud), or (d) to one who would argue against the usual probabilities typical of the circumstances (James & Hazard 1977: 241). What is important about these examples is that they are not sacrosanct, but they can be modified in particular instances by considering the balance of reasons in favor of one policy or another and the nature of the institution in question, although this is more representative of civil litigation than criminal litigation.

Standards of Proof: The law also sets standards for how difficult it is to change the LSQ. This is the specified degree of certainty or standard of proof that must be established before Sam is properly found criminally guilty of assault or Sally may be civilly liable for damaging Jane's garage.

Although both areas of the law recognize the importance of the LSQ, different standards of proof (degrees of certainty) need to be established to find a person guilty in the criminal law compared with finding someone liable in the civil law. A prosecutor must establish Sam's criminal guilt beyond a reasonable doubt, a very high bar to clear. Any uncertainties about Sam's guilt must be removed beyond a reasonable doubt before he can be found guilty. If a jury finds that there is reasonable doubt about Sam's guilt, he would be found not guilty and would be free of those legal charges and the LSQ would remain unchanged.

In civil litigation, Jane needs only to show by a preponderance of the evidence or a balance of the probabilities or that it is more likely than not that Sally caused damage to her garage. Any uncertainty concerning Sally's civil liability need only be removed by a balance of the preponderance of evidence before Sally must compensate Jane for damage to the garage.

Justifications for the standard of proof differ from one area to the other. These ideas have evolved, or been developed, over time to accomplish subsidiary normative goals within different areas of the law. In civil litigation, a mistaken judgment for the plaintiff is [seen as] no worse than a mistaken judgment for the defendant. Thus, the party satisfying the preponderance of the evidence prevails.

In contrast, in the criminal law where the defendant has an interest of transcendent value at stake – life, liberty and good name – the finder of fact must be persuaded of guilt beyond a reasonable doubt (Strong 1992: 957,962). To illustrate these differences, O.J. Simpson, a former professional football player, criminally charged with the murder of his ex-wife, Nicole Brown Simpson, was found not guilty by a jury, because it judged there was too much uncertainty about the prosecution's case to find him guilty beyond a reasonable doubt. Shortly after that trial, in a nearby jurisdiction, Simpson was found liable in civil litigation for the wrongful death of his wife and was required to pay compensation to his children for her death.

It is worth noting that because of the criminal law's harsh sanctions, i.e., individual punishment, substantive normative considerations to protect defendants alleged to have violated the law. Thus, one must have voluntarily chosen to have harmed another and may not be criminally prosecuted for harms inflicted, if, for example, the defendant acted in the midst of an epileptic seizure. A defendant may not be punished for acts committed while insane or having the mental state of a child. Thus, while the criminal law is committed to preventing harm to others enforced by serious punishments, these subsidiary considerations limit criminal actions against those who lack some basic human capacities for criminal responsibility. Analogously, in the criminal law there is a doctrine called the 'presumption of leniency': when a statute may be read in a manner unfavorably or favorably to a criminal defendant, that is, when the law does not provide a clear answer, the presumption of leniency requires that the statute be read in a lenient manner because of the need to protect defendants from the harsh sanctions of the criminal law. Support for this presumption rests on the same normative commitment that people alleged to have committed a crime are presumed innocent until proven guilty by legal procedures. The ethical/normative judgments about differences in personal costs to the party that violated the law are substantially different – quite serious and potentially devastating consequences for many criminal offenses compared with the typical payment of damages to the injured party under civil litigation.

Legal Presumptions: The law also utilizes legal presumptions to address uncertainties on different topics. Legal presumptions are legally created inferences that must be followed in appropriate circumstances. For example, in the criminal law, an adult who kills another is presumed sane, unless evidence to the contrary is established. In other legal areas, a person may be declared legally dead if he or she has disappeared and has not been seen or heard from for seven years; this is a presumption of death unless there is contrary evidence. 'Rebuttable presumptions' obtain unless an opponent produces evidence to the contrary. Historically, there have also been irrefutable or conclusive presumptions in the law, but those have nearly disappeared in US law. In the United Kingdom, a conclusive presumption is that a child below the age of ten cannot be legally responsible for his/her actions (Parliament of the Kingdom of Great Britain and Northern Ireland 1933, 1963, setting the minimum age at ten). No evidence to the contrary can overcome this presumption.

Concerns about Actions that May Harm Others: Although vast areas of the law are concerned with preventing and sanctioning of agents that harm institutions or other persons, there are also prohibitions and sanctions for actions that discourage agents from even risking harm to others.

For instance, killing another person with an appropriate state of mind is a crime (and wrongful death is a tort), but in the criminal law attempting to kill is also a crime even if the targeted victim is not harmed or even aware of the attempt. Conspiring with others to kill or commit another felony, such as bank robbery, is also a crime even if the attempt does not succeed (the bank is not successfully robbed) or not all steps are carried out. Even preparation to commit a crime, such as arson, is a crime even if a fire does not successfully burn the targeted structure. Thus, actual harm to an institution or person need not occur to be a criminal offense. Finally, *reckless endangerment* – the act of being aware of creating "a substantial risk of serious physical injury to another person" (USLegal) – or negligent endangerment – negligently engaging in conduct that creates "a substantial risk of death or serious bodily injury to another – are crimes" (Montana Code Annotated 2019). A person acts negligently with respect to a material element of an offense if one should have been aware of a substantial risk of harm to another.

Attempted offenses, conspiracies, preparation for an offense along with reckless and negligent endangerment authorize protective actions before harm has occurred because a major aim of the criminal law is to prevent harms to others or to some institutions. The law authorizes some degree of prevention (beyond mere threats of deterrence) rather than waiting for harm to occur and then sanctioning the agent after the fact. There are potential costs to individuals from such laws, so such laws must be administered carefully.

To foreshadow concerns about global warming by analogy with the law, scientists need not wait until actual harm to the planet Earth occurs before taking actions, but take action based on recognizable risks to the planet from human or natural forces, e.g., the escape of methane from deep underground, well before harm comes from that source. One might say the global community and the scientists informing them should not recklessly or negligently endanger the Earth if they are aware of risky sources of harm that could be addressed without waiting to satisfy high standards of proof. It seems appropriate to protect the Earth by not negligently or recklessly failing to take actions that would prevent serious global warming.

The law is a humanly created institution to prevent harm, to facilitate a variety of human activities, to adjudicate disagreements between community members, and to serve other purposes in doing so. The instances above illustrate the plasticity or flexibility in the design of these human institutions and often enough, they may differ between legal jurisdictions, areas of law, or between countries. Humans have created them over a period of time with some modifications in how the issues are conceptualized.

Important for our purposes is that normative considerations support the goals and justifications of the institutions as well as how subsidiary aims should be achieved and are central to the designs and functioning of the law. This is also true when we consider regulatory policy and regulatory science as well as international agreements to reduce global warming. The norms that guide an institution and how it functions are important components for understanding it.

Value Assumptions in Regulatory Regimes

Regulatory institutions that have been created to protect the public from exposures to toxic substances such as carcinogens or reproductive toxicants (while permitting companies to develop and market chemical products) utilize analogues to some of these legal ideas. The laws governing exposures to toxicants do not uniformly regard chemical creations as 'toxicologically innocent until proven guilty' as people alleged to have committed a crime or a tort are assumed innocent until proven to have violated the law.

Laws protecting the public from toxic chemical creations (or pollutants) fall into two generic categories: premarket laws and postmarket laws. Premarket testing and scientific review laws for certain classes of substances proposed for commercialization are considered as possibly causing risks to human health, and thus seek to identify risks from products before they enter commerce and people are exposed. Under these laws just because nothing is known about a chemical substance proposed as a pharmaceutical or pesticide, it is not presumed to be toxicologically innocent to humans.

In the criminal law and torts, people are free to carry out their activities and have no difficulty with law until they are alleged to have violated it. In contrast, the mere fact that a company proposes a pharmaceutical or pesticide for commercialization does mean that it is free to immediately prepare it for sale and sell it; the law requires a showing that the product is legally safe under the appropriate statute. For both pharmaceuticals and pesticides chemical creations proposed for sale must undergo a legally specified battery of tests (different for each class of substances) and independent scientific review at the appropriate administrative health agency – the Food and Drug Administration (FDA) for pharmaceuticals or the Environmental Protection Agency (EPA) for pesticides – before they are permitted to be sold in the market. Once they have been approved for marketing, legal conservatism kicks in; they typically remain in commerce until the authorizing agency satisfies a legal and scientific burden of proof to alter their legal status, which typically would involve showing that they posed risks previously not identified (Cranor, 2017:22).

In contrast, postmarket laws typically permit chemical substances into commerce typically without routinely requiring toxicity data about products before they may be sold. Substances subject to most postmarket laws are typically presumed to be toxicologically innocent until they have been shown to pose risks or cause harmful toxic effects. The vast majority of environmental health laws in the United States are postmarket laws. Under an older, largely postmarket law, the Toxic Substances Control Act of 1976 (Cranor 2011, 2017), if evidence is established that a substance such as polybrominated diphenyl ethers (PBDEs) or perfluorinated compounds (PFCs) likely cause cancer or adverse developmental effects to people, then the US Environmental Protection Agency is authorized to take action to reduce exposures to them or to remove them from commerce. If there is credible evidence of toxicity at the time products are proposed for commercialization (but companies are not required to test them), the EPA can demand more data about them. However, if there is no evidence of toxicity when a product is proposed for commercialization, and no studies have been conducted on it, it may enter commerce and is presumed not to pose risks to the populace.

The Federal Insecticide, Fungicide, and Rodenticide Act legally requires that 'pesticides' must undergo premarket toxicity testing by the company, have the data reviewed by the Environmental protection Agency, and be authorized to commercialize it before it is permitted into the environment and the American public exposed to it. The Food, Drug, and Cosmetic Act for pharmaceuticals and a few other categories of products requires extensive premarket testing of a product for its safety and efficacy in test tube and animal studies, as well as numerous randomized control studies with humans. This body of data would then be reviewed by the FDA, before a pharmaceutical may be commercialized and the American public exposed. The FDA must find the product to be both 'safe' and 'effective' before commercialization and exposures occur (Cranor 2011).

Thus, postmarket laws for most chemical substances are legally presumed not to pose risks to human health, that is, to be toxicologically innocent, unless and until there is evidence to the contrary (typically, but not quite always, determined after the product is in commerce). In this, postmarket laws somewhat resemble the criminal law or torts because a company can sell its products that may have toxic effects until a legal case is made to reduce exposures or remove it from commerce. A slightly misleading aphorism capturing this might be 'no evidence, no harm (and no legal interference).'

In contrast, the laws for pharmaceuticals recognize that biologically active prescription drugs will be ingested by or injected into humans and may pose some risks to the health of persons using them. Consequently, they must be tested for possible adverse effects before the general public is exposed. The rationale for the pesticide law is somewhat different. Since insecticides, fungicides, and rodenticides are deliberately created to be biologically toxic to living entities such as insects, fungi, or rodents, compounds toxic to those species may also cause harm to humans (or other living organisms in the environment). Consequently, before exposures occur, a company must show that there are no or insufficient risks from products by testing them in order to have a 'reasonable certainty of no harm' to people and no unreasonable risks to the environment.

The status quo ante for pharmaceuticals or pesticides proposed for commercialization does not necessarily presuppose risks of harm to humans. Instead, it is the possibility that biologically active pharmaceuticals can have adverse side effects and that pesticides, designed to cause harm to other living things, provide normative justifications for imposing a burden of proof for requiring scientific review of the products before commercialization and human exposures.

Value Assumptions Needed in Risk Assessments

However, often the science needed to assess the risks of products in commerce under postmarket laws or products commercialized under premarket laws is too uncertain to address some safety concerns the law requires. There are several issues, including: (1) no scientific studies may have been conducted to determine the seriousness of risks below the level that can be observed in existing studies; and (2) the shape of a risk assessment curve at low levels of exposures may not be determinable by any scientific data. In this instance, some people might urge the agency not set standards at levels that cannot be scientifically identified with certainty, but this could leave some people at risk. Or, there may be several plausible models for the shape of curves at low, but realistic levels, but no scientific data which identifies the correct one; which one should be chosen? All of these options may be sufficiently controversial that they could put EPA actions and the public's health legally at risk. How have our human institutions been modified to address these issues?

Legal procedures for the regulation of chemicals in commerce differ from the criminal law and torts. Burdens of proof and standards of proof in those two areas of the law prohibit successful legal action if the respective standards of proof are not established. However, under regulatory health law, there are different approaches to absence of empirical/scientific evidence. If there are insufficient facts to determine just how toxic a substance is, especially at low doses, the affected company likely may not escape legal consequences. Instead, scientific presumptions that are rational and not 'arbitrary and capricious' may be invoked to assist in conducting the toxicological assessments.

Default Assumptions in Regulatory Science at the US Environmental Protection Agency

The US EPA developed the equivalents of rebuttable presumptions consistent with its overall charge under various laws and consistent with the known science at the time a decision is needed to regulate the toxicity and exposure level of a toxicant. This discussion is restricted to some EPA risk assessment policies for efficient discussion (addressing and interpreting scientific uncertainties to evaluate the toxicity of substances may be somewhat different for pharmaceuticals and pesticides, which is not discussed here).

Rebuttable scientific presumptions for risk assessments are called 'scientific defaults.' Defaults are assumptions or presumptions utilized in science-policy judgments to draw inferences beyond the data or beyond what may be directly observable. They may be used when the experimental scientific data either cannot fully resolve factual issues between different models, or current science simply cannot establish a scientific case to be made to address low levels of exposures to assess the toxicity of a product (EPA 2005; NRC 2009: 192).

The defaults must be consistent with the available science as well as protecting the public's health. The guidelines have been developed to justify and select, from among available options, the assumptions to be used for agency risk assessments. They are needed because distinguishing the available options on purely scientific grounds often is simply not possible, thus either the agency might not be able to protect the public from a toxicant or a scientific policy was needed to select the options for general use. They provide uniform inference guidelines for risk assessment. A number of background considerations support the use of scientific defaults.

Defaults are efficient, making explicit existing assumptions or developing new assumptions needed for decisions (NRC 2009: 241, 264–265). They avoid full *ad hoc* examinations of data. They also provide important starting places, help to clarify discussions, provide predictable and consistent risk assessment structures, and a basis for the public assessment of its actions (NRC 2009: 241, 264–265). The National Academy of Sciences acknowledges that defaults are 'value judgments' needed in scientific policies for risk assessments (NRC 2009: 196–197).

Current science cannot decisively resolve all the issues that need to be addressed in risk assessments. On some issues, the 'correct' scientific answer simply cannot be determined; scientific data simply does not provide a sufficient answer that needs to be addressed. But the public should still be protected from toxicants. In the face of such scientific ignorance what should the EPA do?

In its 2005 revised cancer guidelines, EPA resorts specifically to an explicit normative consideration,

> [T]hese cancer guidelines adopt a view of default options that is consistent with EPA's mission to protect human health while adhering to tenets of sound science. ... The primary goal of EPA actions is protection of human health; accordingly, ... risk assessment procedures, including default options that are used in the absence of scientific data to the contrary, should be health protective.
>
> *(EPA 2005: 1–7)*

Consequently, and quite importantly, while permitting companies to commercialize chemical products, because the EPA's major mission is to protect the human health, the defaults for conducting risk assessments must be consistent with and serve this institutional goal at the same time being sufficiently consistent with the science to carry out the protective task. That is, if there are legitimate scientific concerns that there are risks from exposures to carcinogens, but the exact risks cannot be specified, the agency utilizes scientific defaults to assist risk assessment in protecting the range of exposed individuals.

Examples of Defaults in EPA Guidelines

Children explicitly receive greater legal protection from pesticide exposures as required by the 1996 Food Quality Protection Act (FQPA) (US Congress 1996),

> The 1996 Food Quality Protection Act (FQPA) mandates the use of a safety factor of 10 [to protect infants and children] unless EPA has sufficient evidence to determine that a different value is more appropriate.
>
> *(US Congress 1996: Sec. 408(b)(2)(c)]; EPA 2005: 1–7; NRC 2009: 202)*

Protecting developing children *in utero* or during early development is required by this law and is incorporated into pesticide risk assessments. Developing children are much more susceptible to adverse effects than adults at the same exposure levels. They have lesser defenses to toxic invasions. Some organ systems have 'one chance' to develop correctly, i.e., the brain and immune system.

And, if children are exposed to toxicants at early ages, they have the rest of their lives for any diseases that might be triggered to develop (Cranor 2011: 102).

In addition, a number of specific scientific defaults have been developed for cancer risk assessments, because this disease is of substantial concern to the American public. The goal is to ensure for EPA's risk assessments that the public is appropriately protected from exposures to carcinogens. For example,

> When cancer effects ... in ... humans are attributed to ... to an agent [which might be from a comparatively high occupational exposure], the default ... is that ... data are predictive of cancer in ... other exposed human population[s] [where exposures might be lower].
>
> *(EPA 2005: A-2)*

This default recognizes that studies conducted on occupational populations should be taken seriously as indicating other people might be at risk even though exposures to the general population might be lower. Studies of working populations provide, "no opportunity to observe subpopulations who are likely to be under-represented, such as fetuses, infants and children, women, or people in poor health, who may respond differently from healthy workers"; and occupational studies could, "underestimate the response of certain human subpopulations" (EPA 2005: A-2).

Another default recognizes the limitations of scientific data from human epidemiological studies: they often have "low power to detect and attribute responses and typically evaluate cancer potential in a restricted population (e.g., by age, healthy workers)" (EPA 2005: A-2). Thus, even if there are no data for the effects of a carcinogen on a wider population, this normative guidance requires EPA to address it. The agency must be sensitive to the limitations of epidemiological studies to detect rare diseases or that have too few participants in exposed and control groups.

Normative epistemic consideration can also guide statistical models, so they better discover harms and protect the public as required by the law. Consider epidemiological studies as representative. Conventional views urge that epidemiological studies must be "statistically significant" (Neutra et al. 2018: 138). This is typically understood to mean that an observed finding from a study or, one more extreme, "should be highly unlikely under the non-causal hypothesis with a probability < 0.05" (Neutra et al. 2018: 151–153). However, despite this convention, it is widely recognized that other measures of statistical significance are appropriate.

For example, leading epidemiologists recognize that as a matter of good statistical or scientific practice, statistical significance can be greater than 0.05, even up to 0.10 (Neutra et al. 2018: 152–153). For instance, the US EPA utilized 0.10 for statistical significance for identifying harms from secondhand tobacco smoke. Also, a number of cancers from nuclear radiation fallout were identified with statistical significance at 10 (Cranor 2016: 229). The importance of this point? Permitting a slightly higher level of statistical significance can facilitate the identification of a risk that might not be detected by insisting that significance be.05, while only slightly increasing the chances of a false positive error. The reason? In a statistical study, there is an inverse, but not a linear relationship between studies that risk false positive and false negatives. The lower the level insisted upon for statistical significance to prevent false positives, the less likely a study would be able to detect an adverse effect even if it were present – the study would be unable to detect it.

In addition, supported by the ethical concern to protect the health of the US population from carcinogens, the EPA recognizes that scientific studies might not have had time or good fortune to detect cancer effects in various susceptible subpopulations or life stages. For example, the latency period from *in utero* exposure to the carcinogen diethylstilbesterol to the clinical appearance of vaginal cancer is twenty years. The latency for cancers to appear resulting from exposure to asbestos to clinical appearance is up to 40–45 years (Cranor 2016). Researchers must allow sufficient time for the effect to occur to detect risks of concern. Otherwise, a study may well fail to detect an adverse effect when it has been triggered by an exposure. The default to address this issue is,

> When cancer effects are not found in an exposed human population, this information by itself is not generally sufficient to conclude that the agent poses no carcinogenic hazard to this or other populations of potentially exposed humans, including susceptible subpopulations or life stages.
>
> *(EPA 2005: A-2)*

Because humans are mammals, a biological fact sometimes forgotten or deliberately contested by companies subject to regulation, evidence of carcinogenic effects in animals can provide good evidence of carcinogenic effects in humans, "Positive effects in animal cancer studies indicate that the agent under study can have carcinogenic potential in humans" (EPA 2005: A-3). This default is supported by the public health-protective policy, and "nearly all of the agents known to cause cancer in humans are carcinogenic in animals in tests that have adequate protocols" (EPA 2005: A-3).

Sometimes advocates promoting products considered for regulation argue that benign tumors should not provide evidence that a product is carcinogenic, but this would be both scientifically incorrect and not support EP's and Congressional health protective commitment.

> The default is to include benign tumors observed in animal studies in the assessment of animal tumor incidence, if such tumors have the capacity to progress to the malignancies with which they are associated. This default is consistent with the approach of the National Toxicology Program and the International Agency for Research on Cancer and is more protective of public health than not including benign tumors in the assessment…
>
> *(EPA 2005: A-5, emphasis added)*

At one time before data about the mechanisms of chemical compounds was better understood, this information had quite limited use in risk assessment. A body of data has now been developed that supports a greater use of mechanistic data,

> There is a similarity of the basic pathways of metabolism and the occurrence of metabolites in tissues in regard to the species-to-species extrapolation of cancer hazard and risk [e.g., from animals to humans].
>
> *(EPA 2005: p. A-6)*

Moreover, the EPA has had to create a scientific default for extrapolating from observed exposure levels in human or animal studies to exposures likely to cause adverse exposures levels at much lower doses that cannot be observed or detected by scientific studies.

Consequently, the agency utilizes linear extrapolations from known high-dose effects to unknown low-dose consequences:

> When the weight of evidence evaluation of all available data is insufficient to establish the mode of action for a tumor site and when scientifically plausible based on the available data, linear extrapolation is generally considered to be a health-protective approach.
>
> *(EPA 2005: 3–21)*

This commitment to linear extrapolations from data in the observed range to data points in the unobserved range is sufficiently strong that a departure from it requires special justification. Where the EPA cannot determine the mechanisms by which a carcinogen causes tumors, it must presume that a linear approach should be used, "Nonlinear approaches generally should not be used in cases where the mode of action has not been ascertained" (EPA 2005: 3–22).

Congress and the EPA could have followed the criminal law or tort law by adopting the view that if the toxicity of a substance cannot be identified by scientific studies in accordance with some prescribed standard of proof, e.g., beyond a reasonable doubt or by a preponderance of the evidence, it may not take regulatory actions, thus preserving the legal and public health status quo, in instances in which there is good but not totally decisive scientific evidence of carcinogenicity. This approach was not chosen; it would leave the health of the public at risk. Instead, the EPA used the equivalent of rebuttable presumptions for protecting the public, even though not every scientific 'i' is dotted and every 't' is crossed. Scientific studies simply cannot yet, and probably for some considerable period of time, supply every data point needed for health protections. However, to serve the overall institutional goals, supported by pertinent scientific evidence, they have adopted various normative considerations to protect the public health.

Ethical/Normative Issues in the United Nations Framework Convention on Climate Change

The United Nations Framework Convention on Climate Change, adopted in 1992, is an international (legal) treaty with 197 signatories that aims to prevent "'dangerous' human interference with the climate system …" (UNFCCC 1992). Its collective duties, "bound member states to act in the interests of human safety [of the world and member states] even in the face of scientific uncertainty," stabilizing greenhouse gases at a level that would prevent humanly caused interference with the climate system (United Nations 1994).

On its face, there are several explicit normative considerations that are part of the agreement that each country has agreed to:

1 Member nations have duties from the international agreement to prevent dangerous atmospheric interference with the climate system (UNFCCC 1992: Art. 1) and in particular to adopt policies that will seek to prevent warming from rising no more than 1.5°C but certainly it should be kept below 2.0°C (Paris Agreement 2015);
2 There is also a 'do-no-harm' principle to the environments of others' states or areas beyond a country's national jurisdiction (UNFCCC: Preamble). These first two duties are legal and moral duties agreed to by all countries in the 1992 UNFCCC and the 2015 Paris Agreement (UNFCCC 1992; Paris Agreement 2015), others may be less obvious, but still important;
3 There is an applicable fairness (equity) principle that because different nations make different contributions to Global Warming (GW): Those that make greater contributions to GW warming have greater responsibilities to reduce them in order to reduce or prevent harm to the most vulnerable countries. This is a basic principle of fairness: in a cooperative effort, those who contribute to greater potential harm to common goals should bear a greater responsibility to help reduce them. And, those who make minor contributions have lesser responsibilities to reduce any collectively caused harms; and
4 The Parties should take precautionary measures to anticipate, prevent or minimize the causes of climate change and mitigate its adverse effects: where there are threats of serious or irreversible damage, lack of full scientific certainty should not be used as a reason for postponing

> such measures, taking into account that policies and measures to deal with climate change should be cost-effective so as to ensure global benefits at the lowest possible cost (UNFCCC 1992, Art. 3.3).

There are different ethical issues implicit in the precautionary principle (PP). This provision anticipates that increased climate change in excess of 1.5°C or worst case 2°C will in fact cause adverse effects and probably irreversible damage to planet Earth for many centuries. What is envisioned by the idea of 'full scientific certainty?'

Borrowing from the law, this term might mean (a) that if scientific certainty is less than scientific 'proof beyond a reasonable doubt,' countries need not take cost-effective actions to protect the Earth. This seems much too stringent an intellectual barrier before nations' take action to better ensure a more livable Earth, would likely block achievement of the collective goal, and might be collectively disastrous.

Or perhaps, (b) the PP requires that if the science is not 'more likely than note true,' countries need not take cost effective actions to reduce or prevent GW. However, because both interpretations (a) and (b) are taken from legal litigation, they may be misleading. That is, in both cases, there is the suggestion that the status quo is OK or not unacceptable if a legal defendant is not found guilty or liable, respectively. This further suggests there is a comparatively fixed time for a decision, and, if the status quo was OK before litigation, if there is no legal reason to change it, it's OK post-decision.

Global warming is quite different. It is an ongoing, non-static process. If nothing is done to modify the warming, it is highly likely that substantial harm to the Earth and its inhabits will occur. Inexorable mechanisms of warming likely will continue for some time unless collective actions are taken to slow or reverse it.

For example, there is strong modeling evidence supported by some data observations that abrupt thawing of ice caps (versus gradual thawing) will accelerate

> Mobilization of deeply frozen, ancient carbon, increasing 14°C- depleted permafrost soil carbon emissions by ~125–190% compared to gradual thaw alone. These findings demonstrate the need to incorporate abrupt thaw processes in earth system models for more comprehensive projection of the PCF this century.
>
> *(Anthony et al. 2021)*

What would a near doubling of carbon emissions from under ice caps do to increase global warming? For another example, "self-reinforcing feedbacks could push the Earth System toward a planetary threshold that, if crossed, could prevent stabilization of the climate at intermediate temperature rises and cause continued warming on a 'Hothouse Earth' pathway even as human emissions are reduced" (Steffen et al. 2021).

Perhaps in such circumstances, a better scientific-policy approach would be to utilize presumptions or default assumptions analogous to those adopted for addressing risks to human health in regulatory science and law. Given what the current best sciences show or strongly suggest, what are the best defaults for addressing increased methane or CO_2 emissions that have not been incorporated into climate models to date? The science in these areas may not be certain beyond a reasonable doubt or even highly certain short of that demanding standard, but scientists and decision makers need not do nothing when there is good, possibly not totally decisive scientific findings, but overwhelming moral, policy, and human self-interested imperatives to protect the planet Earth. Choice of good and appropriate defaults to augment scientific findings might be the way to go.

An advantage of defaults is that, since they are deliberately chosen to serve particular purposes, they can be modified and updated with additional data or better understandings of the warming mechanisms as has occurred with EPA defaults in protecting the public's health. Consequently, just because not all of the scientific 'i's' are dotted or 't's' crossed does not mean countries need to do nothing to combat global warming. One can use the best science augmented by good policy presumptions to protect the planet.

There are other analogies to which one could appeal as well. Should the nations of the world and scientists that advise them recklessly to disregard current trends or mechanisms of warming of which they have good evidence, or currently disregard the likelihood of 'tipping points' in feedback loops that would likely cause permanent planetary damage?

We can learn much from how other institutions cope with uncertainties or limited science data, but it seems we should keep our eyes on the normative goal: protecting Mother Earth from becoming unlivable for many humans.

References

Anthony, K.W., von Deimling, T.S. and Nitze, I. et al. (2021) "21st-Century Modeled Permafrost Carbon Emissions Accelerated by Abrupt Thaw Beneath Lakes," *Nature Communications*, 9: 3262. DOI: 10.1038/s41467-018-05738-9, located at www.nature.com/naturecommunications.

Breakthrough - National Centre for Climate Restoration. (2017, revised 2018) Melbourne, Australia. Located at https://docs.wixstatic.com/ugd/148cb0_a0d7c18a1bf64e698a9c8c8f18a42889.pdf.

Cranor, C.F. (2011) *Legally Poisoned: How the Law Puts Us at Risk from Toxicants*. Cambridge: Harvard University Press.

Cranor, C.F. (2016) *Toxic Torts: Science, Law and the Possibility of Justice*, 2nd Edition. New York: Cambridge University Press.

Cranor, C.F. (2017) *Tragic Failures: How and Why We Are Harmed by Toxic Substances*. New York: Oxford University Press.

James, F. Jr. and Hazard, G.C. (1977) *Civil Procedure*, 2nd Edition. Boston: Little, Brown and Company.

Montana Code Annotated. (2019) *Assault and Related Offenses, "Negligent Endangerment," 2019*. Located at https://leg.mt.gov/bills/mca/title_0450/chapter_0050/part_0020/section_0080/0450-0050-0020-0080.html.

National Research Council. (2009) *Committee on Improving Risk Analysis Approaches Used by the US EPA, Science and Decisions: Advancing Risk Assessment*. Washington, DC: National Academy Press.

Neutra, R.R., Cranor, C.F. and Gee, D. (2018) "The Use and Misuse of Bradford Hill in US Tort Law," *Jurimetrics: The Journal of Law, Science and Technology*, 58: 127–162.

Paris Agreement to the United Nations Framework Convention on Climate Change, December 12, 2015, T.I.A.S. No. 16–1104.Mar 18, 202

Parliament of the Kingdom of Great Britain and Northern Ireland, *Children and Young Persons Act 1933, 1963* (23 & 24 Geo.5 c.12), amended in 1963.

Steffen, W., Rockströma, J. and Richardson, K. et al. (2021) "Trajectories of the Earth System in the Anthropocene," *Proceedings of the National Academy of Sciences*. Located at www.pnas.org/cgi/doi/10.1073/pnas.1810141115

Strong, J.W. (ed.) (1992) *McCormick on Evidence*, 4th Edition. Eagan: West Publishing Co.

United Nations. (1994) *Climate Change, What Is the United Nations Framework Convention on Climate Change?* Located at https://unfccc.int/process-and-meetings/the-convention/what-is-the-united-nations-framework-convention-on-climate-change.

United Nations Framework Convention on Climate Change (UNFCCC), May 9, 1992, S. Treaty Doc No. 102-38, 1771 U.N.T.S. 107.Mar 18, 2020

US Congress. (1996) "The Food Quality Protection Act (FQPA)," *Public Law*, 104–170.

US Environmental Protection Agency. (2005). *Guidelines for Carcinogen Risk Assessment*. EPA/630/P-03/001F. Risk Assessment Forum, US Environmental Protection Agency, Washington, DC. March 2005 [online]. Located at http://cfpub.epa.gov/ncea/cfm/recordisplay.cfm?deid=116283.

USLegal, *Legal Definitions*, "Reckless Endangerment," located at https://definitions.uslegal.com/r/reckless-endangerment/#:~:text=Reckless%20Endangerment%20Law%20and%20Legal%20Definition.%20Reckless%20endangerment,disregard%20for%20the%20foreseeable%20consequences%20of%20the%20actions.

11

THE ETHICAL RESPONSIBILITY OF DEVELOPED COUNTRIES TO HELP FINANCE ADAPTATION COSTS IN POOR DEVELOPING COUNTRIES

Workineh Kelbessa

Introduction

The climate crisis is affecting all countries in different ways, and we all need to work together to reduce its impact. In particular, climate change is already threatening lives in the least developed countries. It has had negative impacts on poor people in developing countries rather than on rich people in developed countries, "Poor people suffer more than rich people from climate-related impacts, wherever they live, but poor people in poor countries suffer most of all" (Jamieson 2010: 270; see also Attfield 2014: 203–204). Although Africa's contribution to climate change is insignificant, Africa has been suffering from rising temperatures and evaporation, widespread water stress, increased frequency and severity of droughts and floods, harvest failure, rising sea level, a decline in biodiversity, changes in the spatial distribution of diseases (such as malaria, dengue fever, and cholera) from lower altitudes and latitudes to higher altitudes and latitudes, and conflicts over access to land and water (Kelbessa 2015; Attfield 2020). Africa is already facing harsh weather conditions (e.g., high average temperature, scarce and erratic rainfall) in some places. Local temperatures get warmer in various parts of the continent.

This chapter will discuss the ethical responsibility of developed countries to help fund the adaptation costs of poor developing countries. This does not suggest that developing countries only need adaptation to climate change impacts. What should be emphasized is that adaptation or mitigation is not enough alone to address climate change impacts (Jamieson 2010: 267; Attfield 2014: 210; 2019). Adaptation policy should be supplemented by a climate mitigation policy (Jamieson 2010: 267). These two strategies should work together. As Dale Jamieson stresses, a policy of "adaptation only" will have the moral risk of hurting the poor the hardest.

This chapter relies on the definition of "adaptation" given by the Intergovernmental Panel on Climate Change (IPCC). The IPCC defined adaptation as "initiatives and measures to reduce the vulnerability of natural and human systems against actual or expected climate change effects" (IPCC 2007: 809).

UNFCCC Provisions on Mitigation and Adaptation

Article III of the United Nations Framework Convention on Climate Change (UNFCCC) states that Parties to the Convention should

DOI: 10.4324/9781003039860-14

> ...protect the climate system for the benefit of present and future generations of humankind, on the basis of equity and in accordance with their common but differentiated responsibilities and respective capabilities. Accordingly, the developed country Parties should take the lead in combating climate change and the adverse effects thereof.
>
> *(UN 1992)*

This passage identifies fairness between generations, as well as fairness between nations, although there is no reference to non-human generations.

The cardinal principle of "common but differentiated responsibilities and respective capabilities" (CBDR+R) of the UNFCCC suggests that developed countries should lead in mitigating greenhouse gas emissions (Article 3.1) and supporting adaptation in vulnerable countries (Article 4.4), including providing financial assistance with "new and additional" monies (Article 4.3). Article 4.4 of the UNFCCC indicates that developing countries expressly agreed: "[t]he developed country Parties included in Annex II[1] shall also assist the developing country Parties that are particularly vulnerable to the adverse effects of climate change in meeting costs of adaptation to those adverse effects."

It is also worth noting that in the Preamble to the UNFCCC, all nations agreed to the "no-harm" principle. In particular, developed countries agreed that they have a duty not to harm the environment of other States or areas beyond the limits of national jurisdiction.

Different ethical principles can be used to support the principle of "common but differentiated responsibilities." It can be defended, among others, by requiring countries that can afford to pay the costs associated with climate change. The argument is that developed countries should assist developing countries to cover mitigation and adaptation costs, as they are capable of doing so. According to Henry Shue, "among a number of parties, all of whom are bound to contribute to some endeavor, the parties who have the most resources should contribute the most to the endeavor" (1999: 537).

However, it can be objected that a country's ethical obligations should not rely on its wealth, "[t]his is because being wealthier than others would not confer any special responsibility on a country if it only has enough resources to meet the basic needs of its own citizens" (Page 2008: 561). Edward A. Page suggests that this objection can be addressed by amending Shue's articulation of the "ability-to-pay" approach. "It is not 'the nations with the most resources' that should lead the way in climate policy but rather 'the nations with excess capacity'" (Page 2008: 561).

Also, as P. Baer notes, there are some discrepancies between historical emissions and current wealth. Some countries that have historically caused the most pollution "have remained (or become) relatively poor"; some countries have developed their economy "with relatively low emissions" (Baer 2010: 252–253). Therefore, despite their historical emissions, some countries cannot pay for the adaptation costs of developing countries. Robin Attfield also persuasively argues that the "polluter pays" principle should be related to capacity, as poor countries are incapable of sharing the burden and responsibility for adaptation costs (Attfield 2014: 216). Peter Singer for his part suggests that those who have the power must aid the global poor: "if it is in our power to prevent something bad from happening, without thereby sacrificing anything of comparable moral importance, we ought to do it" (Singer 1972: 231).

The beneficiary pays principle suggests that countries that benefited the most from past emissions should support adaptation efforts in developing countries (Page 2008: 562; see also Keating 1997 cited in Baer 2010: 250). The idea here is that developed countries should help poor countries to adapt to climate change because of their privileged inheritance. Dale Jamieson (2010) also suggests that those who are rich by global standards and benefit from excess emissions should reduce their emissions and finance adaptation.

Page stresses that the above-mentioned approaches converge on some version of "common but differentiated responsibility." He argues that only a principled and philosophically robust reconciliation of three approaches to burden-sharing "('contribution to the problem', 'ability to pay,' and 'beneficiary pays') can generate a satisfactory mix of theoretical coherence and practical application" (Page 2008: 556).

Climate Adaptation Finance

While the articles of the 1992 UNFCCC and subsequent decisions adopted by the Conference of Parties (COP) referred to adaptation finance, in the first decade of climate negotiations, the main focus was on mitigation rather than adaptation. Developed countries were expected to lead the process of preventing global warming based on their Common but Differentiated Responsibilities and Respective Capabilities.

Three reasons can be mentioned for the lack of interest in adaptation: first, lack of motivation on the part of developed nations to address adaptation. Annex 1 Parties[2] (essentially the OECD) did not want to commit themselves to acknowledge responsibility and liability based on historical emissions by focusing on adaptation. Second, some actors and civil society in both developing and developed countries suspect that a focus on adaptation will marginalize mitigation efforts (Anderson 1997 and Burton 1996, cited in Ciplet et al. 2013: 51). Third, there is disagreement within the UN-based group of developing countries, the G77, on Article 4(8) of the UNFCCC. Among others, this Article refers to how the implementation of response measures affects "[c]ountries whose economies are highly dependent on income generated from the production, processing and export, and/or on [the] consumption of fossil fuels and associated energy-intensive products." Countries belonging to the Organization of Petroleum Exporting Countries (OPEC) want compensation to diversify their oil-dependent economies. The EU and other Annex 1 countries did not accept this requirement (Ciplet et al. 2013: 51).

Since the 2011 Marrakech Accords, adaptation has been recognized as an important strategy for addressing the impacts of climate change. Two factors influenced COP's focus on adaptation in the negotiations. First is a marked increase in climate-related catastrophes. Second, the failure of developed countries to take adequate action to deal with anthropogenic climate change and developing countries' request to pay attention to adaptation (Ciplet et al. 2013: 52–53).

Three new funds were established during the 2001 Marrakech Accords to support adaptation efforts in developing countries: The Least Developed Countries Fund (LDCF), The Special Climate Change Fund (SCCF) under the UNFCCC, and The Kyoto Protocol Adaptation Fund under the Kyoto Protocol.[3] Unlike the first one, the second one was designed to assist the adaptation efforts of all developing countries. Concrete adaptation projects and programs are expected to be financed by the Kyoto Protocol Adaptation Fund. It is "resourced by an adaptation levy placed on transactions under the clean development mechanism, the program under which greenhouse-gas reductions are traded between companies in the developed and developing world. The two other funds are supported by voluntary contributions" (Jamieson 2010: 269). Contributions to SCCF and the LDC Fund from industrialized countries are voluntary and small. Only countries that have ratified the Kyoto protocols will contribute to these funds. Other rich countries do not need to do this.

The Bali Action Plan also recognized the role of adaptation in 2007. During the 2009 Copenhagen Accord, developed countries pledged a "fast-start" of \$30bn in 2010–2012 and to reach \$100bn a year by 2020 to cover adaptation and mitigation costs (Ciplet et al. 2013: 52). In 2010, the Cancun Agreements emphasized the need to balance adaptation and mitigation. As a result, the COP agreed on key elements of the umbrella funding agency structure known as the Green Climate Fund (GCF) introduced in the Copenhagen Accord in Durban in 2011 (Ciplet et al. 2013: 63).

Article 9.2 of the Paris Agreement also states that developed countries are under obligation to support developing countries' mitigation and adaptation efforts.[4] Developed countries have committed to helping developing countries with technology, strengthening institutions, and also providing financial support. The Paris Agreement favors a more voluntary approach to national commitments. Countries are expected to provide their nationally appropriate mitigation reductions (NAMAS) and make their activities transparent to national measurements and international assessments.

However, this does not encourage powerful countries to contribute to the cost of adaptation in poor developing countries in proportion to their share of global cumulative greenhouse emissions. There is a huge gap between rhetoric, commitment, and action on climate justice in the UNFCCC. Although the European Union has put concrete actions to meet its Paris Agreement commitments, certain wealthy countries have not yet fulfilled their promises. They have continued to use their economic muscle through a combination of high-handed and tactful diplomacy to impose their will on developing countries during climate negotiations. In addition, the US government decided to withdraw from the Paris Agreement, citing unsubstantiated claims that it would hurt the US economy and jobs. US President Donald Trump announced on June 1, 2017, his intention to withdraw the United States from the 2015 Paris Agreement on climate change mitigation. Jan Burck and his co-authors conducted the 2020 annual climate change performance index and concluded that the Paris Agreement has not helped humanity to reduce greenhouse gas emissions (Burck et al. 2020). Conversely, greenhouse gas emissions have increased significantly since the Paris Agreement. Some countries have failed to meet their ethical responsibilities to address climate change. As a result, the world does not appear to be on track to meet its climate goals. Even if every country delivers on its Paris commitments, the world will fall short of the two-degree ceiling proposed by the Paris Agreement. This chapter suggests that urgent action is needed for all countries to reduce greenhouse gas emissions beyond their Paris commitments.

Developing countries need international funding and technical assistance to deal with the effects of climate change and to prevent future climate-related harms. Adaptation finance justice is required to address these issues,

> Adaptation Finance Justice requires that developed country Parties take precautionary measures to assist developing countries to adapt to climate impacts by providing adequate, scaled up, predictable, balanced and new and additional adaptation finance, with priority to those Parties that are particularly vulnerable. Action should be taken on the basis of equity and in accordance with the principles of common but differentiated responsibility and respective capabilities and should follow a country-driven, gender-sensitive, participatory and fully transparent approach, taking into consideration vulnerable groups, communities and ecosystems.
>
> *(Ciplet et al. 2013: 55)*

At COP26 in Glasgow, which ended on the 13th of November 2021, developed country parties recalled that all developed countries agreed in 2009 to provide \$100 billion a year to support the adaptation and mitigation needs of developing countries, but they have not been able to deliver. As a result, developed countries agreed in Glasgow to fully meet the \$100 billion target and to increase the transparency of their adaptation commitments (UK Government 2021).

Attfield (2014) suggests that developed countries should provide grants rather than loans to developing countries, as they bear the primary responsibility for climate change. The most vulnerable countries also favor grants over loans "to enhance their adaptive capacity and avoid greater indebtedness" (Khan et al. 2020). Developed countries also have a moral obligation to compensate poor developing countries for losses and damages.

Conclusion

Climate change affects everyone differently – wealthy and poor countries. Above all, poor countries are victims of climate change. Although developed countries are responsible for past- and present-day emissions, they have ignored ethical norms, such as distributive justice, the polluter pays principle, and the like. They have failed to attend to their "equitable" and "ethical" obligations under the Paris agreement when they set their GHG emission reduction targets. To make matters worse, they have broken their commitments and duty of justice for more than two decades.

The ethics of adaptation demands that developed countries should assist developing countries to adapt to climate change "because of their greater wealth and capacities" (Markowitz et al. 2015; see also Attfield 2019; Khan et al. 2020). Sustained commitment by the developed world is required to finance adaptation to climate change. Nations should strive to balance their economic interests and duties, responsibilities, and obligations to the health of their people and mother Earth. Developed countries have a legal and moral duty to cut their emissions and to continue to support the poorest, and most vulnerable countries to reduce their emissions, prevent climate-induced harms and enhance their capacity to adapt.

Justice also requires that Africa and other developing countries should be compensated for the exploitation of the past. Some European countries have exploited Africa's human and natural resources since the 15th century. The endless line of Swedish, Danish, Dutch, Portuguese, British, French, and more castles that mark the "Ivory" and "gold" coasts of western Africa is just one example of the European exploitation and export of Africa's rich resources to Europe. European colonization devastated African indigenous communities and their resources. Many Western and Eastern companies, including Chinese and Indian companies, continue to extract natural resources from Africa. So, although developed countries are not ready to recognize climate debt to developing countries, developing countries have the right to demand reparations from the former for the damage done to their people and environment to fight climate change.[5]

Developed countries should also support poor developing countries to address the consequences of climate-induced risks and vulnerability out of enlightened self-interest, as they cannot continue to pursue a decent life when the people in the latter are suffering because of climate change. Building a more sustainable and safer world requires the cooperation of all countries, without which humanity cannot solve problems related to climate change. As Eric S. Godoy notes, social connection responsibility can help humanity effectively respond to climate change, "Social connection responsibility is met by acting with other agents to transform harmful structures rather than by acting as an isolated, individual agent" (Godoy 2017: 699). Iris M. Young states that power, privilege, interest, and collective capacity are important parameters that can be used to fulfill social connection responsibility (2011: 144–147). Those who are socially connected can achieve their goal, as they share these four parameters. Different social groups in developed countries should cooperate and put pressure on their governments and transnational corporations to reduce greenhouse gas emissions and support poor developing countries.

Adaptation is important for intergenerational justice. Many of the impacts that the current generation is experiencing today were caused by past generations. What humans do today will have an impact on humans and non-humans in the future. Reducing fossil fuels will improve the lives of current and future generations of both human and non-human beings alike.

Developing countries should also reduce their emissions as some of them, such as China, India, Brazil, and South Africa, have been emitting a relatively large percentage of global greenhouse gases although as developing countries have a lower responsibility to reduce their emissions under law. They indeed need electricity to improve the living conditions of their people. They must be converted into renewable energy generation. They ought to contribute their fair share to reduce the effects of climate change. We should also recognize that some emerging economies are increasingly providing support to the most vulnerable nations. Therefore, international cooperation is the key to reaching a global climate solution.

Notes

1 Annex II countries are developed countries which pay for costs of developing countries.
2 Annex 1 countries are the industrialized countries of North America and Europe, Japan, Australia, and New Zealand.
3 Decisions 5/CP.7, 6/CP. 7, 7/CP.7 and 10/CP.7.
4 195 countries agreed to the world's first universal action plan to reduce greenhouse gas emissions and adapt to the unavoidable impacts of climate change at the Paris climate conference (COP21) in December 2015. They promised to limit global warming to "well below 2°C above pre-industrial levels and to try to limit the increase to 1.5°C".
5 For further discussion of this issue see Chapter 7.

References

Attfield, R. (2014) *Environmental Ethics: An Overview for the Twenty-First Century*, Second ed. Malden, MA: Polity Press.

Attfield, R. (2019) "Vital Needs and Climate Change: Inter-Human, Inter-Generational and Inter-Species Justice," in P. G. Harris (ed.), *A Research Agenda for Climate Change*, Cheltenham: Edward Elgar Publishing, pp. 15–26.

Attfield, R. (2020) "Africa and Climate Change," *Utafiti* 14(2): 281–294.

Baer, P. (2010) "Adaptation to Climate Change. Who Pays Whom?" in S. M. Gardiner, S. Caney, D. Jamieson and H. Shue (eds.), *Climate Ethics: Essential Readings*, Oxford: Oxford University Press, pp. 247–262.

Burck, J., U. Hagen, N. Hohne, L. Nascimento, and C. Bals. (2020) *Climate Performance Index: Results 2020*. Berlin and Beirut: Germanwatch, New Climate Institute and Climate Action Network.

Ciplet, D., J. T. Roberts, and M. Khan. (2013) "The Politics of International Climate Adaptation Funding: Justice and Divisions in the Greenhouse," *Global Environmental Politics* 13(1): 49–68.

Goody, E. S. (2017) "Sharing Responsibility for Divesting from Fossil Fuels," *Environmental Values* 26: 693–710.

IPCC. (2007) *Climate Change 2007: Impacts, Adaptation and Vulnerability. Contributions of Working Group II to the Fourth Assessment Report of the Intergovernmental Panel on Climate Change*, edited by M. L. Parry, O. F. Canziani, J. P. Palutikof, P. J. van der Linden and C. E. Hanson. Cambridge: Cambridge University Press.

Jamieson, D. (2010) "Adaptation, Mitigation, and Justice," in S. M. Gardiner, S. Caney, D. Jamieson and H. Shue (eds.), *Climate Ethics: Essential Readings*, Oxford: Oxford University Press, pp. 263–283.

Kelbessa, W. (2015) "Climate Ethics and Policy in Africa," *Thought and Practice: A Journal of the Philosophical Association of Kenya* 7(2): 41–84.

Khan, M., R. Stacy-ann, R. Weikmans, D. Ciplet, and T. Roberts. (2020) "Correction to: Twenty-five Years of Adaptation Finance through a Climate Justice Lens," *Climatic Change* 161(2): 251–269.

Markowitz, E., M. Grasso, and D. Jamieson. (2015) "Climate Ethics at a Multidisciplinary Crossroads: Four Directions for Future Scholarship," *Climatic Change* 130(3): 465–474.

Page, E. A. (2008) "Distributing the Burdens of Climate Change," *Environmental Politics* 17(4): 556–575.

Shue, H. (1999) "Global Environment and International Inequality," *International Affairs* 75(3): 531–545.

Singer, P. (1972) "Famine, Affluence and Morality," *Philosophy & Public Affairs* 1(2): 229–243.

UK Government. (2021) "COP26: The Negotiations Explained." *COP26 Negotiations Explained*.docx (ukcop26.org).

United Nations. (1992) *United Nations Framework Convention on Climate Change* (UNFCCC). New York: United Nations, conveng.pdf (unfccc.int).

Young, I. M. (2011) *Responsibility for Justice*. New York: Oxford University Press.

12

CLIMATE JUSTICE IN NIGERIA

A Pathway to Defending the Rights of Disproportionately Affected Persons

Ngozi Unuigbe

Introduction

Climate change is a justice issue. Those who have contributed the least in the emission of greenhouse gases (GHGs) will be the most impacted by, and more vulnerable to, the impacts of climate change. Climate vulnerability is the degree to which geophysical, biological, and socio-economic systems are susceptible to, and unable to cope with, adverse impacts of climate change. Climate change is constantly undermining the enjoyment of human rights, especially of the people on the frontline of the climate crisis who have contributed least to the causes of climate change. These vulnerable persons comprise indigenous peoples, women, children, and people living with disabilities. Therefore, the impacts of climate change, and on the most vulnerable persons, have made the promotion of climate justice extremely urgent.

Climate justice is based on the basic principle of equal treatment and fairness in society. Equal treatment is a basic human right; equity and fairness are required for society to function well. The voice of vulnerable persons, which is the subject of this chapter, is a fundamental aspect of a just framework for climate change action. This is because climate justice builds on the platform of equitable development, human rights, and political voice for everyone. It is an agenda that seeks to redress global warming by reducing disparities in development and policies that lead to climate injustice and discrimination in society. This requires transformative change and the need to look beyond national boundaries to what is good for humanity.

In Nigeria, climate change has negatively impacted people's rights to health, housing, water, and food, among others. Vulnerable groups such as indigenous peoples, women, children, and people living with disabilities have been the worst hit. In addition to its direct impacts on the enjoyment of human rights, climate change has also brought about indirect impacts by diverting resources from crucial public services such as health, education, and agrarian support. Moreover, mitigation, adaptation, and geoengineering measures have adversely affected the exercise of human rights. There is also a high risk of human rights violations resulting from the implementation of resettlement programs for those who are displaced or at risk of displacement due to climate change.

Nigeria has taken several measures to tackle the challenges of climate change. It is a party to the United Nations Framework Convention on Climate Change (UNFCCC) and the Kyoto Protocol, and the Nigerian Federal Government aims to make Nigeria carbon neutral by 2025. In 1999, the Federal Ministry of Environment was established and included the Department of Climate Change, which coordinates national climate change efforts. Through this department, Nigeria

DOI: 10.4324/9781003039860-15

put into effect the National Policy on Climate Change and Response Strategy (NPCC-RS) and established the Science Committee on Climate Change to study the connection between scientific knowledge and climate policy. In 2011, the government also established a National Adaptation Strategy and Plan of Action on Climate Change for Nigeria (NASPA-CCN). Furthermore, it formulated strategies for each of the following priority sectors: agriculture (crops and livestock); freshwater resources, coastal water resources, and fisheries; forests; biodiversity, health, and sanitation; human settlements and housing; energy; transportation and communications; industry and commerce; disaster, migration, and security; livelihoods; vulnerable groups; and education.

While the Nigeria Policy on Climate Change is expected to be implemented soon, this chapter is an opportunity to highlight blueprints on how to make the policy actionable, especially in favor of vulnerable persons in the country. It emphasizes that well-conceived climate actions should reinforce social welfare systems and ensure support is directed to those persons and communities most vulnerable to the negative impacts of climate change.

The chapter is divided into three main sections, the impacts of climate change in Nigeria, the trajectory of climate change politics and policies in Nigeria, and the policy blueprints for effective implementation of the country's extant policy on climate change. The chapter concludes that it is imperative for climate change responses in Nigeria to transition from market-based and scientific approaches to people centered/human rights-based approaches.

Impacts of Climate Change on Vulnerable Groups in Nigeria

This section will examine the impact of climate change on vulnerable groups in Nigeria; namely, indigenous peoples, women, children, and people with disabilities.

Indigenous Peoples

The United Nations Declaration on the Rights of Indigenous Peoples (UNDRIP) and International Labor Organization Convention No. 169 articulate the individual and collective rights of indigenous peoples. UNDRIP affirms that "indigenous peoples are equal to all other peoples, while recognizing the right of all peoples to be different, to consider themselves different, and to be respected as such" (Article 2). It calls for States to "provide effective mechanisms for prevention of, and redress for" actions which would undermine indigenous peoples' rights including their rights to culture, land and resources, and self-determination (Articles 28 & 31). UNDRIP specifically recognizes indigenous peoples' rights to sustainable and equitable development and management of the environment (Preamble), to conserve and protect the environment (Article 29), and to mitigation of "adverse environmental, economic, social, cultural or spiritual impact" (Article 32). The Agreement at the 21st Conference of the Parties (COP 21) of the UNFCCC in Paris, December 12, 2015 (Paris Agreement) acknowledges the rights of indigenous peoples and calls for participatory adaptation that takes into consideration the traditional knowledge of indigenous peoples.

In Nigeria, climate change threatens these rights. Climate change is gravely harming the health of indigenous peoples' traditional lands and waters, as well as the plants and animals upon which they depend for survival. It endangers the delicate ecosystems upon which many indigenous peoples rely for housing, livelihoods, and spiritual and cultural practices. It also contributes to a loss of land and resources, discrimination, and unemployment (Akpodiogaga & Odjugo 2010; Idowu et al. 2011).

Indigenous peoples in Nigeria are the most vulnerable to climate change impacts because they heavily depend on the natural resources affected by climate change, such as trees, water, rivers, and mountains, for their day-to-day living. Despite these profound impacts, however, they are often excluded from decision-making processes on climate change. This exclusion of indigenous peoples in decisions that ultimately affect them, and more-so than any other affected group, is a breach of Nigeria's human rights obligations (Haider 2019). Some of these international legal instruments that Nigeria is party to include: the International Covenant on Civil and Political Rights (ICCPR) adopted and opened for signature, ratification and accession by General Assembly resolution 2200A (XXI) of 16 December 1966, entry into force 23 March 1976, in accordance with Article 49; the International Covenant on Economic, Social and Cultural Rights (ICESCR) adopted and opened for signature, ratification and accession by General Assembly resolution 2200A (XXI) of 16 December 1966, entry into force 3 January 1976, in accordance with article 27; the UN Declaration on the Rights of Persons belonging to National or Ethnic, Religious and Linguistic Minorities, adopted by General Assembly resolution 47/135 of 18 December 1992; the Indigenous and Tribal Population Convention 1957 (Convention 107) of the ILO; the Indigenous and Tribal Peoples Convention of 1989 (Convention 169) of the ILO; and UNDRIP 2007. On the regional level, we have the African Charter on Human and Peoples Rights 1981 (ACHPR).

Women

Gender equality and non-discrimination are fundamental human rights principles. The ICESCR, ICCPR, Convention on the Elimination of All Forms of Discrimination against Women (CEDAW), Beijing Declaration and Platform for Action, and other such instruments to which Nigeria is a party, affirm that Parties and other relevant actors must prioritize gender equality in all their actions and recognize that individuals who are part of certain groups— notably, women, indigenous groups, and children—are entitled to special protections.

The UNFCCC has also highlighted the obligation of Parties to fulfill the rights of women and promote gender equality. The UNFCCC has emphasized the need for gender mainstreaming through all relevant targets and goals in activities under the Convention as an important contribution to increase their effectiveness (2019). The 2010 Cancun Agreements include a detailed call for Parties to address the impacts of climate change on people who are vulnerable to climate change because of geography, gender, age, indigenous or minority status, and disability (Decision 1/CP.16). The agreements reached on December 11 in Cancun, Mexico, at the 2010 United Nations Climate Change Conference (Cancun Agreements) also recognize that gender equality and the effective participation of women are important for effective action on all aspects of climate change (Decision 36/CP.7).

In its Concluding Observations on State Party reports and in a number of General Recommendations, the CEDAW underscores States parties' and other stakeholders' obligations under the CEDAW to adopt targeted and country-specific policies, strategies, legislation, budgets, and other measures to address the gender dimensions of disasters and climate change (UNDP 2009). In 2009 at its 44th session, the Committee underlined that all stakeholders should ensure that climate change and disaster risk reduction measures are gender-responsive, sensitive to indigenous knowledge systems, and respect human rights. Women's right to participate at all levels of decision-making must be guaranteed in climate change policies and programs (CEDAW 2009).

In Nigeria, ensuring gender equity and non-discrimination is critical because climate change impacts, including droughts, extreme weather events, and flooding, affect women, girls, men, and boys differently. Statistics on casualties of climate-related hazards in Nigeria show that women and girls are among the most impacted. They not only provide a significant contribution to the agricultural workforce, but in some places, when men migrate, wives are left behind because they cannot abandon their children. In other places, women can walk up to 10 kilometers in search of water – time that could be used for caregiving and other income-generating activities

(Nnadi 2019). When combined with economic and social discrimination, climate change threatens women's right to education, information, water, food, health care, and freedom from violence. This, no doubt, magnifies gender inequities (Akpodiogaga & Odjugo 2017; Haider 2019).

The two most important resources which are most vulnerable to climate change are drinking water and fuel, and women are the ones who collect and utilize drinking water in most of the rural areas, as well as collect the resources for fuel required for cooking. In addition, women are more dependent on agriculture for food and basic requirements (Onwutuebe 2019). This is one of the reasons that women have more need for access to energy and its related services for housework in both rural and urban areas. In some areas, women are responsible for forest resources management and contribute extensively to farming or agriculture practices (Haider 2019; Onwutuebe 2019).

In the context of a changing climate, food sources are becoming less predictable, and production is decreasing, which also affects women disproportionately. In many places, women participate in crop harvesting as their only source of income, which makes them even more vulnerable. It is also reported that the rise in food prices impacts women severely, and particularly girls, because their health condition deteriorates more and faster than males in the times of food shortage. This only adds to the obligation of all nations, and particularly Nigeria, to include women in decision-making; it is critical to their lives and livelihoods, and necessary to achieve equity and justice in climate change policy-making (Shiru et al. 2018).

Children

Climate change poses an immediate and far-reaching threat to the enjoyment of many, if not all, of the rights enshrined in the Convention on the Rights of the Child, including those related to survival and wellbeing, health, food security and nutrition, water, and access to education. Article 24 of the Convention on the Rights of the Child directs States to account for the "dangers and risks of environmental pollution" to ensure full implementation of the right to health for children. It further outlines how States should ensure full implementation of the right. For example, States "shall take appropriate measures" to "combat disease and malnutrition, including within the framework of primary health care, though, inter alia, the application of readily available technology and through the provision of adequate nutritious foods and clean drinking water, taking into consideration the dangers and risks of environmental pollution." The Cancun Agreement further recognizes the need to "fully account" for the adverse effects of climate change on children (Decision 1/CP. 16).

Children are highly vulnerable to climate change and climate justice can effectively secure their rights. In the United Nations Children's Fund (UNICEF) report, children with poor nourishment are more at-risk to common air-borne infections and illness (2020). And according to the International Food Policy Research Institute, by 2050, there will be a 20 percent increase in malnourished children, even without considering the impacts of climate change (Nelson et al. 2009; Gillespie et al. 2013). This is why it is crucial to understand the higher mortality rate in children as related to climate change health impacts, as much or even more-so than in adults. The highest causes of death among children are malaria, diarrhea, and malnutrition, and climate change is increasing the risk of these diseases. And like other vulnerable groups, children are a large portion of the population, but are rarely prioritized in climate change policies and debates.

In Nigeria, the impacts of climate change on children need to be considered in relation to wider development pressures affecting the country. Challenges such as international economic shocks and stresses, high levels of poverty and inequality, population changes, effects of diseases like lassa fever, yellow fever, HIV and AIDS, and recently, the Coronavirus disease of 2019 (COVID 19), the management of scarce natural resources, and rapid urbanization will each interact with climate change in different ways, and on different groups. The results of those interactions will affect how far the effects of climate change are transmitted to children and households at the local level (Ilevbare 2019). Furthermore, the role of climate change in exacerbating and compounding drivers of insecurity is causing more children to be at risk of violence, exploitation, and abuse, in both conflict and non-conflict settings (Amanchukwu 2015; Duru & Emetumah 2016).

Persons with Disabilities

The Convention on the Rights of Persons with Disabilities (CRPD) calls for an end to harmful stereotypes and to discriminatory policies and practices against persons with disabilities (A/RES/61/106). Further, it recognizes the rights of disabled persons to an adequate standard of living, social protection, gender equality, health, and education, among others. Other instruments further guarantee the rights of persons with disabilities, including the right to non-discrimination, including the Convention on the Rights of the Child, the CEDAW, and the ICESCR. The preamble to the Paris Agreement also calls for states to respect, promote, and consider their human rights obligations to all people, including persons with disabilities, when taking climate action.

Climate change in Nigeria poses unique challenges to persons with disabilities. They lack equitable access to employment and education and face barriers to participation in decision-making processes. These situations mean that persons with disabilities have inadequate access to the resources, information, and services necessary to adapt to the impacts of climate change.

Many persons living with disabilities live in rural areas with limited access to education, employment, health care, and social services, and they face additional challenges as related to disaster preparedness and resilience. Urban climate change-related risks are also increasing, and this results in widespread negative impacts on people and their health, livelihoods, and assets (Ezegwu 2014). For instance, climate change exacerbates extreme weather events, which is one of the factors driving increased migration in recent years. Because the ability to migrate often depends on resources and mobility, marginalized populations – such as people with disabilities – might be unable to travel and so forced to remain in degraded environments without housing, employment, support networks, or health care services. People with disabilities also experience poverty at more than twice the rate of people without disabilities. This puts people with disabilities at heightened risk, as the world's poorest people continue to experience the most severe impacts of climate change through lost income, displacement, hunger, and adverse impacts on health.

The situation for women and girls with disabilities is even more difficult as they have less access to education, employment, and resources, and face greater barriers to participation in decision-making than both women generally and men with disabilities. These situations often mean that women and girls with disabilities face inadequate access to information, including early warning and evacuation measures, the inability to access emergency response food, supplies, or other benefits, and greater risk of violence after disasters (Onwutuebe 2019).

Climate Change Politics and Policies to Protect Vulnerable Groups in Nigeria

Nigeria joined global efforts toward mitigating climate change by becoming a signatory to the UNFCCC in 1992 and the ratification of the Kyoto Protocol of 1997. As a Non-Annex 1 signatory, Nigeria is not required to take any actions to curtail its emission of GHGs. However, Nigeria is obliged to issue four key National Communications: four in-depth review summaries; a progress report; a National Adaptation Program of Action; and a Global Climate Observing System Report. The Federal Ministry of Environment was then given the responsibility for the implementation of the UNFCCC and the Kyoto Protocol in the country. A Special Climate Change Unit (SCCU) was then created within the Ministry to develop a short to long-term national plan to enable Nigeria to respond to its obligations as specified by the UNFCCC and the Kyoto Protocol, and also to coordinate the activities of the Inter-Ministerial Committee on Climate Change (IMCCC) (Federal Republic of Nigeria [FRN] 2008).

In Nigeria, climate change policies are part of the political framework conditions and an important outcome of the political process of the country. Sustainable climate change policies are therefore prerequisites for a strong adaptive capacity as they will go a long way in reducing climate change impacts and vulnerability in the country. Nigeria's response to climate change threats in the context of policy development framework, however, remains a major challenge. There are a number of existing policies that could be adapted and implemented in order to reduce climate change impacts and vulnerability in the country (Ati et al. 2018).

In 1989, the Nigerian government elaborated a National Policy on the Environment in order to meet the challenges of addressing key environmental problems and land degradation (deforestation, desertification, and coastal and marine environment erosion), air and water pollution, urban decay, and municipal waste, as well as hazards of drought, coastal surges, floods, and erosion. The policy was revised in 1999 to accommodate new and emerging environmental concerns. The goal of the revised policy was to achieve sustainable development in Nigeria in order to: (1) secure a quality of environment adequate for good health and well-being; (2) promote the sustainable use of natural resources; (3) restore and maintain the ecosystem and ecological processes and preserve biodiversity; (4) raise public awareness and promote understanding of linkages between environment and development; and (5) cooperate with government bodies and other countries and international organizations on environmental matters (Federal Republic of Nigeria (FRN) 2010).

The country has also enacted a number of specific policies and action plans for the implementation of the National Environment Policy. These policies, which can support national climate change mitigation and adaptation response efforts, include:

- The National Policy on Drought and Desertification: This policy recognizes that climate change could intensify drought and desertification, especially in the northern part of the country that is very prone to these environmental problems. The policy emphasized the need to equip relevant agencies, institutions, and citizens adequately to collect, analyze, and use climate data effectively to ameliorate and combat drought and desertification (FRN 2010);
- The National Forest Policy: This policy seeks to ensure sustainable forest management, promoting participatory process of development, facilitating private sector-forestry development, and adopting an integrated approach to forestry development. The government is currently embarking on several afforestation programs. These include the "Green Wall Initiative", in which a 'green wall' of trees will be planted across the Sudano-Sahelian zone of the country (National Forest Policy 2006);
- National Biodiversity Strategy and Action Plan: This policy aims to develop appropriate framework and program instruments for the conservation of Nigeria's biological diversity and enhance its sustainable use by integrating biodiversity considerations into national planning, policy, and decision-making processes (National Biodiversity Strategy and Action Plan 2016–2020);
- National Erosion and Flood Control Policy: This policy aims to ensure coordinated and systematic measures in the management and control of the climate-related hazards and risks of erosion and floods to reduce their impacts on the people and the environment (National Environmental Soil Erosion and Flood Control Regulations 2011); and
- Other policies related to agriculture, water, coastal resources, and energy. For example, the National Policy on Drought and Desertification 2000; Nation Policy Guidelines on: solid waste management; market and abattoir sanitation; excreta and sewage management; and pest and vector control; and the National Environmental Standards and Regulations Enforcement Agency (NESREA) Act 2007.

In addition to the National Policy on the Environment discussed above, Nigeria has many laws and regulatory measures to promote sustainable environmental management in many sectors of the economy. In 2011, as part of a project titled, Building Nigeria's Response to Climate Change (BNRCC), the National Adaptation Strategy and Plan of Action on Climate Change for Nigeria (NASPA-CCN) was prepared (BNRCC 2011). This document also recommended policies, programs, and measures at the federal, state, and local Government. The recommendations, when acted upon, will increase the resilience of local communities, and of Nigeria as a whole, to the impacts of climate change. These recommended policies, programs, and measures, and their goals, include:

- Agriculture (Crops and Livestock): The goal is to ensure that vulnerable communities and groups alter their agricultural practices to adapt to the changing climate, including predicted temperature and rainfall changes and extreme weather events;
- Freshwater and Coastal Water Resources and Fisheries: The goals are to implement Integrated Water Resource Management (IWRM) in watersheds and coastal regions to reduce the impacts of climate change; plan for and adapt to current and expected impacts of climate change on water resources, coastal resources, and inland and coastal fisheries by addressing water supply, water demand and water management infrastructure; implement strategies for promoting water body enhancement for inland and coastal fisheries; and plan for and adapt to the expected impacts of sea level rise and storm surge on coastal resources and fisheries. The goals of forests are to maintain and restore healthy forest ecosystems through sustainable forest management in order to increase the resilience of human communities in the face of changing climate; improve the understanding of the impact of climate change on the abundance and composition of "closed" forest resources; and increase the extent and diversity of forest cover in Savanna and Sahel in order to address increased aridity caused by higher temperature and greater rainfall variability;
- Biodiversity: To intensify efforts to conserve biodiversity in Nigeria, in order to reduce the threatened impact of climate change on biodiversity;
- Health and Sanitation: The goals of health and sanitation are to strengthen and improve public health, disease prevention, and environmental sanitation under the framework of the Primary Health Care Program; reduce human exposure to climate change-related health risks; and better understand the implications of increasing temperature for human and animal health;
- Human Settlements and Housing: The goal is to develop housing and settlement patterns/practices that enhance climate change adaptation and are resilient to climate change. Energy: To take pre-emptive measures to reduce vulnerability of critical energy infrastructure to climate change impacts;
- Transportation and Communications: To take pre-emptive measures to reduce vulnerability of critical transportation and communication infrastructure to climate change impacts;
- Industry and Commerce: The goals are to protect the Nigerian industrial & commercial sectors from the adverse effects of climate change; explore new markets and opportunities arising from climate change; and enhance Nigeria's competitiveness in this era of global green economy;
- Disaster, Migration, and Security: To strengthen the capacity of the governments and people of Nigeria to carry out disaster risk and vulnerability assessments; prepare for and respond to climate change hazards; anticipate and prepare for climate change induced migrations and ensure security of lives as well as public and private property in the face of climate change impacts;
- Livelihoods: To reduce the vulnerability of people and increase their adaptive capacity to climate change impacts through improved livelihoods development. Vulnerable Groups: Goals

are to develop programs that support and assist the adaptation to climate change by vulnerable groups; and harness opportunities arising from differentiated roles, responsibilities, representation, and experiences of vulnerable groups for climate change adaptation; and

- Education: Goals are to develop a good public understanding of the impacts of climate change and the need for people of all ages to learn about and contribute to environmental protection in general and climate change adaptation in particular; reduce exposure of school children to the impacts of climate change, including extreme weather events; and increase the resilience of educational infrastructure to the impacts of climate change.

The above strategies are profound and carry the potential to effectively protect vulnerable groups in Nigeria from the impacts of climate change. Therefore, to promote climate justice in the country, the Strategic Framework for Voluntary Nationally Appropriate Mitigation Action (NAMA) and National Climate Change Policy for Nigeria, which are yet to be implemented, must expressly protect vulnerable groups. This recommendation includes incorporating a right to climate justice in the Nigerian constitution, as this is the basic guarantee for the defenseless – all other laws are based on the spirit of this fundamental law of the country.

Critically, promoting climate justice in Nigeria will require a multi-national, multi-sectorial, and multi-dimensional approach. The National Policy on Climate Change has been signed by President in 2008. However, the Bill to Establish the National Climate Change Commission – The Bill to an Act to Establish the National Climate Change Commission and for Other Matters Connected Therewith – has only been passed by the National Assembly and is yet to be signed by the President. Policy statements have been written and made, but the political will to advance proper financing and implementation of the various policies is critical for effective response to the climate change challenge, especially for its impacts on vulnerable groups. Clearly, the NASPA-CCN has protective provisions for vulnerable groups in the country. However, it is one thing to have the policies on paper, and yet another to operationalize them effectively. Consequently, the next section will examine how the existing policies can be deployed to effectively protect the interests of vulnerable people in the country.

Policy Directions for Effective Climate Justice in Nigeria

As discussed above, vulnerable groups in Nigeria have experienced compounded impacts of climate change through levels of discrimination. Understanding the impacts of climate change on these persons and groups, learning how to better engage them as active agents of their own rights, and making targeted recommendations therein is critical to fulfilling human rights obligations nationally and internationally, and to further informing and enacting a global climate action plan.

An essential step in the effective operationalization of policies on climate justice is to address the research gap in promoting climate justice. Currently, there is a large gap in understanding the intangible and long-term effects of natural disasters or climate-related events on people's behavior and well-being in Nigeria, including mental health and trauma.. For example, a person's mental health is negatively affected due to the stress of a flooding event. Adding to this, the long-term flood impacts at the neighborhood or community level are poorly documented. As such, it is difficult to determine the effect on human development and the mental state of groups in a region after a flooding happens.

The identification of vulnerable groups is also very critical. It is not sufficient to map vulnerability or identify vulnerable households and individuals. Steps should be taken to provide a rapid and targeted list of vulnerable people, communities, and groups with cost-effective measures. This would be a very useful tool in order to have a long-term strategy and emergency response to any situation and better prepare the country.

Also, climate-related impacts on business-related activities, such as from flood damage, are largely under-explored. Studies should reveal the impacts on the productivity of local businesses, the impacts on food production, the social vulnerabilities in relation to heat and flood, and the ability of communities to cope with potential impacts and what would be the response.

Information, in style and approach, can also be a creative and valuable tool in reducing the risk to vulnerable people from climate change harms. Social networks can both increase and decrease the resilience of vulnerable people, and the use of mobile technology (even in rural areas), interactive games, and software are all important ways to reach people's pockets, educate them, and provide them with accurate and timely information, such as vulnerability maps, hazard maps, and at times locally available maps. The use of Information and Communications Technology (ICT) can also be very helpful, but needs more research to identify the key impacts and results.

It is important to now offer specific policy recommendations for each vulnerable group:

Indigenous Peoples

Human rights law requires that indigenous peoples be meaningful participants in their own development, including climate dialogues and actions. Not only are indigenous peoples particularly harmed due to climate change, but indigenous peoples often respond to climate change in innovative and effective ways, which can help the global community mitigate and adapt to climate change. The IPCC has found that "indigenous, local and traditional knowledge systems and practices, including indigenous peoples' holistic view of community and environment, are a major resource for adapting to climate change" (IPCC 2019). By empowering indigenous peoples and guaranteeing them control over their traditional knowledge, lands, territories, and resources, as called for in UNDRIP and other international and regional instruments, states can simultaneously improve climate mitigation and adaptation efforts and the situation of indigenous peoples.

Women

Realizing women's human rights implies recognizing and addressing the underlying and immediate causes of women's human rights violations, challenging structural constraints to the equal rights and choices of women and girls, and putting in place appropriate policy and programmatic responses in line with human rights principles. A human rights-based approach calls for the participation of marginalized, disempowered, and discriminated against groups of women in decisions that affect their livelihoods and overall sustainable development. Climate action at all levels must integrate a strong human-rights based approach, taking into account not only women's vulnerability to climate change, but their crucial capability as leaders in disaster preparedness and management.

Women and girls are not passive victims of climate change impacts. They are agents, actors, and leaders in crafting climate responses. Women already play key roles in the protection and management of natural resources and disaster response and are at the forefront of actions aimed at reducing human contributions to global warming. Indigenous women, for example, are stewards in knowledge and the use of natural resources and may already serve as community leaders. Climate change mitigation and adaptation should, therefore, not only be gender responsive, but also inclusive of women's rights to full and active participation at all levels of decision-making.

Children

While the Paris Agreement represents a significant advance in terms of formal recognition of children's rights in the framework of global climate action, consideration of the Convention on the Rights of the Child remains largely absent from climate-related policies, action, investments, and dialogue. There is an urgent need for the best interests of the child to be systematically applied in shaping national and international responses to climate change. Children's right to access information and to be heard in decision-making should be upheld, as well as their access to effective and timely remedies. This would also entail the provision of tailored information and mechanisms that are child-friendly and language-appropriate.

Persons with Disabilities

Persons with disabilities are active participants in society and agents of their own rights. In order to protect the rights of persons with disabilities against the harmful impacts of climate change, more must be done to mainstream their rights in climate policies and actions and to ensure their meaningful participation. It is critical that this group be included in climate action. The rights of people with disabilities should be taken into consideration when developing climate policies in order to secure their meaningful, informed, and effective participation during the process.

One important step is to ensure that information about climate risks, plans, and policies are made accessible to everyone – this is in line with the report published by the Office of the High Commissioner for Human Rights (OHCHR) on the Promotion and protection of all human rights, civil, political, economic, social, and cultural rights, including the right to development. Particularly, at the panel discussion on promoting and protecting the rights of persons with disabilities within the context of climate change, it was resolved that all persons, including persons with disabilities, must be recognized as rights holders and barriers to their inclusion, such as discrimination, must be eliminated. An intersectional approach that considered the specific requirements of persons with disabilities should include accountability and data collection disaggregated by disability and other layers of identity. It required the active participation of and consultation with persons with disabilities in the planning of climate action, review processes and decision-making. Persons with disabilities must be supported and empowered to fully participate as agents of change in climate change-related prevention and response processes. The accessibility of venues, information, and communications, as well as the implementation of accessibility standards, in post-emergency planning and reconstruction efforts must also be ensured (HRC 2020).

Conclusion

Climate justice requires that climate action is consistent with existing human rights agreements, obligations, standards, and principles. Unfortunately, it is often the case that those who have contributed the least to climate change, including, women, children, indigenous peoples and peoples living with disabilities, unjustly and disproportionately suffer its harms. Climate-related hazards magnify gender inequalities and widen existing socio-economic and political gaps. Despite the challenges faced, many persons and groups in vulnerable situations play key roles in their households and communities, in the protection and management of natural resources, and at the forefront of actions aimed at reducing human contributions to global warming. For these reasons, these individuals and groups must be meaningful participants in, and primary beneficiaries of, climate action, and when their interests are harmed, they must have access to effective remedies.

Climate justice was embedded in international legal frameworks of fairness and equity. It was included with the common and differentiated responsibility principles of the UNFCCC and in the provisions of the Kyoto Protocol. World development models, laws, and governance are built around the notions of fair justice without discrimination and must now include this objective of fairness in the well-being of people and ecological systems as related to climate justice. There is an important need of evening out the imbalances that are not sustainable in nature and the urgent need for the development of tailored policy responses for groups which are the most vulnerable to the impacts of climate change.

Given Nigeria's status as a fossil-fuel dependent economy with a large climate sensitive agricultural sector, the operationalizing of its current climate change policy and response strategy is critical. Climate justice is just one potential, yet crucial, response to the continued impacts of climate change on vulnerable persons in Nigeria, and it is a response that benefits all of Nigeria. The recommendations included herein bridge the gaps in the theory and application of climate justice, in which the impacts of climate change and social vulnerability must be addressed simultaneously and provide a strong tool for vulnerable groups in Nigeria. Climate justice recognizes the shared responsibility for all to understand the importance of the environment, and that human well-being is directly proportional to the environment.

References

Akpodiogaga, P. & Odjugo, O. (2010). "General Overview of Climate Change Impacts in Nigeria." *Journal of Human Ecology* 29(1): 47–55.

Amanchukwu, R. N. et al. (2015). "Climate Change Education in Nigeria: The Role of Curriculum Review." *Education* 5(3): 71–79.

Ati, O., Agubamah, E. & Abaje, I. (2018). "Global Climate Change Policies and Politics: Nigeria's Response." *Fudma Journal of Politics and International Affairs* 1(1): 106–120.

BNRCC (Building Nigeria's Response to Climate Change). (2011). *National Adaptation Strategy and Plan of Action on Climate Change for Nigeria (NASPA-CCN).* Prepared for the Federal Ministry of Environment Special Climate Change Unit. Available from: http://csdevnet.org/wpcontent/uploads/NATIONAL-ADAPTATION-STRATEGY-AND-PLAN-OF-ACTION.pdf

CEDAW. (2009). *Initial Report on the Convention on the Elimination of All Forms of Discrimination against Women, Timor Leste.* New York: CEDAW.

Duru, N. P. & Emetumah, C. F. (2016). "Evaluating the Effects of Information Literacy on Climate Change Awareness Among Students in Imo State University." *Archives of Current Research International* 4(3): 1–10.

Ezegwu, C. (2014). *Climate Change in Nigeria: The Impacts and Adaptation Strategies.* Available at SSRN: https://ssrn.com/abstract=2543940 or https://doi.org/10.2139/ssrn.2543940

Federal Republic of Nigeria [FRN]. (2008). *Nigeria and Climate Change: Road to Cop15: Achieving the best outcome for Nigeria.* Abuja: Federal Ministry of Environment.

Federal Republic of Nigeria [FRN]. (2010). *National Environmental, Economic and Development Study (NEEDS) for Climate Change in Nigeria.* Federal Ministry of Environment (Special Climate Change Unit). Abuja: Federal Ministry of Environment.

Gillespie, S., Haddad, L., Mannar, V., Menon, P., Nisbett, N. & the Maternal and Child Nutrition Study Group (2013). "The Politics of Reducing Malnutrition: Building Commitment and Accelerating Progress." *The Lancet* 382(9891): 552–569.

Haider, H. (2019). *Climate Change in Nigeria: Impacts and Responses.* K4D Helpdesk Report 675. Brighton: Institute of Development Studies, 1–38.

Human Rights Council Forty-Sixth Session 22 February–19 March 2021, (A/HRC/46/46). New York: United Nations General Assembly.

Idowu, A., Ayoola, S., Opele, A. & Ikenweiwe, B. (2011). "The Impact of Climate Change in Nigeria." *Iranica Journal of Energy & Environment* 2: 145–152.

Ilevbare, F. (September 27th, 2019). *Investigating Effects of Climate Change on Health Risks in Nigeria, Environmental Factors Affecting Human Health.* Available from: https://www.intechopen.com/books/environmental-factors-affecting-human-health/investigating-effects-of-climate-change-on-health-risks-in-nigeria

IPCC. (2001). *Synthesis Report 2001. Contribution of Working Group I, II, and III to the Third Assessment Report of the Intergovernmental Panel on Climate Change.* Cambridge: Cambridge University Press.

IPCC. (2014). *Contribution of Working Group III to The fifth Assessment Report of the Intergovernmental Panel on Climate Change.* Cambridge: Cambridge University Press.

IPCC. (2019). *Special Report on Climate Change and Land.* Cambridge: Cambridge University Press.

National Biodiversity Strategy and Action Plan 2016–2020. Abuja: Federal Ministry of Environment.

National Environmental (Soil Erosion and Flood Control Regulations. (2011). Abuja: Federal Ministry of Environment.

National Forest Policy of Nigeria. (2006). Abuja: Federal Ministry of Environment.

Nelson, G. C., Rosegrant, M. W., Koo, J., Robertson, R., Sulser, T., Zhu, T., Ringler, C., Msangi, S., Palazzo, A., Batka, M., Magalhaes, M., Valmonte-Santos, R., Ewing, M. & Lee, D. (2009). *Climate Change: Impact on Agriculture and Costs of Adaptation. Food Policy Report.* Washington, DC: International Food Policy Research Institute.

Nnadi, O. I. et al. (2019). "Impacts of Variability and Change in Rainfall on Gender of Farmers in Anambra, Southeast Nigeria." *Heliyon* 5(7): 1–14.

Onwutuebe, C. J. (2019). "Patriarchy and Women Vulnerability to Adverse Climate Change in Nigeria." *Sage Open* 9(1): 1–7.

Shiru, M. S., et al. (2018). "Trend Analysis of Droughts during Crop Growing Seasons of Nigeria." *Sustainability* 10(3): 871.

UNDP. (2008). *Final Report of the Gender and Climate Change Finance Workshop, November 2008.* New York: UNDP.

UNICEF Annual Report. (2019). New York: UNICEF.

13

THE RESPONSIBILITY OF NATIONS TO ADDRESS CLIMATE REFUGEES AND DISPLACED PERSONS

Donald A. Brown

Climate Change Impacts Are Already Creating Millions of Refugees and Displaced Persons and Are Threatening to Cause Many Millions More

The chapter begins by describing climate change harms that are already creating millions of climate change refugees and threaten to cause hundreds of millions more. The chapter will use the term 'refugees' to apply to all climate induced displaced persons although under international law the term 'refugee' does not connote all displaced persons, but only those who flee their nation because of fear of persecution or violence. A 'refugee' is defined as a person who has crossed an international border "owing to well-founded fear of being persecuted for reasons of race, religion, nationality, membership of a particular social group or political opinion" (Convention on the Status of Refugees 1951: Art. 1). In some contexts, the definition extends to persons fleeing "events seriously disturbing public order." All climate change refugees are not covered by the 1951 Convention Relating to the Status of Refugees, which protects people who have a well-founded fear of persecution on racial, religious, or other grounds. Therefore, all people displaced by climate change impacts cannot apply to be treated as a refugee by countries they approach, under current law. Climate change affects people inside a nation's own countries, and typically creates internal displacement before it reaches a level where it displaces people across borders. However, there may be situations where the refugee criteria of the 1951 Convention or the broader refugee criteria of regional refugee law frameworks could apply. People may have a valid claim for refugee status, for example, where the adverse effects of climate change interact with armed conflict and violence.

Climate change impacts have created and will continue to adversely affect refugees around the world. These harms include sea level rise, glacial melting, changes in precipitation which causes flooding and drought, famine caused by drought, storm damage from extreme weather events, unbearable heat waves, wildfires, increases in the intensity of tropical storms, increases in degradation of ecological systems including plants and animals on which people and animals depend, the melting of ice masses which affect water supplies, and the degradation of carbon sinks including forests, plants, and the acidification of oceans.

A 2022 report of the Intergovernmental Panel on Climate Change found human-induced climate change has caused more frequent and intense climate extreme events, and thus caused widespread adverse impacts (IPCC 2022) and related losses and damages to nature and people, beyond natural climate variability. Vulnerability of ecosystems and people to climate change differs substantially among and within regions (very high confidence), driven by patterns of intersecting socio-economic development, unsustainable ocean and land use, inequity, marginalization,

DOI: 10.4324/9781003039860-16

historical and ongoing patterns of inequity such as colonialism, and governance (high confidence). Approximately 3.3–3.6 billion people live in contexts that are highly vulnerable to climate change (high confidence). Human and ecosystem vulnerability are interdependent (high confidence). As global temperatures continue to rise, hundreds of millions of people could struggle with floods, deadly heat waves and water scarcity from severe drought, the report said. Mosquitoes carrying diseases like dengue and malaria will spread to new parts of the globe. Crop failures could become more widespread, putting families in places like Africa and Asia at far greater risk of hunger and malnutrition. People unable to adapt to the enormous environmental shifts will end up suffering unavoidable loss or fleeing their homes, creating dislocation on a global scale (IPCC 2022).

Poor nations are far more exposed to climate risks than rich countries. Between 2010 and 2020, droughts, floods and storms killed 15 times as many people in highly vulnerable countries, including those in Africa and Asia, as in the wealthiest countries:

> The number of those forcibly displaced due to climate disasters, conflicts, and violence – both within their countries and internationally – has continued to climb, from 82.4 million at the end of 2020 to more than 84 million by June 2021.
>
> *(UNHCR 2021)*

While serving as Program Manager for UN Organizations at US EPA during the US Clinton Administration, I was invited in 1997 to participate in war games being conducted by the US Army War College about parts of the world that could raise national security threats triggered by social disruption from climate change caused destabilization. One region the Army War College identified during these war games as being a potential source of global disruption was the drought-prone regions of Syria. In 2007, a climate change induced drought began in Syria, which lasted past 2010 and created 1.3 million refugees who eventually destabilized large parts of Europe (Kelley et al. 2015).

In recent years, tens of thousands of people have fled Central American countries because of drought and extreme weather conditions that have made it difficult to grow crops. Yet when refugees from these countries have fled to the US-Mexican border, the US press almost always has failed to connect this crisis to climate change, nor the significant role of the US in creating the climate change impacts which have been making agriculture difficult or impossible as the largest national emitter in historical GHG emissions.

Some parts of the world are particularly vulnerable to climate impacts such as sea level rise, exacerbated by increasingly severe storms and even the economic fallout from the COVID-19 pandemic. From the South Pacific to the Caribbean, many small island developing states are also particularly vulnerable to rising seas. Along with loss of homes, roads, and other infrastructure, rising seas pose serious legal implications for small island states, including imperiling their territorial rights.

NOAA recently issued an ominous new prediction of expected sea level rise which should increase concern about sea level caused migrants even if the ice sheet collapse concerns do not happen. This report prophesized one foot of sea level rise by 2050 and two feet by 2100 (NOAA 2022).

In 2008, the Army War College published a 466-page report which extensively described parts of the world that might create national security threats for the world due to their vulnerability to climate change impacts and their potential to cause conflict (Pumphrey 2008). This report, after reviewing and evaluating the conclusions of the mainstream scientific community, including the Intergovernmental Panel on Climate Change, describes why expected impacts of climate change would create national security threats to governments around the world (Pumphrey 2008). Among other climate change impacts on the global environment, the 2008 report

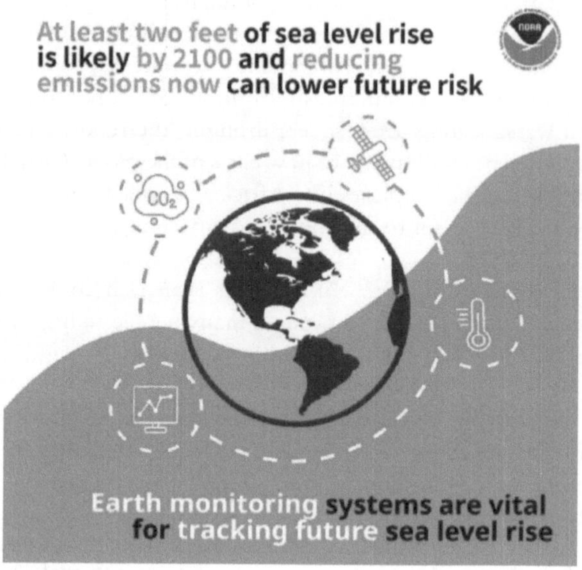

Image 13.1 NOAA 2022

described how climate change impacts on water resources, infectious diseases, and other causes of population movements would likely lead to destabilization and conflict in parts of the world. The report also described how climate change impacts will likely create national security threats to the United States of various types. The report claimed that several parts of the world that have a proclivity to produce violence against the United States are likely to become more violent as they struggle with dwindling natural resources such as water.

Throughout the report, the Army War College quotes and treats some of the direst predictions of IPCC and the mainstream scientific community as if they are entitled to scientific support. For instance, the report expressly says, "the changes predicted to come, the heat waves and droughts and floods, rising sea level and stress to countless species are already being observed" (Pumphrey 2008: 41). And so, several years before members of the Trump administration and some members of the US Congress were claiming that human-induced climate change is a 'hoax,' the 2008 Army War College Report claimed that the predictions of the mainstream scientific community on climate change impacts were entitled to scientific respect. And now, 22 years later, many of the Army War College's predicted harms have come into being.

In October 2021, the White House issued a report on climate change and migration (US White House 2021). This report concluded that the climate crisis is reshaping our world, as the Earth's climate is now changing faster than at any point in the history of modern civilization as defined by changes in average weather conditions that persisted over multiple decades or longer. These include changes in temperature, precipitation patterns, the frequency and severity of certain weather events, and other features of the climate system. When combined with physical, social, economic, and/or environmental vulnerabilities, climate change can undermine food, water, and economic security. Secondary effects of climate change can include displacement, loss of livelihoods, weakened governments, and in some cases create political instability and conflict (US White House 2021: 4)

The United Nations High Commissioner for Refugees (UNHCR) reported that an average of 21.5 million people were forcibly displaced each year by sudden onset weather-related hazards between 2008 and 2016, and thousands more from slow-onset hazards linked to climate change impacts. Hazards resulting from the increasing intensity and frequency of extreme weather events, such as abnormally heavy rainfall, prolonged droughts, desertification, environmental degradation, sea level rise, and cyclones are already causing an average of more than 20 million people to leave their homes and move to other areas in their countries each year (Rigaud 2018).

The Current Major Policy Preoccupation of Nations to Achieve the Paris Agreement's Warming Limit Goals Is Appropriate but Takes Attention Away from Policies Relating to Climate Refugees

The major climate policy preoccupation of the international community has been limiting warming to the 1.5°C–2.0°C warming limit goals established in the Paris Agreement. This is understandable because warming in excess of these amounts can trigger catastrophic warming capable of leading to mass extinctions such as those that have been experienced at least five times in Earth's history.

The recent international focus on policies needed to limit warming to no more than 2.0°C is also warranted because the international community is running out of time to prevent catastrophic warming that becomes much more likely if global temperature rise is more than 2.0°C above pre-industrial levels. The parties to the 2021 Glasgow COP climate negotiations under the UNFCCC explicitly reaffirmed the Paris Agreement temperature goal of holding the increase in the global average temperature to well below 2°C above pre-industrial levels and pursuing efforts to limit the temperature increase to 1.5°C above pre-industrial levels (Glasgow Climate Pact: para. 20).

The current CO_2 atmospheric concentration of approximately 420 ppm is lower than levels that existed during times which GHG triggered mass extinctions. Although current CO_2 atmospheric concentrations are lower than levels that triggered catastrophic warming in Earth's history, there are tipping points in the Earth's system that if destabilized could dangerously accelerate the rise in atmospheric CO_2 concentrations by destabilizing other tipping points and accelerate warming.

The destabilization of some tipping points may potentially lead to ominous consequences caused by destabilization of other tipping points creating a domino effect until atmosphere concentrations are at levels that previously caused mass extinctions (Leahy 2019). Several of the following tipping points are showing early signs of destabilization making needed reductions extremely urgent. Tipping points of concern include the Greenland and West Antarctic Ice Sheets, the Atlantic Meridional Overturning Circulation (AMOC) system, the El Niño-Southern Oscillation (ENSO) System, the Amazon rainforest, and melting permafrost, each briefly explained below.

The AMOC is one of the main global ocean currents and is critical to regulating climate. Cold salty water, which is dense and heavy, sinks deep into the ocean in the North Atlantic, and moves along the bottom until it rises to the surface near the equator, usually in the Pacific and Indian Oceans. But as glaciers and ice sheets melt, they add fresh, less dense water to the North Atlantic, which prevents the water from sinking and impedes circulation. This may be why AMOC has slowed 15 percent since the 1950s. A recent study found that the AMOC is in its weakest state in 1,000 years. Moreover, the latest climate models project that continued global warming could weaken the AMOC by 34 percent to 45 percent by 2100 (Caesar et al. 2021).

El Niño and La Niña are the warm and cool, naturally occurring weather patterns across the tropical Pacific—the El Niño-Southern Oscillation, or ENSO. Every two to seven years, the pattern alternates, bringing disruptions in temperature and precipitation. El Niño causes impacts around the world, such as more drought in India, Indonesia and Brazil, and flooding in Peru. As the ocean warms, it could push ENSO past a tipping point, which would make El Niño events more severe and frequent and could increase drought in the Amazon.

The Amazon rain forest, the world's largest tropical rain forest, stores 200 billion tons of carbon – equal to about five years of global carbon emissions from the burning of fossil fuels – and is home to millions of species of plants and wildlife (Lovejoy & Nobre 2018). If 20–25 percent of the Amazon were deforested, its tipping point could be crossed, according to one study (Lovejoy &

Nobre 2018; see also Guston 2022). Fewer trees would mean less evapotranspiration, and without enough rainfall to sustain itself, the Amazon could start to die back. In other words, parts of the rainforest could transition into a savanna, a drier ecosystem characterized by grasslands and a few trees. In the process, it would potentially release 90 gigatons of CO_2, exacerbating climate change. Crossing this tipping point would also result in the loss of biodiversity and ecosystem services, affect global weather patterns, and threaten the lives of 30 million people, including many indigenous people, who depend on the rainforest to survive. One study found that dieback would occur if we reached 3°C of warming. Recent studies show worrisome signs of Amazon degradation (Lovejoy & Nobre 2018).

The Amazon is already feeling the effects of climate change, as over the last century, temperatures in the region have increased 1°C–1.5°C. The Amazon is experiencing longer and hotter dry seasons that make it more vulnerable to wildfires, reduced evapotranspiration in response to higher levels of CO_2, and there are now more drought-tolerant tree species.

Scientists are unsure whether the Amazon has a single overall tipping point, or when exactly it might be reached, and the ecosystem has some ability to adapt to changing conditions. But fires and drought could cause local changes that spread drying conditions to other regions because of an overall reduction of moisture. Twenty-eight percent of the eastern part of the Amazon is already losing more carbon than it is absorbing due to deforestation. And some climate models predict that by 2035, the Amazon will be a permanent source of carbon (Lovejoy & Nobre 2018).

Permafrost is ground that remains frozen for two or more consecutive years and is composed of rock, soil, sediments, and ice. Some permafrost has been frozen for tens or hundreds of thousands of years. It is found in northern hemisphere lands without glaciers, including parts of Siberia, Alaska, northern Canada, and Tibet. In the Southern Hemisphere, there is permafrost in parts of Patagonia, Antarctica, and the Southern Alps of New Zealand (Resnick 2019). Fourteen hundred billion tons of carbon are thought to be frozen in the Arctic's permafrost, which is twice as much carbon as is currently in the atmosphere. But the Arctic is warming two times faster than the rest of the planet – it has already warmed 2°C above pre-industrial levels. As it warms and thaws the permafrost, microbes come out of hibernation and break down the organic carbon in the soil, releasing CO_2 and methane, which then trigger even more warming and melting. For a discussion of the significance of permafrost melting, see 2019 Arctic Report Card from NOAA found that the Arctic's thawing permafrost could be releasing 300–600 million tons of carbon per year into the atmosphere (NOAA 2019).

Methane stored in ice-like formations called hydrates are also found in permafrost in ocean sediments. This methane may be released as hydrates are thawed by warming seawater. Scientists recently discovered methane leaking from a giant ancient reservoir of methane below the permafrost of the Laptev Sea in the East Siberian Arctic Ocean. Scientists do not know exactly how much carbon could ultimately be released by thawing permafrost or when. According to one report, 2°C of warming could mean the loss of 40 percent of the world's permafrost (NOAA 2022).

Under the 1992 treaty, nations agreed that:

> Nations have duties to adopt policies to prevent dangerous climate change and to take steps toward stabilization of GHG concentrations in the atmosphere at a level that would prevent dangerous anthropogenic interference with the climate system.
>
> *(UNFCCC 1992: Art. 2)*

Although nations agreed they had a duty to prevent dangerous climate change in 1992, and despite almost two decades of efforts to further define a nation's responsibility, no progress was made on defining 'dangerous' until the 2015 Paris Agreement when nations agreed that they have a duty:

> Holding the increase in the global average temperature to well below 2°C above pre-industrial levels and pursuing efforts to limit the temperature increase to 1.5°C above pre-industrial levels, recognizing that this would significantly reduce the risks and impacts of climate change; (b) increasing the ability to adapt to the adverse impacts of climate change and foster climate resilience and low greenhouse gas emissions development, in a manner that does not threaten food production; and (c) making finance flows consistent with a pathway towards low greenhouse gas emissions and climate-resilient development.
>
> This Agreement will be implemented to reflect equity and the principle of common but differentiated responsibilities and respective capabilities, in the light of different national circumstances.
>
> *(Paris Agreement 1015: Art. 2)*

Although the 1992 UNFCCC did not define dangerous climate change, under the 2015 Paris Agreement, 197 nations agreed to adopt policies to keep global temperature rise in this century to well below 2°C above pre-industrial levels and to pursue efforts to limit the temperature increase to 1.5°C (Paris Agreement 2015).

The Intergovernmental Panel on Climate Change (IPCC) has warned that exceeding 2°C could have catastrophic consequences and that we ideally need to keep global warming to 1.5°C. The world has already exceeded 1.1°C warming as of the recently concluded UNFCCC COP26 in Glasgow, Scotland and is on a pathway to exceed 1.8°C by 2100 if all commitments made thus far are achieved thus exceeding the 1.5°C warming limit goal. To limit warming to 2.0°C, developed nations will need to take the requirement that they base their emissions reductions target on the remaining carbon budget for the world that must constrain global emissions and their 'equitable' share of that budget. Although in my experience, most climate scientists and environmental NGOs have no understanding how to apply equity to their GHG emissions reduction target. Moreover, without an understanding of reasonable interpretations of 'equity,' proponents of climate change strategies are unable to respond effectively to inevitable predictable false claims of opponents of climate policies about a nation's fair share compared to another country such as China.

If high emitting nations do not reduce their GHG emissions to levels required of them by equity, the low-emitting developing nations must go to zero immediately if the world has any hope of achieving the 2°C warming limit goal. And so developed nations must determine its GHG emissions reduction target by adjusting the emissions reduction amount needed of the entire world by a serious consideration of equity. This means that if the entire world must reach net zero by 2075, for instance, to achieve the 2.0°C warming limit goal, developed nations after taking equity into consideration may need to reduce CO_2 emissions by –125 percent by 2075. In this writer's experience, few US environmental NGOs nor academics engaged in climate issues have made recommendations on the US Nationally Determined Commitments after adjusting globally needed emissions reductions levels on the basis of a reasonable interpretation of 'equity.' Although reasonable people may disagree on what equity expressly requires of a nation to reduce its GHG emissions, IPCC said its 5th Assessment report that despite some ambiguity about what equity means:

> There is a basic set of shared ethical premises and precedents that apply to the climate problem that can facilitate impartial reasoning that can help put bounds on plausible interpretations of 'equity' in the burden-sharing context. Even in the absence of a formal, globally agreed burden sharing such are important in expectations of what may be reasonably required of different actors.
>
> *(IPCC 2014: 317).*

The IPCC went on to say that:

> (T)hese equity principles can be understood to comprise four key dimensions responsibility, capacity, equality, and the right to sustainable development.
>
> *(IPCC 2014: 317)*

Responsibility is understood to mean historical responsibility for the current problem, not emissions levels per year. For the last several decades, when proponents of climate action proposed serious greenhouse gas reduction goals and strategies, opponents of climate action frequently claimed that such action would be unfair to the United States as long as countries such as China don't do the same; rarely if ever has anyone among the US climate activists responded by saying the total national historic greenhouse gas emissions levels are not a valid criteria for determining a nation's reduction responsibility.

Making matters worse, when respected climate scientists are asked whether the world has enough time to deal with harms of climate change, they usually claim we do if we act aggressively. But this answer is usually in response to the question of whether we have enough time to achieve the Paris Agreement warming limit goal of 2.0°C. The atmospheric concentrations which are already causing refugees are existing concentrations. In this writer's experience, few climate activists seem to understand that CO_2 has features that are different than other air polluting substances that have profound policy implications. These features include CO_2 mixes well in the atmosphere and is very long lived. Although approximately 80 percent of CO_2 emissions are removed by carbon sinks in 100 years, yet some stay in the atmosphere for tens of thousands of years contributing to climate change harms everywhere for a very long time. The following image depicts the reality that atmospheric concentrations of CO_2 rise as CO_2 emissions rise. The policy implication of this fact is all CO_2 emissions are making the problem worse and just some reductions of emissions levels will not stop ice from melting, dangerous killer storms from occurring or other climate effects on the climate system from creating damage to the global system even if global CO_2 emissions are decreasing at rates necessary achieve the 1.5°C and 2.0°C warming limit goals.

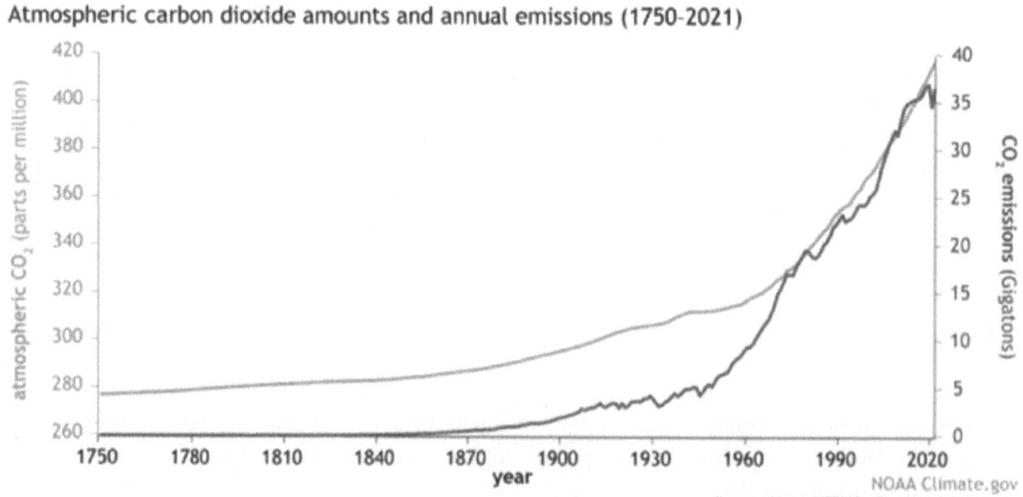

Image 13.2 NOAA 2022

Notice GHG emissions from around the world are raising atmospheric concentrations globally thereby increasing harms everywhere. Therefore, the response by some scientists that 'we have time' may be misleading although perhaps accurate in regard to keeping the world below the 2.0°C warming limit goal of the Paris Agreement. Yet unless governments reduce CO_2 emissions to net zero, CO_2 atmospheric concentrations will rise and harms, including refugee problems, will continue to worsen. For a discussion of this phenomenon, see, "Seven Features That Citizens and The Media Needs to Understand to Critically Evaluate Their Nation's Response to Climate Change" (Brown 2022).

The dominant climate policy discourse has focused on actions necessary to stay within the Paris Agreement's warming limit goals has resulted in little public attention on the reality that all GHG emissions raise atmospheric CO_2 concentrations and resulting harms. Therefore, unless atmospheric CO_2 concentrations are stabilized, all CO_2 emissions will raise atmospheric CO_2 concentrations increasing refugee numbers and associated problems globally. More alarming, current signs of several tipping point destabilization is ominous evidence that refugee problems may plague the world even if CO_2 emissions are reduced to net zero. This is so because warming from current atmospheric CO_2 concentrations is already destabilizing some tipping points.

COP 27 in Egypt Has Finally Agreed to Establish a Loss and Damage Fund to Help Countries Vulnerable to Climate Change

At COP 15 in Copenhagen in 2009, all developed nations agreed to provide USD 100 billion a year to support adaptation and mitigation needs of developing countries. Yet nations have failed to deliver on the money. In 2021, at COP 26 in Glasgow, developed nations agreed to fully deliver on the USD 100 billion goal and provide more transparency on their adaptation pledges. The adaptation and loss and damage funding may help with climate change harms that are creating refugees, but it is too early to tell.

Although all nations agreed under the 1992 UNFCCC that they had duties to prevent activities within their jurisdiction from harming others outside their jurisdiction, the international community had not developed a mechanism for operationalizing the 'no harm' rule despite pressure from developing countries to do so for many years. Yet COP 27 in Egypt has finally agreed to create and fund a loss and damages mechanism and to create a committee to make recommendations to COP 28 in the United Arab Emerites on the details of the loss and damages mechanism (see UN Press Release 2022). COP 28 is also expected to review issues about funding and use of the adaptation mechanism.

Nations also agreed in the 1992 UNFCCC that:

> States have, in accordance with the Charter of the United Nations, the responsibility to ensure that activities within their jurisdiction or control do not cause damage to the environment of other States or areas beyond the limits of national jurisdiction.
>
> *(UNFCCC 1992: Preamble)*

This principle is referred to as the 'no harm' principle.

Climate change is already creating enormous misery and desperation around the world from flooding, drought, sea level rise, forest fires, extreme heat, more intense tropical storms, the spread of tropical diseases. These harms are being caused by atmospheric CO_2 concentrations at or below current CO_2e atmospheric concentrations of 421 ppm, the new historic level of CO_2 atmospheric concentration. The current public debate which has focused mostly on keeping national emissions below levels that represent the nation's fair share of maximum global emissions that are needed to achieve the Paris Agreement's warming limit goals of between 1.5°C and 2.0°C has failed to raise public awareness that any new CO_2 emissions, as we have seen, because they are long-lived and mix well in the atmosphere will likely raise atmospheric CO_2 concentrations thus increasing some climate harms and refugees.

The climate harms to some parts of the world from climate change include sea level rise, droughts, destructive storms, and floods even if some tipping points are not triggered, are already devastating for some parts of the world and will create astronomical costs and suffering particularly to those places which are particularly vulnerable. Most of these harms can create refugees.

Although nations had agreed to adopt policies that prevent activities within their jurisdiction from creating climate change harms beyond their borders, developed nations have finally agreed to fund a mechanism for loss and damages at COP 27 in Egypt, the details of which are scheduled to be finalized at COP 28 in the United Arab Emerites. Other legal changes may be necessary to appropriately deal with refugee issues; for example, to assure that refugees have freedom of movement and rights of other citizens.

References

Brown, D. (2022) "Seven Features That Citizens and The Media Need to Understand to Critically Evaluate Their Nation's Response to Climate Change," *EthicsandClimate.org*. Available at https://ethicsandclimate.org/2020/08/19/why-getting-nations-to-comply-with-ethical/.

Caesar, L. et al. (2021) "Current Atlantic Meridional Overturning Circulation Weakest in Last Millennium," *Nature Geoscience*, 14: 118–120. Available at https://doi.org/10.1038/s41561-021-00699-z.

Convention on the Status of Refugees. (1951) Available at https://www.ohchr.org/en/instruments-mechanisms/instruments/convention-relating-status-refugees.

Glasgow Climate Pact. (2021) *UNFCCC COP 26*. Available at https://unfccc.int/sites/default/files/resource/cop26_auv_2f_cover_decision.pdf.

Guston, G. (2022) "Is the Amazon Approaching a Tipping Point? A New Study Shows the Rainforest Growing Less Resilient." *Inside Climate News*. Available at https://insideclimatenews.org/news/07032022/amazon-rainforest-tipping-point-resilience/.

Intergovernmental Panel on Climate Change. (IPCC 2014) *5th Assessment Report, Contribution of Working Group III to the Fifth Assessment Report of the Intergovernmental Panel on Climate Change*. Cambridge University Press.

Intergovernmental Panel on Climate Change (IPCC). (2022) *6th Assessment Report, Summary for Policymakers*. Available at https://www.ipcc.ch/report/ar6/wg1/#SPM.

Kelley, C. et al. (2015) "Climate Change in the Fertile Crescent and Implications of the Recent Syrian Drought," *PNAS*, 112(11): 3241–3246. Available at https://www.pnas.org/doi/10.1073/pnas.1421533112

Leahy, S. (2019) "Climate Change Driving Entire Planet to Dangerous 'Tipping Point,'" *National Geographic*. Available at https://www.nationalgeographic.com/science/2019/11/earth-tipping-point/.

Lovejoy, T. and Nobre, C. (2018) "Amazon Tipping Points, Science Tipping Points, Science Advances," *Science Advances*, 4(2). Available at https://www.science.org/doi/10.1126/sciadv.aat2340.

NOAA. (2022) *Sea Level Rise Technical Report*. Available at https://oceanservice.noaa.gov/hazards/sealevelrise/sealevelrise-tech-report.html.

Paris Agreement. (2015) Available at https://unfccc.int/process-and-meetings/the-paris-agreement/the-paris-agreement.

Pumphrey, C. (2008) "Global Climate Change National Security Implications," *Monographs*. Available at https://press.armywarcollege.edu/monographs/15.

Resnick, B. (2019) "Scientists Feared Unstoppable Emissions from Melting Permafrost. They May Have Already Started," *Vox*. Available at https://www.vox.com/energy-and-environment/2019/12/12/21011445/permafrost-melting-arctic-report-card-noaa.

Rigaud, K. et al. (2018) *Groundswell: Preparing for Internal Climate Migration*. Washington, DC: World Bank. Available at https://openknowledge.worldbank.org/handle/10986/29461.

UN Framework Convention on Climate Change (UNFCCC). (1992) Available at https://unfccc.int/resource/docs/convkp/conveng.pdf.

United Nations High Commissioner for Refugees (UNHCR). (2021) *Climate Change Is An Emergency for Everyone Everywhere*. Available at https://www.unhcr.org/en-us/news/stories/2021/11/618a301d5/climate-change-emergency-everywhere.html2021e.

UN Press Release. (2022) "COP27 Reaches Breakthrough Agreement on New "Loss and Damage" Fund for Vulnerable Countries," *UNFCCC*. Available at https://unfccc.int/news/cop27-reaches-breakthrough-agreement-on-new-loss-and-damage-fund-for-vulnerable-countries.

US White House. (2021) *Report on the Impact of Climate Change on Migration*. Available at https://www.whitehouse.gov/wp-content/uploads/2021/10/Report-on-the-Impact-of-Climate-Change-on-Migration.pdf.

14

TAKING MORAL RESPONSIBILITY FOR CLIMATE CHANGE BY CITIES

Johan Hattingh

Introduction

"Cities are where the climate battle will be won or lost." This statement by Patricia Espinosa, Executive Secretary of the UNFCCC in 2018 (Suliman 2018) represents a widely held view that what cities will do, or will neglect to do in the very near future, will crucially determine whether humanity will succeed in mitigating climate change to reasonable levels, and adapting to its effects that cannot be escaped. The facts and figures in support of this claim are well documented, and neatly summarized by the United Nations Development Program (UNDP Website 2021) in its motivation of Sustainable Development Goal 11: Sustainable Cities and Communities:

- In 2018, 4.2 billion people, 55 percent of the world's population, lived in cities. By 2050, the urban population is expected to reach 6.5 billion.
- Cities occupy just 3 percent of the Earth's land but account for 60 to 80 percent of energy consumption and at least 70 percent of carbon emissions.
  828 million people are estimated to [currently] live in slums, and the number is rising.
- In 1990, there were 10 cities with 10 million people or more; by 2014, the number of mega-cities rose to 28, and was expected to reach 33 by 2018. In the future, 9 out of 10 mega-cities will be in the developing world.
- In the coming decades, 90 percent of urban expansion will be in the developing world.
- The economic role of cities is significant. They generate about 80 percent of the global GDP.

Conversely, cities are not only significant drivers of climate change; they are equally and significantly exposed to the risks of climate change. *The Special Report of the IPCC on Global Warming of 1.5°C* (2018), for instance, warns that 136 port cities with a population of more than 1 million in 2005 "are at risk from flooding" due to sea level rise under the scenarios of 1.5°C or 2°C global warming in the 21st century unless adaptation is undertaken. The *Report* further warns that cities will also be exposed to extreme heat events, fires, water stress because of droughts, damage of infrastructure (energy, water, transport, and buildings), salinization of groundwater, as well as severe impacts on vulnerable populations such as the poor and those living in informal settlements if no adaptive measures are taken (2018: 231, 241–242).

Many world cities, however, are well aware of these risks and have started to prepare for them in terms of mitigation, or adaptation to the effects of climate change that cannot be escaped, while many other cities have yet to start with their preparations (IPCC 2018: 231; Nicholls et al. 2018).

DOI: 10.4324/9781003039860-17

For instance, ICLEI (Local Governments for Sustainability), a global network of 2,500 local and regional governments committed to sustainable urban development, is active in more than 125 countries influencing policy and development, among other goals, to "help cities, towns and regions anticipate and respond to complex challenges, from rapid urbanization and climate change to ecosystem degradation and inequity" (ICLEI Website 2021a). Similarly, C40 Cities is an alliance of 97 world cities committed

> To take bold climate action, leading the way towards a healthier and more sustainable future. Representing 700+ million citizens and one quarter of the global economy, mayors of the C40 cities are committed to delivering on the most ambitious goals of the Paris Agreement at the local level, as well as to cleaning the air we breathe.
>
> *(C40 Cities Website 2021)*

While initiatives such as these are to be welcomed at face value, they beg the important question to what extent they are really effective, having real results and making a real difference to the readiness of cities and to mitigate global climate change. To be fair, it may be too early to judge, since many of the programs emerging from these and similar alliances are very young and of recent date. Building on earlier programs, C40 Cities (C40 Annual Report 2021), for example, pledged to double down on its efforts to cut carbon emissions by half by 2030, and toward the end of 2020 launched the 'Cities Race to Zero' campaign "which looks to enlist 1,000 cities to sign up to [their] target commitments as part of the multi-stakeholder Race to Zero initiative of the High-Level Climate Champions" (C40 Annual Report 2020).

ICLEI, in turn, launched an ambitious new strategy for climate action and sustainability in June 2018 that included a pledge to

- Promote 100 percent renewable energy;
- Develop urban low emission, climate resilient, mobility friendly development strategies and monitor their implementation;
- Connect climate action to urban development and engage local and regional governments in making national climate action plans (Nationally Determined Contributions or 'NDCs') fit for the Paris Agreement; and
- Support island communities at the front line of climate challenges and address land-based ocean pollution (UN Climate Change News, 21 June 2018).

While the juries are still out on determining whether these initiatives, strategies and action plans will be strong enough to make good on their promises in cities 'on the ground,' a question we need to ask is whether they, in their current format and conceptualization, (a) recognize and respond to the moral dimensions of climate change, and (b) whether recognizing these moral dimensions change the nature and quality of the responses to climate change in any way.

In order to answer these questions, this chapter will start by briefly summarizing what is regarded as the moral dimensions of climate change, why it is so important to spot these moral dimensions, and why it is so important to respond to them as moral challenges. In this regard, it will take as a given that these moral dimensions of climate change have been discussed at length in this volume and elsewhere. It will then proceed to scrutinize seminal documents from ICLEI and C40 Cities, asking to what extent they provide evidence of recognizing the moral dimensions of climate change, and to what extent they provide guidance, if any, to cities to respond to climate change as a moral challenge. Given the finding that ICLEI and C40 Cities at best hold a latent view of climate change as a moral challenge, meaning that they recognize the moral dimensions of climate change on an implicit rather than an explicit level, the chapter will bring the discussion to a close with a reflection on the questions that need to be asked by cities in order to *explicitly* recognize the moral dimensions of climate change and to *explicitly* respond to them as such.

The Moral Dimensions of Climate Change

Climate change is usually seen as nothing but a physical problem that can be understood on the basis of scientific studies, and accordingly, responses to climate change, in so far as they are deemed to be humanly possible, are usually seen as technical challenges, the implementation of which are typically determined on the basis of the financial affordability of the steps that need to be taken to do so, and the political will to tackle these problems. To elaborate: the latest advances in earth system science (for instance Steffen et al. 2018) tells us about the planetary boundaries within which the earth can operate indefinitely as a functional system, sustaining life on earth and providing a basis for its flourishing; it tells us about the unpredictable, uncontrollable and irreversible changes that can occur in this earth system when certain thresholds are surpassed; it tells us about greenhouse gasses and the human causes of climate change, that our only chance to mitigate climate change is to cut carbon emissions, and that we will have to reach carbon neutral economies by 2050 if we want to prevent global warming from spinning totally out of control. Science also tells us that even at the target of keeping global warming from rising above 1.5°C above pre-industrial levels will not mean that we are safe; there will still be adverse effects from climate change to which humanity and other living species will have to adapt in so far as they survive even a 1.5°C rise in average global temperature (IPCC 2018).

As valuable as science is in understanding climate change and its trajectory if left unmitigated, and as valuable as science is in helping us to understand what can be done to mitigate climate change and to adapt to its effects that cannot be escaped, science cannot tell us whether the changes brought about by climate change are acceptable or not; science cannot tell us whether winning or losing from climate change is good or bad; and science cannot tell us what we should do to live good lives with humans and other living beings in the face of climate change that is undermining the very preconditions for living good lives on earth. These are all 'value-rational' questions, and they are asked and answered from a moral point of view. In fact, all questions about the harms of climate change, or the inequities and injustices around climate change are moral questions, and when recognized as such, all of the scientific and technical dimensions of climate change are placed in a different light, requiring a different kind of response than those informed by 'science alone,' or 'pure financial calculation,' or 'clinical technical reason.'

What is at stake when recognizing a problem as a moral concern is a moral value or a moral reason; and moral reasons and values are important because they articulate the distinctions we make between right and wrong, good and bad, and what deserves respect and what not. In our personal lives, we usually make these distinctions to ensure that others are not harmed, not treated unfairly or unjustly, or without the dignity and respect they deserve. As such, morality presupposes an ability to be moved by moral reasons and to act upon them. Bluntly formulated, it means at the least to 'draw a line,' i.e., to stop, prevent or avoid things that are wrong, harmful, unfair, unjust, disrespectful, or undignified – toward humans and toward the wider circle of 'other' non-human living beings. At best, acting on a moral reason may entail positively to promote or enhance what is right, good, respectful, dignified, respectful, fair, or just. Conversely, to ignore moral reasons and to refuse to act upon them is to willfully choose to either harm others, or to treat others unfairly, unjustly or without respect.

Within the context of everyday human interaction, such a deliberate choice to ignore moral reasons is typically met with disgust, and provide grounds for justifiable moral outrage, prompting demands for an explanation and some kind of remedial action. However, within the context of collective action such as global energy use causing climate change, or in the governance of a city or a nation for the well-being of its citizens, it is not as clear-cut to register moral outrage, or to take action on it for moral reasons. Cities, for instance, are characterized by different interest groups existing side by side or overlapping with one another, so that a moral outrage for one we-group may not be a moral outrage for another we-group. At the same time, the governors of cities would typically be loath to take sides between different interest groups if they open their flanks to choices that could be branded as preferential treatment or as arbitrary, or not consistent with the mandate they were elected to govern the city in the first place. In this wider context, however, the basic structure of acting upon a moral reason does not fundamentally change; moral reasons provide compelling reasons also for collectives to act upon, and *not* to act collectively upon moral reasons would constitute a moral outrage of that collective in its own right.

Against this background, some of the more important moral dimensions of climate change can be articulated in terms of concerns about harm, justice, and respect:

- Manifesting in sea level rise, extreme weather events such as droughts, floods, and cyclones, and in associated disasters such as wildfires, pests (such as locusts), changes in disease vectors, and loss of biodiversity and coral reefs, climate change directly harms people by endangering their lives and/or livelihoods;
- Climate change brings about concerns around justice, fairness, and equity in that: those least responsible for causing climate change (typically poor and marginalized people and nations) are the first and the most exposed to the impacts of climate change; those least capable of adapting to climate change (typically poor and marginalized people and nations) are those that will have to adapt the most to effects of climate change that cannot be avoided; those most responsible for climate change (on the basis of historic emissions) have the biggest obligation to act first and the most in mitigation of and adaptation to climate change, and taking the lead in assisting others that are not in a position to do so, while facts tell us that those most responsible tend to shirk away from this obligation; and generations living in the future will be exposed to unfair burdens of climate change if the present generation does nothing in mitigation or in preparation of adaptation; and
- Climate change brings about concerns around dignity and respect in that victims of climate change (especially the poor and the vulnerable) are typically exposed to inhumane conditions when they suffer the consequences of extreme weather events, lose livelihoods, and/or have to migrate in order to start new lives elsewhere.

On the basis of the well-established scientific knowledge that the human causes of climate change predominantly lie in the use of carbon-based energy, the concerns listed above provide compelling moral reasons for mitigating and adapting to climate change and doing so with special obligations to and in solidarity with those most vulnerable to the effects of climate change, or the least capable of adapting to them. The point, as discussed above, is that if my choices and actions contribute to harming others, even if I do so collectively with a vast number of others, I have a moral obligation to stop, prevent or avoid that action, and to put things right where those harms have already taken place. This further presupposes an obligation on those with the capabilities and means to studying the trajectory of climate change on a scientific basis, to continue doing so, and to share this knowledge with the rest of the world on a timely and an equitable basis. It also implies and obligation on the whole of the world's population to act on this knowledge individually and collectively. To deny climate change, or to constantly try to find reasons to neutralize the moral reasons for acting on it, would thus amount to willfully and deliberately exposing others to harm, injustice and disrespect.

Taking it as a given that cities exist and are supposed to be governed with the ultimate aim of the well-being of their citizens, it makes sense to hear from cities themselves to what extent they recognize climate change as a moral challenge and respond to it as such. However, since a comprehensive empirical study of cities with this question in mind is a matter of logistical impossibility, the chapter will turn, in the section that follows, to two high-level alliances such as ICLEI and C40 Cities that have taken a leadership role in guiding cities in their responses to climate change. In this regard, it will be guided by the question to what extent ICLEI and C40 Cities are aware of the *moral* dimensions of climate change in the context of cities, and to what extent the leadership and coordination they provide enable cities to take moral responsibility for climate change.

Global Alliances Leading Cities in Climate Change Responses

From a close inspection of the websites and seminal documents of ICLEI and C40 Cities respectively, it is clear that both of these global alliances are well-positioned to lead cities in the recognition of the moral dimensions of climate change, and to help them design climate change responses that are sensitive to these moral dimensions. In both of these alliances there is a strong emphasis on the harmful impacts of climate change on urban and regional populations, and there is an equally strong focus on equity and justice issues related to impact on the more vulnerable and marginalized of their citizens. In both of these alliances, adaptation on an equitable basis forms part of their response programs, as well as early implementation of measures to mitigate harm from extreme weather events such as droughts, cyclones, and floods, as well as disasters such as wildfires and diseases. With regards to mitigating the causes of climate change, both alliances strongly promote transformation in energy use with an emphasis on re-designing roads and transport systems, water and electricity provision, building, and waste management. Evidence also reveals that both alliances are working within the international frameworks of the UNFCCC and the 2015 Paris Agreement on climate, and that they are well aware of targets such as cutting carbon emissions by half by 2030 and reaching net zero carbon economies by 2050. The question of course is to what extent this leadership, i.e., this high-level consciousness, translates into effective action 'on the ground' in particular and in all cities in the world. Keeping the latter question in mind, this section will be devoted to an overview of the 'moral consciousness' of ICLEI and C40 Cities respectively with regards to climate change and climate change responses.

ICLEI

Founded in 1990 in Canada, and from there spreading to 125 countries all around the world, the broad agenda of ICLEI is the promotion of urban sustainability through systemic change. It does so by creating "connections among the local, regional, national and global governmental levels" through "peer exchange, partnerships and capacity building" (ICLEI Website 2021a). Responding to climate change is part of this agenda with ICLEI's focus on five integrated pathways cutting across different sectors to promote "low emission, nature-based, equitable, resilient and circular development" (ICLEI Website 2021b). Its most recent thinking on climate change is articulated in *ICLEI's Climate Neutrality Framework: Accelerating integrated climate action for sustainable urban development* (2020).

In this publication, ICLEI emphatically states that everyone should "take responsibility for their carbon emissions," and that an "absolute net zero emissions level" should be achieved at the latest by 2050. To reach this target, in parallel with adaptation, ICLEI (2020: 3, 6) calls on

> Local and regional governments [to] set a clear goal and advance rapidly following a holistic and integrated approach that leads to a wide range of co-benefits for sustainable development, such as creating socio-economic opportunities, reducing poverty and inequality, and improving the health of people and nature.

It is worth quoting the five integrated pathways through which ICLEI aims to accelerate climate change action in order to meet the 2050 target of net zero emissions/carbon neutrality (ICLEI 2020: 6):

- "Low emission development pathway: Curbing climate change, creating new economic opportunities and improving the health of human and natural systems";
- "Nature-based development pathway: Protecting and enhancing biodiversity and urban ecosystems, which underpin key aspects of local economies and the well-being and resilience of our communities";
- "Circular development pathway: Inspiring, guiding, co-creating, and scaling up the services and policies of regional and local governments to create better living conditions for their residents, using redesign, reload, reuse, reduce and recover strategies. These strategies can address the untapped hidden emissions in products and services";
- "Resilient development pathway: Anticipating, preventing, absorbing, and recovering from shocks and stresses, especially those brought about by rapid environmental technological, social and demographic change"; and
- "Equitable and people-centered development pathway: Equitable and people-centered development building more just, livable, and inclusive urban communities and addressing poverty."

In order to reach the 2050 emission target in cities, ICLEI (2020: 9) furthermore identifies the importance of (a) removal/sequestration of greenhouse gasses; (b) divestment from fossil fuels and repurposing and reinvesting funds in climate and sustainable energy action; and (c) mechanisms to offset greenhouse gas emissions that cannot be immediately removed, reduced, or avoided. In support of its *Climate Neutrality Framework*, ICLEI launched its *Green Climate Cities Program* (GCC), in alliance with its *Bonn Center for Local Climate Action and Reporting* (also known as The carbon*n* Center). Central to this program is a methodology (reviewed by the Bonn Climate Center) of measuring, monitoring, reporting, and verification of GHG emission reduction, and monitoring and evaluation of climate change adaptation and resilience. All of these data are captured in the carbon Climate Registry, and "used by ICLEI to track local climate action, ensuring transparency, accountability and credibility of the ICLEI network and its collective (aggregated) impact on climate change and sustainable development" (ICLEI 2020: 8).

From this it is clear that ICLEI is informed by a moral consciousness of the harmful impacts of climate change and responding to these impacts within the framework of urgency and on the basis of equity and justice, even if the terms of 'morality' or 'ethics' do not appear in their literature or websites that have been reviewed. Indeed, there is undeniably a normative structure to ICLEI's diagnosis of the crisis of climate change, its commitment to the target of net zero emissions at the latest by 2050, the integrated package of measures it proposes to reach that target, and the mobilization of cities in conjunction with other levels of government across sectors. The question that arises in this context, however, is whether this latent moral consciousness can be made explicit, and to what extent such an explication can make a substantive difference to the responses of cities to the crisis and the risks of climate change. This is as much a philosophical question as it is an empirical one, to which will be returned to in a later section of this chapter.

C40 Cities

Launched in October 2005, C40 Cities currently represents 97 mayors of the world's biggest and most influential megacities, pursuing the mission "to halve the collective carbon emissions of our member cities within a decade, while improving resilience and equity and creating the conditions for everyone, everywhere to thrive" (C40 Cities, 2021: 4). Cognizant of the dangers of a 2°C rise in global temperatures above pre-industrial levels, C40 Cities (2021: 8) aims to "keeping global heating to 1.5°C, while increasing climate resilience and equitable outcomes." The methodology of C40 Cities entails collaboration and knowledge sharing, starting with identifying the world's best climate practice in pioneering cities and then replicating these actions as rapidly as possible in other cities all around the world (2021: 5). This evidence-based strategy led C40 Cities to launch its flagship *Climate Action Planning Program* in 2016. Starting with a pilot of eight leading cities (Boston, Durban, London, Los Angeles, Melbourne, Mexico City, New York City, and Paris), robust climate action plans were drafted aligned with the 1.5°C target of the Paris Agreement, and subsequently these plans were consolidated in a *Climate Action Planning Framework* (2018). Member cities of C40 have since been challenged to adopt this framework for their own climate action planning, and by 2020 a number of 88 cities have committed to do so. While the main aim of this framework is to keep global warming below 1.5°C, it at the same time strives to protect residents, improve resilience, address inequalities, and, as quoted above, "creating the condition for everyone, everywhere to thrive" (C40 Cities 2021: 4, 7).

The *Climate Action Planning Framework* is supported by a comprehensive web-based *Climate Action Planning Resource Center* (see details in the Bibliography), and the Framework itself consists of four key components. Two of these are outcome oriented, entailing emissions neutrality and resilience to climate hazards; while another two are process oriented, aiming at inclusivity and benefits, and ensuring good governance and collaboration respectively. It is worth quoting the summaries of all four of these key components as they appear on the C40 Cities website (2021):

- "Emissions neutrality: Develop a pathway to deliver an emissions neutral city by 2050 at the latest and set an ambitious interim target and/or carbon budget (e.g., 2030)";
- "Resilience to climate hazards: Demonstrate how the city will adapt and improve its resilience to the climate hazards that may impact the city now and in future climate change scenarios";
- "Inclusivity and benefits: Engage with the community to inform the plan, outline the social, environmental and economic benefits expected from implementing the plan, and establish ways to ensure equitable distribution of these benefits to the city's population"; and
- "Governance and collaboration: Detail the city's governance, powers, and capacity, as well as identify the partners who need to be engaged in order to accelerate the delivery of the city's mitigation targets and resilience goals."

The *Climate Action Planning Framework* and its key components are furthermore evidence-based, and transformational in its intent. As described on the C40 website, proper climate action planning requires that evidence is considered about "existing city conditions, including baseline emissions, 2050 emissions trajectory, climate risk, and socioeconomic priorities." Proper climate action planning also "Defines the transformational action and implementation plan, including the development and prioritization of actions and the processes of monitoring, evaluation, reporting and revision."

Without having to go into all of the details of the *Planning Framework*, it is already evident from the above that in its intent and conceptualization, the Framework clearly displays a consciousness of most of the moral dimensions of climate change and climate change responses, even if it does not articulate this consciousness explicitly as a 'moral' or 'ethical' position. In broad outline, the

Planning Framework recognizes the harm and justice concerns related to climate change as sound moral reasons to act upon in a firm commitment to set bold emission reduction targets and adaptation goals on the basis of equity, justice, engagement, inclusivity, and collaboration within the framework of international treaties such as the UNFCCC and the Paris Agreement.

If there is anything at all to be questioned about this *Planning Framework*, it is its normative formulation as a standard or an aspiration, and that it accordingly gives expression to future oriented goals. In itself, there is nothing wrong with such bold goal-setting and an urgent call for action, as long as the heavy lifting and hard task of moving from the present to achieving these goals are also acknowledged and engaged with seriously. It is exactly at this point that the problem of the so-called Action Deficit emerges, i.e., the gap between well-articulated and ambitious goals and targets, on the one hand, and what is actually done in order to reach those goals and targets progressively over time.

Aware of this Action Deficit, C40 Cities launched *Deadline 2020: How cities will get the job done* (2016), in which the urgency of setting bold climate targets in line with the Paris Agreement is emphasized, and cities challenged to develop and start implementing their climate actions plans before the end of 2020. Offering a "blueprint for delivering on the Paris Agreement" and emphasizing at the time of its launch (2016) that what is done by cities in the next four years in collaboration with each other to 'plug the gap' between Climate Ambition and Climate Action will be crucial to make a real difference in the race to a net zero carbon world economy by 2050, *Deadline 2020* demonstrated that it is indeed possible to move out of a carbon-energy based world, and that cities can make a substantial contribution in this regard. However, it is very disappointing to read in the *2020 Annual Report* of C40 Cities that only "54 cities are on track to help keep global heating below 1.5°C," and while it is encouraging to read in this Report that C40 member cities are working hard on achieving reductions in Green House Gasses by implementing measures to clean transport, building and industries, it is equally disappointing to see that this only happens in 30 cities (C40 Cities 2021: 7, 8).

Against this background, it is clear that in spite of the well-articulated ambitions, goals and targets, of ICLEI and C40 Cities, we still have a huge problem with regards to achieving actual results in climate change action. Many cities set good examples of best practice in recognizing the moral challenges of climate change, and many cities consider moral principles in developing and implementing their responses to climate change. These examples of (moral) best practice, however, are found in but a handful of world cities, while the majority of the more than 4,000 world cities with populations more than 1,00,000 appear to seriously lag behind.

This brings us to the point where we have to ask what cities in the latter category need to know and need to ask in order to recognize the moral responsibility they have toward their citizens in the first place, and others wider afield, to respond to climate change through measures of mitigation and adaptation that not only promote health and protect their citizens from climate change risks, but also improve equity and justice for them all. We also need to ask what difference it will make to climate change responses if climate change is recognized as a moral problem. In the next section, the chapter will first focus on the difference ethics can make in understanding and responding to climate change, and then it will turn to some of the most important questions that need to be asked by citizens and governors of cities to assess climate change and climate change action in moral terms.

Discussion: The Difference Ethics Can Make

From the overview above, it is clear that there is a normative structure at work in both of the high-level responses to climate change of ICLEI as well as C40 Cities, leading world cities as they do to take responsibility for climate change. In both cases, this normative structure can be interpreted as a latent moral position, in so far as moral reasons (for example preventing harm, ensuring equity, sharing knowledge) are accepted as a point of departure in the climate change responses they help to develop and support in their implementation. It needs to be asked, though, what difference it would make to climate change action if it is explicitly recognized as a challenge with moral dimensions.

Staying for the sake of this discussion about cities in the realm of human ethics, it can be stated in general terms that an ethical (or moral) view has a negative face in that it focuses on the principles and standards of conduct accepted in civil society in order to prevent or stop such things as harm, inequities, injustices, and circumstances eroding the dignity of individuals and communities. The positive face of an ethical view, in turn, will entail doing something about harms, inequities, injustices etc., with a view to enabling people to live good lives with and for others in just institutions (Ricoeur 1992: 172). While views may differ on what exactly a harm or inequity or the good life or justice is, there is a common consensus that an explicit and active refusal to take harms, inequities, injustices, and the erosion of human dignity seriously is a scandal and a moral outrage. The reason for this is that such a denial not only takes away the humanity of those at the receiving end of such denial, it also diminishes the humanity of society and everyone standing idly by and doing nothing about it.

From this point of view, an ethical stance is the opposite of taking things trivially or lightly; an ethical stance rather places us in that serious realm of scrutiny and questioning whether our choices and actions contribute to, or take away from, the humanity and human dignity of others, and accordingly, also from who we are as human beings. To wit: in the ethical realm, we recognize, scrutinize, question, and debate to what extent our choices and actions place our humanity at stake. Seen in this light, an ethical view entails a process of weighing and deliberation. If this shows that our humanity is put in jeopardy, an ethical view would base this assessment on reasons such as harms, inequities, injustices or disrespect imposed on others or on us. Conversely, an ethical view would require of us to act positively with care and compassion and in solidarity with others, promoting what could be justified as living good lives with and for others in just institutions. As such, ethics is never a process of neutral deliberation, moral scrutiny is always value-based, it takes sides: always for care, equity, justice, dignity, and the good life, and always against harm, inequity, injustice, and the degradation of others and communities.

With this in mind, the importance of taking an ethical/moral view of climate change and climate change action lies in placing climate change and responses to it in the scales of deliberating to what extent they impose harms on individuals and communities, how they do or may intensify already existing inequities and injustices in society, how they may not only undermine the conditions of life on earth in general, but above all, the conditions for living dignified lives as human beings with and for others. Conversely, taking an ethical/moral view of climate change action compels us on the basis of moral reasons to act with the utmost care when developing and implementing climate responses; first, taking into account how the poor, the marginalized, and the most vulnerable in society are affected, making sure that they do not suffer more harms, inequities, and injustices than they already do, but rather less.

The difference that an ethical or moral view of climate change can make to understanding and responding to it thus lies in bringing climate change and climate change responses into a field of scrutiny and assessment in which moral reasons stand central, something that is neglected or totally forgotten if climate change is only conceptualized in scientific, technical, or financial terms. In the latter three frameworks, ethical questioning typically does not occupy the foreground; instead 'hard' facts, instrumental rationality, and efficiency equations dominate. While this 'neutralization' or 'objectification' of the conversation on climate change action is in itself based on value assumptions (that are problematic and must be questioned), it is important to recognize this tendency as such, and also to identify some of the important questions citizens and governors need to ask in order to make a moral understanding of climate change by cities possible and explicit.

Recommendations and Concluding Remarks

The most important questions that can be articulated in order to foreground the ethical dimensions of climate change and climate change action in cities, whether these cities are at the forefront of the fight against climate change or late starters, cluster around five intersecting areas of inquiry.[1] Under each one of these areas, a number of sub-questions can be articulated, each one of which probing for an ethical dimension of climate change or climate change actions.

1 How is climate change conceptualized and problematized in/by the city?

- a Is climate change viewed as primarily a scientific, technical, financial, or moral problem? Is this conceptualization justifying and perpetuating business as usual, locking us into a carbon-based world and lifestyle, or is it providing us with a vision of a world beyond carbon energy and transformative pathways to achieve it?
- b Is climate change problematized as a crisis? If yes, a crisis of what, for whom or what, in which time frame, and what is proposed to address this crisis, if anything? If not, how is the urgency of acting on climate change articulated?
- c Who participated and who are excluded in determining this problematization?
- d Who wins and who loses from this problematization? By which mechanisms?[2] And what does the city propose in order to close the gap between winners and losers?

2 What exactly is the value-basis of this problematization of climate change?

- a Climate harms and risks?
- b Inequities and injustices?
- c Human dignity?
- d Survival? Of whom or what?
- e Is the value basis only related to humans, or is the net cast wider to include environmental issues?
- f Who participated and who are excluded in determining this value basis?

3 What is the knowledge base of the city on the climate system, climate change, its impacts and risks, and climate change action?

- a Does the city use the latest scientific advances in earth systems and climate science as basis for mitigation, adaptation, and building resilience?
- b Does the city use an integrated, systems-oriented approach to understand climate change, and to develop and implement climate action?
- c Does the city have a regularly updated, transparent, and accessible carbon inventory, tracking its:
  - i Current and historic greenhouse gas emissions, per sector, and per capita;
  - ii Direct emissions from the city itself, and indirect emissions generated in other areas due to imports of food, water, energy, building material, and raw material for industries;
  - iii Carbon budget and what is left of it for future use on an equitable basis;
  - iv Emissions reduction trajectory in aggregate, and per sector and per capita over time with important milestones such as 2030 and 2050?
- d Does the city have a climate risks/harms inventory, tracking:
  - i Detrimental impacts of climate change over time on every sector population, activity, precinct, and neighborhood in its jurisdiction;
  - ii The location and needs of the most, and those more vulnerable to the impacts of climate change;

e Does the city disseminate this knowledge to its citizens, making it accessible and concrete, and does it share this knowledge with peer-cities, and other sub-national and national authorities?

4 Does the city have a climate action plan that is:

a In ambition and timeframes at the least well-aligned with UNFCCC guidelines and the highest ambitions of the Paris Agreement, aiming for at least reaching carbon neutrality by 2050, if not earlier;
b The result of inclusive and participatory processes within the city;
c Pursued visibly, transparently, and with vigor, engaging actively in dialogue with those who choose not to participate in climate action;
d Clear in its vision and action plan of all of the systemic transformations required across sectors to deliver on the commitment to become carbon neutral by 2050 on the basis of equity and fairness, and is this vision and plan effectively disseminated and shared with citizens and stakeholders;
e Explicit, pro-active, and dedicated to address climate change mitigation, adaptation, and resilience in every investment it makes and every project it embarks on in providing services and infrastructure to its citizens.

5 Does the city have a strategy to combat climate skepticism and climate change denial in its numerous manifestations?

These questions, when asked by citizens or governors of cities and vigorously pursued in the ongoing conversation of cities on the future we want (or the future we want to avoid), may help cities to recognize the moral dimensions of climate change and climate action, and may help to improve the nature and quality of climate action. These questions may help and enable cities to discover what it means to be moved by the weight of moral reasons in climate action: to stop, prevent, or avoid harm, inequity, injustice, or the erosion of dignity; to live good lives with and for others in just institutions in the face of climate change. We are moved by moral reasons in our daily lives; nothing prevents us to do the same collectively in the spaces and ambit of cities grappling with the challenges of climate change and its impacts on urban spaces. In fact, if we never stop asking moral questions in our daily lives, it would be irrational to suspend the process of moral questioning when it comes to the very spaces we live our daily lives in. In short, these questions may help and enable cities to live up to their *raison d'etre*, their reason for being for centuries, and that is to ensure that their citizens, all of their citizens, can live well together, for each other, in just institutions. Of late, one important qualification was added to this conversation, namely, to do so within the limits of planetary boundaries.

Notes

1 These questions are inspired by Wouter Actherberg (1994) who formulated similar (but not identical) questions in order to rescue the radical, critical edge of the notion of sustainability from its stale over-utilization.
2 Bent Flyvbjerg (2001, 2012) used this question in his Progressive Phronetic approach to social science and case studies in applied ethics in order to probe power relations and justice issues.

References

Achterberg, W. (1994) *Samenleving, natuur en duurzaamheid. Een inleiding in de milieufilosofie*. Assen: Van Gorcum.
C40 Cities. (2016) *Deadline 2020: How cities will get the job done*. Available at: https://www.c40.org/researches/deadline-2020

C40 Cities. (2020) *C40 annual report 2020*. Available at: 2827_C40_annual_report_2020_16April2021.original.pdf (c40-production-images.s3.amazonaws.com)
C40 Cities. (2021) *Website*. Available at: https://www.c40.org/about
C40 Cities. (not dated) *Climate action planning resource centre*. Available at: https://resourcecentre.c40.org/
Flyvbjerg, B. (2001) *Making social science matter: Why social inquiry fails and how it can succeed again*. Cambridge university press.
Flyvbjerg, B., Landman, T., & Schram, S. (Eds.) (2012) *Real social science: Applied phronesis*. Cambridge University Press.
ICLEI. (2020) *ICLEI's climate neutrality framework: Accelerating integrated climate action for sustainable urban development*. Available at: https://e-lib.iclei.org/publications/ICLEIs_Climate_Neutrality_Framework.pdf
ICLEI Website. (2021a) *About us*. Available at https://www.iclei.org/en/About_ICLEI_2.html
ICLEI Website. (2021b) *Our pathways, our approach*. Available at: https://iclei.org/en/our_approach.html
IPCC. (2018) *Global warming of 1.5°C*. Available at: https://www.ipcc.ch/site/assets/uploads/sites/2/2019/06/SR15_Full_Report_High_Res.pdf
Nicholls, R.J. *et al.*, 2018: "Stabilization of global temperature at 1.5°C and 2.0°C: Implications for coastal areas." *Philosophical Transactions of the Royal Society A: Mathematical, Physical and Engineering Sciences*, 376(2119), 20160448. Doi:10.1098/rsta.2016.0448.
Ricoeur, P. (1992) *Oneself as Another*. University of Chicago Press.
Steffen, W. *et al.* (2018) "Trajectories of the earth system in the anthropocene." *Proceedings of the National Academy of Science* (PNAS), 115(33), August 2018. Available at: https://www.pnas.org/content/pnas/115/33/8252.full.pdf
Suliman, A. (2018) "Interview-climate battle will be won or lost in cities, says U.N. climate chief." *Thomson Reuters Foundation News*. Available at: https://news.trust.org/item/20180426135359-0uj8q
UN Climate Change News. (21 June 2018) *ICLEI presents new strategy for climate action and sustainability*. Available at: https://unfccc.int/news/iclei-presents-new-strategy-for-climate-action-and-sustainability
UNDP. (2021) *Website*. Available at: https://www.za.undp.org/content/south_africa/en/home/sustainable-development-goals/goal-11-sustainable-cities-and-communities.html

List of Further Readings

UNESCO. (2017) Declaration of ethical principles in relation to climate change. Available at: https://unesdoc.unesco.org/ark:/48223/pf0000260129
World Commission on the Ethics of Scientific Knowledge and Technology (COMEST). (2010–2015) Ethical principles for climate change: Reports of the World commission on the ethics of scientific knowledge and technology (COMEST) of UNESCO (2010–2015). Available at: https://unesdoc.unesco.org/ark:/48223/pf0000245280_eng

15

THE LISBON CASE STUDY

The Ethical Obligations of Urban Policymakers to Support Policies Entailed by Mainstream Climate Science

Karla Matos

Introduction

Climate change can be conceptualized in scientifical, political, and ethical theories, addressed in numerous policies and regulations related to different aspects of human–nature relationships, and translated into calls for action, along with the implementation of official environmental programs. This issue matters for every region, every country, and every city around the globe as societies are watching the dangerous growing gap between their still weak understanding of the political, economic, scientific, and social aspects of climate change and the lack of official political actions in setting effective rules, regulation, educational programs, and participatory policy mechanisms. Creation and enforcement of climate change policies is urgent and must be properly addressed by governments. And this must be done in a collaborative way, with mainstream climate science, and understanding the gravity of the crisis before us, for life on Earth is at stake.

Several scientific developments can produce solid ethical, political, and legal environmental challenges but they require long-term actions through the engagement of a variety of stakeholders, in a participatory dialogue, aiming at having their different views brought up as well as developing tools to contribute to or assist policymaking processes in assessing the impacts of science on society. According to Barry et al. (2013), policy strategies and action repertoires have strong normative and ethical undertones by nature, although the threats of climate change are constantly changing politics, policymaking, and many of the policy innovations developed over the last decade. Due to that, policies frequently and quickly get old, having to be periodically reviewed after new scientific discoveries (like on climate change), which takes time, pace, and political will.

Barry et al. (2013) raise questions like: 'Who is included in and excluded from consultations and (non)decision-making on climate change?' or 'Who is held accountable by whom for lack or insufficiency of solid action?' And we can add, 'What are the best practices that highlight ethical issues and how they can inspire a greater number of governments and civil society?' The attempt here is to advance a local framework, analyzing if the ethical obligations of urban policymakers can produce outcomes that support climate science policies; and, vice-versa, if scientific information can reach policymakers – and societies – in a way that city policies shall be efficiently designed, redesigned, and implemented to meet climate scientific advice, and be resilient enough to evolve.

DOI: 10.4324/9781003039860-18

But Why Cities?

According to the United Nations, more than 50 percent of the world's population live in urban areas, and this is expected to reach up to 68 percent by 2050 (United Nations 2018). Cities seem to be natural co-leaders of the agenda; they are territories that have unique chances to raise sustainable living standards at a faster speed by following low carbon paths, as they also account for a staggering 70 percent of GHG emissions.

The matter of the challenge is that of mitigation and adaptation to climate change. The widely publicized *Intergovernmental Panel on Climate Change - IPCC Special Report on Global Warming of 1.5°C* became one of the most important and proven public scientific reports of the 21st century as it reveals why the rise of every tenth of a degree matters (IPCC 2018). About the role of cities, the most recent *Summary for Policymakers* (IPCC 2021) affirms with very high confidence that cities intensify human-induced warming locally and further urbanization, responsible for intense emissions, together with more frequent hot extremes, will considerably increase the severity of heat waves. Consequently, mean and heavy precipitation and resulting runoff intensity are a lot more probable.

On one hand, in cities "one has to agree with critics since the current urban planning models are not ecologically inclined and not adapted to the current needs" (Allam et al. 2020: 7). On the other hand, "there are concerted citizenry calls for revising urban planning models and policies to better accommodate practices and projects that promote increased resilience against climate change" (Allam et al. 2020: 7). Yet, cities can host successful ethically guided responses to help address some of the most difficult issues related to climate change; for example, consumption behaviors and waste recycling,

> Our consumption, (...) demand for resources like foodstuffs, water and building materials exacerbated by increasing urban populations, also plays a key role in climate change because of their extraction and fabrication needs, prompt unhealthy practices including wholesale vegetation clearance and firing of more land to facilitate agricultural productivity.
>
> *(Allam et al. 2020: 6)*

Going further, in cities the necessary reduction of emissions is difficult to achieve, not only technically but also politically. They demand unprecedented innovation of lifestyles and international cooperation (Warszawski et al. 2021). Considering that the construction of lifestyles occurs primarily in cities, environmentalists and scientists are discussing evidence that cities may be the solution to several environmental problems, including climate change (Lee 2014). However, the relation between a city and climate change is considered highly complex as it is related to complicated social issues. This is perhaps the vital reason why mainstream climate science must be accessible to urban policymakers, so that cities can have accurate plans and proper laws to lead real and intrinsic changes with societal support and responsibility.

Also, according to Lee (2014), there is evidence that applied ethics concerning reducing inequity in cities helps societies to have better environmental outcomes, as the consequences of climate change are the result of unethical human practices and unsustainable standards of living. Lack of enforcement of internationally agreed-upon climate change policies, for instance, leads to the unfair social and economic scenario where all can witness the rich countries – and their cities within – getting richer while the less developed countries continue to get poorer, suffering the growing consequences of a changing climate, mostly induced by the former.

Unfortunately, the demands for climate friendly cities also create a paradox. According to Goldsby (2020), as calls for greener lifestyles grow, lower-income families and communities do not manage to follow the trend. As a result, they end up stuck in the previous century, forced to adopt lifestyles that lead to greater greenhouse gas emissions. In general, as inequity rises, climate negative impacts grow, which "means that our moral obligations with regard to the climate include addressing social inequity" (Goldsby 2020: 396). This is certainly the main problem of inequity related to climate and cities and a strong reason why the response to climate change must be just and inclusive or we will go beyond 1.5 percent, and quickly; in other words, the climate of political and scientific agendas need to meet and belong to every citizen and every city development program around the world. This means that our ethical obligations to the climate agenda must include global cultural and social inequity issues.

Tools for Making the City Climate Agenda Work

It is right to say that there is no time to lose. We need solutions to reduce inequities and halt the melting ice at the same time, and one step forward should be to have science-based regional and international policymaking encourage and guide cities to have their plans focused on interactions between the Sustainable Development Goals (SDGs) and Climate Change Adaptation (CCA) guidelines. This could be achieved, for instance, by unpacking SDG 11, to make cities and human settlements inclusive, safe, resilient, and sustainable, SDG 13 to take urgent action to combat climate change, and CCA strategies altogether, so that cities can have common indicators to review and measure their local climate change action plans, in a way they clearly show how they contribute to national and regional targets. Also, as IPCC's Adaptation Planning and Implementation documentation systematically addresses its calls for local governments, city plans should be reviewed at every recent report, in order to keep on updating them with effective science-based information. Moreover, according to Lubchenco et al, those recommendations for scientific guidance to follow the SDG agenda and build CCA plans should refer to the importance of active participation of the social sciences and a deeper engagement of decision-makers and practitioners, leading to an understanding that these agendas need to reach individuals and groups, as they can help and are key to the construction of multidimensional operational actions on the ground (Rodriguez et al. 2018).

Uncertainty about many of the changing climate impacts vis-à-vis desired benefits may, in certain cases, put decisions on hold. However, mainstream science-based literature and arguments integrating common and participatory local fora knowledge and efforts may be able to point out these areas of uncertainty and eliminate distorted perceptions. Perhaps, only with political will a less costly and more efficient change in city climate policies will be possible. In short, in the cases of uncertainty, it is recommended that multi-sector local groups identify losses and gains resulting from environmental and climate impacts and all possible mitigating and compensatory actions should be included in city plans and actions.

As in the European Union, for instance, Europeans face environmental and sustainability challenges from an economic development matrix. There is a scenario in which the model is being reviewed, as sustainable alternatives for consumption habits are being launched and improved at higher scales than ever. Society and the scientific community have also started to introduce concepts and practices of Circular Economy,[1] with the purpose of getting the most out of production's full cycle, as well as managing stocks. So, in order to remain strong in the policymaker's agenda, environmental sustainability – having climate issues in its chore – has to be understood as an integral part of the development process; it must not be considered in isolation anymore (Mendes 2009), which is just one of the reasons why such indicators, tools and agendas are so relevant.

Choosing an Adjective for the City and the Role of Regional Programs

Several 'brands' of sustainable urban development – green cities, resilient cities, low carbon cities, sustainable cities, friends of the climate cities, healthy cities, smart cities – have allowed different groups, political parties, and governments to create opportunities, dynamics, and pathways for climate actions. Nevertheless, these different programs have also created confusion and competition for and among local politicians and decision-makers, even though it looks as if they connect their guidelines to SDGs and climate change's scientific documents and concerns. The matter seems unethical as it can indicate the political-electoral agenda still surpasses the others, and cities end up taking the long road, spending twice – or more – as much time and resources on never-ending local plan designs.

Fortunately, there are many positive aspects concerning regional and national programs and contests designed to attract cities to the climate agenda. Due to these, a high number of cities have created CCA plans, environmental programs, SDG campaigns, and strategies over the last decade, leading us to wonder: are many of these climate plans just words on paper? Are mitigation and adaptation issues being co-opted by traditional growth agendas? Do they depend on active citizenry and non-governmental groups for local governments to become aware of and highly committed to their implementation, or is it a convenient cover for business as usual?

There are numerous initiatives all over the world, but let's take the example of the European Union Green Capital (EGC) Award. It was initiated in 2010 by the European Commission to promote environmental performance within European cites at a time when environmental concerns, and specifically climate change, started to be at the top of several urban agendas (Davoudi 2014). Since its first edition, the EGC Award has been fairly successful in celebrating both the host city's environmental achievements and its plans to develop a sustainable future (Gulsrud et al. 2017). In order to win, the candidate city must demonstrate that it has a consistent record of achieving high environmental standards; is committed to ongoing ambitious goals for further environmental improvement toward sustainable development; and can act as a role model to inspire other cities and promote best practices to other European cities (European Commission 2018). This Award establishes a two-stage evaluation process. First, an international independent Expert Panel assesses each city's application based on 12 environmental indicators: Climate Change Mitigation; Climate Change Adaptation; Sustainable Urban Mobility; Sustainable Land Use; Nature and Biodiversity; Air Quality; Noise; Waste; Water; Green Growth and Eco-innovation; Energy Performance; and Governance. Second, the Panel selects a shortlist for the next stage of the competition.

These topics seem to demonstrate that the EGC Award is not simply given to cities with outstanding parks, biodiversity, and water quality, but to those that consider sustainability at their core. Actually, a very recent addition to the list, "Eco-Innovation and Sustainable Employment," clearly shows that the EU Commission means to emphasize economic development to be an important quality of a sustainable city. Still, candidate cities are asked to provide three types of data for each category: the existing environmental performance of the city; the city's performance facing its stated plans and strategies; and an outline of their future plans and strategies, enhancing that host cities will be committed to ongoing and ambitious environmental action. At the last stage, finalist cities formally present their commitment to an international panel in which the city environment and their capacity to act as a role model to others is evaluated.

The City of Lisbon

In the city of Lisbon, Portugal, local sustainability policies and laws have led it to become the European Union Green Capital 2020, awarded by the Karmenu Vella, EU Commissioner for Environment, Maritime Affairs, and Fisheries at a ceremony in Nijmegen, the Netherlands, the former EGC. In addition, the city was also awarded the European Entrepreneurship Region prize – EER 2015, was host of the Web Summit 2016-2018, and is a leader in the networks Sharing Cities 2020 and Smart Cities and Communities. It was also the first European capital city to sign the New Covenant of Mayors for Climate and Energy in 2016, which helped leaders to formalize policies and actions, such as changes in the city's master plan to allow green corridors to be implemented.

Lisbon, the capital and largest city of Portugal, has a population of 5,47,800 people and its metropolitan area is of 2.8 million, with 18 municipalities. It is one of the world's oldest capitals, visible in its rugged facade, Roman ruins, run-down aesthetics, and even traces of the 1755 Lisbon earthquake, with subsequent fires and a tsunami that almost destroyed the entire city. At first glance, besides recycling bins, green areas, and parks located throughout the city, it is not easy to see Lisbon as a sustainable city. However, the city has been proving it has an ample vision and ambitious targets to become green and a 'city of neighborhoods' in which, for instance, anyone, anywhere in the city can have easy and clean means of transportation, integrating bicycle and walking paths in between boroughs.

Ana Teixeira Pinto (2018) describes some ancient features of Portugal's national identity: they are discoverers, adventurers, and explorers of many of the world's regions; they are a coastal nation, with archipelagos in the Atlantic Ocean: the Azores and Madeira; and, more recently, they are economically dynamic (there was an economic boom in the country about three years after the 2008 crisis that had severely hit the country). Together, these three characteristics could actually produce a cultural commodity that seems to be part of the reason why Lisbon has features of a resilient city that can walk fast on the low carbon path.

The EU Expert Panel (European Commission 2018) has explicitly considered Lisbon a particularly strong city in the fields of sustainable land use, sustainable urban mobility, green growth, climate change adaptation, and waste management because Lisbon has achieved a 50 percent reduction in CO2 emissions (2002–2014) and reduced energy consumption by 23 percent and water consumption by 17 percent from 2007 to 2013. In addition, Lisbon has strategies for sustainable urban mobility, with actions to restrict car use and prioritize cycling, public transport, and walking paths and has one of the world's largest networks of electric vehicle charging points. 39 percent of the municipal public buses electric, and 93.3 percent of Lisbon's population live within 300 meters of a public transport service. The majority of the population, 76 percent also lives within 300 meters of green urban areas.

Furthermore, one of the strategic projects for Lisbon is to commission a municipal solar power plant for charging their public e-fleet of buses and waste management vehicles, as well as provide renewable energy for communities.[2] In addition, their Solar Platform – Solis 3 – is a free web tool developed by the Environment and Energy Agency to inform and educate people about solar energy, and works as a tool for citizens to determine the solar energy potential of their building and how to install solar panels.

Lisbon's Actions as the EU Green Capital 2020

It was a very promising year for Lisbon but carrying out the city's journey as 'European Green Capital 2020' was not an easy task due to the Coronavirus pandemic that started in March of the same year.[3] Nevertheless, before that, in January, the population had answered to the city call for a 'trees planting day,' when around 5,400 volunteers planted 20,000 trees and shrubs in one single day. Because of the greater commitments to the European Commission as an EGC,[4] green growth was actually a crucial goal for the city of 1,553 hectares of public green areas. Lisbon then presented a Local Action Plan for Biodiversity for the use of its four green corridors, and for the implementation of other four, by 2022. At the end, these green corridors will run a total of 45 km through the city.

According to Goldsby (2020), there are many aspects of a city that can make it easier for one to adopt a carbon-friendly lifestyle. One of them is by having more open areas, mainly parks, gardens, and green fields for outdoor activities. He also says that because cities are typically geographically concentrated, the population can emit fewer greenhouse gases simply by getting around the city, walking, biking, or using public transportation. These types of urban mobility actions in Lisbon are key to meeting its CO_2 emissions reduction target. In order to embody this strategy, starting in 2019, Lisbon interconnected its public transport services – buses, trams, trains, subway, funiculars, and lifts – in a single ticketing method across the metropolitan area, accessible to 2.8 million people. This system has also been enhanced by 90 km of cycle paths, with a further 60 km currently under construction and 50 km more planned. Also, there is a municipal shared bike system of around 3,000 bikes that are rented approximately 20,000 times a day, and the city has developed one of the largest European networks of electric vehicle charging points with 540 public stations, having now a 91 percent municipal car e-fleet. Electric buses, and some that use biofuel, serve Lisbon's bus routes and there are ongoing investments in natural gas buses in attempts to increase the use of biofuel from collected used cooking oil.[5]

According to Donald Brown (2017) in his paper "571 Strategies to reduce GHG emissions adopted by 44 cities,"[6] while national responses to climate change are still inadequate, many local governments around the world have started to reduce their emissions at very ambitious levels. Although Lisbon and the 17 municipalities of its Metropolitan Area are still developing a common Climate Adaptation Plan, adaptation actions have been considered ambitious as they crosscut the city plans with the Lisbon Commitment - Climate Action 2030 charter and consortium of more than 200 organizations.

Circular economy strategies are playing a key role in the commitment, with initiatives to reduce consumption and improve waste management, reuse, and upcycling. Single use plastic is banned at public events, plastic bags are not for free anymore at all stores and markets since early 2021, and, concerning non-recyclables, a project has been developed for biowaste reduction, as the city is putting forward free composting bins and use training for families.

Another accomplishment was doubling the local participatory budget for 5 million euros in 2020, calling citizens to have a voice, campaign, and vote on projects related exclusively to urban sustainability like community gardens. Goldsby (2020) affirms that community gardens could provide low-income families access to healthier foods, improving their overall well-being. Lisbon has 750 land plots for community gardens in place – ready to increase to 1,000 plots – and are being looked after by 750 families and community groups.

It looks as if Lisbon's citizens are likely to be engaged in civil actions. Citizenship and bottom-up strategies and policies play an important role in several of their green choices, with the UN 2030 Agenda for Sustainable Development and the recently approved European climate targets acting[7] as guides for city policymaking. There are also several other ongoing local bills on water issues, green energy, land use, and eco-innovation projects and policies, involving universities as well as citizens, businesses, and national and foreign entrepreneurs.

At a broader view, it is indeed a very important moment for societies, taking into account that local governments are realizing their potential – that they must legislate and ought to participate in scientific and political national and international debates about the climate emergency. And by doing so, they create better plans and policies as a matter of justice and safety – and climate.

When starting the Stop Ecocide movement,[8] Higgins (2015) highlighted that

> The rules of our world are laws, and they can be changed. Laws can restrict or they can enable. What matters is what they serve. Many of the laws in our world serve property – they are based on ownership. But imagine a law that has a higher moral authority... a law that puts people and planet first. Imagine a law that starts from first do no harm, that stops this dangerous game and takes us to a place of safety....
>
> *(Higgins 2015)*

Perhaps traditional policymakers find it hard or even not possible to make profound changes in the system as a few powerful world leaders and corporations rule the global market, leaving little room for a low-carbon economy to flourish. Things have changed though; the planet is in serious danger of an unprecedented collapse and societies are bearing witness to dramatic changes in our global climate. What else can be done if ethical obligations are not met through global, national, and local science-based actions, policies, and laws?

Does Lisbon Incorporate Ethics in Policy and Support Climate Science?

According to a regional review provided by the most recent Eurobarometer[9] Climate Report 513 (European Commission 2021), climate change is the second single most serious problem facing the world for the 27 EU Member States, including Portugal. Also, the majority of respondents answered that climate change is a "very serious problem" at this moment, and Portugal, with 91 percent leads this opinion but with only 35 percent believing that local governments are responsible for tackling climate change.

At Portugal's capital level, from the various official documents researched, it is likely that mainstream climate science has been fairly considered by Lisbon when implementing public policies. As they are constantly calling for social and academic participation and supporting science and education, their current obligations may be having an important role in changing city laws and vision toward a sustainable approach. In addition, it appears that the city′s ongoing environmental and climate policies are intrinsically living principles of ethics through their actions, such as this call for citizens' participation in the EGC 2020 campaign

> Choose to evolve! We need the indifferent, the conformed and the skeptical. We need those who care too much about the car. And those who do not turn off the light. We need those who let the water run. And those who linger in the bath. We need those who trash the sea. And those who throw into the air. We need the pessimists and the consumerists. Those who want a straw. The plastic bag. And the disposables. We need those who recycle apologies and nothing else. Those who do not want it and those who do not believe it. We even need those who mean no harm. We need the indifferent. We can no longer save the world without them.
>
> *(European Union 2021)*

Residents and visitors could – and still can – read these words electronically in sites and ads and on fliers, magazines, and huge banners all over the city. It is a call for people to think about what is right and what is wrong, an invitation for the ones who care and do not care. This can be considered a message to make people stop and think about ethics, values that rarely change, as they are intrinsic and somewhere inside everyone.

Further analysis is necessary for their committed challenges to sustain, understanding that policymakers must continuously take steps in support of policies entailed by climate science, making further decisions "in the face of scientific uncertainty about the timing and magnitude of climate change impacts" (Brown 2010). It seems that, for local authorities, climate science is more useful as a guide for social, cultural, and political choices than as a method to follow, scientists must keep looking for ways to maintain its undeniable importance, namely in the understanding of complex phenomena and disruptive situations. It is crucial that climate science maintains its impartiality and accuracy; in other words, its strength. Climate science tends to lean toward politics by simplifying conclusions and ignoring uncertainties, and politics tends to get closer to science by justifying decisions with scientific facts. This article understands that science should stay within its hard core of competence and also be open for information exchange with policymakers, just as public policymakers should promote open and inclusive decision-making processes and support science, taking into account its uncertainties and specific domain.

Conclusion

Recent international calls for a global urban science with new scientific knowledge that can answer complex urban challenges, like tackling climate change by different stakeholders, emphasize the importance of systemic multidimensional city action. The topics raised here are a modest contribution to this end, particularly when adding mainstream climate science to cities' decision-making arenas.

Timing is an essential factor for strengthening climate change adaptation and mitigation plans and their implementation in cities. The dynamics of urban growth in this century creates a limited timeframe to develop adaptation strategies at an affordable cost, but insufficiently fast responses for emerging climate change impacts seem to already be putting sustainable development in jeopardy. Urgent calls for citizens to see city actions and plans through ethical lenses can enable a systemic view to flourish, governance to work, city networks to improve, and sustainability to become real.

City policies and climate science, along with innovation, seem to be able to produce some effective answers for the climate crisis. Moreover, when a local climate plan is built within a framework that considers the SDGs and Climate Change Adaptation – or mitigation – strategies, they naturally end up having approaches for tackling social inequality growth as well as for policymaking and mainstream climate science integration. SDGs and climate change issues are actually catching people's attention, bringing new opportunities for different stakeholders' engagement to escape from the trap of 'business as usual' planning.

Promoting policies for reducing vulnerability, facilitating universal access to basic services, investing in green lifestyles, and education for all are indeed able to accelerate adaptation at the local level, as long as they become trustworthy and society cares for them. Although accessible clean water, local empowerment, job training, and inexpensive basic health care are crucial for sustainability, they are not sufficient for a city to lower its carbon emissions. Ethical premises and actions must be crosscutting in order to change unsustainable policies and practices.

Through Lisbon's case study, we could see that focusing on building green strategic planning and policies can be a long and prosperous project for the city. They have won an EU award for innovations that required city actions in partnership with citizens, businesses, and the scientifical community. Decision-makers and stakeholders enabled actions while promoting local sustainable development policies, as well as influenced the urban economy, transport infrastructure, and behavioral patterns. Yet, the development of green areas and community gardens, protecting ecosystem services, and developing nature-based solutions also seem to be able to increase the set of urban adaptation options.

This article has also attempted to illustrate how ethics can contribute to civic understanding of climate change threats and opportunities for a city to grow toward zero GHG emissions. Of special interest is the place of public policy within wider approaches to climate ethics and social inequalities reduction. In general, urban action and policy must always serve as a hub for transitions in production and consumption systems and, consequently, in lifestyles. Lisbon's climate plans and actions will be used in the future as a model for others that are planning to go green, explicitly showing that an ethical framework – alongside a lot of work and creativity – are needed to lead the integrative implementation of public policies in synergy with mainstream climate science.

Notes

1 According to the European Parliament, circular economy is a model of production and consumption, which involves sharing, leasing, reusing, repairing, refurbishing, and recycling existing materials and products in a way that the life cycle of products is extended. In February 2021, the Parliament adopted a resolution on the new circular economy plan with additional measures to achieve a carbon-neutral, environmentally sustainable, toxic-free and fully circular economy by 2050, including targets for the use of materials by 2030. <www.europarl.europa.eu/news/en/headlines/economy/20151201STO05603/circular-economy-definition-importance-and-benefits>.

2 Located on the Atlantic coast of Portugal, the city has an average of eight-hour sunshine per day, one of Europe's capital cities with the highest exposition to the sun.

3 Located on the Atlantic coast of Portugal, the city has an average of eight-hour sunshine per day, one of Europe's capital cities with the highest exposition to the sun.
challenge. The city then started to plan things for its legacy to last.

4 Every city has to write a post-assessment report for their laureate year documenting its highlights. After five years, another report has to be submitted to the Commission. The report summarizes the year and describes outcomes.

5 Further information in <www.sustaineurope.com/lisbon-european-green-capital-2020-20201228.html>.

6 In <https://ethicsandclimate.org/2017/09/01/the-enormous-potential-of-local-governments-to-reduce-ghg-emissions-a-paper-that-identifies-571-strategies-adopted-by-44-cities-around-the-world/>.

7 The EU has set itself two climate targets for the future. The first EU climate target is set for 2030, while the other one is set for 2050. The initial target for 2030, based on the previous climate and energy framework in 2019, was set at a 40 percent reduction in greenhouse gas emissions compared to 1990 levels. Later, with the introduction of the European Green Deal to make Europe a climate-neutral region by 2050, this EU climate target was updated to 55 percent, including emission reduction and removals <https://eucalls.net/blog/eu-climate-targets>.

8 <www.stopecocide.earth/what-is-ecocide>.

9 Eurobarometer is the polling instrument used by the European Commission, the European Parliament and other EU institutions and agencies to regularly monitor the state of public opinion in Europe on issues related to the European Union as well as attitudes on subjects of political or social nature. The aim of the Eurobarometer is also to provide quality and relevant data for experts in public opinion and the public alike. <https://europa.eu/eurobarometer/about/eurobarometer>.

References

Allam, Z., Jones, D. & Thondoo, M. (2020) *Cities and Climate Change – Climate Policy, Economic Resilience and Urban Sustainability*, Cham: Springer Nature.

Barry, J., Mol, A.P.J. & Zito, A.R. (2013) "Climate Change Ethics, Rights, and Policies: An Introduction," *Environmental Politics* 22 (3), pp. 361–376.

Brown, D.A. (2010) *Have We Been Asking the Wrong Questions About Climate Change Science? Why Strong Climate Change Ethical Duties Exist Before Scientific Uncertainties Are Resolved*. <https://ethicsandclimate.org/2010/07/>.

——— (2017) *571 Strategies to Reduce GHG Emissions Adopted by 44 Cities*. <https://ethicsandclimate.org/category/cities-and-climate-change>.

Davoudi, S. (2014) "Climate Change, Securitization of Nature, and Resilient Urbanism", *Environment and Planning C: Government and Policy* 32(2): 360–375.

European Commission. (2018) *European Green Capital Award 2020: Expert Panel Technical Assessment Synopsis Report*. <https://ec.europa.eu/environment/europeangreencapital/wp-content/uploads/2013/02/EGCA_2020_Technical_Assessment_Synopsis_Report.pdf>.

European Commission. (2020) *European Green Capital: Rules of Contest*. <https://ec.europa.eu/environment/europeangreencapital/wp-content/uploads/2020/Rules_of_Contest_EGC_2022_EGL_2021_Web.pdf>.

European Commission (2021) "Eurobarometer," *Climate Change Series*. <https://europa.eu/eurobarometer/surveys/detail/2273>.

European Union (2021) "Lisbon European Green Capital 2020," *Sustain Europe*. <www.sustaineurope.com/lisbon-european-green-capital-2020-20201228.html>.

Goldsby, M. (2020) "Paradox in the City: Urban Complications Regarding Climate Change and Climate Justice," in Meagher, S.M., Noll, S. & Biehl, J.S. (eds.), *The Routledge Handbook of Philosophy of the City*, Chapter 34, Milton Park: Routledge, pp. 390–398.

Gulsrud, N.M., Krajter **Ostoić,** S., Faehnle, M., Maric, B., Palomieni, R., Pearlmutter, D. & Simson, A.J. (2017) "Challenges to Governing Urban Green Infrastructure in Europe: The Case of the European Green Capital award," in Pearlmutter, D., Calfapietra, C., Samson, R., O'Brien, S. Krajter **Ostoić,** S., Sanesi, G. & Alonso del Amo, R. (eds.), *The Urban Forest: Cultivating Green Infrastructure for People and the Environment*, Cham: Springer International Publishing, pp. 235–258.

Higgins, P.H. (2015) *Quote in.* <www.stopecocide.earth/polly-higgins>.

IPCC. (2018) *Special Report Global Warming of 1.5%.* <www.ipcc.ch/sr15/>.

IPCC. (2021) *Climate Change 2021: the Physical Science Basis, Summary for Policy Makers.* <www.ipcc.ch/report/ar6/wg1/downloads/report/IPCC_AR6_WGI_SPM.pdf>.

Lee, T. (2014) *Global Cities and Climate Change: The Translocal Relations of Environmental Governance*, New York: Routledge.

Mendes, M.C. (2009) *Desenvolvimento Sustentável.* <www.tratamentodeagua.com.br/artigo/desenvolvimento-sustentavel>.

Pinto, A.T. (2018) "The Art of Gentrification: The Lisbon Version," *Afterall A Journal of Art Context and Enquiry* 45, pp. 88–97.

Rodriguez, R.S., Ürge-Vorsatz, D. & Barau, A.S. (2018) "Sustainable Development Goals and Climate Change Adaptation in Cities," *Nature Climate Change* 8: 181–183.

United Nations. (2018) *Department of Economic and Social Affairs News.* <www.un.org/development/desa/en/news/population/2018-revision-of-world-urbanization-prospects.html>.

Warszawski, L., Kriegler, E., Lenton, T.M., Gaffney, O., Jacob, D., Klingenfeld, D., Koide, R., Costa, M.M., Messner, D., Nakicenovic, N., Schellnhuber, J., Schlosser, P., Takeuchi, K., Van der Leeuw, S., Whiteman, G. & Rockström, J. (2021) "All Options, Not Silver Bullets, Needed to Limit Global Warming to 1.5°C: A Scenario Appraisal," *Environmental Research Letters* 16: 064037.

16

THE ETHICAL RESPONSIBILITY OF INDIVIDUALS TO REDUCE GREENHOUSE GASES

Rainier A. Ibana

Introduction

Aldo Leopold defined the parameters of environmental ethics in an essay entitled *The Land Ethic.* According to him, "A thing is right when it tends to preserve the integrity, stability, and beauty of the biotic community. It is wrong when it tends otherwise" (1949). Leopold's distinction between right and wrong extends the scope of ethics from interhuman relationships to include the environment and the other species that dwell therein. The surrounding environment, moreover, can be extended from local ecological niches to include the whole planet as a 'biotic community.' This extension is urgent today because the Earth's biosphere is being threatened by global warming due to climate change. Former UN Secretary General Ban Ki-Moon noted that "Climate change is, simply, the greatest collective challenge we face as a human family" (Ki-Moon 2009).

The United Nations Framework for Climate Change (UNFCCC) attributed the problem of climate change as anthropogenic in nature. It defines this problem in terms of "human activity that alters the composition of the global atmosphere and which is in addition to natural climate variability observed over comparable time periods" (UNFCCC 1992: Art. 1.2). Global warming was precipitated by the processes of industrialization when greenhouse gas emissions, most of them carbon dioxides, were released to the environment, which then absorbed, trapped, and reemitted the sun's life-giving energies back to the Earth's surface. This led to global warming and its adverse consequences: melting ice-caps, unpredictable weather conditions, violent typhoons, salt-water intrusions on agricultural lands, sinking islands, desertification, and floods, among others.

But if climate change is rooted to the collective social transformations that beset humanity as a whole, it will be very challenging, if not overwhelming, for individuals to address this problem. Only human beings, with their capacity for reasoning and decision making, however, are empowered to redirect this impending planetary catastrophe. Humans cannot abdicate their responsibility to protect and save the planet and their fellow "earthlings" (Rolston 2006: 65). Effectively addressing this responsibility, however, requires understanding the scientific background that it entails and the special location of human beings in the natural world.

The Basic Principles of the Ecological Sciences: The Laws of Thermodynamics and Biology's Ten Percent Law

Greenhouse gas (GHG) emissions are about energy and its transformations. The technological innovations brought about by industrialization, however, led to the exponential increase of the human being's power to harness nature's latent energies. The physical laws of thermodynamics

 DOI: 10.4324/9781003039860-19

tell us that energy is neither created nor destroyed. It is merely converted from one form to another.

In the process of converting energies, however, the recipient can only receive the potential energy of the resource in accordance with its limited capabilities. Since nature's resources are greater than the human beings' capacity to receive and utilize them, the excess energies are released to the environment. Such excessive energies are multiplied exponentially by industrial innovations to harness the seemingly infinite resources of the planet.

This surplus energy can be traced in the record of expenses charged by electricity franchisers to their consumers. Electricity bills itemize such charges as transmission and distribution costs along with 'system loss.' The Philippine Distribution Code of 2016 defines system loss as "the difference between the electric energy delivered to the distribution system and the energy delivered to the end-Users and other entities connected to the system" (PDC 2016). This charge is added to the actual energy used by the consumers. These expenses, just like most environmental contributions to the production and consumption processes, are 'hidden costs.' They are usually not accounted for because they are not located within the scope of the production and consumption systems. Electric meters measure only the wattage that are consumed, not the energies that were lost in the process of converting and transferring energies.

The reduction of energy is also evident in the broad pyramidic structure of feeding levels in an ecological niche. Biologists estimate that secondary consumers, such as herbivores, consume only 10 percent of energies from feeding on vegetation. Carnivores, food consumers at the third level that feed on herbivores, consume only 10 percent of energies from animal meat. The latter, therefore, receives only 1 percent of the 10 percent of the energies that were initially consumed by herbivores. Vegetarians are more efficient in terms of their absorption of energies from the biotic community because they get their nutrients directly from plants instead of the indirect route taken by carnivores. This can be illustrated by a pyramid of biology's 'Ten Percent Law,' with carnivores at 1 joule of energy (the top of the pyramid); herbivores at 10 joules of energy; vegetation at 100 joules of energy; and soil and other producers at 1,000 joules of energy (the bottom of the pyramid).

These feeding stages are technically called trophic levels (Molles 1999: 136). The energies that were not absorbed into the pyramid of feeding stages are described as 'entropic.' The notion of entropy refers to the degradation and decay that accompanies the emergence of life. As life becomes more complex and intensified in its ascension toward higher levels of organization, chaos surrounds this organizational complex until a new level of equilibrium is obtained between the emergent organism and its surroundings.

The introduction of a predator or a new technology into a feeding system, for example, disrupts a prevailing equilibrium. Previous predators are deprived of energy while the available sources of energy are reduced proportionately in relation to the new predators. The relationships among the members of the niche are reorganized until a new balance is obtained. The so-called 'Tragedy of the Commons' illustrates how additional consumers such as cows or fishermen, will reduce the share of resources of the other consumers in an ecological niche (Hardin 2001). They have to agree on a new distribution scheme if equal sharing is to be maintained among them.

This process of balancing energies within and outside ecological systems follows from the laws of thermodynamics. Energies that are not transformed within an ecological niche must go somewhere else. This excess energy is usually in the form of heat. But the earth must maintain a stable temperature where life can survive and flourish. Climate scientists are worried because they estimate that the tipping point for the irreversible damage on the planet's delicate biotic equilibrium will be breached if heating levels are reached beyond the 1.5°C rate of increase from pre-industrial levels.

Human beings, with their intellectual prowess, nevertheless, can become the most voracious creatures on earth if their consumption patterns are not mitigated by their inherent capacities for intelligence, freedom of choice, and compassion. These faculties must be exercised judiciously if they are to be worthy of their stature as *homo sapiens* (Rolston 2006: 66).

When humans realize that their own survival is also at stake in their mindless relationship to their surroundings, they can be led to cultivate ecological virtues such as frugality, prudence, and a reverential attitude toward the natural world. These virtues correspond to ethical principles such as sustainable development, the precautionary principle, and respect for the integrity of the natural world. These virtues and principles are consistent with the norms enumerated in Leopold's *Land Ethic*: "the integrity, stability, and beauty of the biotic community" (1949).

Duties of Individuals to Reduce Green House Gases

On the basis of the scientific context described above, the duties of individual human beings to reduce greenhouse gas emissions can be enumerated in terms of the following set of duties: (1) duties to conserve energy, (2) duties to delay the onset of entropy, (3) duties to act in accordance with nature's cycles of production, and (4) duties to participate in collective actions that demand the reduction of GHGs. These duties complement, support, and follow one another.

Duties to Conserve Energy: Lessons from the COVID 19 Pandemic

The first and foremost duty of individuals to lower GHG emissions is to live closer to the ground. The natural world flourished when the level of consumption of the Earth's resources was minimized to that of bare necessities during the COVID-19 pandemic. The use and emissions from fossil fuels were reduced when people and products stood still to prevent the spread of the virus. The air became fresher and more habitable, wildlife thrived, and some animals even ventured to intrude into human habitats.

Backyard and community gardens sprouted as food consumers took the task of food production literally into their own hands. Supply chains were shortened when purchases were limited to local goods and services. The transfer of energies was reduced significantly. Patronizing local products reduced transportation costs as consumer goods need not be imported from abroad and people merely walked or ride bicycles to reach their destinations. Using the stairs instead of elevators and escalators became a source of physical exercise while reducing the consumption of fossil-fuels that energize automated machines.

Fossil fuel by-products, such as plastic wrappers and containers, also became unnecessary. Fresh fruits were packaged in biodegradable materials such as leaves and paper instead of non-biodegradable ones. One need not worry about the shelf-life of commodities because the periods of transportation were shortened. Consumer preferences also shifted to fresh fruits and vegetables over meat products in order to enhance the immune system. The feeding level went down to that of herbivores as vegetarianism became a popular lifestyle option.

These practices of frugality that were implemented during the pandemic can be exercised also under normal circumstances. They do not only conserve the energies extracted from the environment. They also contribute to human health and well-being. Health, livelihood, and extending lifespans are existential concerns of every individual.

Duties to Delay the Onset of Entropy

The simplest way to reduce GHGs is to cultivate more vegetation. Plants sequester carbon dioxide emissions, while producing oxygen to cool down the environment. If every individual person can grow plants to at least make up for their carbon footprints, humanity can be on their way to restoring the required temperature for life to flourish on the planet. But since GHG pollution is exponentially multiplied by engines that empower industrialization, plant cultivation must be matched by other technical innovations that reduce GHG emissions.

Farmers have employed the technique of agroforestry not only to increase their productive capacity but also to strengthen their ecological niche. By designing farms that imitate the symbiotic relationship of plants, agricultural products, and animals, small farm holders and gardeners can optimize their production while increasing the resilience of their farms from the contingencies brought about by harmful impacts driven by the volatility of market forces.

It is therefore not enough to plant anything anywhere, although such activities can reduce the harm caused by air pollution. The rapid evaporation of water from arid lands is dumped somewhere else by virtue of the laws of thermodynamics as mentioned above. Humans can construct irrigation and water systems that follow the contours of the land. Flood waters can be channeled to aquifers by carefully planting trees that guide rainwater to be absorbed by the land and used productively in ponds that create habitats for water fowl and fishes.

Employing the principle of biodiversity, as defined by the UN Convention on Biological Diversity, is not a matter of the number of individuals or species within an ecological niche. It is about the quality of relationships among the organisms that dwell in it. These relationships can be arranged in such a way that they feed and enrich the members of the biotic community. Such arrangements are not limited within the boundaries of farmlands but must encompass weather conditions, the surrounding terrain, quality of the soil, sunlight, and even seasonal animals that visit the region where the area is located. Sustainable development, therefore, does not happen instantly and accidentally. It must be carefully planned and collectively deliberated upon from the synoptic vision of the environment.

These prescriptions are understandable from the perspective of Natural Law ethics, which presupposes that all beings tend to fulfill their purpose or telos. When this orientation is thwarted by external conditions, however, the organism reacts in a negative way: repulsed and revolted by the evil that blocks them. Plants and animals, for example, try to surmount the obstacles that prevent their self-development. Roots eat up pavements to reach the earth and animals cry in pain when hurt by others. To a large extent, humans also learn about good and evil through their senses. In *The Definition of Moral Virtues*, ethicist Yves Simon describes "knowledge by inclination in moral matters by comparing its feelings of attraction and repulsion to the sense of smell" (1999).

Simon's insight is relevant when applied to the environmental ethics of greenhouse gas emissions. Communities and business corporations frequently succumb to the lure of financial profitability when evaluating power plants, mining projects, and water dams. But natural individual persons are the ones affected by issues of health and livelihood in relation to the surroundings from wherein they depend for their daily subsistence.

Local knowledge claims arise from the lived experiences of individuals who actually reside in vulnerable areas. They serve as valuable sources of information on how to identify and solve ecological problems. It is well known that indigenous peoples are the best custodians of the environment because of their actual valuation of the resources that abound in their surroundings.

It is important for individual citizens to voice their concerns and be allowed to participate in deliberations and decision-making processes that affect their living conditions. UNESCO's *Declaration of Ethical Principles in Relation to Climate Change* (2017) states that local, traditional, and indigenous knowledge must be taken into account when appropriate. UNESCO's Commission on the Ethics of Scientific Knowledge and Technology (COMEST), in its formulation of the precautionary principle, also recommends that the choice of actions that pertain to human activities that may lead to unacceptable harm "should be the result of a participatory process" (COMEST 2005: 14).

In light of their potential contributions, individuals, as citizens of localities, regions, and nation states, should be able to assert their rights to participate and be involved in decision-making processes that concern their own health and well-being. This participation is also warranted by the need to secure a better future for the next generations.

Duties to Abide by the Cyclical Patterns of Nature

Another set of duties for individuals to reduce GHG emissions is to abide by nature's cyclical patterns of production. The basic elements of nature move in cycles. The Earth itself revolves around the sun. In this revolution, seasonal cycles are affected by the length of the sun's rays as it falls on the surface of the earth. Arid and hot climates are located near the center of the Earth, moderate temperatures are felt in the middle, while cold weather is the norm for those who live at the opposite end of the Earth's poles.

Excess energies from the sun could have bounced back from the earth had it not been for GHGs that blocked its release from the atmosphere. The heat generated by the sun empowers the wind and ocean currents while affecting the rate of evaporation and condensation of water. Excessive heat speeds up the movement of these elements, thereby creating more frequent and stronger typhoons.

To mitigate these climatic conditions, humans can reduce the production of excess energies by recycling and reusing materials whose production has already affected nature. This will help reduce the rate of extraction of energies from the natural world while minimizing wastes at the end of the production line.

Waste segregation can also be practiced at the household level. Biodegradable materials can be composted as fertilizers, while non-biodegradable ones can be reused or repurposed as pots, water containers, and wrapping materials. Linear methods of production can be tempered by creating feedback loops within production systems. These loops regenerate energies by returning excess energies into the production system. Appliances, such as refrigerators and air conditioners with inverters, are examples of this feedback loop. Reusing utensils instead of patronizing disposable plastic products is another energy saving practice that individual consumers can become mindful of.

The mass production of solar-powered energy systems has enabled consumers to afford using them. Even low-income groups have innovated solar-powered bottles filled with chlorinated water and inserted on their roofs have lighted hovels in low-income communities. Innovations that make use of wind energy may also be deployed by entrepreneurs by reconfiguring discarded appliances. There is no limit to the capacity of the human mind to imagine the extension of the lifespan of material products by recycling, reusing, and refurbishing them for other purposes.

Such reflective and creative deployments of human innovation also exhibit a cyclical pattern. The act of reflection is not only oriented to the object of knowledge but returns to the reflective agents themselves in order to affirm the veracity of their conceptions of reality. Such reflexivity is intensified within the context of engaging others in collective decision-making processes that involve deliberation, dialogue, and consensus-building.

Duties to Participate in Collective Actions that Demand the Reduction of GHGs

The advent of mass communication technologies and the social media has empowered individuals to participate in the advocacy for the reduction of GHGs. Greta Thunberg's heroic boycott of her own schoolwork to call on world leaders to pay attention to the global problems of climate change and global warming is a testament to the power of individuals to intervene in the mindless activities of a consumerist culture that forebodes the possibility of a bleak future for humanity.

In a landmark case filed in the Philippines by minors represented by their parents in *Oposa v. Factoran*, the Supreme Court ruled in favor of the petitioners to (1) "Cancel all existing timber license agreements in the country" and to (2) "Cease and desist from receiving, accepting, processing, renewing or approving new timber license agreements" (1993). Timber, felled trees that could have sequestered carbon dioxide emissions from the environment, weakens the concerted effort of environmentalists to reduce global warming.

Aside from invoking the principle of 'intergenerational solidarity' among the petitioners and their constitutional right "to a balanced and healthful ecology in accord with the rhythm and harmony of nature" (Philippine Constitution 1987: Section 16, Article II), the ruling pointed out that such basic rights "need not even be written in the Constitution for they are assumed to exist from the inception of humankind."

This pronouncement is in accord with the first precept of Natural Law in St. Thomas Aquinas' ethical theory. According to this 'Angelic Doctor':

> The order of the precepts of natural law is the order of our natural inclinations. For there is in man a primary and natural inclination to good, which he has in common with all things, inasmuch as everything desires the preservation of its own being (esse) according to its nature.
>
> *(Clark 2000: 299)*

The second precept of Natural Law extends this injunction for self-preservation toward caring for others, specifically of children. This is significant by virtue of the principle of intergenerational solidary that justified the personality of children and future generations to file a class suit against the government. According to this Thomist precept:

> There is in man a second inclination to more specific ends according to the nature he has in common with other animals. According to this inclination, those things are said to be of natural law that "nature has taught all animals," instincts such as the union of husband and wife, the education of children, and so forth.
>
> *(Clark 2000: 300)*

More recently, in 2015, the citizens of the non-profit Urgenda Foundation obtained a ruling from the Supreme Court of the Netherlands "to protect its citizens' human rights in the face of climate change and must reduce emissions by at least 25% compared with 1990 levels by the end of 2020." This ruling was followed by other environmentalists who filed similar cases against their governments "across the world, from Pakistan to New Zealand" (Kaminski 2019).

Collective actions by organized non-state actors can serve as venues for individuals to influence nation-states and businesses to reduce GHG emissions. Climate change and global warming entails a structural transformation of the public sphere dominated by government and businesses. Individuals must therefore influence this arena by cooperating and organizing with like-minded individuals in order to effectively make a dent in this global problem.

In this regard, civil society, an alternative social system fueled by ideas and voluntary action, is emerging in the public sphere as a direct venue for individuals to express their sentiments and arguments in favor of reducing GHGs emissions. Civil society affords individuals with the possibility of participation in common concerns that affect them as citizens and stakeholders Instead of the representations made by the state's powerbrokers and the mediation of stockholders fueled by the financial interests and profits. In this regard, civil society movements offer a broader understanding of social problems on the basis of arguments and rational deliberation. They offer themselves as viable alternatives to the collective regimes of power and money.

Conclusion

The list of duties enumerated above was derived from nature's patterns of consumption, production, and regeneration. Human beings, as rational and emancipated participants in the natural world, however, are able to intervene in economic and political processes that violate the life-giving systems of nature and introduce new ways of living that are consonant with the natural world. As Aldo Leopold counsels above, a human intervention is "right when it tends to preserve the integrity, stability, and beauty of the biotic community. It is wrong when it tends otherwise" (1949). The broadest context of human actions is the Earth's 'biotic community.' This community is alive, interactive, and intricately connected. Human action, therefore, has wide-ranging affects to the rest of this community.

Energy, the source of greenhouse gases, flows through these webs of connections from the deepest bowels of the earth to the cloud formations that hover above the Earth. As the "wise species" (Rolston 2006: 66), humans are beginning to realize their foibles. But they can only correct their mistakes when they begin to act as stewards and not taskmasters of the earth's destiny.

References

Clark, M. (2000) *An Aquinas Reader*, New York: Fordham University Press.

Hardin, G. (2001) *The Tragedy of the Commons*. Available at https://www.garretthardinsociety.org/articles_pdf/tragedy_of_the_commons.pdf. Accessed on January 26, 2022.

Kaminski, I. (2019) "Dutch supreme court upholds landmark ruling demanding climate," *The Guardian*. Available at https://www.theguardian.com/world/2019/dec/20/dutch-supreme-court-upholds-landmark-ruling-demanding-climate-action. Accessed on January 20, 2021.

Ki-Moon, B. (2009) *Secretary-General's Remarks to Korea University [as prepared for delivery]*. Available at: https://www.un.org/sg/en/content/sg/statement/2009-08-17/secretary-generals-remarks-korea-university-prepared-delivery.

Leopold, A. (1949) *The Land Ethic*. Available at https://rintintin.colorado.edu/~vancecd/phil308/Leopold.pdf. Accessed on January 25, 2022.

Molles, M. (1999) *Ecology: Concepts and Applications*, New York: WCB McGraw-Hill.

Rolston III, H. (2006) "Intrinsic values on earth: Nature and nations," In Henk, A.M.J. ten Have (ed.), *Environmental Ethics and International Policy* (pp. 47–67), Paris: United Nations Educational, Scientific and Cultural Organisation (UNESCO).

Oposa v. Factoran G.R. No. 101083 July 30, 1993, Manila: Supreme Court. Available at https://www.lawphil.net/judjuris/juri1993/jul1993/gr_101083_1993.html. Accessed on January 25, 2022.

Philippine Distribution Code of 2016 (PDC) (2016), Republic of the Philippines, Energy Regulatory Commission. Available at https://www.erc.gov.ph/Files/Render/media/PhilippineDistributionCode-2016Edition.pdf. Accessed June 5, 2023.

Simon, Y. (1999) *The Definition of Moral Virtues*, New York: Fordham University Press.

UNESCO. (2017) *Declaration of Ethical Principles in Relation to Climate Change,* Paris: UNESCO.

United Nations Framework Convention on Climate Change (UNFCCC). (1992). Available at https://unfccc.int/resource/docs/convkp/conveng.pdf. Accessed on January 25, 2022.

World Commission on the Ethics of Scientific Knowledge and Technology (COMEST). (2005) *The Precautionary Principle*, Paris: UNESCO.

PART III

Ethical Issues Raised by Arguments Frequently Made in Opposition to Climate Change Policies

17

INSTRUMENTAL RATIONALITY AND CLIMATE POLICY

J. Timmons Roberts and Donald A. Brown

Introduction

In 1991, the nations of the world worked to draft early versions of what would become the U.N. Framework Convention on Climate Change (UNFCCC). That agreement has been signed by 195 countries, essentially every nation on the planet, even reluctant ones like the United States and Saudi Arabia.

In that agreement were key ethical principles, which technically remain in force today:

1 That the nations of the world would work together to 'avoid dangerous climate change';
2 That the wealthy nations, who had created most of the problem of climate change and who had the resources to do so, would act first, under the 'principle of equity'; and
3 Those same nations pledged to reduce their emissions of the gases that create global warming, and act in a way to help the poorer nations prepare for future climate-related disasters.

Concrete proposals along these lines were put forward, including one from the tiny Pacific ocean nation, Vanuatu. The nation put forward its proposal in the name of the Association of Small Island States, calling for an insurance mechanism to be set up to help these countries pay for the costs of climate impacts that can't be avoided or adapted to.

That proposal never was given consideration. The proposal of equity in expected emissions reduction – proposed by India and supported by many developing nations – was never properly considered in the Kyoto round of negotiations in 1997. Mandatory reductions at a level that would actually 'avoid dangerous climate change' have never been adopted.

This chapter argues that the reason for the failure of the world to respond in these fundamental ways to the crisis of climate change is because *instrumental rationality* prevails instead of ethically-based principles, like precaution and equity. Instrumental thinking should sound familiar, e.g., 'everyone only acts in their own interests,' 'the market will solve it,' 'the powerful get their way,' 'cost-effective solutions are the right ones,' or 'the realistic thing to do is to accept things the way they are.' Here, the principles underlying global climate negotiations are introduced, but the chapter spends most of its time on understanding why instrumental rationality is so powerful and difficult to overcome.

DOI: 10.4324/9781003039860-21

Ethical Principles in UN Climate Talks and Agreements

Governments around the world have agreed under the 1992 United Nations Framework Convention on Climate Change (UNFCCC), and agreements by parties under this treaty since then, including the Paris Agreement (2015), to adopt national climate change policies on the basis of several ethical principles.

First, developing nations feared their own efforts to pull themselves out of poverty would be blocked by efforts to cut the use of fossil fuels and stop them from using their forest lands for agriculture and timber. But since emissions anywhere on Earth warm the planet everywhere, wealthy nations realized that climate change could only be solved with action by all countries. Thus, they agreed in 1992 that the duty to establish national policies had to be in accordance with "equity" and "common but differentiated responsibilities and respective capabilities" (UNFCCC 1992: Art. 3.1). That phrase is used so often in discussions of climate politics it is referred to as simply CBDR, or CBDR+RC. That is, all nations are responsible for climate change ('common responsibilities'), but some are much more responsible than others ('differentiated responsibilities'). Like kindergarten ethics, if you make a mess, you should be the one to clean it up. Likewise, some nations have much more money ('capabilities') than others and should do the most, the soonest. And those with both responsibility and capability should act first.

This principle continues in documents all the way through the latest, the Paris Agreement, and continues to be debated about. Some say it has been gutted and revised to the point of being essentially meaningless. For example, since the contentious Copenhagen negotiations held in 2009, all nations are only expected to offer their own voluntary pledges for how much emissions reductions they choose to undertake. This is called 'pledge and review,' and is seen in the Paris negotiations language that nations should offer their own Nationally Determined Contribution (NDC).

Reviews of those NDCs, sadly, show that very few nations have made pledges that meet any reasonable criteria for acting according to their respective responsibilities and capabilities. A group called EcoEquity.org has led the development of an 'Equity Reference Calculator,' by which different criteria can be inputted to the site and expectations for each nation can be compared with actual pledges/NDCs (Climate Equity Calculator 2019).[1] The 'mitigation gap' is massive, between what emissions reductions are required for just action and what is actually being pledged. And on top of that, there is a massive 'adaptation gap' (UNEP 2022).

Second, countries pledged to 'avoid dangerous climate change,' therefore applying what is called the 'precautionary principle.' That ethical principle explicitly prohibits nations from using scientific uncertainty as an excuse for not taking action to prevent dangerous anthropocentric interference with the climate system (UNFCCC 1992: Art 3.3). In spite of thousands of scientific studies showing the reality of the phenomenon, industrial interests, like the 'Big Oil' companies, have continued to argue for decades that the science was uncertain (PBS 2022). Finally in Paris, the nations of the world agreed to work to keep the global temperature rise under 2°C (United Nations 2015: Art. 2). The other part of that article was even more precautionary, seeking to stay as close to 1.5°C warming as possible, a goal advanced by the Least Developed Countries group and the Association of Small Island States.

In addition, there are numerous other non-controversial ethical norms that are understood to apply to nations as a matter of international law on global environmental problems such as climate change. These include the 'no harm principle,' which obligates nations to prevent people or entities within their jurisdiction from harming people and nations outside their borders (UNFCCC 1992: Preamble). Another is the 'polluter pays principle,' which requires those nations causing harm from pollution to pay for the damages they cause (Rio Declaration 1992: Principle 16). Unfortunately, again, these ethical principles are stated, but rarely acted upon by the nations of the world when it comes to climate change.

Why Nations Are Not Following Basic Ethical Principles

From its very beginning, climate policy action has been quite divergent from the principles that were laid down in the 1992 Framework Convention. Most nations around the world have set their national targets and negotiating positions based on perceptions of their economic self-interest, and what is believed to be politically feasible.

The failure of ethical principles to get traction in guiding policy is a much broader problem than in regard to climate change policy formation alone. Despite the emergence of the academic sub-discipline of environmental ethics in the late 1970s, ethical principles are failing to influence environmental policymaking for most environmental problems (Hargrove 2003; Frodeman 2006). Unfortunately, the academic discipline of environmental ethics has failed to pay adequate attention to actual controversies that arise in environmental policymaking debates.[2]

This chapter will now explain why a first order task that needs to be addressed before ethical principles can play their appropriate role in shaping environmental public policy is to open policymaking arguments on environmental issues including climate change to ethical reflection. This is a first order task, because throughout the world those responsible for environmental policymaking are following instrumental reason, a mode of thinking which hides or ignores ethical questions. Governments routinely do this to determine the acceptability of environmental policies. It is a first order problem because before one can consider what ethical principles should guide policy formation, policymaking must be made open to ethical critique and reflection. If policymakers do not see and respond to the ethical issues that are implicitly raised by arguments raised against proposed policies, they cannot apply the appropriate ethical rules.

Instrumental rationality is a mode of rationality that is exclusively concerned with the search for efficient means or scientific facts which, consequently, is not concerned with assessing the goals – or ends – that policies should pursue. This form of rationality has existed throughout history but has become more dominant in post-Enlightenment liberal democratic capitalist societies (Cruickshank 2014).

Ethics rationality, on the other hand, is concerned about what the goals of society *should be*. *Ethical reasoning* seeks to determine what should be the goal of human behavior by examining what is right or wrong, what is permissible or impermissible, what actions are obligatory or non-obligatory, and how burdens of preventing harm should be justly distributed.

Instrumental rationality, because it focuses on means, usually ignores ethical questions about what the goals of policy should be, despite the fact that every argument against a proposed environmental policy already contains an unstated norm. For instance, we often hear the claim that a proposed climate change policy should not be adopted because it imposes unacceptable costs. However, this claim rests on the unstated norm that the government should not adopt policies that impose significant costs on the economy or specific industries. The nature of climate change, of course, means that when powerful industries manage to avoid costs to their businesses through intensive lobbying, the long-term wellbeing of the most vulnerable and powerless members of society are imperiled and completely ignored.

Scientific and economic reasoning, which have increasingly dominated public policymaking from the beginning of the Enlightenment, almost always focuses on how to achieve goals, not on what goals or ends should be desired. Economic rationality often focuses on how to maximize human preferences. Ethics asks a different question of economic activity, namely, what preferences humans *should have*. Scientific reasoning usually tests hypotheses to determine what 'is.' Moral philosophers believe that determining what 'is,' which is the proper domain of science, cannot determine what 'ought' to be, which is the domain of ethics.

Instrumental rationality that scientists and economists deploy in their search for scientific and economic facts has dominated public life and higher education for several centuries. That instrumental rationality dominates environmental policymaking is clear, given that most government environmental agencies are staffed nearly exclusively by engineers, scientists, economists, and lawyers. Very infrequently are there any employees trained in ethics. This is a huge problem because very few employees of environmental agencies or scientific organizations that make policy recommendations can spot problematic ethical issues that should be acknowledged in policy debates, and particularly ethical issues that are ignored or hidden when instrumental rationality is deployed. Although employees of government agencies responsible for policy formation often understand they should apply policy rules entailed by relevant laws, many relevant laws do not contain clear rules on how to respond to economic and uncertainty arguments against proposed environmental policies.

Instrumental rationality dominates public policy formation for at least two reasons. Sociologists, including Max Weber, have predicted that instrumental rationality would over time crowd out ethical rationality in modern societies, because increasingly complex human problems would be relegated to bureaucracies run by technical experts, whose expertise depends on the use of instrumental rationality (Thomas 2017). Since the power of experts depends, in part, on maintaining the fiction that their expertise is the central key to solving modern problems, these experts are reluctant to acknowledge that their analytic tools for solving problems are often ethically inadequate, and sometimes ethically inappropriate (Thomas 2017). Moreover, particularly in capitalist societies, wealthy interests are able to hire experts and frequently do so to fight government action that would reduce profits.

Second, opponents of proposed environmental policies usually frame opposition to these policies on the basis of excessive costs to governments or specific industries, or to a lack of scientific certainty about the harms the policy seeks to prevent. These arguments very frequently hide controversial normative assumptions. For instance, cost arguments made in opposition to environmental policies often rest on the very ethically dubious idea that any policy which creates significant cost to a nation, regional economy, or to a specific industry should not be adopted, even when the problematic behavior causes serious harm to people or nations who have not consented to be harmed. The public debate in response to these claims often narrowly focuses on the magnitude of the costs, or whether the regulatory action will create jobs. In so doing the debate ignores several serious ethical problems with these arguments. Cost-benefit analysis (CBA) is riddled with these ethical problems.

The Rise of Cost-Benefit Analysis

Nearly every policy issue raises issues of whether the expense of instituting it is worth the benefits of doing so. In climate change politics, we are bombarded with apocalyptic arguments from forcing families, companies, and nations to stop or sharply reduce burning of gas, oil, and coal. Ads or talking points conjure images of aging pensioners freezing in their dark homes, because they are unable to buy a tank of heating oil or pay their electric bill. 1970s images of lines at gas stations are rerun as gas prices soar. Solar and wind power are painted as unreliable and too expensive, and the damages from climate change is painted as uncertain or far in the future. In such times, and in all times, policymakers need to make decisions on what to do. Should a radical new approach be taken? Or should the status quo system be continued?

Just days into his presidency, on February 17, 1981, Ronald Reagan signed Executive Order 12291, required that all federal "regulatory action shall not be undertaken unless the potential benefits to society from the regulation outweigh the potential costs to society" (Shabecoff 1981). The order turned "what had been a useful economic tool into an imperative of Federal decision making" (Shabecoff 1981). Several US presidents went on to issue executive orders which require that any proposed regulation must satisfy a CBA before it may be put into effect (Congressional Research Service 2014).

CBAs seemingly put all major decisions on a level footing and potentially avoid needlessly burdensome approaches to solving social problems from being adopted. Yet CBAs hide important ethical issues. If, for instance, a CBA concludes that government action to protect vulnerable people or ecological systems should *not* be taken because costs of taking action to reduce an environmental threat outweigh the economic value of harms avoided, controversial ethical assumptions may be hidden. These assumptions can hide in factual assertions about the magnitude of the costs, or in the value of benefits. This is particularly true if:

1 Potentially (but not fully proven) catastrophic harms were ignored in the CBA;
2 The costs of taking action would be imposed upon parties that are harming others, yet the victims of the harm have not consented to be harmed;
3 Things that were believed to be sacred by one culture are valued in the CBA as if they were commodities whose value can be measured adequately by 'willingness-to-pay' monetary measures. CBAs usually commodify all human values and thus value is restricted to monetary value while ignoring other values including sacred value or beliefs that certain entities should not be for sale. Thus, in CBAs, usually the value of things that could be harmed are measured by human preferences measured in monetary values. Yet ethics is concerned with what preferences people *should* hold, not simply what preferences people hold;
4 Human rights will be violated if regulatory action is not taken;
5 The proposed government action implements the ethical duty of people to not harm others on the basis of self-interest. The assumption that self-interest will solve environmental problems, especially in the context of profound inequalities of wealth and poverty, is highly suspect; and
6 The benefits of government action to protect the environment are discounted too greatly compared to current costs to those who must act to prevent harm (Brown 2008).

Thus, if a decision to take no government action on a potential environmental problem is justified only as a matter of imbalance between costs and benefits, dubious ethical assumptions are frequently hidden in the CBA calculations. Meanwhile, ethical principles, including those that have been widely acknowledged as valid and applicable to government policy formation, are often ignored.

In our experience, proponents of environmental policies rarely identify the ethical problems with the use of CBAs. When faced with cost-based arguments made in opposition to their proposed policies, proponents almost always respond to the cost-based arguments by making counter cost claims. And so public debate about proposed policies usually focuses on economic 'factual' claims while ignoring ethical principles. This is seen in decades of environmentalists adopting economic arguments, even while these arguments are seen to not effectively mobilize the public (Wetts 2020).

CBAs in US policymaking routinely exclude damage to other nations, violating the ethical norm against causing such harm. Using only the cost to those causing harm to others as a justification for failing to abate that harm violates well-established principles of international environmental law, including the 'polluter pays principle' (Rio Declaration 1992: Principle 16) and the 'no-harm principle' (UNFCCC 1992: Preamble). Yet the US continues to very frequently base the acceptability of (lax and harmful) environmental regulations on the results of a CBA.

In 1997, while working as the Program Manager for United Nations Organizations in the U.S. Environmental Protection Agency (US EPA) Office of International Affairs, Brown closely observed the US debate about whether the country should agree to the Kyoto Protocol under the UNFCCC. This debate focused exclusively on two different kinds of CBAs. One was completed by the US EPA, the other by the US Department of Energy. The two CBAs reached slightly different conclusions about the likely magnitude of negative impacts on the US Gross Domestic Product if the country agreed to be bound by the Kyoto Protocol. Amazingly, both CBAs examined costs and benefits to the United States alone, while completely ignoring potentially harsh climate impacts on poor and vulnerable people and nations around the world. Yet no one in the US government nor even the major non-governmental organizations participating in the debate raised any ethical problems with the US reliance on these selfishly myopic CBAs.

In most Western capitalist countries, corporations and their industry associations have huge political power to frame public policy questions, and they do not hesitate to exercise their power to prevent any government action that could lower corporate profits (Brulle 2023). So, the public debate on proposed policies often focuses on economic 'facts,' not ethical duties, despite the almost universally accepted ethical norm agreed to by almost all religions and nations that people should not harm others on the basis of self-interest.

Opposition arguments against proposed environmental policies often rest on the unstated and very dubious norm that regulatory action limiting commercial activities should not be taken unless the harms are proven by the government with high degrees of scientific certainty. This is claimed even in cases where achieving high levels of certainty is scientifically difficult, or very prohibitively expensive.

For over thirty-five years now, opponents of US action on climate change have frequently based their opposition on scientific uncertainty about human-caused climate change harms, despite the fact that the United States agreed to the 'precautionary principle' when it agreed to the UNFCCC in 1992 (Art 3.3). This principle states that governments will no longer fail to take action on the basis of scientific uncertainty. Yet, advocates of national action on climate change, in response to opponents' scientific uncertainty arguments, almost always simply claim that the scientific 'facts' of harm have been sufficiently scientifically demonstrated. They fail to argue the ethical rule that precaution is required once it is scientifically established that significant harm will likely be created by certain human activities.

If a government decides not to act to reduce the threat of environmental harm on the basis of lack of proof of harm, such a decision can hide important ethical questions. This is particularly true if:

1 The government assumes that the proponents of government action to prevent environmental harm should shoulder the burden of proof of demonstrating harm, particularly in matters where proof is very expensive, difficult to demonstrate, or cannot be fully demonstrated before potential harms are experienced;
2 There is credible but uncertain evidence that the current activity may be approaching thresholds that could trigger very serious consequences ('tipping points');
3 If the government waits until all uncertainties are resolved, it may be too late to prevent serious harm;
4 Some very serious potential harm is judged to be low probability just because the mechanism for causing serious harm is not completely understood. This means that the probability of the serious harm cannot be confidently evaluated; and
5 The victims of potential harm have not consented to put at risk.

All of these considerations are relevant to climate change; yet the United States has failed to decisively act on the issue since international climate change negotiations began 35 years ago. Opponents of US climate change policies have repeatedly claimed that there is insufficient proof of human-induced climate change caused harms, even pointing to the benefits it might bring. This is so despite the fact that the international community in 1992 in the UNFCCC determined and agreed that the precautionary principle should guide national responses to prevent dangerous climate change.

Conclusion: Where Is the Science?

The most prestigious scientific institutions in the world, including most national academies of science and the Intergovernmental Panel on Climate Change, have concluded with high levels of confidence that humans are causing and threatening great harms from human-induced climate change. Even conceding, for the sake of argument, that great harms from human-induced climate change are not yet fully proven, ethical principles require that action should be taken to reduce the threat of climate change. Yet the ethical basis for requiring action is almost never discussed in the US public debate about whether scientific uncertainty about human-induced climate change is an appropriate justification for an unwillingness to act.

Scientists employed by environmental agencies usually focus on understanding the environmental harms and risks of various human activities and whether proposed government action will acceptably reduce threats to human health and the environment. The goals of environmental regulatory action are usually given to these scientists and bureaucrats by law or regulation. These laws typically specify that water or air pollution should be reduced to a point where unreasonable harm to humans or ecological systems is avoided. Yet, in the face of scientific uncertainty about whether human actions may cause harm, scientists cannot determine by scientific methods alone *who should have the burden of proof* or *what quantity of proof* should satisfy that burden. These are fundamentally ethical questions.

An understanding of the ethical problems with instrumental rationality leads one to an understanding of why nations often ignore even well-established ethical principles in policy formation. The ethical principle that no nation should harm others outside their jurisdiction on the basis of national economic interest should be at the center of such an understanding. For this reason, we believe that a first-order problem on the road to a world which formulates policies guided by ethical principles is to open policy formation controversies to expressed consideration of ethical issues. This goal requires that those engaged in policy formation identify the ethical issues frequently hidden in economic and scientific arguments. These are often hidden in arguments against proposed policies addressing environmental issues around the world.

Unfortunately, most professionals engaged in environmental policy formation have no training that would help them identify the hidden ethical issues embedded in arguments made against environmental and sustainable development policies. Nor do NGOs, academic observers, or journalists have the training to spot ethical problems made by opponents of proposed policies that are derived from various forms of instrumental rationality. Training in ethical principles and understanding why they are required for humanity to address the crisis of climate change would be an important start. It will also be required that there be a rebalancing of the authority to make decisions, toward those who will or have suffered the devastating harms from climate change.

Notes

1 A research project led by Widener University Commonwealth Law School and the University of Auckland found that despite express national promises under the Paris Agreement to base national climate commitments known as Nationally Determined Contributions (NDCs) to reduce the threat of climate change to prevent warming as close as possible to 1.5°C but no more than 2°C, on the basis of equity and common but differentiated responsibilities, all 24 nations studied actually set their NDCs on economic self-interest. Yet this conclusion was not determinable from the documents that nations submitted to the UNFCCC Secretariat when the nations submitted their NDCs (Brown & Taylor (nd) *National Climate Justice, Research Project on Ethics and Justice in Formulating National Climate Policies, Lessons Learned*). The study also found that environmental NGOs in the country that supported national action on climate change did not seem to understand how to critique the failure of the nation to set its NDC on the basis of the nation's ethical obligations including ethical obligations that the nation expressly agreed to.

2 Academic environmental ethics since its inception in the late 1970s has been almost exclusively focused on theoretical issues, such as how to ground a biocentric or ecocentric ethics, while completely failing to help policymakers understand what is ethically wrong with specific arguments made by opponents of environmental policies who almost always rely on arguments derived from instrumental rationality which hide or ignore dubious unstated norms on which the arguments are based.

References

Brown, D. (2008) *Ethical Issues in the Use of Cost-Benefit Analysis of Climate Change Programs.* Available at https://ethicsandclimate.org/2008/06/01/ethical-issues-in-the-use-of-cost-benefit-analysis-of-climate-change-programs/, accessed December 16, 2017.

Brown and Taylor. (nd) *National Climate Justice, Research Project on Ethics and Justice in Formulating National Climate Policies, Lessons Learned.* Available at https://nationalclimatejustice.org, accessed December 24, 2017.

Brulle, R.J. (2023) "Advocating Inaction: A Historical Analysis of the Global Climate Coalition," *Environmental Politics*, 32(2): 185–206.

Congressional Research Service. (2014) *Cost-Benefit and Other Analysis Requirements in the Rulemaking Process'.* Available at https://fas.org/sgp/crs/misc/R41974.pdf, accessed December 18, 2017.

Cruickshank, J. (2014) "Democracy Versus the Domination of Instrumental Rationality: Defending Dewey's Argument for Democracy as an Ethical Way of Life," *Humanities,* 3: 19–41. Available at https://www.mdpi.com/2076-0787/3/1/19/htm, accessed December 20, 2017.

Frodeman, R. (2006) "A Policy Turn in Environmental Ethics," *Environmental Ethics*, 28: 3–20.

Hargrove, E. (2003) "What's Wrong? Who Is to Blame?" *Environmental Ethics*, 25 (1): 3–4.

Paris Agreement. (2015) Available at https://unfccc.int/files/meetings/paris_nov_2015/application/pdf/paris_agreement_english_.pdf.

Rio Declaration on Environment and Development (Rio Declaration). (1992) Available at https://www.un.org/documents/ga/conf151/aconf15126-1annex1.htm, accessed December 24, 2017.

Shabecoff, P. (1981) "Reagan Order On Cost-Benefit Analysis Stirs Economic and Political Debate," *New York Times*, November 7.

Thomas, W. (2017) *Max Weber on Rationality in Social Action, in Sociological Analysis in Modern Life, Rational Action.* Available at http://www.rational-action.com/?s=Weber, accessed December 22, 2017.

United Nations Environment Program (UNEP). (2022) *Adaptation Gap Report.* Available at https://www.unep.org/resources/adaptation-gap-report-2022.

United Nations Framework Convention on Climate Change (UNFCCC). (1992) Available at https://unfccc.int/resource/docs/convkp/conveng.pdf.

Wetts, R. (2020) "Models and Morals: Elite Orientated and Value-Neutral Discourse Dominates American Organizations' Framing of Climate Change," *Social Forces*, 98(3): 1339–1369. DOI: 10.1093/sf/soz027.

Additional Reading

Intergovernmental Panel on Climate Change (IPCC). (2014) *5th Assessment Report, Contribution of Working Group III to the Fifth Assessment Report of the Intergovernmental Panel on Climate Change*, Chapter 4. Sustainable Development and Equity. Sec 4.6. 2.1, p 4. Available at http://www.ipcc.ch/report/ar5/wg3/, accessed December 23, 2017

Oxfam. (2020) *Oxfam Report on Climate Financing.* Available at https://www.oxfam.org/en/research/climate-finance-shadow-report-2020

18

ETHICAL PROBLEMS WITH COST ARGUMENTS MADE IN OPPOSITION TO CLIMATE CHANGE POLICIES

Stephen M. Gardiner and Donald A. Brown

Cost and Its Rivals

Government consideration of climate law and policies at the national and international level became more frequent in the early 1990s as the international community began creating serious international legal obligations in response to the scientific consensus of the enormous risks of human caused climate change. Since then, opponents of climate policies have often successfully framed the public debate on government responses to climate change so that it focused on scientific uncertainty and claims of unacceptable costs. Typically, such cost arguments take a very simple form: the basic idea is that if a given policy will entail notable costs for a certain actor (e.g., a firm, sector, a generation, or the country as a whole), that policy is untenable, and the government is justified in rejecting it. In other words, simple costs arguments are often assumed to be decisive. The main actors, and sometimes even their opponents, assume that demands to do more are unreasonable.

One place where simple cost arguments are common is in the political domain, where opponents of climate action frequently emphasize the economic costs of making transitions away from fossil fuels and other kinds of emissions-intensive activities. Typically, such objections are very one-sided. For one thing, they often fail to take into account the benefits of climate action, whether these are direct benefits (e.g., in terms of reductions of climate risk or harms) or indirect benefits (e.g., the health benefits of improving air quality). For another thing, the emphasis is usually only on the short-term and narrowly economic costs which accrue to the current population or particular groups within it. Given such problems, cost arguments made in political contexts are often prejudicial and ethically objectionable, even in narrow terms.

More nuanced kinds of cost arguments are presented by economists and other analysts employing conventional policy tools, and in particular traditional cost-benefit analysis (CBA). Within these circles, cost arguments have a more mixed record, and are subject to much internal and external controversy. Still, if one looks at the field historically, since the late 1980s, it seems clear that those pushing standard market CBA have tended to downplay the climate threat relative to other groups within the climate community, including especially climate scientists, climate activists, and those interested in climate ethics and justice. For example, the climate economist most celebrated within the policy community appears to be Nobel Prize winner William Nordhaus. Yet his approach has long been the subject of intense criticism (from outside economics and also sometimes from within) for calling only for modest, incremental change. As such, it is seen by many as contributing to political inertia on climate change, especially in earlier decades.

DOI: 10.4324/9781003039860-22

Our central concern about the role of cost arguments, whether in politics, economics, or policy, is a general one. It is clear on both ethical and legal grounds that to regard cost arguments as necessarily constituting a decisive consideration against climate action is a mistake. Indeed, there is a strong presumption against it. The reason is that other ethical considerations matter, are often weightier, and arguably also have a much stronger claim to be decisive. In what follows, we elaborate on this idea, focusing on the notions of harm, rights, and responsibility as our examples, and exploring the ways in which they apply to nation states.

Context

As a preliminary, let us remind ourselves of the context and what is at stake. In general, climate change is already causing loss of life and increasing health problems from intense storms, droughts, floods, intense heat, and rising seas. For example, a 2019 Special Report of the UN General Assembly found that climate change was causing 150,000 premature deaths, a number which is sure to increase as temperature rises (UN General Assembly 2019). Moreover, the number of those forcibly displaced due to climate disasters disruption, and violence – both within their countries and internationally – has continued to climb, from 82.4 million at the end of 2020 to more than 84 million by June 2021 (UNHCR UNRefugee International 2021).

As horrific as these current climate impacts are, they are at the moderate end of the scale compared to what is projected in the future. First, while the planet has seen a rise in average global temperature of around 1.1°C so far, the internationally agreed targets for "avoiding dangerous anthropogenic interference" with the climate system are much higher. Under the Paris Agreement, they call for restricting the rise in average global temperature only to "well below" 2.0°C, and for pursuing efforts to get close to 1.5°C (UNFCCC 2015). In other words, conventional temperature targets envision allowing many more climate impacts to occur, and at levels likely to be much higher than those being experienced today.

Second, we are currently far from being on track to meet the mainstream targets. For example, for intermediate emissions scenarios, the IPCC projects a 2.1–3.5°C rise by 2100 and 2.3–4.6°C by 2200; and for a very high emission scenario, 3.3°C–5.7°C by 2100 and 6.6–14.1°C by 2200 (IPCC 2021). Crucially, global temperature rises of 4.5°C or higher are comparable in magnitude to an ice-age like shift, except in the warmer direction, and would be occurring much more quickly, over decades, not thousands of years. Yet human civilization has no experience of living through such radical global change. As a result, it is widely thought in the scientific community that the impacts would strain the ability of human and other systems to cope, and thereby threaten catastrophe, including social collapse (Gardiner 2011; Anderson 2012). One reason for this (among many) is the concern that at around 4°C global agriculture may begin to unravel.

A third point is that the mainstream projections are in many respects optimistic, including in the way they handle probabilities, climate sensitivity, and assumptions about scientific progress (Gardiner: Chapter 4, this volume). One particularly striking example is that additional warming may cause the global system to run into various 'tipping points' that could dangerously accelerate warming, yet these are typically marginalized or neglected within mainstream projections (e.g., Lenton et al. 2019).[1]

Crucially, the upper warming limit goal of 2°C in the Paris Agreement was based on an informal scientific consensus in 2015 that the tipping point feedbacks would not likely be triggered until warming exceeded warming of 1.5°C–2°C (Cho 2021). However, this is a questionable assumption. For instance, recently there has been some evidence that several tipping points of concern are already showing signs of destabilization, perhaps to a point of no return (Lenton et al. 2019). These include methane permafrost (Anthony et al. 2018), arctic summer ice sheets, and the Greenland ice sheet (Morgan McFall-Johnson 2020). Moreover, breaching some of these tipping points could trigger a domino effect that unleashes other feedbacks, potentially creating an existential crisis for much of life on Earth (Leahy 2019).

Importantly, this context already casts doubt on the relevance and decisiveness of simple cost arguments. To flesh out this concern, let us examine three ethical considerations: harm, human rights, and responsibility.

Harm

When it comes to the notion of harm, and in particular obligations not to cause harm, there are prior international agreements, principles and norms which appear to take precedence over simple considerations of cost. For instance, in the foundational international treaty on climate change – the United Nations Framework Convention on Climate Change (UNFCCC 1992) – all nations committed themselves to preventing "dangerous anthropogenic interference with the climate system," and subsequently, under the 2015 Paris Agreement to interpreting this as avoiding a rise in global temperature of more than 2.0°C, while striving to get as close as possible to 1.5°C (UNFCCC 2015).

Countries also specifically agreed in the framework convention that:

> States have, in accordance with the Charter of the United Nations, the responsibility to ensure that activities within their jurisdiction or control do not cause damage to the environment of other States or areas beyond the limits of national jurisdiction.
>
> *(UNFCCC 1992: Preamble)*

This principle is often referred to as the 'no-harm principle.' In accepting it, each nation explicitly acknowledged that it had a duty to adopt climate change policies that would keep climate change from inflicting damage on others outside its jurisdiction. In the wider international context, the acceptance of this principle is not surprising. The more general principle that states have a responsibility not to cause damage beyond their borders has independent standing as a customary principle of international law (Bernie et al. 2008).

The relevance of the no-harm principle to simple cost arguments should be clear. First, and most generally, the principle applies even if complying with it involves costs. If a state could excuse itself from this kind of duty simply by asserting that compliance would involve some cost to it, this would undermine the point of the principle: a no-harm principle that applied only if compliance were costless to those subject to it would be intolerably weak and barely amount to any constraint at all. Given this, very simple cost arguments against climate action cannot succeed. On the contrary, there is a strong presumption that damage should be prevented even if doing so is costly to those responsible.

Second, the no-harm principle has immediate application in the policy world. Climate change is already creating substantial misery and desperation around the world from flooding, drought, sea level rise, forest fires, extreme heat, more intense tropical storms, the spread of tropical diseases, and so on. These harms are being caused by atmospheric CO_2 concentrations at or below 419 ppm and a global temperature rise of less than 1.5°C. One notable upshot of this is that the no-harm principle is relevant to policies that prevent loss and damage, and fund climate adaptation. For instance, unfortunately there is currently no legal mechanism under international law to adjudicate the economic value of damages for the harms created. Given this, the no-harm principle suggests that the nations of the world are obligated to support the creation of a loss and damages mechanism. Notably, the international community agreed to negotiate on this issue at COP 26 in Glasgow in 2021, on the basis of an agreement that are responsible for the harms they create (Brown 2022).

Third, the no-harm principle is also relevant in the longer term. For instance, as was said above, breaching the Paris targets for 'dangerous climate change' will almost certainly produce even more pronounced levels of harm than is currently being experienced. Moreover, prolonged increases in anthropogenic climate forcing are likely to accelerate the risks of triggering tipping points in the climate system that could cause widespread devastation for human and nonhuman systems, including social collapse, or other hard or crash-landing scenarios (Gardiner 2011). More generally, the effects of emissions will be with us for a long time. For one thing, some of the geological processes being triggered by climate change (such as the melting of the West Antarctic Ice Sheet) will play out over centuries but are now likely to be irreversible. For another thing, CO_2 emissions are different than most other air pollutants in that they mix well in the atmosphere and are very long lasting. For example, while on average 80 percent of CO_2 is removed in 100 years by carbon sinks, some percentage lasts for thousands of years in the future, thus contributing to climate harms worldwide for a very long time.

Fourth, as the last point suggests, there is a strongly intergenerational aspect to the most serious climate harms (Gardiner 2011). While some may arise in the next few decades, many harms are most likely to occur late this century and in those to follow. Given this, there are good reasons to believe that the worst risks do not fall predominantly on current decision-makers (who are generally in their fifties or older), but rather on the young and those yet to be born, who must rely on the present policies to protect them. This suggests that the duty not to cause damage in the future should be understood in an especially stringent manner, and that provisions for protection, compensation, and redress (including loss, damage, and climate adaptation) should be understood in a forward-looking, and not just a backward-looking way (e.g., Shue 2022).

Human Rights

A second ethical consideration beyond costs is the relevance of rights, and in particular basic human rights. Importantly, it is widely believed that climate change is now violating the most basic human rights, including the rights to life and health (e.g., ICHRP 2008; UN OHCHR 2009; Caney 2010), and the newly recognized human right to a healthy environment. This tends to undermine simple cost arguments, since rights considerations usually trump a narrow concern for costs. Crucially, national responsibilities to protect human rights cannot be overridden simply because fulfilling those responsibilities would impose costs on a particular government responsible for preventing human rights violations. Consequently, countries may not refuse to adopt climate policies necessary to prevent climate impacts that violate basic human rights simply on the basis that such policies involve costs to the nation.

One reason for this is that responsibilities around basic human rights are foundational responsibilities, and on many views vital to the justification of the legitimacy of nation-states themselves (see Lewis and Brown: Chapter 5, this volume). As such, they constitute a background or baseline against which costs are to be assessed. They are prior to simple considerations of costs, in ways which both constrain the relevance of those costs and constitute the very context in which those costs must be understood.

More generally, while basic human rights may not be decisive over all other considerations, they have an especially high status, imposing a strong burden of proof on those seeking to override them. Simple arguments about costs do not meet that burden. Moreover, more complex arguments that might do better would probably need to appeal to costs of a special kind, such as those that are themselves related to either basic human rights or similar foundational concerns. For example, cost arguments that are ultimately about preserving subsistence or the basic integrity of particular communities might succeed, since those issues are foundational; but costs arguments that appeal merely to market preferences for carbon-intensive luxury goods do not suffice. Of course, it is worth noting that the former arguments – focused on subsistence or cultural survival – are no longer simple cost arguments, but appeal to a deeper range of ethical values.

Responsibility

Our third ethical consideration is responsibility. Here our concern is that simple cost arguments are not an acceptable justification for a nation's refusal to adopt climate policies necessary to prevent horrific climate impacts both because they have made prior commitments and because they have more general ethical responsibilities. One example of a prior commitment is that, under the Rio Declaration in 1992, each country agreed to the 'polluter pays' principle, which states:

> National authorities should endeavor to promote the internalization of environmental costs and the use of economic instruments, taking into account the approach that *the polluter should, in principle, bear the cost of pollution,* with due regard to the public interest and without distorting international trade and investment.
>
> *(Rio Declaration 1992: Principle 16, emphasis added)*

Under subsequent agreements, nations also agreed on how the objectives of international climate policy established under the UNFCCC should be understood. One of the few achievements of Copenhagen was officially to recognize the idea that avoiding dangerous anthropogenic interference meant staying below a 2°C rise in the global average temperature. Paris confirmed this goal, and then strengthened it in two ways: the objective became to hold the increase to "well below" 2°C, while "pursuing efforts" to limit it to 1.5°C (UNFCCC 2015: Article 1(a)]. In addition, the Paris Agreement reiterated the aim of reaching a global peak in emissions "as soon as possible" (UNFCCC 2015: Article 4), and included the idea that each country's national contribution "reflect its *highest possible ambition*" (UNFCCC 2015: Article 4.3, emphases added). These statements and others like them make it difficult to excuse lack of appropriate action on the basis of simple cost arguments.

In summary, we have argued that simple cost arguments face a considerable burden of proof when deployed against climate action by nation states and other entities intent on avoiding or minimizing their contributions. Cost is only one normatively relevant consideration, and is often trumped by other considerations, such as harm, basic human rights, and responsibility. These other considerations are not only ethically more important in most settings, but they are also explicitly included and prioritized in legal frameworks and precedents that are relevant to climate policy. Ignoring such normative concerns by deferring to simple costs arguments is therefore ethically irresponsible. It may also signal a deeper moral corruption (Gardiner 2011: Chapter 9). By this we mean roughly that, given our positionality, the current generation, and especially the affluent in more powerful countries, is susceptible to ways of thinking and talking about the climate problem that are distorted and self-serving, and obscure many of the problem's most important features. In particular, we might expect framings to emerge that gloss over, omit, or otherwise diminish important issues of justice and ethics.

On Cost-Benefit Analysis

Our concerns apply both to simple political arguments that focus on cost alone, and also to more complex arguments based on standard economic CBA. However, due to its special influence in some circles and its undoubted complexity, it is worth adding some additional points about CBA itself.

What Is Cost-Benefit Analysis?

To guard against possible confusions, let us begin with some background (see also Gardiner 2011; Brown 2018). The label 'cost-benefit analysis' is often used in at least three different ways (following Gardiner in Gardiner and Weisbach 2016: 76–82). On the first usage, it can refer to any kind of analysis that tries to *pick out the positive and negative aspects* of a given course of action or policy, or, more broadly, identifies its salient features.[2] To mark this usage and distinguish it from the others, we propose the term 'pros and cons analysis' (PCA). In our view, this form of 'cost-benefit analysis' is deflationary. In particular, PCA contains no reference to anything distinctively economic, nor even to any notion of 'cost' or 'benefit.' Notably the 'pros' and 'cons' might take any form; for example, it could be a 'pro' that a given policy promotes justice and a 'con' that it violates rights. More generally, arguably all normative and policy analyses involve identifying 'pros and cons.' Hence, endorsing PCA merely amounts to endorsing analysis of some kind or another. Importantly, this is not the exclusive domain of economics, but should also include ethics. Indeed, one might argue that ethical analysis is primary here (e.g., Gardiner in Gardiner and Weisbach 2016).

Consequently, we propose that if by 'cost-benefit analysis' one really means the deflationary sense, one should adopt the terminology of 'pros and cons analysis' and stop muddying the waters with the phrase 'cost-benefit analysis,' since that suggests something more specific and theoretically loaded. In particular, continuing to use the phrase 'cost-benefit analysis' encourages confusion, and is thereby at risk of facilitating moral corruption. One sign of this is that it is plausible to argue that conventional neoclassical economics is too limited to cover the domain of PCA. For instance, the IPCC has concluded that while the methods of economics are suited to measuring and aggregating the well-being of humans, they are not appropriate for taking account of justice and rights (IPCC 2014: 224). In our view, such values are essential to any sensible analysis of climate change. Hence, if a PCA is to be serious, it must move beyond limited economic methodology to other approaches that can include other key ethical dimensions (such as justice and rights). A PCA should analyze options that are ethically appropriate.

A second use of the phrase 'cost-benefit analysis' refers to a method for choosing between policies based on which yields the highest net benefits. Again, to mark this usage and distinguish it from the others, we endorse Gardiner's revised terminology and call this 'net benefit analysis' (NBA). NBA goes beyond pros-and-cons analysis. It assumes that the 'pros and cons' now being dealt with are of a specific sort. Most importantly, they are directly comparable, and in such a way that it makes sense to speak of a 'net' benefit: that which is left over when the costs are subtracted from the benefits.

Importantly, NBA is controversial and for familiar reasons. Take, for instance, the fact that many rights theorists would object to analyzing a situation involving a rights violation by counting the violation as a 'cost' to the victim which is then to be weighed against the 'benefits' to the perpetrator. Rights theorists would typically not, for example, analyze the torture of an innocent person by weighing the cost to the victim against the benefits to the sadists enjoying it, in order to produce an overall calculation of net benefit. Instead, they would say that no 'weighing' should take place: the rights of the victim are decisive, and whatever joy the sadists may get has no status.

A third usage of the label 'cost-benefit analysis' refers to a specific conception of NBA that relies on a set of distinctive assumptions and techniques within contemporary economic theory. These include that costs and benefits are to be understood in terms of 'willingness to pay'; that they should be expressed in standard market prices (or proxies for them); that they should be assessed in terms of their net present value; and that they should be mediated through a standard positive discount rate. To mark this usage and distinguish it from PCA and NBA, we call this 'standard market cost-benefit analysis' (or 'market-CBA' or MBA, for short). Importantly, market-CBA is very far from both PCA and NBA. For example, most forms of PCA and NBA that ethicists take seriously would not endorse understanding ethical concepts such as justice and rights in terms of either willingness to pay or market-prices, nor would most philosophical utilitarians understand well-being in this way. Similarly, few ethicists would accept discounting the future using a standard positive discount rate (e.g., 3–7 percent).

Limits of Market-CBA

The distinctions just offered between different usages of the label 'cost-benefit analysis' are useful both in themselves and also because they help us to make a more fundamental point about cost arguments in the climate context. Specifically, in the world of policy, the dominant form of CBA is market-CBA, and this is the view that is most controversial and most likely to come into conflict with core ethical concepts. Let us offer just two concerns.

First, given its reliance on market prices and willingness to pay, market CBA is very likely to come into conflict with more philosophically robust and attractive conceptions of *well-being*. For example, consider the outrage in the 90s when economists evaluated lives lost due to climate change in terms of foregone income, with the predictable result that lives (and especially deaths) in poorer nations counted for much less in their market CBA than those in richer nations (Broome 2012). The outrage here was justified and based on ethics. Comparing lives simply in terms of foregone income in order to reach policy conclusions is theoretically bizarre and morally reprehensible.

Second, and more generally, market CBA also threatens other values, such as justice, human rights, duties to avoid harm, responsibilities for damages, liberty, and so on. To illustrate, let us identify two issues that arise from market CBA's focus on satisfying consumer preferences. The first issue is that while market CBA is concerned with satisfying preferences as such, ethics has a crucial role to play in determining *which preferences* should be included and excluded (Dworkin 1977: 234; Sagoff 1982). For example, it is not ethically appropriate for public policy to aim to maximize the satisfaction of sadistic preferences, such as those that some might have that specific others suffer or be deprived of their liberties. Similarly, it is not appropriate to allow market-CBA to determine policy when questions of basic human rights and justice are central. So, for example, few would propose that the government should use market-CBA to determine whether to decriminalize child prostitution, or to decide when elder abuse is permitted.

The second issue surrounding preferences is that, arguably, market CBA misunderstands the nature of ethical judgments. One concern is that attempts to reduce all values to measurements of willingness-to-pay risk confusing mere preferences, whose significance might perhaps be measured in terms of the intensity with which they are held, with values, whose import should be assessed in terms of the reasons that support them (Sagoff 1982). For example, just as we should not evaluate mathematical claims (such as $2 + 2 = 4$) by asking how strongly mathematicians feel about them, or, more specifically, how much they would be willing to pay for the rest of us to accept them, so we should not evaluate ethical claims in these ways. To do so misunderstands the point of both ethics and mathematics.

A central problem here is that market-CBA requires that government policy-making translate all values into commodity value. For market-CBA to operate, whatever pros and cons might be associated with climate change must be reduced to their economic value. So, for instance, we are quickly led to discussions of the economic value of people (e.g., valuing lives in terms of foregone income), the economic value of health (e.g., the costs of disease to GDP), the economic value of preserving treasured ecological systems, plants and animals, and so on.[3] Yet ethicists (and most other people) typically think that the value of such things cannot be adequately reduced to narrowly economic terms, as if they were items to be traded in the same way as commodities in a market.

How do these problems manifest themselves in practice? One place is in the use of the concept of the 'social cost of carbon' (SCC). This is a number employed by governments to estimate in dollars the long-term damage caused by the emission of one extra ton of carbon dioxide into the atmosphere, with the aim of showing how much it would be worth to us today to avoid that damage (e.g., EPA 2017). The SCC raises many issues. Here we note two.[4]

First, in determining a SCC, any analysis must face enormous challenges in determining what future harms will be created by emissions. For example, consider the difficulty of offering reasonable estimates of costs for the long-term future, given that a significant percentage of carbon dioxide will likely last in the atmosphere for hundreds or even thousands of years (Archer 2005). How is market CBA to deliver a reliable evaluation of costs and benefits over such a time-frame, particularly based on current market prices?

Second, analysts must confront further deep issues in placing a value on future harms. For example, in principle, a high atmospheric CO_2 concentration is capable of leading humanity and many other species to extinction, particularly if a cascade of tipping points occurs. But why would we think that the 'willingness to pay' of the current generation's consumers provides an acceptable foundation on which to build an answer?

Third, the SCC must take a position on whose interests count. Politically, this has had great resonance in the United States, where successive administrations have taken different views on whether the interests of only Americans need to be taken into account, or the interests of humanity more generally. Those hostile to climate action have often insisted on restricting the cost assessment to Americans alone. Yet this is ethically indefensible. Doing so ignores the threat that climate change poses to humanity as such, and the harms being imposed by high-emitting nations on others around the world (e.g., Mintz-Woo 2018).

Notably, this issue is not limited to the national scale. Many subnational governments, including US States like Pennsylvania, for example, have used CBA to determine whether proposed climate policies are justified by comparing the costs of the policies to the economy of the government implementing the policy, such as Pennsylvania, to the economic value of the harms avoided by the policy only on the sub-national government. Yet this approach is ethically problematic because such comparison ignores the harms to the rest of the country – and the rest of the world – that are caused by emissions from the subnational government. In essence, the level of permissible harm is being determined by asking what the perpetrators of the harms are willing to pay to avoid the harms – all while ignoring many of the victims, who are either in the future or in other countries.[5] Both moves are strange analytically, and unacceptable ethically. Sadly, hiding such issues behind the veneer of 'cost' arguments is all too predictable given the background risk of moral corruption.

We conclude that though some opponents of ambitious climate action have argued that a nation's response to climate change must satisfy market CBA, such arguments should be contested. While market CBA may be a useful tool for investigating particular economic issues in certain specific contexts, ethics ask different questions. For example, market CBA cannot determine what justice requires of our choices; similarly, while market CBA is concerned with preference-satisfaction, ethics asks us to consider which preferences are acceptable to have. All of this suggests that rather than standard market CBA, what is needed is what we have called pros and cons analysis (PCA), and ethical concepts are central to that enterprise.

Ethical CBA?

Let us turn now to an important debate. Some advocates of standard market-CBA argue that their view is ultimately descriptive, and does not aim to address ethical issues, which (they say) fall outside the scope of their ambitions for their theory. However, others adopt an explicitly ethical approach to CBA, and argue that crucial parameters within it are themselves based on ethics, including through ethical thought-experiments of the sort often employed by philosophers (Stern 2006). What should be said about this dispute and its implications? Let us make three quick points.

First, it should be emphasized that the dispute here occurs within economics and policy as well as outside. Numerous prominent economists, including climate economists, find themselves on different sides of the question. Thus, this is as much an intra-disciplinary dispute as one that plays out between disciplines.[6]

Second, we are more sympathetic to the ethical approach, and in particular to the claim that ethical judgments are central to climate economics, including market-CBA and net-benefit analysis. Most notably, many of the most important ingredients in economic models for climate change involve variables that are either essentially ethical, or strongly influenced by ethical considerations. At least, we believe that this must be true for any responsible kind of CBA that it would be reasonable to use to guide major policy decisions.

One area where this is especially clear is in discounting the future, where decisions about basic parameters dominate the models, in the sense of largely determining the results. Crucially, these decisions are largely based on what values one assigns to benefits to future people, to the importance of intergenerational equality, and to the risk of human extinction (Stern 2006; Gardiner 2011; Zuber and Fleurbaey in press). Economic modeling tends to make bold assumptions about these matters. Unfortunately, at the time of writing, these assumptions do not appear to reflect mainstream ethical discussions, including those based on incorporating the values suggested above, such as preventing harm, protecting basic human rights, recognizing responsibilities, and promoting justice.

The third point is that support of market-CBA is justified by some economists on the basis of utilitarian ethical theory, which claims that society should develop policies that maximize human well-being, understood in terms of preference-satisfaction. However, this raises two key problems. The first is that most philosophers are not utilitarians and indeed hold that maximizing utility is not an ethically supportable justification for sacrificing other key values, such as basic human rights. The second problem is that most actual utilitarians – and especially philosophical utilitarians – would not endorse market-CBA, but rather reject it on the grounds that it does not provide a sound methodology for understanding or implementing utilitarianism. For instance, most philosophical utilitarians would reject the willingness-to-pay approach. Similarly, many would question the basic methodology of a strong reliance on direct calculations of costs and benefits (traditionally, an 'act-utilitarian' approach). Instead, they would favor rule-utilitarianism, which proposes that we should put in place social rules and practices which maximize utility over the long term. Consequently, they would likely not try to evaluate particular policies – such as climate policies – directly by summing up their projected costs and benefits, as market-CBA does. Indeed, notably, rule utilitarians are typically sympathetic to human rights, and so would probably accept a strong presumption that conflicts between simple cost arguments (such as those generated by market-CBA) and human rights violations ought to be resolved in favor of human rights.

There is much to be discussed at a theoretical level. Nevertheless, we conclude that at the very least, there should be a strong presumption that there is a fundamental lack of fit between the problems we face with climate change and employing conventional market-CBA as a solution. This undermines the relevance of the simple cost arguments. It suggests that instead of bowing to such arguments, we must take seriously a wider range of values.

Ethics in Climate Agreements

Fortunately, there are straightforward ways to make wider values visible. One source of climate responsibilities is climate agreements themselves. Consider several relevant examples.[7]

To begin with, under the UNFCCC, developed country Parties have already agreed to provide financial resources to assist developing country Parties in implementing the objectives of the Convention (UNFCCC 1992: Art. 4, 3), and agreed that they have the responsibility to assist the developing country Parties that are vulnerable to the adverse effects of climate change in meeting their costs of needed adaptation to adverse effects (UNFCCC 1992: Art 4, 4). The Paris Agreement also states that the developed countries shall provide financial resources to assist developing country Parties with respect to both mitigation and adaptation in continuation of their existing obligations under the Convention (Paris Agreement 2015: Art. 9.1). Yet many cost arguments simply ignore these and similar responsibilities. In particular, the arguments made by opponents of climate change based on the costs to a nation of needed climate policies typically have not even considered the costs that developed countries may be responsible for covering, given that they have already agreed that they must contribute to financing the mitigation and adaptation costs of climate change to poor developing countries.

Additional responsibilities emerge from considering responsibilities of harms caused by a nation to people outside the nation's boundaries in terms of loss and damage. For instance, the no-harm principle recognized in the UNFCCC makes nations responsible for climate losses and damages to other nations caused by activities within their jurisdiction.

Unfortunately, there are pressing problems in operationalizing this principle. For one thing, there is an absence of legal rules in the international legal system that prescribe how the value of damages should be allocated among all nations responsible for the climate change harms. For another thing, the fact that all nations have contributed to rising atmospheric CO_2 levels makes it unlikely that a court will find any country financially legally liable for a specific amount of losses or damages in another country.

Nevertheless, because nations have agreed in the UNFCCC that they have a duty to prevent activities in their jurisdiction from harming countries and people beyond their borders, many of the most vulnerable countries have been pushing for the creation of a financial mechanism under the UNFCCC that would compensate vulnerable countries for climate losses and damages that adaptation cannot remediate.

Importantly, at the 2012 Doha Conference of the Parties under the UNFCCC, the international community agreed to establish a formal mechanism for compensation for losses and damages, the Warsaw Mechanism for Loss and Damages (WMLD). Article 8 of the 2015 Paris Agreement made the WMLD an official negotiating body of the UNFCCC. Since the beginning of negotiations of the WMLD, negotiations have gotten bogged down over how to finance compensation for losses and damages in developing countries as developed nations have stressed that any agreement on compensation should not be understood as establishing legal liability for the developed nations to compensate for losses and damages. Although developed nations will likely prevail in avoiding any language that could be construed as establishing their clear legal liability for losses and damages in developing nations, some believe that developed nations will likely agree eventually to create some mechanism, such as an insurance fund, to compensate vulnerable developing countries for some kinds of losses and damages in developing countries which developed countries will be expected to provide financing for.

Notably, there are practical as well as ethical and legal reasons to follow through on such commitments. Financial support of developing nation's mitigation obligations under the UNFCCC is not only legally required under the UNFCCC but also practically important because large-scale investments by developing countries are required to significantly reduce their emissions and very dangerous climate change will not likely be avoided unless developing nations reduce their GHG emissions to their fair share of safe global emissions. Financial support for developing nations by developed nations is also both legally and ethically required to meet the adaptation needs of developing countries.

The issue of loss and damage is pressing. Climate impacts, such as sea-level rise and more frequent droughts and floods, are already causing devastating effects to communities and individuals in developing countries. These impacts to developing nations are already affecting developed nations because, for instance, between 2008 and 2011, approximately 87 million people were displaced due to extreme weather events which have caused a mass migration of refugees which are already destabilizing many developed nations, particularly in Europe (Brookings 2019). Since 2014, serious drought in and severe weather in Central America have caused large migrations of refugees which have put pressure on the US southern border (Wernick 2018). Because climate change caused refugees are already destabilizing developed countries who have been fleeing vulnerable areas of poor developing nations that have become inhabitable due to climate change-induced droughts, floods, loss of drinking water, and rising seas, developed nations have a strong practical incentive to assist developing nations with adaptation. If developed countries do not help finance adaptation needs in developing countries, they will experience growing conflict and stress caused by vulnerable people's problems including the 150 million refugees that the World Bank predicts will be created by a 2°C temperature rise by the end of this century, a temperature rise that now appears to be almost inevitable (World Bank 2018).

Conclusion

In this chapter, we have argued against the common assumption that cost arguments – and particularly those made on behalf of the current generation – either straightforwardly trump, or else provide a very weighty presumption against, robust climate action. We did so by identifying other salient values and ethical constraints, including ones already embodied in law and policy. These include duties not to cause harm, to protect basic human rights, and to accept various kinds of climate responsibility. We also argued that conventional, market CBA is severely limited in its usefulness, given the nature of the wider ethical concerns. By contrast, we claim that ethics plays fundamental roles within any responsible PCA. One way to help keep ethics central would be to stop using the term 'cost-benefit analysis' in different ways, which helps to facilitate ambiguity and confusion. Thus, we recommend revising our terminology as well as our practices.

Acknowledgments

The authors are grateful to Kathryn Gwiazdon for helpful feedback on earlier drafts.

Notes

1 A tipping point can be understood as a critical threshold in the earth climate system – such as major ocean and atmospheric circulation patterns, the polar ice sheet, and the terrestrial and ocean carbon stores – the passing of which produces a steep change in the system (Breakthrough Institute 2018). Progress toward triggering a tipping point is often driven by positive feedbacks, in which a change in one component of the climate system leads to further changes that eventually 'feedback' onto the original component to amplify the initial effect. A classic global warming example is the ice-albedo feedback. This happens when warming causes ice sheets to melt, which then prevents that ice reflecting heat energy from the sun back out into space, which then causes more heat energy to warm the Earth, which then melts more of the ice sheets.

2 For instance, the salient features may include not just the "pros and cons" themselves, but also the conditions under which they are salient, how they relate to each other, and further contextual factors.

3 For instance, some plants and animals are sacred in some cultures, such as cows in India and elephants in Thailand, so that using willingness to pay to determine the value of climate harms avoided requires transforming sacred value into commodity value.

4 There are numerous other ethical problems with the use of CBA to determine the acceptability of climate policies. See, Brown, D. (2008) Ethical Issues in the Use of Cost-Benefit Analysis of Climate Change Program, available at https://ethicsandclimate.org/2008/06/01/ethical-issues-in-the-use-of-cost-benefit-analysis-of-climate-change-programs/.

5 This is usually calculated by governments without the acceptance of those whose interests will be harmed by determining the 'willingness to pay' for protecting things that will be harmed that have no market value and by determining the present value of things that will be harmed in the future by discounting the values of things harmed in the future by judging what discount rate should apply, a decision for which there is no value-neutral way of proceeding.

6 There are signs that the descriptive view dominates in some circles. For example, consider the granting of the Nobel Prize to Nordhaus – and Nordhaus alone – rather than awarding it jointly to include more ethically-minded economists, such as Nicholas Stern or Martin Weitzman.

7 Notably, the arguments made by opponents of climate change policies based on the cost to a government of adopting climate policies tend to ignore commitments that countries have made in the past and which bring with them ethical responsibilities.

References

Anderson, K. (2012) "Going Beyond Dangerous – Brutal Numbers and Tenuous Hope," *Development Dialogue*, 61: 16–41.

Anthony, et al. (2018) "21st-Century Modeled Permafrost Carbon Emissions Accelerated by Abrupt Thaw Beneath Lakes," *Nature Communications*. Available at https://www.nature.com/articles/s41467-018-05738-9#author-information

Bernie, P., Boyle, A. and Redgewell, C. (2008) *International Law and Environment*, Nairobi: UNEP.

Breakthrough Institute. (2018) *What Lies Beneath, On the Understatement of Existential Climate Risk*. Available at https://docs.wixstatic.com/ugd/148cb0_a0d7c18a1bf64e698a9c8c8f18a42889.pdf.

Brookings Institution. (2019) *Climate Crisis, Urban Migration, and Refugees*. Available at https://www.brookings.edu/research/the-climate-crisis-migration-and-refugees/.

Broome, J. (2012) *Climate Matters: Ethics in a Warming World*, New York: Norton.

Brown, D. et al. (2018) "A Four-Step Process for Formulating and Evaluating Legal Commitments Under the Paris Agreement," *Carbon & Climate Law Review*, 12(2): 98–109.

Brown, D. (2022) "Why Developed Nations Should Support the Development of a Loss and Damages Mechanism for Refugees," *EthicsandClimate.org*.

Caney, S. (2010) "Climate Change, Human Rights and Moral Thresholds," in S. Humphreys (ed.), *Human Rights and Climate Change*, Cambridge: Cambridge University Press. Reprinted in S. M. Gardiner, S. Caney, D. Jamieson and H. Shue (eds.), *Climate Ethics: Essential Readings*. New York: Oxford University Press.

Cho, R. (2021) *State of the Planet, How Close Are We to Tipping Points*, Columbia University. Available at https://news.climate.columbia.edu/2021/11/11/how-close-are-we-to-climate-tipping-points/.

Dworkin, R. (1977) *Taking Rights Seriously*, London: Duckworth.

Environmental Protection Agency (EPA). (2017) *The Social Cost of Carbon*. Available at: https://19january2017snapshot.epa.gov/climatechange/social-cost-carbon_.html

Gardiner, S.M. (2011) *A Perfect Moral Storm: The Ethical Tragedy of Climate Change*, New York: Oxford University Press.

Gardiner, S.M. (this volume) "The Ethics of Climate Targets: Extinction Rebellion v. The Climate Establishment" in D. Brown and K. Gwiazdon (eds.), *The Routledge Handbook of Applied Climate Change Ethics*, London: Routledge.

Gardiner, S.M. and Weisbach, D. (2016) *Debating Climate Ethics*, New York: Oxford University Press.

Intergovernmental Panel on Climate Change (IPCC). (2014) *Contribution of Working Group III to the Fifth Assessment Report of the Intergovernmental Panel on Climate Change*, Cambridge University Press.

Intergovernmental Panel on Climate Change (IPCC). (2019) *Special Report on 1.5 C*. Available at https://www.ipcc.ch/sr15/.

International Council on Human Rights Policy (ICHRP). (2008) *Climate Change and Human Rights: A Rough Guide*, Geneva: International Council on Human Rights.

Leahy, S. (2019) "Climate Change Driving Entire Planet to Dangerous Tipping Point," *National Geographic*. Available at https://www.natiTonalgeographic.com/science/2019/11/earth-tipping-point/.

Lenton, T.M. et al. (2019) "Climate Tipping Points: Too Risky to Bet Against," *Nature*, 575: 592–595.

Mintz-Woo, K. (2018) "Two Moral Arguments for a Global Social Cost of Carbon", *Ethics, Policy & Environment*, 21.1: 60–63.

Morgan McFall-Johnson. (2020) "Greenland's Melting Ice Sheet Has Passed The Point of No Return," *Science Alert*. Available at https://www.sciencealert.com/greenland-s-melting-ice-sheet-has-passed-the-point-of-no-return-scientists-say.

Paris Agreement. (2015) Available at https://unfccc.int/files/meetings/paris_nov_2015/application/pdf/paris_agreement_english_.pdf.

Rio Declaration on Environment and Development. (1992) Available at https://www.un.org/en/development/desa/population/migration/generalassembly/docs/globalcompact/A_CONF.151_26_Vol.I_Declaration.pdf.

Sagoff, M. (1982) "At the Shrine of Our Lady of Fatima or Why Political Questions Are Not All Economic", *Arizona Law Review*, 23: 1281–1298.

Shue, H. (2022) *The Pivotal Generation*, Princeton: Princeton University Press.

Stern, N. (2006) *The Economics of Climate Change: The Stern Review*, HM Treasury. Available at https://webarchive.nationalarchives.gov.uk/ukgwa/20100407172811/https:/www.hm-treasury.gov.uk/stern_review_report.htm.

United Nations Framework Convention on Climate Change (UNFCCC). (1992) Available at https://unfccc.int/resource/docs/convkp/conveng.pdf.

United Nations General Assembly. (2019) *Special Report on Human Rights and Climate Change*. Available at https://www.ohchr.org/Documents/Issues/Environment/SREnvironment/Report.pdf.

United Nations Office of the High Commissioner for Human Rights (UN OHCHR). (2009) *Report on the Relationship Between Climate Change and Human Rights*, UN Doc. AHRC/10/61.

Wernick, A. (2018) *Climate Change Is Contributing to Migration of Central American Refugees, The World*. Available at https://www.pri.org/stories/2018-07-15/climate-change-contributing-migration-chttps.

World Bank. (2018) *Climate Change Could Force Over 140 Million to Migrate Within Countries by 2050*. Available at https://www.worldbank.org/en/news/press-release/2018/03/19/climate-change-could-force-over-140-million-to-migrate-within-countries-by-2050-world-bank-report.

Zuber, S. and Fleurbaey, M. (2023) "Discounting and Intergenerational Ethics", in S.M. Gardiner (ed.), *Oxford Handbook of Intergenerational Ethics*, New York: Oxford University Press.

19

THE CLIMATE CHANGE DISINFORMATION CAMPAIGN

Attacking the Common Good, Advancing the Self, and Destroying Democracy

Kathryn Gwiazdon and Donald A. Brown

Introduction

> The words of truth are simple, and justice needs no subtle interventions, for it has a fitness in itself; but the words of injustice, being rotten in themselves, require clever treatment.
>
> – *a diary entry by Thomas Jefferson, citing Euripides (Ricks 2020:79)*

For over 30 years, through a sophisticated and well-funded disinformation campaign, the fossil fuel industry has been successful in framing the public debate surrounding climate change on issues of nationalism, excessive costs, and scientific uncertainty (if not outright, 'hoax!'). This is despite the fact that the international community, including the United States, has already agreed to be legally bound by ethical principles that counter these narratives under the 1992 UN Framework Convention on Climate Change (UNFCCC) and the 2015 Paris Agreement. These include the 'no-harm principle,' which makes governments responsible for harms to others from activities within their borders, de-legitimizing both a nationalist and cost approach; the 'precautionary principle,' which requires states to act with precaution in the face of uncertainty, making claims of scientific uncertainty an illegitimate excuse for inaction; and the 'principle of equity,' which makes fairness (i.e., justice and accountability) a requirement for the prevention, mitigation, and adaptation of harms, countering once again both nationalist and cost arguments.

This chapter deals with the failure to get traction for ethical principles in claims about what nations should do to achieve the common good in the face of climate change, and argues that this failure is the intentional result of deceptive disinformation tactics by the fossil fuel industry, reaching and manipulating all institutions of a democracy. These principles – rooted in state responsibility for the common good, justice, and truth – provide states with a counter narrative and language to defend their indefensible inaction (or not enough action) on climate change. For purposes here, it is understood that all such claims implicitly have the form, 'because of facts A, B, and C (factual premise), governments should do D (normative/ethical conclusion).' Here, 'normative' means right or wrong, an ethical duty, or a prescriptive conclusion in light of facts. It is important to also note that the United States and other nations have agreed to other related ethical principles embedded in human rights doctrine which have also been ignored in public debates about climate change.

DOI: 10.4324/9781003039860-23

This chapter will open with the evolution of the disinformation campaign, beginning with the 1971 *Powell memo* and evolving into the climate-specific tactics seen today. It will then explore the common threads that carried its evolution: its attacks on the common good, justice, and truth, and through arguments rooted in neoliberal ideology and nationalism, or 'the self.' The last major component of the chapter will be to discuss the intentionally targeting of education in order to leave society – present and future – without the language, skills, and places (i.e., without the ability) to counter the narrative that the disinformation campaign asserts. Much of this discussion focuses on critical acts within the United States that have created and advance the disinformation campaign, with national and global repercussions.

The ultimate purpose of this chapter is for states and civil society to realize the power and potential of ethics to climate change action by understanding that it is *intentionally* targeted because of that power. And although it is understandable (not ethical, not just) why the fossil fuel industry, as entities that exist to create profits for its shareholders, seek to undermine a power that directly affects its future profits, it is inexcusable for nations, who exist to create and defend and provide the opportunities for the common good, to undermine that power. Ethics is the inquiry into right and wrong behavior, to what is just and unjust, and when responsibility, or accountability, attaches – it is the practice of states. And when states and elected officials allow the tactics of the disinformation campaign to not only exist, but thrive, they are betraying the purpose of a state and their duties to the people.

From the Powell Memo to the Climate Change Disinformation Campaign

The climate change disinformation campaign is decades old. It is powerful, it is deadly, and it is successful: it is the underlying cause of the failure to get traction for ethical principles in climate change discourse – of the intentional disempowerment of ethical principles. The well-funded, sophisticated spread of disinformation is rooted in denialism, scientific uncertainty, and economic 'costs' as a viable defense to preventing or being held accountable for harm (i.e., to ensuring justice). Denialism is the subject of another chapter in this *Handbook*, so most of the focus here is on scientific uncertainty and 'costs' arguments – although denial, or rejection, of the purpose of a nation is certainly unpacked here.

There are often two origin points when discussing the climate change disinformation campaign. The first is the targeted attack on science in the 1950s with the tobacco industry (although, when considering religion, the attack on science is centuries old in human history, reconnecting again when the disinformation campaign aligned with the far-right evangelical movement in the United States). The tobacco industry, through its own scientific research, learned that its profitable product was killing people, and so, instead of warning the public, they hid the science and distorted the truth, deliberately "manufacturing uncertainty" to "[delay] regulation and victim compensation for decades" (Mann & Toles 2016, citing Michaels 2020: 69–70; see also Oreskes & Conway 2011). This model has since been replicated almost exactly by the fossil fuel industry (see Michaels 2020; see Oreskes & Conway 2011). The second is where we begin, in 1971 in the United States, with a confidential memorandum from Lewis Powell, Jr. (*Powell memo*), then Vice President of the US Chamber of Commerce, to Eugene Snydor, Jr., then Chairman of the Education Committee of the US Chamber of Commerce (Powell 1971). It was titled, *Attack on American Free Enterprise System*. It helped to create a widely accepted, unquestionable cultural narrative that the government is the problem, not the solution, to many of society's most troubling problems, thereby disempowering the most powerful actor in international law: the nation-state, as well as the only authority powerful enough to counter domestic state interests in the United States: the federal administrative state, which holds the power of regulation. In addition, if the state is seen as the problem, and the state is responsible for justice and the common good, how can any state respond to global or shared crises by appealing to notions of justice and the common good? The disinformation campaign neuters the very power and purpose of a democratic nation.

Cultural narratives, like those we see with the arguments coming from the climate change disinformation campaign – 'we should care only for us,' 'we should not pay for our harms if we have the power to not pay for our harms,' 'we should not believe experts if it does not suit our interests,' 'regulation is an attack on my freedom,' 'climate change is too expensive' – become so accepted by a society that citizens often become afraid to challenge them. They are assumed as true because they have been ingrained in our worldviews, which then closes or stifles critical discourse on them. And what is a disinformation campaign if not a campaign against truth, and against discourse on right and wrong behavior?

The *Powell memo* birthed the climate change disinformation campaign (see Whitehouse 2021). Powell claimed that the free market was under attack, as evidenced by the successful civil and political movements in the 1960s that advanced social and environmental protections (yes, the very movements that placed the United States as the global leader in human rights, democracy, and the environment). Yet the *Powell memo* construed these victories on rights, inequality, and injustice as a threat to corporate power and the neoliberal ideology and free-market enterprise that undergirds that power. The *Powell memo* criticized corporations for their lack of vigor in responding to these 'challenges' and called for a much more aggressive response from the business community to protect free enterprise from criticism from college campuses, the pulpit, the media, the intelligentsia, literary journals, the arts and sciences, and politicians.

And now we begin to see the formation of the core of the climate change disinformation campaign's methodology: a multi-faceted attack on education, the media, and the branches of governance. Two months after the *Powell memo* was released, President Nixon nominated him to the US Supreme Court – the highest court in the nation – where he directed and influenced the law and governance of the most powerful nation on Earth for 15 years, further embedding 'the Powell narrative' as 'the American narrative.'

Since it was written, the recommendations of the *Powell memo* have all been well-funded and aggressively pursued (see Brown 2015). He called for the explicit designation of a member of senior management who would have the responsibility to fight attacks on the free enterprise system. He called for the expansion of the role of business organizations to fight the threats of the free market; most importantly, the US Chamber of Commerce, which has the time, finances, and organizational capacity to powerfully unify the response of the business community.

To counter criticism of the business community from college campuses, he asked that business organizations, including once again the US Chamber of Commerce, to support scholars who will defend the free enterprise system; develop speakers and support speakers' bureaus that will counter the liberal rhetoric coming from college campuses; subject textbooks to ideological review; insist on equal time for campus speakers exposing the views of the business community; insist that college faculties be balanced by those who will defend the free enterprise system; and request that graduate schools include courses that support the free enterprise system. He also encouraged local chambers of commerce to provide the views of the business community in high schools. He also called for scholars to publish in scholarly journals and that the business community should establish incentives for scholars to publish defenses of free enterprise in books, papers, and pamphlets.

To counter criticism of the business community in the media, he recommended that they establish staff who work with the media to communicate to the general public the views of the business community; monitor and criticize television programs that 'unfairly' criticize the free enterprise system and, where appropriate, file complaints with the Federal Communications Commission. He also called to monitor radio and other media and pressure them to cover the views of defenders of the free enterprise system and to spend more money on advertising that expressly supports the free market system.

In his most powerful move, he targeted politicians and the judiciary, calling for a much more aggressive and deliberate campaign to support politicians who support the interests of the business community and that they need to become much more involved in the judicial system in, among other things, filing litigation and *amicus curiae* briefs in important cases. Today, we saw this tactic explicitly referenced in *West Virginia v. EPA* (2022), a case before the US Supreme Court that the fossil fuel industry *intentionally* brought, and intentionally brought before this *specific* court, to determine whether a federal regulatory agency had the power to regulate greenhouse gas emissions as related to climate change. In an *amicus curiae* brief filed by US Senator Whitehouse and others, he raises some inconvenient truths, and the resulting decision by the Court only gave those truths more credence,

> Regrettably, there was another parallel effort, one targeting the judiciary. To achieve the political goal of 'deconstruction' of the so-called 'administrative state,' judicial appointments were made part of the 'larger plan' to eradicate regulatory agencies... In the last administration, former White House Counsel Don McGahn... stated plainly that 'the judicial selection and the deregulation effort are really the flip side of the same coin.' More than $400 million was spent on this deregulatory effort targeting the judiciary, with much of it coming in large donations from anonymous sources, while fossil fuel interests donated added millions to the then-president's reelection campaign.
>
> *(Whitehouse et al. 2022)*

In the very next section, he links that campaign to the case before them,

> Now this industry machine turns to the judiciary to push its policy agenda, asking the Court yet again to do what Congress has consistently refused to do: gut EPA's regulatory authority under one of our preeminent environmental protection laws. The anti-regulatory arguments advanced here would shield deep-pocketed and politically-powerful coal, oil, and gas interests from regulation under the statutory EPA authority challenged by the petitioners, and give polluters new opportunities to tangle and delay regulation of their emissions. It is no surprise that most, if not all, of the amici supporting the petitioners in these cases receive substantial funding from this industry.
>
> *(Whitehouse et al. 2022)*

Whitehouse asked the Court to "refuse to participate in this industry-driven project" that "[reeks] of politics, and [is] advanced by thinly-disguised but highly-motivated industry front groups."

Unfortunately, however, the Court decided in favor of corporate interests, as the machine had desired, instructed, and *constructed* the Court to do; indeed, to do otherwise would have required them to acknowledge the existence of the machine that put them into their seats to do the very job that they were being asked to do in that very case (see Feldman & Kott 2021, regarding the power of the Federalist Society; see also Whitehouse 2022). And the Court came to their decision with their own 'thinly-disguised' efforts, gutting decades of regulatory powers through a novel and bizarrely constructed 'major questions' doctrine (*West Virginia v. EPA* 2022). The result was exactly what Whitehouse had warned: the majority justices had "[undermined] the public's remaining faith in the Court" (Whitehouse et al. 2022). A success for the disinformation campaign, another loss for democracy.

But to those following the disinformation evolution, this result was no surprise. For decades, those espousing 'free market' ideology had already been targeting the judiciary. The Olin Foundation, another organization central to the larger disinformation campaign, had been "central to the Federalist Society's success" (Aronoff 2020). A "stealth political attack" that "the country's best law schools didn't grasp," they worked to introduce 'Law and Economics' in US law schools as a tool to advance free market ideology. And it worked. Following the *Powell memo*, they flooded schools with grant money, "taking advantage of students financial need to indoctrinate them with a particular ideology" (Mayer 2016: 109). In addition, they funded luxurious seminars for judges, at one point accounting for 40 percent of the federal judiciary (Mayer 2016: 110). Topics included, "The Moral Foundations of Capitalism;" and climate, listed alongside terrorism, as "Challenges to Liberty and the Rule of Law" (Mayer 2016: 110).

On wider issues, Powell wanted to harness the power of corporate shareholders to advance the interests of the business community and dramatically increase finances in support of opposition to those threatening unfettered markets, including increasing the staff of organizations like the Chamber of Commerce to engage in this work. As recent as January 12, 2023, we see the continued advance of the US Chamber of Commerce pushing the narrative that government, the source of regulation to protect the common good, is the source of their problems. Suzanne Clarke, President and CEO of the US Chamber of Commerce, gave their State of American Business remarks, titled "Agenda for American Strength." She argued that "government itself is seen as a source of risk for businesses," and "the only risk that businesses say is rising... is the risk that comes from our own government" (Clarke 2023). She continues that "regulatory overreach" threatens "a fragile world economy" and that "It is not the role of government to direct the behavior of business, redistribute power in our economy, or undermine the competition that fuels free enterprise" (Clarke 2023). 'Strength,' fear [of a fragile economy], and the savior under attack, 'free enterprise' – the seedy, authoritarian-leaning under-current of the disinformation march continues.

Importantly, Powell also wanted to defend the free enterprise system by linking personal freedom to free enterprise. The notions of liberty engrained in the very ideals of the nation were now to be directly and completely entangled with the free enterprise system. This is how cultural narratives take root: a core facet of a culture already accepted as true, 'liberty,' is then distorted. It is why special interests try to usurp words like 'freedom,' or 'life' even – they know their power to an entire culture, and they try to take that power. And it is the same way they try to demean words and concepts that they cannot take – like ethics, justice, equity, and the common good.

But liberty cannot exist without a concern for the public good. President John Adams believed that,

> ...Public virtue is the foundation of republics. There must be a positive passion for the public good, the public interest, honor, power, and glory, established in the minds of the people, or there can be no republican government, nor any real liberty. And this public passion must be superior to all private passions.
>
> *(Ricks 2020: 119–120, citing From John Adams to Mercy Otis Warren, 16 April 1776)*

Attacking the Common Good

But, unfortunately, "Corruption has... come to be seen as old-fashioned bribery rather than as it was originally conceived, as the act of placing special interests above the common good" (Hart 2007). A decade after the *Powell memo*, the success of this propaganda – to support less government regulation for the common good – was already evident when US President Ronald Reagan proclaimed in his inaugural speech that, "[the] government is not the solution to our problem; government is the problem" (Reagan 1981). Although many, if not most, people would acknowledge the benefits of free markets, and that government is sometimes the problem, it is irrational to conclude that the private sector alone could provide most public goods, such as affordable health care, protection from environmental threats, towns designed to promote social interaction, affordable high-quality education, affordable housing, etc. – the very things that merchants and despots throughout history have ruthlessly schemed against. This scheming is inevitable when the solution to growing global threats requires the regulation of new technologies that have admitted value but dangerous potential for harm. Current potentially beneficial technologies of concern include, for example, artificial intelligence and bioengineering.

This 80s embrace of neoliberalism also came at the same time as the right wing turn against science; a natural consequence to a field that was not benefitting their end goals. There were obvious harms (i.e., pollution) that the free market caused and could not fix, and regulation was a potential solution. Therefore, the aim became to challenge regulation and the science that undergirds it (Oreskes & Conway 2011). So, the "politically-motivated scientists" are not to be trusted and the scientific establishment was controlled by the liberal agenda (Oreskes & Conway 2011: 59–63).

Evolving from Powell, and rooted and realized through deceit, greed, and corruption, the intentional spread of disinformation on climate change is a powerful adversary to the common good, to national and international action on climate change. Yet, the fathers who founded the United States – and later those who carried it – believed that the goal of democracy was to achieve the common good, a government "of the people, by the people, for the people" (President Abraham Lincoln, Gettysburg Address 1863). The oldest state constitution in the United States, that also served as a model for the US Constitution, is Massachusetts, drafted by the man who would later become President John Adams. It also serves as the oldest functioning written constitution in history – a testament to the history and present of the United States. Article 7 of the 1780 Massachusetts Declaration of Rights (i.e., the Massachusetts Constitution) reads, "Government is instituted for the common good; for the protection, safety, prosperity and happiness of the people; and not for the profit, honor, or private interest of any one man, family, or class of men." For its supporting precedent, if offers James Otis, 'The Rights of the British Colonies Asserted and Proved,' (1764), "...But let the origin of government be placed where it may, the end of it is manifestly the good of the whole..." This is the model for democratic governance. And this is what the disinformation campaign actively seeks to dismantle.

The essence of the common good is 'justice' and achieving the common good is the essential role of government. Justice is the belief in equity, or 'fairness': that if is there is a harm, those who caused that harm will be held accountable; that there will be efforts to make that harmed person or thing 'whole' again; and that actions will be put into place to deter future similar harm. Without justice, harm – and those who harm – will continue unabated.

Yet, despite the foundational role of justice in the rule of law and in democratic societies, it is usually not the focus for agendas and during negotiations of treaty-making bodies and conferences of the parties (COPs). However, the differences between developed and developing nations about what fairness requires are usually the most time-consuming issues. This was seen most recently at COP15 of the Convention on Biological Diversity (CBD) in Montréal, where the entire agreement was stalled because biodiversity rich, developing nations in Africa, including Cameroon, Uganda, and the Democratic Republic of the Congo (DRC), argued that they were not receiving enough support from developed nations to support the '30 by 30' plan [to protect 30 percent of the Earth's biodiversity by 2030] (Greenfield & Weston 2022).[1] These contentious issues continue, while corporate interests that are economically threatened by issues under consideration have been largely successful at negotiations because they have generated national political opposition, and often by the dissemination of sophisticated disinformation.

Getting traction for justice in government affairs has become more urgent since the 1970s, when well organized, aggressive, sophisticated efforts undermined governments' central role in ordering society for the common good. While co-author to this chapter, Donald A. Brown, worked for the US EPA on UN international environmental issues, he saw corporate interests lobby the agency to oppose provisions of the biodiversity treaty and climate law that had the support of other countries. Most Americans, including US-based NGOs, seem to be unaware that the United States is widely viewed as an obstructionist on many global environmental issues, although many non-nationals believe there is still hope that the United States can make democracy work for the common good. In the last decade, however, the manipulation of democracy in the United States has made it much easier for money to dominate politics by removing checks or limits on corporate donations, voter suppression, gerrymandering, and allowing candidates to not disclose donors. This, while other countries have often made it easier to vote. In Australia, for example, Australian citizens have a duty to vote, enforced with a routine fine (e.g., NSW Electoral Commission 2022). In Japan, political money is not allowed to be used on television (which also explains why one hears political messages on loudspeakers from small trucks all the time) (see Curtis n.d.; see Nadeau & Fahey 2017).

As much as justice issues are avoided in international negotiations, they are indispensable to the goals of any international environmental treaty. For instance, those drafting the United Nations Framework Convention on Climate Change (1992) had to grapple with what rules would govern each country's GHG emissions reduction target in light of the fact that some nations more than others are responsible for the current problem. Although the treaty negotiations established very general rules about national responsibilities to adopt policies to prevent dangerous climate change, the international negotiations were unable to agree on how to allocate responsibility among nations for emissions except in the most general and abstract terms. This is so despite the fact that climate change is a problem that necessarily requires some guidance on how to allocate responsibility among nations for reducing national GHG emissions.

Some nations have been pushing for more clarity on these issues for decades. The best the initial round of negotiations could agree on is that the developed countries should take the lead on reductions and that each country should reduce GHG emissions to levels required to achieve any warming limit goal in accordance with "equity and common but differentiated responsibilities" (UNFCCC 1992). Most international environmental governance processes have gotten bogged down in strong differences between developed and developing states with differences not fully resolved in the initial negotiations. Thus, initial texts coming out of the first international negotiations often attempt to resolve any conflicts between rich and poor countries with 'weasel words,' or words which give no clear guidance, in the hope that further negotiations will resolve important but ambiguous language on crucial issues. This is exactly what happened at the most recent CBD COP noted above, as their development of how any fund would operate would be left to future negotiations (Greenfield & Weston 2022). Yet, despite over 27 years and 27 COPs, the UNFCCC is still full of such weasel words, even on the meaning of central terms such as 'equity' and 'responsibility.' It seems the process itself – of vague words and future promises – always seems to push action to the future and not to the present, a very happy result for the disinformation campaign.

During international negotiations for the UNFCCC COPs each year, oil, gas, and energy industry lobbyists have been well represented, along with their political counterparts: US congressmen, mostly from US fossil fuel states that are closely monitoring the US position on issues important to them. They are there to ensure that the US should make no commitment on issues that these industries believe will hurt their interests. And although the COPs often make loud pronouncements about the importance of civil society or Indigenous representatives, COP26 in Scotland had more industry representatives than civil society (and even those present had no access to the actual negotiations) (Democracy Without Borders 2021), and COP27 in Egypt was outrightly hostile to any substantive involvement by civil society and NGOs, with claims of surveillance and intimidation (Michealson & Milman 2022). This, for a treaty meant to address a global existential crisis to all humanity. This, the result of decades of intentional work against the role of governments, justice, and the common good. The UNFCCC COPs – and other international environmental treaty-making bodies – are no longer the space for states and civil society, but for neoliberal ideology.

Advancing the Self

It is often said that states are the powerholders in international law, but the true powerholder seems to be neoliberal ideology. Although the chapter on denialism in this publication flips even that premise, for when ideological principles are at odds with self-profit, self-profit always seems to win. But, not-without-irony, taking the deceitful at their word – that this is all ideologically-driven – and building off of the *Powell memo*, neoliberal ideas have gotten traction around the world. The central idea of neoliberal ideology is that market processes should order society for the common good through the operation of the market's invisible hand, even that the law should work for the markets and not the other way around (see the earlier discussion on 'Law and Economy'). Early proponents of neoliberal ideology claimed there was no, or at least a greatly reduced, need for the government to develop rules and regulations to achieve the common good based on justice. As a result, justifications for government regulation on environmental issues that existed in the first 20 years of the modern environmental movement, such as regulation needed to adequately protect human health and the environment or protect the environment for future generations, were gradually replaced by cost-benefit analyses or other economic criteria.

An example. While Brown was working as the US EPA Program Manager to UN Organizations during the Clinton administration, the United States was considering ratifying the Kyoto Protocol. Much of the policy talk in the agency centered on which of two different cost-benefit analyses should guide their decision. During this time, the Global Climate Coalition (GCC), an international lobbying group of businesses who opposed action to reduce GHG emissions, was waging an intense national campaign in opposition to ratification. After President Clinton negotiated a deal in Kyoto acceptable to the United States, many were surprised that he never submitted it to Congress for ratification. Observers, however, attribute this failure of the common good, this failure of democracy, to the successful lobby of the GCC.

But the attacks on the common good go beyond the inversion of the purpose of the state ('to achieve the common good'), or even the neutering of the state in global governance ('the state is the problem'), as described above; as non-sensical as it seems, the disinformation campaign also actively embraces nationalism, and the inherent racism within it, and the avoidance of state responsibility. When nationalism is seen as rooted in greed and the 'self,' and the rallying cries of the disinformation campaign are 'liberty!,' free market, and individual rights, it does not seem so non-sensical.

In the front matter to *States of Denial: Knowing About Atrocities and Suffering* (2001), Stanley Cohen cites George Orwell's 'Notes on Nationalism,'

> All nationalists have the power of not seeing resemblances between similar sets of facts… The nationalist not only does not disapprove of atrocities committed by his own side, but he has a remarkable capacity for not even hearing about them… In nationalist thought, there are facts which are both true and untrue, known and unknown. A known fact may be so unbearable that it is habitually pushed aside and not allowed to enter into logical processes, or on the other hand it may enter into every calculation and yet never be admitted as a fact, even in one's own mind… Indifference to objective truth is encouraged by the sealing off of one part of the world from the other, which makes it harder and harder to discover what is actually happening… If one harbours anywhere in one's mind a nationalistic loyalty or hatred, certain facts although in a sense known to be true, are inadmissible.

Behind a deceitful methodology to spread deceit are deceitful actors who desire 'more for me, and not for thee,' i.e., greed (and additional tenets of authoritarianism). Dark money (see Mayer 2016) is the insertion of hidden (i.e., non-transparent) money in politics that is often legally protected to remain hidden from public view. The money is 'donated' to political actors which then translates into action or inaction on policy issues, whether it be gun violence, women's rights, or climate change ('donation' is in parenthesis because donations by definition do not require a *quid pro quo*).

The 2010 US Supreme Court decision *Citizens United v. Federal Election Commission 2010* opened the floodgates for dark money. In it, the Court decided that corporations have the same rights to 'free speech' as citizens, and that financial 'donations' could be considered speech. With a single decision, and again, a very receptive court, all previous attempts at limiting corruption and money in politics had been reversed. In Justice John Paul Stevens dissent, the decision was "a radical departure from what has been settled First Amendment law," a "rejection of the common sense of the American people who have recognized the need to prevent corporations from undermining self-government since the founding, and who have fought against the distinctive corrupting potential of corporate electioneering..." (Mayer 2016: 229, citing *Citizens United* dissent). To Justice Stevens, the right of free speech was for "individual Americans, not corporations" (Mayer 2016: 229, citing *Citizens United* dissent). This is an important point in the evolution of the climate disinformation campaign. The highest court in the United States just concretely advanced a core tactic of the disinformation campaign: environmental protection through regulation – the very reason corporations donate to political campaigns to slow and stop – was an attack on freedom.

Destroying Democracy

The climate change disinformation campaign seeks to advance individuals, institutions, and worldviews that support the fossil fuel industry by ruthlessly and systematically dismantling democracy and good governance, science, and education. Some tactics include, but are not limited to, lying or reckless disregard for the truth about climate science; cherry-picking' the science by focusing on unknowns while ignoring what is well-settled in climate science; cyber-bullying and ad hominem attacks on scientists and journalists (and climate change policy advocates); manufacturing bogus, non-peer-reviewed climate science through the creation of ideologically motivated conferences and publications; the use of ideological think-tanks to promote the views of climate change deniers through their media outreach, speakers bureaus, publications, and conferences; the use of front-groups[2] and fake grass-roots organizations, known as 'Astro-turf' groups,[3] to promote the views of climate change deniers and hide the true parties and motives; and making specious claims about 'bad science' that are based upon the dubious assumption that no conclusions in science can be made until everything is proven with high levels or even absolute levels of certainty. They learned from the tobacco industry that "if you could convince people that science in general was unreliable, then you didn't have to argue the merits of any particular case" (Oreskes & Conway 2011: 217). In a confidential memo, 'Winning the Global Warming Debate,' Frank Luntz argued that climate deniers had to sell themselves as protectors of the environment, and that they "absolutely must not raise economic arguments first" (Mayer 2016, citing Luntz 209). In other words, "telling the truth about their financial interests was a recipe for losing" (Mayer 2016: 209).

Undermining Truth

Its tactics were designed to not only undermine citizens' faith in mainstream climate science, but in truth itself. And the consequences of these tactics – inaction on climate harms – are deadly. Due to the intent, the methodology, and the harms, it has even been argued that the disinformation campaign is a new kind of crime against humanity (Brown 2010) – due to the intent, the methodology, and the harm. Its purpose has no connection to the indispensable role of skepticism in science and the right of free speech, yet it manipulates those foundations of democracy to exist and persist.

This is why a major challenge for getting traction for ethics in climate policy formation is to get traction for *truth*. This would combat the lies and misinformation increasingly sophisticatedly spread throughout the government's population by increasingly sophisticated technology. Political scientist and philosopher Hannah Arendt (1907–1975) described in her paper 'Truth and Power' (1967) that throughout history, politicians whose power is threatened often respond with lies. And our founders agreed. Alexander Hamilton believed "it is not safe to trust to the virtue of any people," and James Madison stated that "constant experience shows us, that every man invested with power is apt to abuse it... To prevent this abuse, it is necessary from the very nature of things, power should be a check to power" (Ricks 2020: 206, 208). Virtue carried the American Revolution but checks on power is what carried the Federalist Papers and the resulting US Constitution. "If men were angels," Madison wrote in Federalist Paper 51, "no government would be necessary" (Ricks 2020: 212).

And we have been warned of this by economists themselves. Economist and philosopher Adam Smith (1723–1790), most well-known for convincing society of the value of the free market through an 'invisible hand,' warned governments that the 'merchant' class would sometimes ruthlessly scheme against the common good (Sagar 2018). This is why governments, while still acknowledging the benefits of free markets and private sector institutions for some purposes in a democracy, need to establish rules that anticipate the very likely behavior of economically powerful interests to undermine democratic processes that help determine and defend the common good.

Getting traction for truth in the climate debate is not a new political challenge – if anything our own history argues for the institutionalize of stronger anti-corruption measures in government – but it is nevertheless much more challenging given the sophistication and effectiveness of technology to spread disinformation and how accurately and acutely they are able to target, or even create, receptive audiences. Through big data, yes, but also through tried-and-true totalitarian tactics of fear and emotion.

The Climate Troll

The "key was to question the science," and for Penn State climate scientist Michael Mann, this was a "virtual hunting license" (Mayer 2016: 209). And hunt they did. Brown calls this tactic 'false intimidating manipulative smears' (FIMS), aimed at anyone who appears in the media who challenges the culturally unchallengeable narrative of neoliberal, 'free market is liberty' ideology. In common parlance today, these individuals who mockingly call people extreme and offensive names to intentionally trigger emotion and reactions are called 'trolls,' and one of the most infamous climate trolls is Marc Morano, a Republican political operative that runs the climate denialist website, ClimateDepot.com. He argues, "You have to name names and you've got to go after individuals" (Mayer 2016: 211). And the more successful the climate advocate, the more extreme and aggressive their ad hominem attacks. For example, NASA climate scientist James Hansen (who also led the future generations climate case, *Juliana v. United States*) is "a wannabe Unabomber," Mann is a "charlatan," part of – and saying this without irony – a "lavishly funded climate machine" (Mayer 2016: 211), and teenage climate activist Greta Thunberg is "an autistic prophet" (Waldman 2019). He was also named a "climate killer" by Rolling Stone in a 2009 article by Tim Dickinson, alongside such names as Warren Buffett (a billionaire investor and former economic advisor to the Presidential campaign of Barack Obama), Rex Tillerson (who later became US Secretary of State under President Donald Trump), and US Senators Mary Landrieu, John McCain, and James Inhofe, but the article has since been scrubbed from its pages, and apparently, the entire internet.[4]

And the ad hominem attacks were not just for modern day climate advocates; they reached into history to smear environmental and regulatory heroes. Rachel Carson, who saved millions because her work *Silent Spring* (1962) led to regulatory action to ban dichlorodiphenyltrichloroethane

(DDT), became a "mass murderer," with "the blood of millions" dripping from her hands, because she ended the "most valuable chemical ever synthesized" (Oreskes and Conway 2011: 216, 232). They knew if they demonized her, they could "strengthen the arguments against regulation everywhere" (Oreskes & Conway 2011: 217). This type of revisionist history is of course part of a larger anti-democratic governance campaign, unpacked more thoroughly in the chapter on denialism in this publication.

Frequently used FIMS include calling a climate change policy advocate an 'alarmist,'[5] 'socialist,' or a believer in 'entitlements,' and sometimes these smears include harassment by phone and email campaigns and threats of violence, all the more worrisome with their ideological alignments with guns rights advocates, "whipped up to believe that climate action would take away their freedom and their guns" (Mayer 2016: 223). Mann and Toles call these ad hominem attacks "the last line of defense in a losing battle" (2016: 81). They do not have facts on their side, so they turn to personal attack; "It's all about distracting the public, vilifying the enemy, and going straight to that enemy's strength – in this case, trust and credibility" (2016: 82). For Mann, this climate denial machine included the Republican-led Pennsylvania state legislature, who withheld funding to Penn State University unless they investigated him (Mayer 2016: 222). Attack ads were run, protests were organized, death threats were made, and an investigation began (Mayer 2016). And although the investigation ultimately cleared Mann of any wrongdoing,[6] he was concerned about the repercussions for young academics, potential future scientists, who witnessed such powerful state-sponsored, corporate-paid, intimidation tactics on a university climate science professor (Mayer 2016: 223).

The climate advocate is then left responding to these empty smears – or they are silenced under the weight of the harassing, abusive tactics of the climate change disinformation campaign. During the Harper and Trump Administrations in Canada and the United States, respectively, we saw some of these tactics brazenly placed before the public. An arrogant power-play in itself. The science was silenced as industry representatives replaced scientists on agency policy boards, climate change was scrubbed from government websites, and government officials could not say 'climate change' or speak to the press about it (see Learn 2017; see Webb & Kurtz 2022). More recently, the authoritarian leader of Hungary, Viktor Orbán, has put the Hungarian Academy of the Sciences under direct government control, terrified academics into silence, and forced the Central European University out of the country (Applebaum 2020: 51). Hungary is the only European country to have shut down an entire university – to have removed funding from university departments – that the ruling party dislikes for political reasons (Applebaum 2020: 141). That certainly sounds reminiscent of what occurred to Mann: a Republican state legislature threatening to withhold public funding from a state university until they investigated – and with no cause – an influential climate scientist. This is also a good time to note that the Conservative Political Action Conference, the "largest and most influential gathering of conservatives in the world"[7] welcomed Orban to their 2022 conference, where he received a standing ovation with his anti-immigrant, nationalist rhetoric, and just days after saying that Hungary must not become a "mixed-race" country. In his CPAC speech, he proclaimed, "Politics are not enough. This war is a culture war. We have to revitalize our churches, our families, our universities, and our community institutions" (Allison & Johnson 2022). And the packing of boards, the shuttering of institutions – there was nothing anyone could do about it. This is state-sponsored, corporate-funded climate denialism, with no institutional checks and balances available, as democracy requires.

No Place, No Space, No Voice for Democracy

The founders anticipated that a goal for the common good would create disagreements among contending parties about what a government should do to achieve the common good or even what is the common good. Thus, they recommended that citizens should be educated in skills to help them evaluate factual disputes, namely science, history, and ethics, in order to help them critically evaluate disagreements about justice. Without this – the training in critical thought on aspects of justice – the attainment of justice and the common good would not be possible. And we are witnessing this. The *Powell memo* showed how education – and particularly higher education – was a specific target because of its power to advance justice and the common good. In his 2019 book, *The Common Good*, economist, scholar, and public intellectual Robert Reich wrote the following:

> The common good consists of our shared values about what we owe one another as citizens who are bound together in the same society – the norms we voluntarily abide by, and the ideals we seek to achieve... A concern for the common good – keeping the common good in mind – is a moral attitude. It recognizes that we're all in it together. If there is no common good, there is no society.

Reich then connects these important assertions about the common good to education, writing that "Education is a public good that builds the capacity of a nation to wisely govern itself, and promotes equal opportunity... Without an educated populace, a common good cannot even be discerned. This is fundamental." It is the same understanding that for democracy to work, for liberty to exist, there must be an agreed-upon truth on which to debate and decide the actions that impact people's lives. Yet, the climate disinformation campaign denies people that truth, that liberty (and yes, while at the same time claiming that they are the torch-bearers *for* liberty). Liberty is not possible without truth. A common good is not possible without a civically – or scientifically[8] – educated population. And the disinformation campaign seeks to dismantle even that opportunity.

In David Bromwich's introduction to Arendt's *On Lying and Politics*, a single publication that includes both *Truth and Politics* (1954) and *Lying in Politics* (1971), he states that Arendt saw American universities as "the bulwark against institutional lies" and that "the chances for truth were 'greatly improved by the mere existence of such places'" (Arendt 2022: xxxv). And she connected this search for truth as the place to be "truly free" (Arendt 2022: xxxv, Bromwich citing Arendt's *Crises of the Republic* 1972). Today, due to free market fundamentalists and climate denialists such as James Buchanan and Charles and David Koch, it is argued that George Mason University in Virginia is ground zero for de-regulation policy (although, as noted above, law schools around the country have been entangled in their successful campaign). Mayer's *Dark Money* goes into explicit detail how "the Kochs have not enriched academia but corrupted it," where

> Entire academic areas at universities can be bought just like politicians. The difference is that universities are supposed to permit open dialogue and exchange of ideas and not be places for indoctrination of innocent students with dictated propaganda prescribed by outside special interests.
>
> *(Mayer 2016: 156)*

And so, as a result of decades of targeting and manipulation since the *Powell memo*, it comes as no surprise that our educational institutions are not adequately training students to identify ethical issues embedded within notions of justice and the common good so that they may respond effectively when issues of justice and the common good are at stake. The GHG emissions of the United States harms people inside and outside our borders, often the most vulnerable people, and often those who are not contributing to the climate crisis. This is a *justice* issue, but rarely is it framed as such. Those who are causing the harms are not being held accountable for the harms. Those who are not causing the harms are feeling the harms. It is a system of harm to past, present, and future generations without justice, thereby harming the common good. And so, we must ask ourselves, why is this unjust system of harm to the common good being allowed to continue? And not just being 'allowed to continue,' it is being supported – in law, policy, and institutions – so that it continues.

Looking specifically at higher education, we can see that the goals of these institutions have increasingly shifted from teaching students the skills needed to participate in democratic processes to achieve the common good to teaching students the skills needed to make them more attractive to potential employers in science, engineering, and technology. Although claims about what governments should do to achieve the common good have both factual premises and normative conclusions, this shift in higher education's major focus has increased the power of opponents of environmental policies to frame the public debate on facts which usually ignore the relevant ethical considerations. For instance, all governments of the UNFCCC agreed to be bound by the 'precautionary,' 'equity,' and 'no-harm' principles when adopting GHG emission reduction targets, yet opponents of climate change action use denialism ('it's a hoax!'), 'excessive costs,' and 'scientific uncertainty' arguments. This then frames the debate where proponents of climate policies – including passionate and well-intentioned non-governmental organizations – then ignore any critical evaluation of the normative conclusions and instead focus on counter factual claims about cost and uncertainty. After decades of gutting education institutions of the arts and humanities, we have lost the language of humanity to defend action for humanity. This has left even the passionate and well-intentioned proponents of climate change policy bereft of the understanding or language to successfully advance their own policies for climate action. The harm of this prioritization by higher education on science and technology is then compounded by the refusal of some technical experts to supplement their work with ethical considerations, as they do not know how to do so, or they are more interested in protecting the position and specialty of their field. Another convenient result of decades of cultural framing, institutional manipulation, and disinformation.

The climate change disinformation campaign has been so thorough in its perversion and subversion of democratic institutions – the spaces, places, and words for ethical action for the common good – that they have not only successfully advanced arguments for inaction on climate change, but they have rendered near impossible arguments for action. If people want to voice the frustration they feel for the harms or injustice that they are bearing witness to: well, they cannot find the words to do so; after all, 'what is the *language* of ethics?' And if they do have the words – for justice, rights, and responsibility, for right and wrong behavior – and they approach scientists and economists and the technology industry with those words: well, 'this isn't the *place* for ethics.' Or, they may have the words but do not have the access to the negotiations or the proceedings or the people making the decisions: well, 'this isn't the *space* for ethics.'

Civil society is not taught the language of civic, ethical engagement. And this has been done with intent. The fields of science, economics, and technology exclude ethics and values-considerations from their work. And this has been done with intent. And powerholders, the gate-keepers of ethics in democratic discourse – who largely have not have been taught in civics and ethical engagement or the meaning of the state, justice, and the common good, who may be the purchased product of *the Powell memo* and the climate change disinformation campaign, and who definitely are the product of decades of neo-liberal ideology – close the gates to ethics in democratic discourse. And this has been done with intent. Timothy Snyder's *On Tyranny: Twenty Lessons for the Twentieth Century* lists "9 – Be Kind to Our Language," warning that "the classic novels of totalitarianism warned of the domination of screens, the narrowing of vocabularies, and the associated difficulties of thought. (2017: 61)." Jefferson was also a proponent of education as a safeguard to tyranny, "This last is the most certain and the most legitimate engine of government. Educate and inform the whole mass of the people... They are the only sure reliance for the preservation of our liberty" (Jefferson 1787). Again, and this cannot be stressed enough, the climate disinformation campaign is more than an attack on climate action, it is an attack on democratic governance. We are not taking this attack seriously enough and we are losing our democracy as a result.

Conclusion

> I love the man that can smile in trouble, that can gather strength from distress, and grow brave by reflection. 'Tis the business of little minds to shrink, but he whose heart is firm, and whose conscience approves his conduct, will pursue his principles unto death.
>
> *Thomas Paine (1737–1809)*

Power often works by spreading a narrative through a culture until it becomes so accepted that citizens are afraid to challenge it. But we have so many courageous, influential leaders – Martin Luther King, Hannah Arendt, John Lewis, Greta Thunberg, Sheldon Whitehouse – who believe, and who have lived and live that belief, that citizens have a duty to call out injustice when and where they we see it. And that is what this chapter aims to do: (1) identify a cowardly, intentional, dangerous methodology that seeks to weaken and dismantle our democratic institutions, our very lives, and society, so that (2) we can more courageously, more intently target their tactics, strengthen our institutions, and hold them to account.

But this is also the responsibility of states. It is their duty to identify these deceitful and dangerous tactics, counter them with appropriate safeguards and regulation, and re-invigorate the ethical principles that they have already agreed to be bound by as a nation and as a community of nations: work and build strong institutions for the common good; search for truth; do no harm; if you do harm, be held accountable. In Martha Nussbaum's *Political Emotions: Why Love Matters for Justice*, she writes that

> [Immanuel] Kant felt that individuals have an ethical duty to join some group that will strengthen the good tendencies they have (tendencies to treat other people well), so that those tendencies will have a greater chance of winning out over the bad.

(2013: 4). Join us.

Notes

1 And highlighting the tension in detailing with issues of equity, although the agreement was finally adopted, the DNC is insisting that they legally objected, Cameroon is calling the proceedings "a fraud," and Uganda believes that there has been a "coup d'état" of the COP (Greenfield & Weston 2022). It is also important to note that the United States and the Vatican are the only two parties to the CBD who did not sign the agreement.

2 See Whitehouse et al. (2022), "These industry front groups and their funders are part of a well-organized and tightly connected web of organizations, which we often see dedicated to promoting climate denial and opposing climate action in Congress, at executive agencies, and in the courts. Amici Competitive Enterprise Institute, Americans for Prosperity Foundation, Cato Institute, Mountain States Legal Foundation, Landmark Legal Foundation, and Southeastern Legal Foundation are all part of this network of groups."

3 See Mayer (2016: 159–161), "a new kind of weapon in the arsenal of several of America's biggest businesses – a fake populist movement secretly manufactured by corporate sponsors – not grassroots, but 'Astro-turf,' as synthetic groups came to be known." This same model has been adopted by the anti-public education advocates in the far-right movement in the United States, with fake 'parents' organizations attacking local school boards with national right-wing attacks (see Graham 2021).

4 Reference to it can still be found via the Environmental Defense Fund and other third-party sites, but any direct links to Rolling Stone are no longer active; see https://blogs.edf.org/climate411/2010/01/29/rolling-stone-calls-out-the-climate-killers/.

5 The George C. Marshall Institute, a "central cog in the denial machine," issued the first attack labeling climate science 'alarmist' (Mann & Toles 2016: 72–73; see also Oreskes & Conway). Founded by Cold War-era scientists, these "free market fundamentalists" also loosely and widely branded climate scientists as socialists, watermelons, green trees with red roots (Mann & Toles 2016: 72–73; see also Oreskes & Conway). It dissolved in 2015 and was re-branded as the CO2 Coalition, another non-profit front group that advances the climate disinformation campaign.

6 See Investigation of climate scientist at Penn State complete (2010), available at https://www.psu.edu/news/university-park/story/investigation-climate-scientist-penn-state-complete/.
7 See https://www.conservative.org/.
8 See Mann & Toles (2016: 3) "The fact of the matter, however, is that there is a weakness in the scientific system that can be exploited. The weakness is in the public understanding of science, which turns out to be crucial for translating science into public policy. Deliberate confusion can be sown under a false pretext of 'skepticism.' And the scientific process is continually under assault by bad-faith doubt mongers."

References

Allison, N. and Johnson, L. (2022) "Orbán Gets Warm CPAC Reception After 'Mixed Race' Speec Blowback," *Politico*. Available at https://www.politico.com/news/2022/08/04/viktor-orban-cpac-00049935.

Applebaum, A. (2020) *Twilight of Democracy: The Seductive Lure of Authoritarianism*. New York: Doubleday.

Arendt, H. (2022) *On Lying and Politics*. New York: Library of America.

Aronoff, K. (2020) "This Supreme Court Was Designed to Kill Climate Policies Polluters Helped Build The Court's Conservative Majority. Would Democratic Laws Stand a Chance Against it?" *The New Republic*. Available at https://newrepublic.com/article/159766/supreme-court-designed-kill-climate-policies.

Brown, D. (2010) "Is Climate Science Disinformation a Crime Against Humanity?" *The Guardian*. Available at https://www.theguardian.com/environment/cif-green/2010/nov/01/climate-science-disinformation-crime.

Brown, D. (2015) "The Seeds of the Corporate Funded Climate Disinformation Campaign, the 1971 Lewis Powell Memo," *EthicsandClimate.org*. Available at https://ethicsandclimate.org/2015/10/31/the-seeds-of-the-corporate-funded-climate-disinformation-campaign-the-1971-lewis-powell-memo/.

Citizens United v. Federal Election Commission. (2010) US Supreme Court, 558 U.S. 310. Available at https://www.law.cornell.edu/supct/html/08-205.ZS.html. Justice John Paul Stevens dissent available at https://www.law.cornell.edu/supct/html/08-205.ZX.html.

Clarke, S. (2023) "Agenda for American Strength," *keynote Remarks Given at the US Chamber of Commerce, State of American Business*, January 12, 2023. US Chamber of Commerce. Available at https://www.uschamber.com/improving-government/suzanne-clarks-2023-state-of-american-business-remarks.

Cohen, S. (2001) *States of Denial: Knowing About Atrocities and Suffering*. Cambridge: Polity Press.

Curtis, G. (n.d.) *The Government of Modern Japan: Japan's Elector Laws – Strict Laws Regulating Election Campaigns*. Columbia University: Asian Topics. Available at http://afe.easia.columbia.edu/at/jp_elect/govjel01.html.

Democracy Without Borders. (2021) *Civil Society Marginalized at COP26 Climate Negotiations in Glasgow*. Available at https://www.democracywithoutborders.org/21159/civil-society-marginalized-at-cop26-climate-negotiations-in-glasgow/.

Feldman, N. and Kott, L. (2021) *Takeover: How a Conservative Student Club Captured the Supreme Court*. Pushkin Industries [audiobook].

Graham, E. (2021) "Who Is Behind the Attacks on Educators and Public Schools? The Manufactured Outrage Perpetuated by Dark Money Networks Is Both a Danger to Educators and a Distraction from Helping Students and Parents," *National Education Association*. Available at https://www.nea.org/advocating-for-change/new-from-nea/who-behind-attacks-educators-and-public-schools.

Greenfield, P. and Weston, P. (2022) "Cop15: Historic Deal Struck to Halt Biodiversity Loss by 2030," *The Guardian*, December 19, 2022. Available at https://www.theguardian.com/environment/2022/dec/19/cop15-historic-deal-signed-to-halt-biodiversity-loss-by-2030-aoe.

Hart, G. (2007) "The Commonwealth: Our Public Legacy," *Daedalus*. Available at https://www.amacad.org/publication/commonwealth-our-public-legacy.

Jefferson, T. (1787) *From Thomas Jefferson to Uriah Forrest, with Enclosure*, 31 December 1787. Available at https://founders.archives.gov/documents/Jefferson/01-12-02-0490.

Learn, J. (2017) "Canadian Scientists Explain Exactly How Their Government Silenced Science," *Smithsonian Magazine*. Available at https://www.smithsonianmag.com/science-nature/canadian-scientists-open-about-how-their-government-silenced-science-180961942/.

Mann, M. and Toles, T. (2016) *The Madhouse Effect How Climate Change Denial Is Threatening Our Planet, Destroying Our Politics, and Driving Us Crazy*. New York: Columbia University Press [Kindle edition].

Mayer, J. (2016) *Dark Money: The Hidden History of the Billionaires Behind the Rise of the Radical Right*. New York: Doubleday.

Michaels, D. (2020) *The Triumph of Doubt: Dark Money and the Science of Deception*. New York: Oxford University Press.

Michealson, R. and Milman, O. (2022) "Civil Society Groups Report Surveillance and Intimidation at Cop27," *The Guardian*. Available at https://www.theguardian.com/environment/2022/nov/10/civil-society-groups-report-surveillance-and-intimidation-at-cop27-egypt.

Nadeau, P. and Fahey, R. (2017) "In Praise of Japan's Election Campaigns," *Tokyo Review*. Available at https://www.tokyoreview.net/2017/10/japan-election-campaigns/.

New South Wales (NSW) Electoral Commission. (2022) *I Didn't Vote*. Available at https://elections.nsw.gov.au/voters/i-didnt-vote.

Nussbaum, M. (2013) *Political Emotions: Why Love Matters for Justice*. Cambridge: Belknap Press.

Oreskes, N. and Conway, E. (2011, Forward 2019) *Merchants of Doubt: How a Handful of Scientists Obscured the Truth on Issues from Tobacco Smoke to Climate Change*. New York: Bloomsbury Publishing.

Powell, L. (1971) *Confidential Memorandum: Attack on American Free Enterprise System* (*The Powell memo*). Available at https://scholarlycommons.law.wlu.edu/cgi/viewcontent.cgi?article=1000&context=powellmemo.

Reagan, R. (1981) *Inaugural Address, US President Ronald Reagan*. Available at https://www.reaganlibrary.gov/archives/speech/inaugural-address-1981.

Reich, Robert. (2019) *The Common Good*. New York: Knopf.

Ricks, Thomas. (2020) *First Principle: What America's Founders Learned from the Greeks and the Romans and How That Shaped Our Country*. New York: HarperCollins Publishing.

Sagar, P. (2018) "Adam Smith and the conspiracy of the merchants," *Global Intellectual History*, 6: 4, 463–483.

Snyder, T. (2017) *On Tyranny: Twenty Lessons for the Twentieth Century*. New York: Tim Duggan Books.

Waldman, S. (2019) "Climate Deniers Launch Personal Attacks on Teen Activist," *Scientific American*. Available at https://www.scientificamerican.com/article/climate-deniers-launch-personal-attacks-on-teen-activist/.

Webb R. and Kurtz, L. (2022) "Politics v. science: How President Trump's war on science impacted public health and environmental regulation." *Progress in Molecular Biology Translational Science*, 188(1): 65–80. Available at https://www.ncbi.nlm.nih.gov/pmc/articles/PMC8793038/.

West Virginia v. Environmental Protection Agency [EPA]. (2022) 597 US ___. Available at https://www.law.cornell.edu/supremecourt/text/20-1530.

Whitehouse, S. (2021) *The Scheme 1: The Powell Memo*. Sheldon Whitehouse: US Senator for Rhode Island. Available at https://www.whitehouse.senate.gov/news/speeches/the-scheme-1-the-powell-memo.

Whitehouse, S. (2022) *The Scheme: How the Right Used Dark Money to Capture the Supreme Court*. New York: The New Press.

Whitehouse, S., Blumenthal, R., Sanders, B. and Warren, E. (2022) *No. 20-1530, Brief as Amici Curiae in Support of Respondents, West Virginia v. EPA*. Available at https://www.whitehouse.senate.gov/imo/media/doc/WV%20v.%20EPA%20Amicus.pdf.

20

ETHICAL RULES THAT 'CLIMATE SKEPTICS' SHOULD ABIDE BY

Donald A. Brown

The Scientific Consensus

This chapter reviews the indefensible tactics and the enormous success of the fossil fuel industry funded climate science disinformation campaign which seeks to undermine citizens' faith in mainstream climate science. The chapter then explains that although skepticism in science is a good thing, climate skeptics should comply with ethically generated rules explained in this chapter when making claims against the consensus view of mainstream climate science.

In order to set the foundation for responding to climate change skeptics, it is important to identify what is normally considered the scientific consensus view established by the most prestigious scientific organizations in the world in order to set the foundation for responding to climate change skeptics. The consensus view is usually understood to be that initially articulated by the Intergovernmental Panel on Climate Change (IPCC). The IPCC was established in 1988 by the World Health Organization (WHO) and the United Nations Environment Program (UNEP) to enable governments to assess the scientific, technical, and socio-economic information relevant to understanding climate change, its potential impacts, and options for adaptation and mitigation (IPCC 2023). The IPCC does not do original research but synthesizes and summarizes the extant peer-reviewed climate science to make recommendations to governments and policymakers about needed climate policies (IPCC 2023).

Any government who is a member of WHO or UNEP may be a member of the IPCC, with current membership at hundred 195 countries (IPCC 2023). Therefore, countries who have sometimes opposed international action on climate change on scientific grounds, including the United States and Saudi Arabia for instance, have the same power as governments which have traditionally strongly supported national action on climate change, including most of the governments in the European Union and many of the nations' most vulnerable to climate impacts due to sea level rise or flooding, such as small island developing states. Given that the IPCC's assessment reports must be unanimously approved by the member countries, including countries who have opposed strong international responses to climate change, one can conclude that there has been a broad consensus about the IPCC's scientific conclusions among nations of the world.

Although there is a broad scientific consensus on most of the critical climate scientific issues, there remain some scientific issues about which there is still some uncertainty, such as 'climate sensitivity.' Climate sensitivity is typically defined as the global temperature rise following a doubling of CO_2 concentration in the atmosphere compared to pre-industrial levels. Pre-industrial CO_2 was about 260 parts per million (ppm), so a doubling would be at roughly 520 ppm (Met

DOI: 10.4324/9781003039860-24

Office 2023). There is also some controversy about whether the suite of models on which the IPCC relied reliably captured the speed at which some of the most worrisome tipping points would start to tip. Both of these uncertainties, if proven to be true, would likely support the conclusion that the IPCC consensus view is not stringent enough, but it is too early to tell. However, in response to arguments from some climate change skeptics, many scientific organizations with relevant expertise, including the most prestigious scientific organizations and academies of science in most countries, have endorsed the IPCC consensus position that most of the global warming in recent decades can be attributed to human activities.

When the founding nations of the 1992 UN Framework Convention on Climate Change (UNFCCC) agreed to the climate treaty, they anticipated that some climate science issues would remain somewhat uncertain. This is why all nations unanimously agreed to the inclusion of the 'precautionary principle,' which is discussed in several chapters in this *Handbook*. Although, in my experience as a practitioner and professor on climate change issues, most of US citizens do not initially understand the reasonableness of the precautionary principle. However, when you explain that there are many environmental threats for which the government practically cannot establish quantitative risks in a timely matter, due to their complexity or time and money constraints, a precautionary approach quickly seems more reasonable.

While serving as Program Manager for United Nations organizations in the EPA in 1997 under the Clinton administration, I was asked by the US State Department to co-chair a negotiation taking place in the UN Commission on Sustainable Development (UNCSD) that was considering whether nations would agree in writing that the 'balance of the evidence suggests a discernible human cause of climate change,' a statement that had already been agreed to by all nation's IPCC scientists in 1995. At the conclusion of these negotiations, every country in the world that was seated in the UNCSD, approximately 160 countries, agreed, including Saudi Arabia and other Organization of the Petroleum Exporting Countries (OPEC). And conclusions increased in confidence in every report over the last 25 years, with the most recent report claiming that the human cause of climate change was virtually certain, meaning at least a 95 percent probability.

The reason for the universal international agreement among nations that humans are responsible for climate change is that the evidence of human causation is extraordinarily compelling, despite the fact that the Earth has experienced warming and cooling cycles throughout history in responses to natural forces. The confidence of human causation is very high because:

1 Scientists can predict how the Earth will warm up differently if a layer of greenhouse gas (GHG) emissions at a known elevation in the atmosphere warms the Earth compared to how the planet warms due to natural forces that have caused warming in the Earth's historical heating and cooling cycles. These differences are referred to as 'human fingerprints.' These fingerprints of human causation include:

   a Night warming faster than days;
   b If warming is due to solar activity than the upper atmosphere (the stratosphere) should warm along with the rest of the atmosphere, but if the warming is due to the greenhouse effect, the stratosphere should cool because of the heat being trapped in the lower atmosphere (the troposphere), and satellite measures show that the stratosphere is cooling;
   c The combination of the warming troposphere and the cooling stratosphere should cause the tropopause, which separates the two, to rise, which has been observed;
   d It was predicted that the ionosphere would shrink, and indeed it is shrinking;
   e Less oxygen in the air as CO_2 combustion removes oxygen; and
   f More heat returning to Earth;

2 Have compared the temperature forcing of human GHGs to forcing of the natural causes of climate variations in 'attribution studies,' and have concluded that only the forcing from human activities can explain the recent rise in global temperatures;
3 Have known precisely since the mid-1880s, the amount of forcing a molecule of CO_2 generates in watts per square meter;
4 Have known that the CO_2 accumulating in the atmosphere is from fossil fuel combustion because of its chemical isotope;
5 Determined that the CO_2 accumulating in the atmosphere is directly proportional to the timing and amount of fossil fuel combustion around the world;
6 Tested these lines of evidence rigorously in computer model experiments since the 1960s; and
7 These models have not only accurately predicted future warming, they have been run backward and accurately described past temperature regimes (Wight 2010).

The global confidence in human causation of warming is derived not only from the fingerprint evidence, but also from scientific tests designed to compare whether the warming being experienced on Earth can be attributed to those natural forces which are known to have driven historical changes in climate, such as regular changes in the sun's energy reaching the Earth. This kind of study is called an 'attribution' study. Image 20.1 below compares the warming expected from human activities in red, to the warming expected by natural forcing in blue, to the actual observed warming in black. Thus, this comparison is very strong evidence for attributing recent warming to human caused forcing.

The scientific confidence in the consensus view of climate change is also extraordinarily strong because of the creation and work of the IPCC. The IPCC's mission is to synthesize the peer-reviewed climate science and socio-economic literature on climate change and make recommendations on policy to the international community (IPCC 2023). Approximately every five years, starting in 1990, thousands of scientists, most of whom have been recommended by member governments for their scientific expertise, produce comprehensive three volume IPCC reports.

The IPCC has issued six reports since 1990, known as 'Assessments,' and produced in three different working groups. Working Group I (WGI) synthesizes the physical climate science literature. Working Group II (WGII) synthesizes the science on climate change impacts, adaptation, and vulnerability, and Working Group III (WGIII) focuses on mitigation. I was a contributing author on a new IPCC Chapter in WGIII on ethics and sustainability in the IPCC 5th Assessment

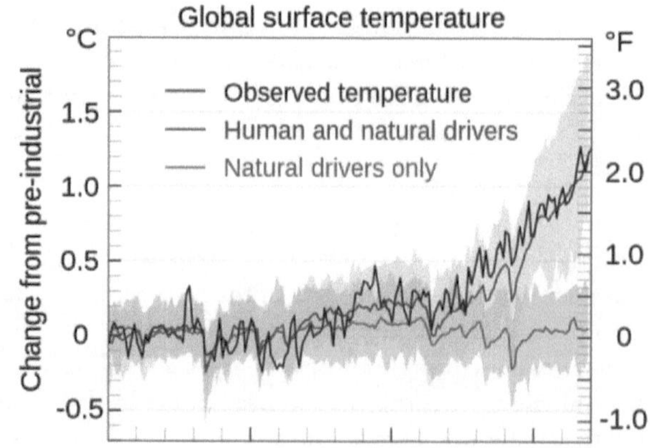

Image 20.1 Global surface temperatures 1850–2020 (relative to 1850–1900) [Efbrazil (CC BY-SA 4.0)]

Report (AR5) and for a chapter on sustainable development in WGII for the 6th Assessment Report (AR6).

In August of 2021, IPCC published AR6. Among many of its major conclusions, it found that:

1. Climate impacts are already more widespread and severe than expected;
2. We are locked into even worse impacts from climate change in the near term than previously;
3. Risk will escalate quickly with higher temperatures, often causing irreversible impacts of climate change;
4. Inequity, conflict, and development challenges will heighten vulnerability to climate risks in the future;
5. Adaptation is crucial;
6. Visible solutions already exist but more support must reach vulnerable communities;
7. The world is set to reach the 1.5°C level within the next two decades and only the most drastic cuts in carbon emissions from now would help prevent environmental disaster; and
8. Some impacts of climate change are already too severe to adapt to; therefore, the world needs action now to address losses and damages (IPCC 2021).

Climate Skepticism

In *Challenging Climate Change: The Denial Countermovement*, sociologists Riley Dunlap and Araon McCright describe how some fossil fuel companies, corporations that depend on fossil fuel, business organizations, and free market fundamentalist foundations have successfully prevented government action on climate change by funding the climate change disinformation campaign which they explain sought to undermine the public's confidence in mainstream climate science (2015: 300). Despite the acknowledgment that skepticism is an essential ingredient of progress in science and should be encouraged, the tactics that of many of the skeptics have engaged in the climate change disinformation campaign cannot simply be understood as an exercise of free speech.

Despite the almost universal agreement among nations that climate change is human-caused and very dangerous, many of US politicians frequently have and continue to argue that human-induced climate change is a hoax. As of March of 2021, there were 139 elected officials in the United States Congress who deny the scientific consensus on human caused climate change (CAP 2021). These members received more than $61 million in lifetime contributions from the coal, oil, and gas industries (CAP 2021). And so, many US federal legislators who have taken a skeptical position on climate change have also received money from fossil fuel interests.

In the United States, state governments control a significant amount of the nation's GHG emissions. Ten US States are responsible for half US GHG emissions. These states include Texas, California, Pennsylvania, Illinois, Ohio, Florida, Indiana, Louisiana, New York, and Michigan (Friedrich et al. 2017). Each level of government controls some activities that produce GHG emissions that other levels do not control. For instance, US states exercise control over some aspects of land use and some forms of public transportation, building codes, and electric power generation that the federal government does not control. Thus, there is a need for all levels of government to adopt climate policies if a national government is going to achieve its GHG-reduction obligations.

In my experience, citizens of subnational governments are almost never aware that emissions from the subnational government are contributing to raising atmospheric GHG concentrations. Therefore, their failure to reduce the subnational government's GHG emissions to zero is contributing to harms around the world such as those that are causing refugees (see Chapter 13 in this *Handbook* on refugees).

After returning from my work with the US Environmental Protection Agency in 1998, I served as the lead staff responsible on climate issues in the Office of Chief Council for the Pennsylvania

Department of Environmental Resources (PA DER). One of my initial assignments was to monitor a cap-and-trade program that was under development among ten northeastern states. This state, regional climate change program was referred to as the regional greenhouse gas inventory (RGGI). The states began RGGI negotiations in 2003 at the prompting of then New York Governor George Pataki. After attending several RGGI negotiations and reporting back to PA DER management, it became clear that there was no appetite at that time in Pennsylvania for a GHG cap-and-trade program because, as it was explained to me by state officials, "Pennsylvania was a coal state." Although Pennsylvania finally joined RGGI in April of 2022, it is still an open question whether the regulations that Pennsylvania will need to implement RGGI and achieve GHG emissions reduction goals, recently announced by Governor Wolf, will survive. The regulatory process itself provides numerous opportunities for the fossil fuel industry to block regulations.

During hearings in the Pennsylvania legislature over the last decade on potential state legislation that would reduce their GHG emissions, hearings were increasingly dominated by testimony of climate skeptics who spread odious disinformation (see Brown 2019). Making matters worse, most of the hearings were in the Pennsylvania House of Representatives, run by a legislator who was a strong vocal climate change skeptic. As a result, the legislative hearings never adequately considered the immense harms from Pennsylvania GHG emissions to the rest of the world, nor any serious counter arguments to the skeptics' claims that challenged the scientific consensus position articulated by the IPCC. This phenomenon offers evidence that the climate change disinformation campaign has worked among a large percentage of US citizens and US legislators.

Frank Lutz, a Republican consultant, is widely attributed as an influential shaper of the GOP message on climate change aptly crystallizes the goal of the fossil fuel disinformation campaign. That is not to get the science right to but to undermine citizens' faith in the consensus climate science,

> The scientific debate is closing… but not yet closed. There is still a window of opportunity to challenge the science… should the public come to believe that the scientific issues are settled, their views about global warming will change accordingly. Therefore, you need to continue to make the lack of scientific uncertainty a primary issue in the debate.
>
> *(Burkeman 2003, quoting Lutz)*

Rules that Climate Skeptics Must Abide By

Skepticism in science is a good thing; in fact, it is the oxygen that allows science to make contributions to the human understanding of how the world works. But skeptics must abide by the rules of science which require that scientific claims be subjected to peer-review and the internationally agreed-upon precautionary principle for issues about which some uncertainty remains. Several sociologists, including Bruelle and Dunlap noted above, have documented in many peer-reviewed sociological papers how some fossil fuel companies, related industrial organizations (such as the American Petroleum Institute), and free market fundamentalists foundations and think tanks have funded and supported efforts to undermine the public's faith in the consensus view of climate science, similar to the way the tobacco industry supported disinformation about the health threats of smoking tobacco. One of the tools in this effort has been to financially support or publicize the claims of climate skeptics whose claims have rarely been subjected to peer review, nor do they acknowledge the precautionary principle.

Based on analysis of the national public debates in several countries, the United States is not the only country in the world that appears to have been influenced by a fossil fuel organized disinformation campaign. *A Study of National Commitments* examined the national GHG reduction commitments of fifteen countries and the explanation for their commitments (Brown & Taylor

2015). The study found problems with all explanations, particularly in regard to express ethical commitments agreed to by nations, such as the promise to base their reductions based on equity.

In 2009, I was deeply surprised when the US State Department, under the conservative administration of President George W. Bush, invited me to make a presentation on climate ethics to the Scottish legislature as they were debating a national GHG emissions target. When I arrived at the Scottish Parliament Building in Edinburgh, the debate was already underway with one parliamentarian arguing that Scotland should set a tough target because Scotland owed it to the rest of the world. This was an argument I had never heard in the United States, but an argument which I believed was ethically sound. It was also required by the 'no-harm' principle which all countries agreed to in the Preamble of the UNFCCC. Shortly thereafter, Scotland set a target of net zero CO_2 emissions by 2045.

Since then, I have learned that another Scotsman, Adam Smith, although known around the world for convincing most of civil society of the benefits of free markets, also energetically warned people that the merchant class would sometimes ruthlessly scheme against the public interest. I still wonder whether Scotland's apparent imperviousness to the disinformation of the fossil fuel industry is explainable because its citizens had a deeper understanding of Adam Smith's warning (Sayer 2018).

The scheming of the fossil fuel industry and related participants in the disinformation campaign includes:

1 Lying or reckless disregard for the truth about mainstream climate change science;
2 Cherry-picking mainstream climate science by focusing on an issue about which there may be some scientific uncertainty, while ignoring a vast body of climate science which is well-settled;
3 Manufacturing non-peer-reviewed climate change science claims by several techniques, including organizing conferences in which skeptics dominate and then publishing the results without peer-review;
4 Creating think tanks, front groups, and Astroturf groups which widely have disseminated untruthful claims about mainstream climate science, and which were sometimes created to hide the real parties in interest, usually members of the fossil fuel industry;
5 Publishing and widely disseminating dubious manufactured climate change scientific claims that have not been subjected to peer-review; and
6 Widely attacking mainstream climate scientists and journalists who have called for action on climate change (Brown 2016).

The tactics of the climate change disinformation campaign have been incredibly detrimental to government progress on reducing the enormous threat to the world that human-induced climate change now clearly poses. Therefore, the international community should agree upon several ethically-based rules that skeptics, seeking to undermine the scientific consensus conclusions articulated by the IPCC, should abide by. These rules should include:

1 Skeptics should agree to peer-review any claims skeptical of the peer-reviewed climate science consensus view before they are published;
2 Climate skeptics should agree to subject their claims to review by scientific organizations whose members are sufficiently interdisciplinary, such as the Academies of Science, so that they can review the interdisciplinary issues that arise in determining the nature of climate risks;
3 Because there is abundant robust evidence that climate change is a major threat to human life, biodiversity, and ecological systems on which life depends, that the consensus view holds

that delay in reducing GHG emissions makes worse and more damaging, skeptics should acknowledge that they have the burden of proof in demonstrating that climate change harms of concern are without scientific warrant;

4 Where proof of harm using normal scientific methods cannot be obtained in a timely manner because of limitations in observing physical processes on which an evidence-based risk assessment can be determined or there is not sufficient time or money to gather the needed information needed to determine a risk using the epistemic norms of relevant disciplines, skeptics should support precautionary science needed to determine the risk;

5 Skeptics should acknowledge that who should have the burden of proof and what quantity of proof should satisfy the burden of determining the risk of a potentially dangerous risk is not a value-neutral scientific question but an ethical matter which is sensitive to the nature of the harm and who is harmed (this is the ethical basis of the precautionary principle);

6 Given the potential catastrophic nature of potential harms from climate change, those who oppose government action on climate change should be seen to have a strong duty to demonstrate, with high levels of proof, that the harm will not occur with the participation of those who will be harmed;

7 All skeptics should be asked to acknowledge that although some scientific uncertainty in the entire body of evidence on which a consensus position on climate harm is based is likely to remain, according to the precautionary principle (which all nations have agreed to under the UNFCCC), the remaining uncertainty is not adequate grounds for suspending a government's responsibility to lower GHG emissions to its fair share of safe global emissions; and

8 Related to (7), governments should support precautionary science for potential risks that cannot be quantified by empirically-used risk quantifications by authorizing institutions to conduct precautionary science with adequate opportunity for entities to engage by shifting the burden of proof to opponents of government action.

References

Brown, D. (2019) "Climate Change Disinformation Comes to Pennsylvania," *Ethics and Climate*. Available at https://ethicsandclimate.org/2019/05/08/climate-change-disinformation-comes-to-pennsylvania/.

Brown, D. (2016) "Why Exxon's and Other Fossil Fuel Companies' Funding of the Climate Change Disinformation Campaign Cannot be Excused as an Exercise in Free Speech but Must Be Understood as Morally Reprehensible Disinformation," *Ethics and Climate*. Available at https://ethicsandclimate.org/2016/07/29/why-the-climate-change-disinformation-campaign-cannot-be-excused-as-an-exercise-in-free-speech-but-should-be-understood-as-morally-reprehensible-disinformation.

Brown, D. and Taylor, P. (2015) *A Study of National Commitments*, Gland: IUCN Press.

Center for American Progress [CAP]. (2021) *Climate Deniers in the 117 Congress*. Available at https://www.americanprogress.org/article/climate-deniers-117th-congress/.

Dunlap, R. and McCright, A. (2015) "Challenging Climate Change: The Denial Countermovement," in R. Dunlap and R. Brulle (eds.) *Climate Change and Society: Sociological Perspectives*, New York: Oxford University Press.

Friedrich, J., Ge, M. and Tankou, A. (2017) "6 Charts Show How 20 Percent Of US States Generate 50 Percent Of Emissions," *Ecosystem Marketplace*. Available at https://www.ecosystemmarketplace.com/articles/6-charts-to-understand-us-state-greenhouse-gas-emissions/.

Burkeman, O. (2003) "Memo Exposes Bush's New Green Strategy," *The Guardian*. Available at https://www.theguardian.com/environment/2003/mar/04/usnews.climatechange.

Intergovernmental Panel on Climate Change. (2021) *Sixth Assessment, Summary for Policy Makers*. Available at https://www.ipcc.ch/report/ar6/wg1/chapter/summary-for-policymakers/.

Intergovernmental Panel on Climate Change [IPCC]. (2023) *About the IPCC*. Available at https://www.ipcc.ch/about/.

Met Office. (2023) "What Is Climate Sensitivity?" *His Majesty's Government*. Available at https://www.metoffice.gov.uk/research/climate/understanding-climate/climate-sensitivity-explained.

Sagar, P. (2018) "Adam Smith and the Conspiracy of the Merchants," *Global Intellectual History*, 6: 4, 463–483.

Wight, J. (2010) "The Human Fingerprint in Global Warming," *Skeptical Science*. Available at https://skepticalscience.com/its-not-us.htm.

21

CLIMATE CHANGE, DIFFICULTY, AND IGNORANCE

Philip Robichaud

Introduction

There are a number of ways in which ignorance might be taken to be relevant to ethical issues raised by climate change. Many might think that individuals living in well-to-do countries who make no efforts to reduce their contributions to climate change are morally responsible and thus blameworthy. But what if they are ignorant concerning the fact that human action is causing climate change? Does their ignorance constitute an excuse? When it comes to moral appraisal, ignorance can sometimes function as an excuse for actions that are wrong or that have bad effects, and there much discussion has aimed at figuring out when ignorance can function as an excuse and when it can't. Plausibly, the ignorance of an individual living in the 1950s regarding the reality of climate change as *exculpating*, given that the evidence regarding the link between emissions and climate change was either non-existent or not well-circulated. By contrast, cases of such ignorance in individuals living currently in epistemic environments in which the relevant evidence is readily available might not be exculpating. Rather, the fact that one can remain ignorant in this context can be considered inculpating, and this would entail that people who act on that ignorance and wrongfully fail to curb their climate change impact would lack an excuse.

This chapter will canvas some of the positions that have been defended in the literature regarding the way in which ignorance regarding climate change should impact our moral appraisal of individuals. It will then examine how the difficulty of 'getting things right' in these domains should affect the analysis, and argue that careful attention to the nature and degree of difficulty will allow for a much more subtle understanding of how ignorance can impact climate ethics.

Ignorance about Climate Change and Blameworthiness

There are different ways of appraising an agent's past conduct in the domain of their climate impacting actions. This includes, for example, characterizing them as virtuous (or vicious, as the case may be) (Jamieson 2007; Sandberg 2011). For purposes here, however, moral appraisal involves assessing whether agents are morally responsible, which is to say whether they are the appropriate targets of our moral gratitude or blame. In cases where agents, say, act wrongly and have no excuse, it is natural to maintain that they are blameworthy for their action and that it is appropriate to actually blame them. The discussion surrounding the impact that ignorance about climate change should have on moral appraisal is part of a general debate concerning what is often called the epistemic condition of moral responsibility. We normally think that an agent's being in

 DOI: 10.4324/9781003039860-25

control of what they are doing is a necessary condition of moral responsibility—agents who fail to exert a certain degree of control over their actions may have an excuse. But another condition of responsibility requires that agents have certain true beliefs about their actions and their likely consequences. Agents who are blamelessly ignorant about certain harms that might result from their actions may also be excused from moral responsibility. However, not all ignorance functions as such an excuse, and a great deal of the literature on the epistemic condition attempts to characterize when ignorance excuses and when it does not.

Initial discussions regarding the question of whether ignorance about climate change could function as an excuse took as a starting point that this ignorance was certainly an excuse prior to the scientific community coming to a widely communicated consensus regarding anthropogenic climate change. Vanderheiden (2007) argues that the situation changed once climate scientists developed such a consensus and were able to distribute their findings via Intergovernmental Panel on Climate Change (IPCC) Assessment reports. At that point, individual ignorance would seem willful and thus inculpating. Vanderheiden appeals to a reasonable person standard in defense of the claim that individuals who don't know about anthropogenic climate change nevertheless ought to have known. Adopting this analysis makes for easy sorting between ignorant agents who meet the epistemic condition of moral responsibility and those who don't. Unless someone remains completely cut off from information sources that contain, for example, the findings of IPCC regarding the anthropogenic climate change, their ignorance is culpable.

More recent discussions (Robichaud 2017; Wündisch 2017) have drawn upon more developments in the literature on the epistemic condition to argue that the case of climate ignorance permits a much more fine-grained analysis. One implication of these discussions is that even up to the present day some, even many, who are ignorant about climate change are excused. Two influential developments of the epistemic condition are particularly instructive to apply in this context. The first maintains that ignorance is culpable only when it has the right kind of etiology, namely that it traces to a witting violation of what Rosen calls "procedural epistemic obligations" (2004). Given the potential negative impact of our actions, we are under all manner of obligations to inquire or investigate prior to acting. For example, parents of peanut-allergic children should inquire into the presence of peanut butter in a free lunch prior to giving it to their child. If the parent opted not to inquire (perhaps because they're ashamed to ask) and the lunch contained nuts, their ignorance regarding the lunch's nut content traces to a witting violation of an obligation to investigate. Since their ignorance traces to a witting violation of one's procedural epistemic obligations, it is inculpating and thus the agent would be blameworthy for any resulting harm.

How this understanding of the epistemic condition applies to the context of climate change depends crucially on the details of the case. If we restrict our attention to the context (such as the current one) in which there is a widely communicated consensus regarding anthropogenic climate change, it is easy to imagine cases in which individuals suspect that they should inform themselves on the matter, but who choose not to do so out of laziness or a fear that they might learn that their current carbon-intensive lifestyle is morally problematic. Such agents would, like the parent discussed above, be ignorant due to a witting violation of a procedural epistemic obligation. The procedural epistemic obligation in this case would be to seek information regarding the causes of climate change and its negative impacts, and the individuals in question can be thought of as 'willfully' ignorant given that they knew, or at the very least suspected, that they should have looked into the relevant matters, yet they chose not to.

However, although there are many other kinds of cases for which this view would suggest that the agent's ignorance is exculpating, two such cases will be discussed here with an eye toward arguing that the view is wrong-headed. Consider the following case, which comes from Wündisch (2017):

> Bob grew up in a rural area, has a conservative peer group, works at an assembly line of a car factory, and mostly tunes into conservative news media. He believes climate change exists but does not believe that it is manmade. Whenever Bob seeks out information from alternative news sources he remains unconvinced...because he finds the persons presented as experts by the conservative news media more trustworthy.
>
> *(289)*

Bob is ignorant about climate change, and we can presume that he regularly acts on this ignorance and fails to curb his emissions, even when it would be easy to do so. In order to assess whether Bob's ignorance is inculpating according to the view under consideration, we need to check whether he wittingly violated a procedural epistemic obligation in coming to be ignorant. Although the details leave this open to interpretation, on two very plausible readings of the case, Bob either already complied with his procedural epistemic obligations or his failure to do so was not a witting failure. Regarding the first, as argued by Wündisch (2007: 282) the strength of one's epistemic obligations to undertake investigations may depend on the epistemic environment that they occupy. The idea here is that our obligations to investigate should only be so strong. Given Bob's starting point for engaging in such inquiry, it would take a particularly strong independence of mind and perhaps strenuous study in order for him to see beyond the biased information that is most readily available to him in order to come to a reliable assessment of the evidence. Wündish argues that given Bob's epistemic circumstances, it may be reasonable for him to cease investigations at a point at which he will fall short of having true beliefs about the causes of climate change (282). Because such ignorant agents do not actually fall short of their procedural epistemic obligations in the first place, their ignorance will not be inculpating.

Another reading of Bob's case yields the same verdict but comes to it by a different avenue, namely via the observation that even if Bob is obligated to conduct more inquiry, his failure to do so is not a witting one and so the ignorance that traces to it is not inculpating. A plausible position regarding our epistemic obligations in the context of climate change is that, given the stakes, the demands of inquiry are quite strong. Thus, they might be strong enough that even Bob, whose epistemic circumstances are less than ideal and make it comparatively difficult for him to come to the correct beliefs regarding climate change, would have procedural epistemic obligations that would require him to continue to (better) inform himself regarding anthropogenic climate change. It is of course difficult to pin down exactly how strong such obligations are, but for our purposes, let us stipulate that Bob's ignorance (and that of many others who inhabit the same milieu) traces to a failure to meet his investigative obligations. It is implausible that this failure is a witting one in the sense that Bob believes he is under such an obligation but simply does not follow through on it. Indeed, one frustrating feature of such cases is that, to the contrary, the relevant agents take themselves to have looked into matters sufficiently. Bob may in fact see himself as having gone *beyond* the call of duty since he looked at what he, probably wrongly, conceives to be a range of information sources. On the view of the epistemic condition under consideration, Bob's ignorance would hereby not have an inculpating origin, and one could argue that, therefore, he would be excused from any wrongful failures to minimize his carbon emissions.

This analysis of Bob's case has a wide application. Anyone for whom it is incorrect to say that they strategically avoided becoming ignorant about climate is relevantly similar to Bob. All agents who inhabit poor epistemic circumstances will have the strength of their epistemic obligations diminished such that they can be somewhat easily met. An even wider set of agents—even those who come from good epistemic circumstances—may simply fail to take themselves to be obliged to the amount of inquiry needed to come to true beliefs. The lesson is that an account of the epistemic condition that requires witting failures to comply with procedural epistemic obligations will entail that all such agents will be non-culpably ignorant of climate change and would have an ignorance-based excuse for their wrongful failures to decrease their emissions.

A completely distinct family of accounts of the epistemic condition of moral responsibility affords an alternative, and, for many, more defensible, analysis of these kinds of cases. These accounts maintain that ignorance is inculpating only if it traces to a substandard 'quality of will.' The idea, roughly, is that morally relevant ignorance can sometimes be explained by an agent's failure to care sufficiently about the matter at hand. Interestingly, this failure to respond to the moral weight of the situation may explain and result in witting failures to comply with one's procedural epistemic obligations, if for example, the reason why the agent does not follow through on what they know or suspect to be an obligation to investigate climate change is that they just do not care about climate harms to current or future generations. However, what is inculpating about this ignorance, according to quality of will views, is not that their ignorance traces to witting failures to comply with procedural epistemic norms. Rather, it is what the ignorance indicates regarding the degree of an agent's concern for the morally relevant features of the situation. Quality of will approaches to moral responsibility abound, but a prominent one that has been discussed in the context of climate change ignorance is developed by Nomy Arpaly. The core of Arpaly's account is that [ignorant] agents are blameworthy for their actions to the degree that they fail to respond to the relevant moral reasons that their situation presents them (2002: 129–132). Responding to moral reasons involves being sensitive to them both in what one believes and does. For example, imagine someone is throwing a dinner party for colleagues. The party planner will count as responsive to relevant moral reasons only if they (a) inquire about any allergies the guests may have, (b) come to have true beliefs regarding these allergies, and (c) avoid serving food with the relevant allergens. An agent who is less responsive to these reasons will not care to look into them, or even if they do, will not bother to change the recipe (using the 'oh, it's probably not that serious' justification). The relevant moral reasons in the climate change context are simply the harms that are occurring and are likely to occur from climate change. Importantly, the current situation with respect to our understanding of the relevant climate harms and the availability of information regarding their nature and causes entails that these moral reasons are accessible to someone who is sensitive or responsive to these reasons. A particular agent's ignorance about climate change would be inculpating according to Arplay's view if the reason why they are ignorant is that they are insensitive or unresponsive to these reasons.

It is easy to see how this would explain why the strategically ignorant agent discussed above would be blameworthy. An agent who fails to be moved by the moral gravity of the current situation and opts to remain ignorant regarding the reality of climate change for the purposes of, say, avoiding cognitive dissonance, reveals themselves to be unresponsive to the relevant moral reasons. Such an agent is ignorant, but plainly meets the epistemic condition for moral responsibility. But, what about someone like Bob who makes some efforts to inform themselves and even recognizes what the moral stakes would be *if* climate change was real, but who nevertheless remains ignorant due to mistrust of, for example, 'the liberal scientific establishment'? The quality of will view allows for a more subtle analysis of these cases, and much will depend on whether the manner in which Bob conducted his inquiry displayed a sensitivity to the relevant moral reasons. Plausibly, Bob was responsive to these reasons to some degree, but unresponsive to another. He did take steps to inform himself which shows some sensitivity to the issue at hand, but his tendency to blanketly discount views that were unpopular in his social milieu may be explained by a cavalier attitude with respect to the relevant climate harms, which indicates a failure to respond to relevant reasons. A counterpart who was more reasons responsive would have perhaps raised more questions and been more open to the countervailing views (which in this case, of course, are the true ones). Thus, according to the quality of will view, Bob's ignorance is to some degree inculpating.

Two further points regarding this analysis bear mentioning. The first is that it allows that some ignorant agents are more or less blameworthy since (un)responsiveness to reasons is something that also comes in degrees. Bob's counterpart who is not at all motivated to explore alternative (for them) sources of information and who is too quickly satisfied that their views are correct is even more unresponsive to the relevant moral reasons than Bob and so would be more blameworthy. Someone else who made even more rigorous efforts while still stopping short of being fully open to countervailing evidence will be less unresponsive to reasons than Bob and so less blameworthy. Arplay's view, thus, has the resources to make informative and intuitively plausible distinctions in the degree to which agents who are ignorant about climate change are blameworthy for acting on their ignorance. A second point is that the quality of will view, such as Arpaly's, allows for a clear way of understanding how the difficulty of avoiding false beliefs about climate change can impact the degree to which agents are blameworthy. Arplay argues that an agent is less blameworthy for their wrong actions if it would have taken a really superb quality of will—understood as reasons-responsiveness—for the agent to avoid wrongdoing (2002: 91). Here, the concern is not that the difficulty presented by one's epistemic circumstances might entail that they have weaker procedural epistemic obligations, as we saw above. Rather, the concern is that an agent's epistemic circumstances that make it [much] more difficult to come to true beliefs about climate change can mitigate blame. If someone inhabits a milieu in which climate change is widely seen as a hoax, then it would take a really superb quality of will to avoid ignorance about climate change (and subsequent ignorant wrong-doing). Such agents are not free from culpability on this view, but they are not as blameworthy as they would have been if they inhabited circumstances that would have made coming to the truth much easier. In what follows, the chapter will focus a bit more on the role that difficulty plays in theorizing about the epistemic condition of moral responsibility and show how recent developments in the relevant literature allow for even more subtle analyses of whether certain agents who are ignorant about climate change can still be blameworthy for ignorant wrong-doing.

Difficulty and Climate Change Ignorance

The literature on how difficulty impacts moral responsibility in the sense of attributions of praise and blame is rich and varied. Much of that literature has focused on how certain factors, such as addiction, or mental disorders, such as depression, can make performing the right actions more difficult, even on the assumption that one lacks ignorance of morally relevant factors (see, e.g., Nelkin 2016). Recently, Bradford (2017) and Guererro (2017) have each defended novel approaches to understanding how the difficulty of 'getting things right' would affect whether ignorant agents are morally responsible. As discussed in the previous section, many of the more complex cases involving agents who are ignorant about climate change are ones in which the agents are not willfully or knowingly shirking their epistemic obligations. Instead, they involve agents who, for some other reason, become and remain ignorant about climate change. As we will see, the factors that make it difficult to avoid ignorance will matter significantly in analyzing such cases.

Bradford (2017) defends a nuanced account of how the difficulty of coming to know some fact of moral relevance, whether it is a moral fact or an empirical one, can mitigate blameworthiness. Bradford begins by noting that when it takes effort to form or maintain true beliefs, there are many different kinds of factors that can explain why [significant] efforts are needed. Two central factors are features of the agents themselves, such as the strength of their cognitive or emotional capacities, and the complexity and quality of the agent's evidential situation. Bradford's core idea is that an ignorant agent's blameworthiness can be mitigated only if some threshold of amount of effort has been made toward 'getting things right.' If the amount of effort exerted is below this threshold, then the agent's ignorance is fully inculpating. How high this threshold should be depends on two things. The first is the moral stakes involved with acting wrongly. If remaining ignorant of some morally relevant fact would result in the performance of some grave moral wrong, then this threshold is rather high—we expect people to do a lot to avoid grave wrongs, and so blameworthiness would only be mitigated once an agent has really given it their all, but still remains ignorant. The second factor that determines how much effort an agent must expend if their blameworthiness is to be mitigated is whether the factor that makes it difficult for them to avoid ignorance reflects badly on the agent. For example, if what makes it hard for someone to know some morally relevant fact is that they are "insensitive, unsympathetic, (and) cranky" (Bradford 2017: 183), then it seems plausible that the threshold of effort that this agent must exert before blameworthiness can be mitigated would be quite high. When the factors that make 'getting things right' do not reflect badly on the agent, as they would not if it came down to the complexity of the issue or some congenital cognitive or attentional shortcoming, then the threshold of effort that these agents must exert before blameworthiness is mitigated may be much lower, such that their blameworthiness would start to be mitigated even after only a small amount of effort.

We can now apply Bradford's framework directly to the climate change case to see what it would entail. If we presume the current situation in which most have ready access to reliable scientific findings concerning climate change, then someone who simply spent no effort attempting to inform themselves would fall below any plausible threshold and so their blameworthiness for, say ignorantly failing to curb their emissions, would not be mitigated. Presumably, this verdict would apply to many cases of climate ignorance, and, importantly, on Bradford's account, their blameworthiness need not trace to a witting failure to investigate. Before analyzing cases where some investigative efforts were undertaken, it is crucial to establish how gravely wrong it would be for an agent to perform some act on the basis of their ignorance regarding climate change. Thus far, the chapter has mentioned the somewhat vague wrong of failing to curb one's emissions. Broome maintains that "each of us has a clear duty to emit no greenhouse gases" (2012: 96). This claim is based both on the harm that our emissions cause and the injustice involved with causing avoidable harm to another just for one's own benefit (2012: 54). Broome explicitly classifies this harm as serious. Though there is significant discussion of the negative impact that say a year or lifetime of emissions has on someone's life, Broome estimates that the annual emissions of persons living in a developed country can shorten the lives of others by several days (2012: 73–75). Thus, according to one influential climate ethicist, the harm of committing an injustice and of shortening lives is far from insignificant. In the context of applying Bradford's framework, this would set the threshold for how much we would have to investigate before blameworthiness for *failing to decrease* our emissions to [effectively] zero would be mitigated would be rather high. Returning then to the case of Bob discussed above, it is plausible that his ignorance would likely fall short of this higher threshold. Though he indeed makes some efforts to inform himself and even ventured out of his comfort zone (somewhat), Bob likely falls short of the requisite amount of effort, given the elevated moral stakes. This verdict is further strengthened if we also take into account the fact that his blanket distrust of non-conservative information sources is indicative of an epistemic vice. It reflects badly on Bob that he is not open-minded with respect to what are the consensus findings of the scientific establishment which are being communicated in neutral media sources. It follows from Bradford's framework that the amount of effort that he would have to make before his ignorance could be mitigated would be even higher, thus increasing the plausibility of the position that his ignorance is inculpating rather than excuse-giving.

As mentioned above, Bob is a stand-in for a segment of individuals who are ignorant of climate change. It is striking, of course, that many probably make even fewer efforts than him. If Bob's blameworthiness is not mitigated even though he has made some efforts to inform himself, then all of those who do even less are certainly blameworthy as long as the assumptions that we have made in Bob's case also apply.

A second, mostly distinct account of how difficulty impacts blameworthiness was developed by Guerrero (2017). Much of the discussion up to this point has focused on cases in which it is difficult to avoid ignorance, given that someone has made some efforts to inform themselves. Various factors can make it the case that such attempts to 'get things right' will take more effort if one is to succeed in having true beliefs about climate change. Guerrero notes that another way in which it would be difficult to have true beliefs that are relevant to moral decision-making involves factors that make it difficult for an agent to even *try* to inform themselves by investigating (2017: 212). For some agents, it more naturally occurs to them that there is morally relevant information that they lack, and so it is not difficult for them to start trying to investigate. For others, this may be really difficult. Here, the relevant factors do not make it difficult to 'get things right' once one is in pursuit of true beliefs, rather they make it difficult for one to *initiate* the pursuit. Examples of such factors mentioned by Guerrero include cases in which the pursuit of true beliefs,

> Would require significant sacrifice according to (one's) values; would require (one) to depart from the norms or beliefs of her society; would require (one) to think or value very differently than she currently does.
>
> *(2017: 211)*

Illustrated in the context of climate change, we can imagine that an agent lives in a society in which the predominant norms and beliefs sanction and even encourage activities with a significant carbon footprint. Such an agent may even value greatly the kind of freedom that comes with carbon intensive activities such as long-distance air travel. If one imagines this agent in a reflective moment, stopping to consider how they could be and do better, it simply might not occur to them to investigate the issue of climate change given the presence of the factors just mentioned. It is in this sense that it would simply be difficult for them to initiate the relevant inquiry.

Guerrero argues that for certain influential theories of blameworthiness for which control is crucial for moral responsibility, this kind of difficulty would mitigate blameworthiness (2017: 212). The reason is that when one's values strongly support a certain set of views on a matter of importance or when one already has one's mind settled, it is difficult for someone to even consider that the matter is worthy of investigation—we are not in control of what occurs to us to investigate the factors listed above make it seem even less under our control. For this reason, it seems like blameworthiness for such agents would be mitigated—such agents have even less control over whether they end up with false beliefs about morally relevant matters. The agents considered above could thus claim that their values and the predominant social norms effectively shut off plausible avenues of inquiry for them, and thus that they had greatly diminished control over whether they were ignorant about climate change.

However, Guerrero is quick to note that the verdict is different for theories of blameworthiness that ground moral responsibility attributions on the fact that the [ignorant] action reflects the moral perspective or their quality of will. The fact that it would be difficult for one to even try to look into the facts relevant to some course of action would actually establish or ground their blameworthiness rather than mitigating it. So, the values and beliefs that would make it more difficult for climate change to appear on an agent's moral radar as something to investigate would indicate a kind of moral insensitivity that quality of will views would regard as inculpating. The fact that it is difficult for such agents to even try to get things right says all one needs to know regarding their blameworthiness.

Thus, there is a certain type of ignorant agent for whom Guerrero's account seems to deliver a split verdict—if it would be difficult for someone to even try to have correct views on climate change given their background beliefs and deeply held norms, then whether this kind of difficulty mitigates blameworthiness depends on which kind of theory of moral responsibility one accepts. Nevertheless, it is important to keep in mind the relative prevalence of this kind of ignorance. In many countries, climate change is a very prominent topic of discussion in the media and in entertainment, and, because of this exposure, it is likely that most people have undertaken some, even just minimal, effort to figure out what they think about the matter, at which point the kind of difficulty that is linked to efforts is the more relevant one. Taken together, Bradford and Guerrero's frameworks allow us to reach subtle verdicts regarding how the difficulty of coming to true beliefs about climate change should affect the moral appraisal of ignorant agents. When evaluating particular cases of ignorance, we must attend to both the source and the kind of difficulty that individuals face.

Conclusion

This chapter has brought together several theories about how ignorance affects moral responsibility. According to the first, which states that for ignorance to be inculpating, it must trace to witting failures to comply with one's procedural epistemic obligations, only strategically ignorant agents would meet the epistemic condition of moral responsibility. By contrast, quality of will accounts of the epistemic condition would classify a larger set of ignorant agents as blameworthy—as long as their ignorance traces to a failure to be sufficiently responsive to relevant moral reasons, then these ignorant agents are blameworthy. In the context of quality of will accounts, the notion of difficulty was considered, and in order to further examine how difficulty in forming and maintaining true beliefs about climate change should impact the blameworthiness of ignorant agents, two distinct approaches were examined. According to Bradford, the difficulty that agents face when trying to determine what they think regarding the reality of climate change can mitigate blameworthiness only if their efforts fall above a particular effort-threshold. We saw that given the moral gravity of the situation and the fact that many who cut off inquiry prematurely do so due to epistemic character traits that reflect badly on them, many agents would not meet this threshold. The blameworthiness of such agents would not be mitigated. Finally, we saw that a distinct kind of difficulty involved in *trying* to be informed about climate ignorance would either mitigate or not, depending on the theory of blameworthiness one adopts. Although many of the responsibility verdicts that were discussed above are tentative, it demonstrates that there are a variety of theoretical resources in moral responsibility literature that can be applied in the context of morally appraising people who are ignorant about climate change. The work of such moral appraisal should be undertaken with care, but we should welcome the fact that the complexities of real-life cases can be captured by several of the theories on offer.

References

Arpaly, N. (2002). *Unprincipled virtue: An inquiry into moral agency.* Oxford University Press.

Bradford, G. (2017). Hard to Know. In P. Robichaud & J. W. Wieland (Eds). *Responsibility: The epistemic condition* (pp. 180–198). Oxford: Oxford University Press.

Broome, J. (2012). *Climate matters: Ethics in a warming world (Norton global ethics series).* WW Norton & Company.

Guerrero, A. (2017). Intellectual Difficulty and Moral Responsibility. In P. Robichaud & J. W. Wieland (Eds). *Responsibility: The epistemic condition* (pp. 119–218). Oxford: Oxford University Press.

Jamieson, D. (2007). When utilitarians should be virtue theorists. *Utilitas*, 19(2), 160–183.

Nelkin, D. K. (2016). Difficulty and degrees of moral praiseworthiness and blameworthiness. *Noûs*, 50(2), 356–378.
Robichaud, P. (2017). Is ignorance of climate change culpable? *Science and Engineering Ethics*, 23(5), 1409–1430.
Rosen, G. (2004). Skepticism about moral responsibility. *Philosophical Perspectives*, 18, 295–313.
Sandberg, J. (2011). "My emissions make no difference": Climate change and the argument from inconsequentialism. *Environmental Ethics*, 33(3), 229–248.
Vanderheiden, S. (2007). Climate change and the challenge of moral responsibility. *Journal of Philosophical Research*, 32(Supplement), 85–92.
Wündisch, J. (2017). Greenhouse gas emissions and individual excusable ignorance after 1990: A study of excusable ignorance in collective action problems. *Environmental Philosophy*, 14(2), 275–315.

22
SAVING DEMOCRACY
Denying the Alternate Reality of Climate Denialism

Kathryn Gwiazdon

Introduction

Climate change denialism, here, simply 'climate denialism,' is a particularly insidious tactic of the climate change disinformation campaign, yet it is also a result of the climate change disinformation campaign. This chapter will focus on the tactic to purposefully misinform through denial in order to stop or stall action on climate change, as opposed to the psychological or sociological underpinnings of denialism that people may use to mentally cope with the existential crisis before us,[1] or those who deny climate change due to ignorance, whose culpability is explored elsewhere in this *Handbook*. Climate denialism is (1) the intentional denial of truth; (2) by largely corporate actors, with and through the assistance of political and academic actors (as 'useful idiots'[2] and as knowing accomplices); (3) through a massively funded campaign rooted in deceitful and strategic entanglement of difficult or impossible to trace donors; (4) with a façade of free market, freedom, and democratic debate; (5) in order to stop or stall action on climate change, particularly regulatory action; (6) with the knowledge that stalling or stopping action on climate change will cause grave harm and destruction to individuals, societies, our planet, and the relations that guide them; and (7) for the creation and maintenance of unequal power, wealth, and white supremacy. This chapter seeks to offer arguments to combat this tactic by exploring how a campaign that rejects truth, equality, justice (i.e., accountability), and notions of society – the ethical underpinnings of a democracy – is a campaign that seeks to destroy democracy itself. This is the threat before us, and those who claim to support democracy must use this language when advocating for action on climate change.

In theory, democracy – its principles, actors, mechanisms, and institutions – does not and should not ever support the creation of unequal power and wealth. This is why denialists actively work to intentionally destroy it. But they know the power of its principles, its history, and its heroes. This is why denialists actively work to distort it. They usurp the narratives of democracy in their arguments and the names of their institutions,[3] creating an alternate reality where freedom is defined by oppression, truth is denying truth, free speech is for paying corporate donors, and the common good is defined by individual rights and private property. And all of these mistruths, half-truths, and alternative truths have an undercurrent of intentionally placed division, discord, and violence. This is why the climate denialism roadmap is a roadmap to authoritarianism. It is not only an 'environment' issue, or an existential life issue, it is a governance issue.

To those fighting climate denialism, the world is the aim because the world is at stake. But to climate denialists, denying and obstructing climate change is just one part of a larger obstruction

DOI: 10.4324/9781003039860-26

to anything that strengthens the common good and the idea that the purpose and duty of a state is to protect and provide for the common good. This is why the climate crisis is the perfect lynchpin for their larger aims. It is a universally shared crisis that only a democratic approach – rooted in truth, justice, and the common good – can address. It is this approach, after all, that would lead to historical accountability for past harms, present-day accountability for current harms, technological changes, and greater regulation. Climate change, and the democracy needed to address it, is an attack on their industry investments and the accumulation of personal wealth; and so, instead of evolving and investing in a sustainable, equitable future, they invest in warfare: attacking those who attack their interests. Yes, climate denialism is simply, and strongly, psychological and political warfare for greed, and with a body count to match. Real harm, with real lives lost, is being committed from this deceit, and they are successfully protecting themselves from accountability with an ideology that they have a right to deceive.

Their tactics – and often-times the same actors – can be found against public education, universal healthcare, public welfare, social security, etc., and yet *for* gun rights and white Christian nationalism. They work against majority will, government oversight, and checks and balances, yet *for* de-regulation, privatization, and an unbridled free market. The common theme: individual rights over the common good, 'my freedom' above and/or at the expense of your freedom. This makes the fight against climate change denialists even more difficult; they are not just 'anti-environmentalists,' but anti-society, anti-rule of law, and anti-democratic governance itself. It is not just a climate denialism campaign; it is a democracy denialism campaign. And the arguments to combat it must acknowledge this.

It is crucial to also understand that the denialists are winning.[4] The tentacles of their anti-government campaign of deceit reach into all aspects and institutions of society: science, academia, non-profit organizations, private life, and public policy. They are creating an alternative reality that is replacing democratic governance, discourse, legislation, and jurisprudence, with a non-working, dysfunctional organization administered by lies, propaganda, oligarchy, and authoritarianism. The very idea, let alone the realization, of a non-working government only helps their arguments that "government is the problem, not the solution" (Reagan 1981).

A democracy cannot exist unless there is a common understanding of truth on which to debate and legislate. If that truth is denied, mocked, and replaced with lies, we have legislators bringing snowballs to the Congressional floor, laughing at the absurdity that anyone could think global warming exists if it is cold outside,[5] or a Senator using traditional totalitarian fear and 'us versus them' tactics that climate advocates "believe that [farmers and commuters] are a fundamental threat to the world, and they have to have the power and the control of those Americans' lives to implement their radical vision for humanity."[6]

This article can talk about their tactics, but the denialists can speak for themselves. Dr. Maya Angelou, an American author and civil rights advocate (1928–2014), argued, "when someone shows you who they are believe them."[7] And climate denialists repeatedly have shown us who they are. An incredible article in 2019 collected the views of several US Congressmen on climate change, and a quick overview is valuable here to provide evidence of the scale and scope of climate denial that will be unpacked below: "The federal government should *stop regulating* and *stomping on our economy* and *freedoms* in the name of a *discredited theory*;" "biggest *scam*... Al Gore's biggest *money-making machine*;" "*alarmists*... no general consensus," "*sound science*... the science is not settled on climate change;" "Every time you have that soil or rock, whatever it is, that is deposited into the seas, that forces the sea levels to rise;" "our climate has always changed... dig up some fossils;" "[not] supported by scientific data, but... by *political agendas*;" "global warming *prophets of doom*... I welcome more offshore energy exploration," "global warming *alarmists*... wants us to unilaterally *handicap... industry with over-burdensome EPA regulations*;" "the *liberal elite*;" "largely *leftist propaganda* to change the way Americans live and create more government *obstruction and intrusion in our lives;*" concerning the Green New Deal, "...the far-left's Santa Claus wish list... ...the philosophies and the ideas behind this textbook *socialism* are not just foolish. They're *dangerous;*" "...harm farmers, small towns, and mining and rural communities with its policies that are rooted in *socialism;*" "... efforts to prop up wave after wave of *job-killing regulations* that are leading to *skyrocketing food and energy costs;*" "global warming *doomsayers... irrational fanaticism;*" "as a *Christian*... ...if there's a real problem, He can take care of it;" "*Radical* activists... Logical people know otherwise," "Today, it's this, but tomorrow it could be the *2nd Amendment or immigration... Not giving the other side a say* and refusing to hear opinions contrary to your own – that's being a real 'denier,' and, it creates a chilling effect on public debate and free speech;"

"This whole global warming *myth* will be exposed as what it really is — a way of *control* more than anything else," "*[like] the policies of Venezuela's Hugo Chavez...* it could lead to people '*starving* to death;'" "tantamount to *genocide;*'" "The climate hasn't warmed in quite a few years... That is proven scientifically;" and "[The federal government's] assertions about climate change are based on *junk science*" (Cranley, emphasis added).

The arrogance and *ad hominem* attacks, the uncertainty and the outright denial, the emotive and exaggerated language of fear and harm and suffering, the threats of control and socialism, the mockery and the politicization – it is all there. And all of this is part of the denialist machine. But the point here is that instead of the necessary, difficult democratic discourse that needs to occur to determine the best path forward to address climate change, we have bought-and-paid-for political theater that stops and stalls climate action, and loyalty tests akin to authoritarianism. It is lies being debated and lies being legislated. Lies being shared in the media. Lies triggering emotions, extremism, and anti-government resentment. And this absurd, alternative reality is strategically and intentionally planned to create the future *they* want, where power and wealth grows unchallenged and is concentrated in the hands of a few.

Ethics is the inquiry into right and wrong behavior and when responsibility attaches. As such, it is the foundation of justice and the rule of law. This may seem a straightforward inquiry when discussing deceitful tactics funded to dismantle democracy, and at the expense of people and the planet, but the disinformation campaign has been so successful in its entanglement with free market ideology that lying and greed have become normalized and "selling one's soul [is] quotidian rather than scandalous" (Brown 2019: 163). This is why a core purpose of this chapter is to raise the seriousness of climate denialism and the language we use to counter it. Let there be no mistake: denialism is a democracy killer. Climate denialism is a planet killer. And there is no hope for a successful response to climate change without the truth, equality, and justice – and for the common good – that democracy demands.

In this inquiry into the ethics of climate denialism, this chapter opens with an introduction to the democratic agitators behind climate denialism, with just a small glimpse given above in the language of US legislators (this article largely focuses on climate denialism in the United States, noting that climate denialism and even climate change is seen as an American export). It will then turn to the anti-democratic tactics of climate denialism, creating an alternate reality where lies are legislated, and authoritarianism is quite literally sold as democracy. This is done by attacking truth and dignity, disempowering and dehumanizing both its 'enemies' and its adherents. Only by knowing its tactics and knowing the seriousness of its end aims, can we hope to address it.

The Anti-Democratic Agitators of Climate Denialism

Denial can come in many forms, including, "...*cognition* (not acknowledging the facts); *emotion* (not feeling, not being disturbed); *morality* (not recognizing wrongness or responsibility); and *action* (not taking active steps in response to knowledge)" (Cohen 2001: 9). It can be individual or collective, personal or public – or a mix of each – and it can lead to a cultural denial, where the culture of a society lives in an alternate reality, created and comforted by the power-holders, where people pretend to believe that which they know is fake and bear allegiance to meaningless slogans (Cohen 2001: 11). Official denial would be considered a highly organized collective act of the state, sustained by massive resources, politicization, 'spin-doctors,' and creative twists of the truth creating their rhetoric and propaganda, "built into the ideological façade of the state" (Cohen 2001: 10). Official and cultural denial are also mutually dependent – each lifting up the other. Telling the truth would amount to treason against your tribe (Cohen 2001: 11), and "an entire language of denial has been constructed in order to evade thinking about the unthinkable" (Cohen 2001: 11).

Climate denial is private and public, individual and collective, and highly organized and well-funded. It is a cognitive act, an emotive act, and a moral act. It is an *act* (action that triggers responsibility) and an act (theater). Lying, after all, is an act that seeks to alter reality and "the blurring of the dividing line between factual truth and opinion belongs among the many forms that lying can assume, all of which are forms of action;" the liar is therefore "an actor by nature" since "he wants to change the world," where "the people will then hold new beliefs in consequence of the stories he told" (Arendt 2022: Introduction, xxii–xxiii). The adherents to climate denialism act by showing their loyalty and bringing snowballs to the Senate floor; they act by creating a language of uncertainty: 'sound science' and 'junk science;' and they act by calling their 'enemies' elitists, socialists, and alarmists who seek to steal your liberty, property, and Americana itself. They "back up and collude in each other's denials," and up against such strong truths, they resort to "minimization, euphemism, and joking," with such "selective blindsight… always politically calculated (Cohen 2001: 64, 163). And their ideological façade is free market fundamentalism, but as this chapter will later show, that is also a lie.

To use lies, tricks, and deception in such a politically calculated manner, where society itself is at stake, makes these consummate actors nothing more than democratic agitators. In Leo Lowenthal and Norbert Guterman's *Prophets of Deceit: A Study of the Techniques of the American Agitator*, there are three devices of the agitator, whose "doctrine thus consists in drawing the ultimate consequences of a totally amoral opportunism:" unseriousness; transformation of meaning; and anti-universality, with each inter-related (1949: 31–33). And we see each in climate denialism. Climate change advocacy, and the ideals that underlie it, is spun as "bunk, hogwash, lies." And while the denialist cloaks himself as a "champion of legality" (in the United States, the conservative party – the key political movement behind climate denialism – refers to themselves as the 'law and order' party), they undermine laws as attacks on "individualism," "[twisting] the meaning of basic ideals in such a way that he infuses them with his own content" (Lowenthal & Guterman 1949: 31). And like the agitator, climate denialists, "explicitly [reject] the ideal of universality," where tolerance, humanitarianism, and universal justice are "opposed to truth:" they are weak, silly, non-Christian sentimentalities that are "contrary to self-interest" (Lowenthal & Guterman 1949: 32). This also sets the stage for religious-supported violence: they "shout 'tolerance' but with their own foul tongues, they would lap up the blood of their own critics" (Lowenthal & Guterman 1949: 33).

The agitator also embraces conspiracy and "seizes on the 'simple folk' theme as a pretext for fostering an aggressively anti-intellectual attitude" (Lowenthal & Guterman 1949: 109). All of the above was on display during and after the armed insurrection on the US government on January 6, 2021. Whipped into emotion by the intentional deceit of private and public actors, embedded with language of warfare, supporters of former President Donald Trump violently attacked an institution of democracy while shouting variations of god, guns, country, and freedom (along with misogynist, racist, and anti-Semitic slurs). This, while far-right media host Tucker Carlson sat on a set that looked like a log cabin and continued to foment this fear and division – pushing white nationalism, lies, and conspiracy to millions (and climate denialism, of course).[8] To Lowenthal and Guterman,

> The agitator gravitates towards malaise like a fly to dung… he grovels in it, he relishes it, he distorts and deepens and exaggerates the malaise to the point where it becomes a paranoic relationship with the external world. For once the agitator's audience has been driven to this paranoic point, it is ripe for his ministrations. The prevalence of malaise in recent decades is reflected in growing doubt with relation to those universal beliefs that bound western society together.
>
> *(1949: 17–18).*

And this is just one example of the power of a lie and those who use it.

Philosopher and political activist Jean Paul Sartre (1905–1980), who spent most of his life fighting violent, racist authoritarianism, also agreed on the use of "unseriousness" in the methodology of those who seek to control through dehumanization and division. In *Anti-Semite and Jew*, like with Lowenthal and Guterman, his agitator is evidenced by the anti-Semite,

> Never believe that anti-Semites are completely unaware of the absurdity of their replies. They know that their remarks are frivolous, open to challenge. But they are amusing themselves, for it is their adversary who is obliged to use words responsibly, since he believes in words. The anti-Semites have the right to play. They even like to play with discourse for, by giving ridiculous reasons, they discredit the seriousness of their interlocutors. They delight in acting in bad faith, since they seek not to persuade by sound argument but to intimidate and disconcert.
>
> *(1948)*

But these often-unserious actors have a very serious end goal: create an alternate reality – a 'democratic' government where their power is not challenged or checked.

'Behind the curtain'[9] of climate denialism is Koch Industries, Inc., "a financial kingpin of climate denial."[10] Their executive, Charles Koch, alongside his brother David Koch, are not only the minds and money behind much of today's climate denial machine, but modern-day 'free market' fundamentalist ideology, as well – a façade that seeks to directly dismantle democratic institutions through appeals to free markets and competition, but whose true aim is to protect personal wealth. To look at the principles of a movement, it is necessary to look at the principles behind the money that funds that movement.

Poison, pay, and deceive: Charles Koch "wants to pull the government out by the root" (Mayer 2016: 374). Entangling all aspects of civil society, they target universities,[11] non-profit organizations, museums,[12] religious institutions,[13] government agencies, and elected politicians. Living by a mantra to "never compromise," and protect their wealth no matter the cost, they seek to "take the *demos* out of democracy," for "If American political institutions render market-oriented reforms too difficult to achieve, then perhaps those institutions should be changed" (MacClean 2017: 127, 223).

And no one compares to the Kochs, with "a pattern of lawbreaking, political manipulation, and obfuscation" that has no parallel (Mayer 2016: 4–5). Many books and organizations track the funding and activity of the Koch brothers, so only a few items will be emphasized here (see Mayer 2016). For example, in 2001, in "one of the most significant cases ever brought under the Clean Air Act," Koch Industries plead guilty for "an orchestrated scheme to conceal benzene emissions – a known carcinogen – from regulators and the community" (USDOJ 2001; Mayer 2016: 125). David Uhlmann, a USDOJ prosecutor, believes that "Environmental crimes are almost always motivated by economics and arrogance, and in the Koch case there was a healthy dose of both" (MacClean 2017: 125). It is important to note that this is just one of several cases before the US-DOJ; a year prior, Koch Industries had to "pay the largest civil fine ever imposed on a company under any federal environmental law to resolve claims related to more than 300 oil spills from its pipelines and oil facilities in six states" (USDOJ 2000). The point is – poison, pay, and deceive is their *modus operandi*, and federal regulatory action is a check on their power.

And 'pay' refers to more than paying court-determined settlements and fines. The Koch brothers not only donate millions to political campaigns for political outcomes, they manipulate and fund scholarship, "providing intellectual ammunition for an autocratic businessman" (MacClean 2017: 209; see also Mayer 2016: 208). Richard Fink, former executive vice president of Koch Industries, economics professor and founder and director of the Mercatus Center at George Mason University,[14] and a "main political lieutenant" for Charles Koch, explains this "three-phase takeover of American politics" as an "investment in intellectuals whose ideas would serve as the raw products," an "investment in think tanks that would turn the ideas into marketable policies," and "the subsidization of 'citizens' groups that would, along with 'special interests,' pressure elected officials to implement the policies" (Mayer 2016: 141–142).[15] Today, the libertarian, evangelical, and conservative movements in the United States have largely become that "instrument of a tyrant," with "no hope that anyone in the think-tanks [will] speak out – they [have] all been bought (MacClean 2017: 209). In the words of Arundhati Roy, "Funding has fragmented solidarity in ways that repression never could" (2014: 37).

The economic entanglements and influence of the oil industry go far beyond the Kochs, however, even into environmental organizations and treaty-making bodies that have explicit missions to protect the human and natural environment. One example: it was just announced (January 2023) that the next Conference of the Parties of the UN Framework Convention on Climate Change would be hosted by Sultan Ahmed Al Jaber, CEO of the Abu Dhabi National Oil Company, the world's 12th largest oil producer. The Sultan will be joined by Her Excellency Razan Al Mubarak, President of the International Union for the Conservation of Nature (IUCN) and UN Climate Change High Level Champion.[16] Ms. Al Mubarak's husband is Badr Jafar, President of Crescent Petroleum and Chairman of Pearl Petroleum, named as one of the 100 most powerful Arabs by Arabian Business and Gulf Business in 2019 and 2020. This begs any rational actor to ask, how can oil executives, oil families, and oil money lead towards realistic solutions on climate harm when it is their business causing the harms? Their presence on these boards, with these organizations, seems to be a mockery of any legitimate law and governance attempt to address this crisis, especially when knowing that part of the denial machine is to diminish the seriousness of the crisis, the undertaking that needs to occur to address it, and the accountability of the industries responsible. Can anyone expect justice when the judges are judging – and sentencing – themselves?

Or perhaps the presence and leadership of oil money in climate 'action' is just a mockery of the climate change law and governance system, another snowball on the legislative floor. Or perhaps it is just arrogance, as noted above. Their involvement is not some hidden treachery, after all, but an out-in-the-open ownership of the dialogue and decisions affecting climate change: the CEO of an oil company, with the help of the leader of the world's oldest and largest environmental organization (a multi-millionaire whose wealth comes in large part from the oil industry), will host the world's largest treaty making body on climate change, a harm caused by oil companies. It is an absurdity. But this absurdity is the reality of climate denialism, to which we are all playing along.

For oil and gas industry advocates, climate change is the perfect scapegoat "because it struck at the very root of economic activity: the use of energy" (Oreskes & Conway 2011: 168). The "coal, oil, and gas magnates [form] the nucleus of the Koch donor network," buying political messages of being 'job creators,' patriots, and freedom fighters responsible for 'American energy independence' (Mayer 2016: 200). And the messaging works, a glimpse in that small sampling of legislator's denialist quotes above. If only there was a successful, and accurate, counter-messaging: how they make America dependent on a limited resource; how they promote stagnation and non-competitiveness against innovative sustainable technology; how they create deadly jobs – poisoning not only the employees but the public; how they limit American freedom by hiding the truth and controlling decision-making; how they hurt an entire country – and world – in order to enrich themselves. Can you imagine if the billions that go into denialism and discord to protect old technologies and 'old money' (i.e., white wealth) went into innovation, lessening the inequality gap, and strengthening democracy?

It is clear to climate denialists that the denialism proffered by Koch and their adherents is not dealing with science or even free market ideology, itself a façade in the alternate reality they are creating,[17]

> Given the overwhelming strength of the evidence, as we have seen, that climate change is (1) real, (2) caused by humans, and (3) a grave threat, one might rightfully ask how it is that some of our most prominent elected officials can still deny that climate change is even happening. The answer, of course, is that climate change denial isn't really about the science; it is instead about the politics. It is about powerful vested interests that find the implications of the science (that there is a need to stop burning fossil fuels) inconvenient. It is about a massive disinformation campaign to justify an agenda of inaction.
>
> *(Mann & Toles 2016: 53).*

These "climate breakdown terrorists" seek to destroy democratic institutions for the dangerous ideology of self: power and greed (Michaels 2020: 183). Even when their mask is to attack science, what is really happening is the intentional erosion of trust (Michaels 2020: 194).

To prevent accountability for industry's harms, they attack regulations and "any sort of mitigation policy effort that detracts from fossil fuel-based economic growth strategies" (Bohr 2021), seeking ultimately to "dismantle the administrative state" (Whitehouse et al. 2022), specifically the US Environmental Protection Agency (EPA) (e.g., *West Virginia v EPA* 2022). Their clear mission is to "mislead large sections of the American public into thinking that the evidence for human-caused warming [is] uncertain, unsound, politically tainted and unfit to serve as the basis for any kind of political action" (Mann & Toles 2016: 78–79). And they have the help of all branches of government behind them, not just the US Supreme Court (see Whitehouse 2022). In a "fifth column assault on American democratic governance," they are attacking from within (MacClean 2017: xxxii).

A powerful point in the evolution of climate denial was President Ronald Reagan's declaration, "[the] government is not the solution to our problem; government is the problem" (1981). He "came to power… on a platform of reducing regulation, decreasing the reach of the federal government, and unleashing the power of the private enterprise" (Oreskes & Conway 2011: 74). Before him, however, the attacks on regulation through language had already begun; the acid rain debates saw industries begin to publicly smear "*irresponsible* environmentalists who insist on *absurdly* strict emission standards at the *expense of jobs and the economy*" (Oreskes & Conway 2011: 74: emphasis added). And after him, in the 1990s, we saw more of their language and methodology take shape, 'sound science' versus 'junk science,' with direct attacks at the EPA "rigging" numbers (Oreskes & Conway 2011: 143). And this exact language and methodology continues to this day, decades later, with any leader or political movement that deceives to sow doubt, cause division, and for personal profit. It has been so normalized that those who dispute it are now the irrational.

The Alternate Reality of Climate Denialism

Perhaps "the most familiar usage of the term denial refers to the maintenance of social worlds in which an undesirable situation (event, condition, phenomenon) is unrecognized, ignored, or made to seem natural" (Cohen 2001: 51). In other words, the creation and maintenance of an alternate reality. In climate denialism, Mann and Toles see "six stages of denial," from (1) the globe isn't warming to (2) it is, but it's natural; (3) it's caused by humans, but it's minimal; (4) it'll be good for us anyway; (5) it's too expensive to do anything about it; and, finally, (6) we'll find a simple or inexpensive technological fix (2016: 68); or, in other words, science denial, economic denial, humanitarian denial, political denial, and crisis denial (Maslin 2019). This denial, in light of such overwhelming evidence, and in the face of such urgency is "the height of irresponsibility in relation to human well-being" (Falk: 76).

In the introduction to Hannah Arendt's collection, *On Lying and Politics*, David Bromwich finds that her writings describe a "larger tendency" about "the kingdom of untruth," that "A falsified present is created by government and its outworks in media, in order to ease the acceptance of a future desired by the powerful" (Arendt 2022: Introduction, xxvii–xxviii).[18] The kingdom is the tentacles and entanglements of the denialist machine, coming from positions of authority, that work in collusion to undermine truth. Lies "become the warrant for other lies" that create "a substitute world in which the truth is just one more opinion" (Arendt 2022: Introduction, xiv, xix). And to Arendt, these lies are a direct attack on democracy, "The more democracy is loosened from standards of truthfulness, reasonableness, accountability, and problem-solving through comprehending and negotiating differences, in turn, the more discredited it becomes" (2022: 87).

Limitless lies are necessary to keep the deceptive image going – that is the power of truth. There is only one truth, and if you work against it, you must always work against it [because the alternative is to acknowledge the truth and face accountability]. In this "image-making," people will deny or neglect any truth that hurts that image. This is when "highly respected statesmen… build their policies on such evident non-facts," and they all "harbor an element of violence" because "organized lying always tends to destroy whatever it has decided to negate, although only totalitarian governments have adopted lying as the first step to murder" (2022: 40). In her foundational text, *The Origins of Totalitarianism*, Arendt argues that totalitarian regimes – or those on that path – offer a "superior realism," "[thriving] on this escape from reality into fiction, from coincidence into conspiracy:"

> [Conjuring] up a lying world of consistency which is more adequate to the needs of the human mind than reality itself... The force possessed by totalitarian propaganda – before the movements have the power to drop iron curtains to prevent anyone's disturbing, by the slightest reality, the gruesome quiet of an entirely imaginary world – lies in its ability to shut the masses off from the real world.
>
> *(1966: 352–353)*

Arendt makes two powerful references here. This consistency – this repetition of lies – in order to create 'illusory truth' is one of the most well-known tactics of fascists, with some variation of 'if you tell a lie big enough and keep repeating it, people will eventually come to believe it' attributed to Nazi propaganda minister Joseph Goebbels (MacClean 2017: 211; Martinez-Conde 2019). In addition, 'iron curtain' is a historical reference to the division between western democracy and Soviet communism, and the complete containment of 'truth' and power by Josef Stalin in the Soviet Union. These liars and deceivers, and the entire network that holds them up, were seen as the greatest threats to democracy; so much so that a world war was fought to end it.

And in order to show power over truth, to "[propagate] a particular worldview" and "motivate and rationalize extreme action" (Lynch 2021), these tactics are used by the climate denialist machine: their big lie is repeated often enough, by numerous actors, and in numerous ways that their audience comes to believe it (and sometimes the liars themselves come to believe it); and their iron curtain is held tightly around their conservative, Christian, anti-government coalitions, creating a safe space of social, political, and moral echo-chambers, where no one is disloyal enough to disagree and all those beyond that curtain are considered enemies. For climate change, "enemies include most climate scientists, but also anyone perceived as an opponent to the lifestyle made possible by access to cheap fossil fuel" (Bohr 2021). And one doesn't compromise – a requirement for democracy – with enemies.

For Pulitzer Prize author Anne Applebaum, an expert on authoritarianism, this iron curtain is an "information sphere" that provides "no easy way to distinguish between conspiracy theories and true stories," and it has help with modern technology and big data, where "social media algorithms themselves encourage false perceptions of the world" (Applebaum 2020: 113). This has led to "no common debate, let alone a common narrative. People have always had different opinions. Now they have different facts" (Applebaum 2020: 113).

Another uncomfortable parallel with authoritarian movements, in particular with 1920's Soviet Russia, are the public demonstrations of support for the alternate reality and its untruths. Like the snowball mentioned above, we often see this with fights for TV airtime: who can make the most absurd statement or quip that will be easily translatable and easily digestible to ensure national news coverage – and repetition. It is important to note that these one-liners are usually pulled from pre-approved party position papers, i.e., documents presented by the national party to elected officials that dictate party-line responses on different issues, often drafted with the help of lobbyists and special interest firms. If the actor sticks to the script, the actor may be awarded with committee appointments (and more airtime).

When discussing one-party rule under Lenin, Applebaum describes how places in universities, government, and advancements were all due to loyalty, not merit, and placements went to those willing to conform, and above all, to those "who most loudly professed their belief in the party… who participated in public displays of enthusiasm" and "engaged with the alternate reality," evidence that they were "willing to defend it" (2020: 23–24, 38, 54). It seems the conservative party in the United States today is already living a one-state system. Climate denial is now a loyalty test for the conservative party and was "very nearly a litmus test for Trump appointments" (Michaels 2020: 184, 194).

The known defense of untruths is a show of loyalty for access to more power. This is where we are in American democracy. But why is the public audience so receptive? Arendt believes that with this fragility of facts, "Lies are often much more plausible, more appealing to reason, than reality, since the liar has the great advantage of knowing beforehand what the audience wishes or expects to hear" (Arendt 2022: 65). Also, when the liar is faced with a particularly obvious truth, they can easily change tactics and pretend that his is not "the gospel truth," but rather an "'opinion' to which he claims his constitutional right" (Arendt 2022: 36). This not only nurtures the 'my rights, my freedom' approach to denialists, but this is also how denialists demand equal airtime in media, appealing to democratic notions of 'fair and balanced' democratic dialogue, to conceal what is actually the truth: a known lie masked as an opinion (see Oreskes & Conway 2011: 19).[19] This is another example of how lies multiply lies to create the limitless lies of an alternate reality. Arendt does then acknowledge that "this is frequently done by subversive groups, and in a politically immature public…" (Arendt 2022: 36).

This approach – the deceit, the masks, the blind loyalty – is more conducive to fascists, who "[deny] objective truth in favor of a glorious myth articulated by leaders who claimed to give voice to the people" (Snyder 2017: 12). Indeed, "There can no longer be any doubt that so-called mass movements of a fascist nature have a very deep structural connection to delusional systems" (Adorno 2020: 16). So then, what are we left with? Impotent truths? Deceitful power (see Arendt 2022: 3–4)? Or, perhaps, a dead truth. In Timothy Snyder's short-but-strong *On Tyranny: Twenty Lessons from the Twentieth Century*, he argues that "Post-truth is pre-fascism" and that truth dies in four ways (all rooted in denialism and all seen in climate denialism machine):

1 The open hostility to verifiable reality, which takes the form of presenting inventions and lies as if they were facts and the creation of a fictional counterworld;
2 Shamanistic incantation, depending on 'endless repetition' designed to make the fictional plausible and the criminal desirable;
3 Magical thinking, or the open embrace of contradiction, which requires a blatant abandonment of reason; and
4 Misplaced faith, seen as the strongman or savior that alone can fix the problems of the world; and once truth becomes oracular rather than factual, evidence is irrelevant (2017: 66–68, 71; see also Bohr 2021, "cast yourself as the savior of traditional American life").

Václav Havel (1936–2011) was the last president of Czechoslovakia, the first democratically elected President of the Czech Republic, and a dissident to the communist regime that led to several years in prison. He lived authoritarianism and then led a democratic nation. Havel believed that, "A world where 'truth' flourishes not in a dialectic climate of genuine knowledge but in a climate of power interests is a world of mental sterility, petrified dogmas, rigid and unchangeable creeds leading inevitable to creedless despotism" (Wilson 1992: 64). There is no path to democracy for those who seek to destroy truth.

And there is no path to freedom. For all the false cries of 'freedom!' by the denialist machine, and amidst the appeals to white Christian nationalism, it must be understood that freedom is only possible in a governance that ensures political equality,

> Political equality is democracy's foundation... [It] alone ensures that the composition and exercise of political power is authorized by the whole and accountable to the whole. When political equality is absent, whether from whether from explicit political exclusions or privileges, from extreme social or economic disparities, *from uneven or managed access to knowledge*, or from manipulation of the electoral system, political power will inevitably be exercised by and for a part, rather than the whole.
>
> *The demos ceases to rule (Brown 2019: 23, emphasis added; see also Arendt 2022: 32)*

To Margaret Thatcher and other 'free market' fundamentalists, there is no such thing as a society[20] (which allows them to ignore systemic harms), and the state (e.g., federal taxation and regulation) and the majority (often pluralistic) will take your freedom. What they fail to understand is that "Freedom without society destroys the lexicon by which freedom is made democratic... freedom without society is a pure instrument of power, shorn of concern for others, the world, or the future" (Brown 2019: 44–45). Their freedom is not something borne of a democracy.

In Snyder's tenth lesson, "believe in truth," he warns us that "To abandon facts is to abandon freedom. If nothing is true, then no one can criticize power, because there is no basis upon which to do so. If nothing is true, then all is a spectacle" and "The biggest wallet pays for the most blinding lights" (2017: 65). Arendt supports this assertion, "Freedom of opinion is a farce unless factual information is guaranteed" (Arendt 2022: 19) and even "Plato believed that demagogues exploited free speech to install themselves as tyrants" (Snyder 2017: 9). Truth checks power; truth allows freedom of opinion and freedom of speech; and truth allows people in a democracy to debate and discuss and make reasoned decisions. And only through truth can free markets move. Otherwise, they are simply a pawn in someone else's scheme, used as a tool to make someone else's end.

As such, freedom is also an issue of autonomy. Denialists not only think that the majority cannot and should not make decisions for a society (largely because they will more often decide for the common good and against market interests),[21] but they also believe that their adherents are incapable of making the right decision, and so *they* should. By offering lies and deceit in order for people to make their decisions, they leave them incapable of making self-made decisions. This act of lying for a particular political end takes away their individual autonomy and their ability to self-govern, ultimately "[weakening] the instinct by means of which free peoples know not what is dangerous or is not dangerous" – a dangerous totalitarian tactic (Orwell 2020: 312).

Christine M. Korsgaard, the Arthur Kingsley Porter Research Professor of Philosophy at Harvard University, explores this instrumentality and the corresponding culpability of these lies. She opens with Kantian ethics, "based on a view of how human beings should treat one another that is deeply democratic," that argues that you "must never treat a human being — either yourself or another — as mere means to an end" (Korsgaard 2022). To treat someone "as a tool rather than as a rational being" violates autonomy:

> What I say to you is just whatever I think will *work* to produce the result I want. What is wrong with my lie does not rest in its consequences, but in the fact that *I have taken the decision out of your hands.*
>
> *(Korsgaard 2022, emphasis added)*

And as such, "force and coercion and lies are… the most fundamental forms of wrongdoing – the roots of all evil" (Korsgaard 2022).

But, of course, responsibility also rests upon the listener to seek truth, "You submit to tyranny when you renounce the difference between what you want to hear and what is actually the case… thus the collapse of any system that depends on individualism (Snyder 2017: 65–66). This is why "consciousness-raising" is important, "to combat the numbing effects of… denial," which would require "radical changes in media and political culture" to counter the normalization of lies by elected officials in the face of known truths and to make it difficult for people to claim ignorance (Cohen 2001: 11).

Taking away individual autonomy – taking away the freedom to self-govern – is authoritarian governance, as evidenced by an authoritarian-aiding free market fundamentalist, James Buchanan:

> How can a citizenry that would, if given the chance, re-elect Bill Clinton by a large majority, how can such a citizenry be expected to understand either the justice or efficiency of markets and the necessity of constitutional constraints on governance?
>
> *(MacClean 2017: 204)*

"People who failed to see," Buchanan argued, "are to be treated as a subordinate members of the species, akin to animals who are dependent" (MacClean 2017: 212). Dehumanization for control; dehumanization for profit – and yet another tactic of both climate denialism and authoritarianism. This "anti-govt rhetoric that continues to saturate our political life is rooted [more] in support for slavery rather than liberty" (MacClean 2017: 8). Think of all of the dehumanization that needs to occur for denialism to work: the denial of injury, the denial of a victim, the denial of responsibility. When reality is denied, or its significance, the harm is also denied, as well as its seriousness, i.e., the suffering and the level of suffering is not acknowledged (Cohen 2001: 22, 52, 60–61). This moral indifference is a denial of a human's moral legitimacy (Cohen 2001: 98).

Conclusion

"The chances of factual truth surviving the onslaught of power are very slim indeed; it is always in danger of being maneuvered out of the world not only for a time, but potentially forever." – Hannah Arendt (2022: 9)

For climate denialists, we know their plans to poison [people, planet, and democracy], pay, and deceive, and we know their weaknesses of truth, justice, and the common good, i.e., democracy. If the end aim of denialism is the concentration of power and wealth, then it is no surprise that decades of their influence and ideologies of deceit, division, racism, and greed have fractured personal, public, and geo-political relations, increased inequity of power, increased inequality, and created a lack of public trust in the integrity of democratic institutions. This is what has been purchased. But they have not won yet.

First and foremost, to counter climate denialism, we must be engaged in culturing democracy,

> Democracy… requires constant vigilance to prevent concentrated wealth from grasping the levers of political power. Wealth – corporate, consolidated, or individual – will never stop reaching for these levers, and once it has a significant hold, there is no limit to its self-serving practices... Neither markets themselves nor winners within them can be permitted to rule if democracy is to prevail…
>
> *(Brown 2019: 26)*

The state must act with intent to make democracy easier and work to deliberately reduce inequalities in power among citizens. We must demand integrity of our political institutions and money out of politics. *Citizens United v FEC* and *West Virginia v. EPA* must be overturned; voting rights legislation must be passed. Strengthening democracy strengthens the common good, which strengthens climate action.

We must use the proper language and identify the denialist for what they are, and the ideology for what it is – self-interested agitators who view democracy as a threat to their personal, unequal gain. And we must identify that they create and foment malaise – distrust, broken relationships, and dehumanization – in order to advance their anti-democratic ends. We must show that their principles of 'free market' and 'freedom' are ruses and refuse to let an alternate reality rule us. We will not be sold authoritarianism under the guise of democracy.

We must defend truth. Like democracy, truth is an action,

> Where everything lies about everything of importance, the truth-teller, whether he knows it or not, has begun to act; he, too, has engaged in political business, for, in the unlikely event that he survives, he has made a start toward changing the world.
>
> *(Arendt 2022: 38)*

Arendt understood that

> No human world… will ever be able to survive without men willing to do what Herodotus was the first to undertake consciously – namely, to say what is. No permanence, no perseverance in existence, can ever be conceived of without men willing to testify to what is and appears to them because it is.
>
> *(Arendt 2022: 4–5)*

Truth checks power – which is exactly why it is targeted. When Grotius "wished to limit the absolute power of the prince," he said, "even God cannot cause two plus two to not make four" (Arendt 2022: 22). And just as two plus two makes four, climate denialism advances and is advanced by wealth that is the product of deceit, predation, exploitation, white supremacy, religious extremism, and authoritarianism.

But we must also understand that sharing truth, i.e., education, may not be enough, which is why we must keep denialists from power,

> To those in the grip of arrogant ideologies… it is unclear, *at best*, that just lobbing more facts at them is going to help at all — if by 'help' we mean 'change their minds.' To this point, we must be clear: in such cases, what matters is not changing their minds but keeping them from power.
>
> *(Lynch 2021)*

And if they do come to power, we must have the courage to not 'play along' with these actors and in their alternate reality, where legislators and justices claim to be exercising deliberative democracy and blind justice, when in reality they are merely producing what has been paid for.

In Havel's *Power of the Powerless*, he offered the story of a greengrocer living the lies and mass indifference of a totalitarian society. In finding his "suppressed identity and dignity," he stops 'playing along' with those in power,

> By breaking the rules of the game, he has disrupted the game as such. He has exposed it as a mere game. He has shattered the world of appearances, the fundamental pillar of the system… He has demonstrated that living a lie is living a lie. He has broken through the exalted façade of the system and exposed the real, base foundations of power. He has said that the emperor is naked. And because the emperor is in fact naked, something extremely dangerous has happened: by his action, [he] has addressed the world.
>
> *(Wilson 1992: 147)*

By choosing not to live a lie, he threatened and fractured the entire system, and "not because of any physical or actual power he had, but because his action went beyond itself, because it illuminated its surrounding and, of course, because of the intolerable consequences of that illumination" (Wilson 1992: 147). Havel continued,

> In the post-totalitarian system, therefore, living within the truth has more than an existential dimension (returning humanity to its inherent nature), or a noetic dimension (revealing reality as it is), or a moral dimension (setting an example of others). It also has an unambiguous political dimension. If the main pillar of the system is living a lie, then it is not surprising that the fundamental threat to it is living the truth. This is why it must be suppressed more severely than anything else.
>
> *(Wilson 1992: 147–148)*

This cleansing and enlightening light of truth also allows us to see and find other truth-tellers; it allows us to build solidarity: the key to action on climate change. Together, we must counter the normalization of for-profit denialism and deceit, "generating an anti-democratic culture from below while building and legitimating anti-democratic forms of state from above" (Brown 2019: 28), but "we must not fight lies with lies, we must not try to be just as clever as it is… we must counteract it with the full force of reason, with the genuine ideological truth" (Adorno 2020: 39–40). And for those who fight climate denialists, that genuine ideological truth is democracy – and a democracy that has the courage and capacity to act upon climate change.

Notes

1 For a discussion on the sociological underpinnings of climate paralysis, see Norgaard (2011).
2 A political pejorative used by left-wing and right-wing ideologists to describe an unwitting accomplice. It does not mean that they are 'stupid,' but rather that they are ignorant of the end aims for which they are being manipulated. See also https://www.merriam-webster.com/dictionary/useful%20idiot.

3 A list of Koch-funded climate denialist groups can be found at https://www.greenpeace.org/usa/fighting-climate-chaos/climate-deniers/front-groups/, including but not limited to Acton Institute for the Study of Religion and Liberty, American Council on Science and Health (ACSH), Americans for Prosperity Foundation, Center for Freedom & Prosperity Foundation, FreedomWorks, Frontiers of Freedom, Independence Institute, James Madison Institute (JMI), John Locke Foundation (JLF), and Students for Liberty. Additional entities are also listed in the amicus brief submitted by Senator Whitehouse et al for *West Virginia v. EPA* (2022); see also Mayer (2016). For conservative organizations in the US, it is a common argument that one can know what they seek to harm by simply looking at their name; for example, the Federalist Society advances principles of anti-federalism, state's rights, and individual liberty (and with their moniker, James Madison – a vision which would "horrify Madison" (MacClean xxxiv)).

4 At the recent COP27 in Egypt (November 2022), Jennie King, a representative from the Institute of Strategic Dialogue, talked with reporters after releasing a report about the impacts of climate disinformation on public perception, "We are now seeing out-and-out denial is making an absolute comeback" (Collins 2022). The article suggests, "The increase in climate denialism can be linked to a wider embrace of conspiracy theories that includes false ideas about everything from Russia's war in Ukraine to the COVID-19 vaccines. Misinformation around politics and the widely debunked notion of election fraud have sowed mistrust in democracy in the US. That willingness to ignore facts and science has meant more audiences are receptive to dismissing the notion of a climate crisis." Many climate change advocates are also concerned about the increase of disinformation on Twitter, a platform that reaches millions, because of the hostile takeover by Elon Musk, a self-professed 'free speech' advocate; see "#ClimateScam: denialism claims flooding Twitter have scientists worried" (Milman 2022).

5 US Senator James Inhofe, a Republican from Oklahoma, brought a snowball to the Senate floor to deny the existence of global warming. US Senator Sheldon Whitehouse responded by referencing NASA and climate scientists, but it is often the snowball that is remembered, showing the power of a denialist's tactic when they use mockery. Inhofe also famously remarked that "global warming is the greatest hoax that has been perpetrated" (Barrett 2015).

6 US Senator Tom Cotton, a Republican from Arkansas (Cranley 2019).

7 Interview with Oprah Winfrey (1997), Oprah Winfrey Network; video available at https://www.youtube.com/watch?v=nJgmaHkcFP8.

8 See, "Tucker Carlson's Manufactured America" (Garber 2021). For his most recent act of climate denialism, see Dicker (2022), "Tucker Carlson Says Global Warming Is Not A Threat But Winter Is."

9 A reference to the Wizard of Oz film in 1939 that refers to deception: the 'wizard' is simply a man hiding behind a curtain, creating the illusion of 'a great and powerful Oz.' From the book, *The Wonderful Wizard of Oz*, "Oh he will see you… although he does not like to have people ask to see him" (Baume 1900: 126).

10 In 2010, Greenpeace named Koch Industries "the kingpin of climate denial." See full report, Greenpeace (2010) *Koch Industries: Secretly Funding the Climate Denial Machine*. Available at https://www.greenpeace.org/usa/research/koch-industries-secretly-fund/.

11 See infra note 15.

12 See Svoboda (2019).

13 E.g., Acton Institute for the Study of Religion and Liberty; see also Schlanger (2017); O'Connor (2017).

14 See Slevin (2022), "…perhaps no academic institution is as potent and dangerous a spreader of climate change disinformation as George Mason University in Fairfax, VA. George Mason hosts a pair of research centers sponsored by Climate Change Counter Movement donors: Institute for Humane Studies and Mercatus Center… A former Koch Industries employee described Mercatus as 'a lobbying group disguised as a disinterested academic program,' in a leaked internal report."

15 Fink also co-founded Citizens for a Sound Economy Foundation, now re-branded as Americans for Prosperity Foundation, a climate denialist 'Astro-turf' group, i.e., fake grassroots organization (see Mayer 2016; Michaels 2020).

16 See Bourke (2021), "IUCN's Al Mubarak: the conservation leader who says we can work with oil majors. While campaigners call for a ban on fossil fuel sponsorship, the head of the planet's biggest nature network takes a different stance."

17 See Mayer (2016: 4), "They said they were driven by principle, but their positions dovetailed seamlessly with their personal financial interests."

18 Hungarian authoritarian ruler Viktor Orbán was asked to explain the secrets of his success. He said, "it was important to not have to share power with other parties… Also, it helps to have the support of the media" (Applebaum 2020: 141).

19 According to Cohen (2001: 168), "The mass media have a near monopoly in creating the cultural imagery of suffering and atrocities," and it does not help address climate change harms when denialists intentionally target them. For example, "Documents uncovered by journalists and activists over the past decade lay out a clear strategy: First, target media outlets to get them to report more on the 'uncertainties' in climate science, and position industry-backed contrarian scientists as expert sources for media. Second, target conservatives with the message that climate change is a liberal hoax, and paint anyone who takes the issue seriously as 'out of touch with reality'" (Westervelt 2019). This leads to the media becoming "unwitting accomplices. By perpetuating the notion that there are two equal 'sides' when it comes to objective matters of fact such as evolution and climate change, mainstream-media outlets have reinforced the perception that there is legitimate doubt about these matters" (Mann & Toles 2016: 9).

20 Margaret Thatcher was the Prime Minister of the United Kingdom 1979–1990 and is an icon for neoliberal ideology. In an interview she gave in 1987, she attacked those made vulnerable by government policy, painting them as people who seek to have the government solve their problems, "'I'm homeless, the government must house me.' They are casting their problems on society. And, you know, *there's no such thing as society*. There are individual men and women and there are families. And no government can do anything except through people, and people must look after themselves first." See also, Samuel Britton's Opinion in the Financial Times, "Thatcher was right – there is no 'society,'" trying to re-write free market fundamentalism as "methodological individualism" against government "free-riders" (2013). A core of this ideology is dividing society into the makers and the takers, i.e., opportunistic free-riders who try to take from the government (and by extension, the wealthy who fund the government). This is also the foundation of a decades-old, racist, sexist conservative mantra in the United States, the 'welfare Queen' (see Sparks 2003).

21 "Unless they are tricked, trained, or effectively disenfranchised, the workers and poor will always fight markets as unfair in their distribution of opportunities and rewards. This class can be tricked… with appeals to white masculinity, especially since liberty, rather than equality, secures those powers. It can be trained… that there is no alternative (TINA)," and neoliberalism is "unchallengeable;" "It can be disenfranchised through voter suppression, gerrymandering, bought elections and legislation, and other ways of insulating legislative power from democratic will or accountability" (Brown 2019: 63–64). Brown sees white supremacy as a direct attack on democracy, "Its subject abhors the democracy it holds responsible for its wounds and seeks to pull democracy down as it goes down. If white men cannot own democracy, there will be no democracy" (2019: 180).

References

Adorno, T. (2020) *Aspects of the New Right-Wing Extremism*, New York: Polity.

Applebaum, A. (2020) *Twilight of Democracy: The Seductive Lure of Authoritarianism*, New York: Doubleday.

Arendt, H. (1966) *The Origins of Totalitarianism*, New York: Houghton Mifflin.

Arendt, H. (2022) *On Lying and Politics*, New York: Library of America.

Barrett, T. (2015) "Inhofe brings snowball on Senate floor as evidence globe is not warming," *CNN*. Available at https://www.cnn.com/2015/02/26/politics/james-inhofe-snowball-climate-change/index.html.

Baume, L.F. (1900) *The Wonderful Wizard of Oz*, Chicago, IL: G.M. Hill Co. Available via the US Library of Congress at https://www.read.gov/books/oz.html.

Bohr, J. (2021) "The structure and culture of climate change denial," *Footnotes: A Magazine of the American Sociological Association*. Available at https://www.asanet.org/footnotes-article/structure-and-culture-climate-change-denial/.

Bourke, I. (2021) "IUCN's Al Mubarak: The conservation leader who says we can work with oil majors," *The New Statesman*. Available at https://www.newstatesman.com/environment/biodiversity/2021/10/iucns-al-mubarak-the-conservation-leader-who-says-we-can-work-with-oil-majors.

Britton, S. (2013) "Thatcher was right – there is no 'society,'" *Financial Times*. Available at https://www.ft.com/content/d1387b70-a5d5-11e2-9b77-00144feabdc0.

Brown, W. (2019) *In the Ruins of Neoliberalism: The Rise of Antidemocratic Politics in the West*, New York: Columbia University Press.

Citizens United v. Federal Election Commission [FEC]. (2010) US Supreme Court, 558 U.S. 310. Justice John Paul Stevens dissent available at https://www.law.cornell.edu/supct/html/08-205.ZX.html.

Cohen, S. (2001) *States of Denial: Knowing About Atrocities and Suffering*, Cambridge: Polity Press.

Collins, K. (2022) "Climate change denial makes an untimely comeback: The science about the climate crisis is unequivocal, but a recent report shows a rise in misinformation about the topic," *CNET*. Available at https://www.cnet.com/science/climate/climate-change-denial-makes-untimely-comeback/.

Cranley, E. (2019) "These are the 130 current members of congress who have doubted or denied climate change," *Business Insider.* Available at https://www.businessinsider.com/climate-change-and-republicans-congress-global-warming-2019-2.

Dicker, R. (2022) "Tucker Carlson says global warming is not a threat but winter is," *Huffington Post.* Available at https://news.yahoo.com/tucker-carlson-says-global-warming-114440143.html.

Garber, M. (2021) "Tucker Carlson's manufactured America: The Fox host has a new daytime show, and he's using it to poison the meaning of patriotism," *The Atlantic.* Available at https://www.theatlantic.com/culture/archive/2021/07/tucker-carlson-today-americana-fake-log-cabin/619411/.

Greenpeace. (2010) *Koch Industries: Secretly Funding the Climate Denial Machine.* Available at https://www.greenpeace.org/usa/research/koch-industries-secretly-fund/.

Korsgaard, C. (2022) "It's for your own good: Is there moral justification for paternalistic lies?" *Religion and Ethics, ABC News.* Available at https://www.abc.net.au/religion/christine-korsgaard-is-there-moral-justification-for-paternalis/11950108.

Lowenthal, L. and Guterman, N. (1949) "Prophets of deceit: A study of the techniques of the American agitator," in M. Horkheimer and S. Flowerman (eds.) *Studies in Prejudice*, New York: Harper & Brothers.

Lynch, M. (2021) "The value of truth," *Boston Review.* Available at https://www.bostonreview.net/articles/michael-patrick-lynch-epistemology-tk/.

MacClean, N. (2017) *Democracy in Chains: The Deep History of the Radical Right's Stealth Plan for America*, New York: Penguin.

Mann, M. and Toles, T. (2016) *The Madhouse Effect How Climate Change Denial Is Threatening Our Planet, Destroying Our Politics, and Driving Us Crazy*, New York: Columbia University Press, Kindle edition.

Martinez-Conde, S. (2019) "I heard it before, so it must be true," *Scientific American.* Available at https://blogs.scientificamerican.com/illusion-chasers/i-heard-it-before-so-it-must-be-true/.

Maslin, M. (2019) "The five corrupt pillars of climate change denial," *The Conversation.* Available at https://theconversation.com/the-five-corrupt-pillars-of-climate-change-denial-122893.

Mayer, J. (2016) *Dark Money: The Hidden History of the Billionaires Behind the Rise of the Radical Right*, New York: Doubleday.

Michaels, D. (2020) *The Triumph of Doubt: Dark Money and the Science of Deception*, New York: Oxford University Press.

Milman, O. (2022) "#ClimateScam: denialism claims flooding Twitter have scientists worried: Many researchers are fleeing the platform, unnerved by the surge in climate misinformation since Musk's chaotic takeover," *The Guardian.* Available at https://www.theguardian.com/technology/2022/dec/02/climate-change-denialism-flooding-twitter-scientists.

Norgaard, K. (2011) *Living in Denial: Climate Change, Emotions, and Everyday Life,* Boston: MIT Press.

O'Connor, B. (2017) "How fossil fuel money made climate change denial the word of god," *Splinter.* Available at https://splinternews.com/how-fossil-fuel-money-made-climate-denial-the-word-of-g-1797466298.

Oreskes, N. and Conway, E. (2011, Forward 2019) *Merchants of Doubt: How a Handful of Scientists Obscured the Truth on Issues from Tobacco Smoke to Climate Change*, New York: Bloomsbury Publishing.

Orwell, G. (2020) *Orwell and Politics*, London: Penguin House.

Reagan, R. (1981) *Inaugural Address, US President Ronald Reagan.* Available at https://www.reaganlibrary.gov/archives/speech/inaugural-address-1981.

Roy, Arundhati. (2014) *Capitalism: A Ghost Story*, Chicago, IL: Haymarket Books.

Sartre, Jean-Paul. (1948) *Anti-Semite and Jew*, New York: Schocken Books.

Schlanger, Z. (2017) "The 'scientists' who support Trump's choice to run the EPA are creationists with opaque funding sources," *Quartz.* Available at https://qz.com/888424/a-climate-denying-evangelical-group-connected-to-the-koch-brothers-supports-trumps-choice-to-run-the-epa.

Snyder, T. (2017) *On Tyranny: Twenty Lessons from the Twentieth Century*, New York: Tim Duggan Books.

Slevin, I. (2022) "Universities are enabling climate change denial. Yes, Brown too," *Brown Political Review.* Available at https://brownpoliticalreview.org/2022/03/universities-enable-climate-change-denial/.

Sparks, H. (2003) "Queens, teens, and model mothers: Race, gender, and the discourse of welfare reform," in S. Schram, et al. (ed.) *Race and the Politics of Welfare Reform*, Ann Arbor: University of Michigan Press.

Svoboda, M. (2019) "New Koch-funded fossil exhibit at the Smithsonian is curiously quiet on fossil fuels," *Yale Climate Connections.* Available at https://yaleclimateconnections.org/2019/07/new-koch-funded-fossil-exhibit-at-the-smithsonian-is-curiously-quiet-on-fossil-fuels/.

Thatcher, M. (1987) "Interview for woman's own ('No such thing as society')," *Margaret Thatcher Foundation: Speeches, Interviews and Other Statements.* Available at https://newlearningonline.com/new-learning/chapter-4/neoliberalism-more-recent-times/margaret-thatcher-theres-no-such-thing-as-society.

US Department of Justice [USDOJ]. (2000) *US Environmental Protection Agency: Koch Industries to Pay Record Fine for Oil Spills in Six States*. Available at https://www.justice.gov/archive/opa/pr/2000/January/019enrd.htm.

USDOJ. (2001) *Koch Pleads Guilty to Covering Up Environmental Violations at Texas Oil Refinery*. Available at https://www.justice.gov/archive/opa/pr/2001/April/153enrd.htm.

Westervelt, A. (2019) "How the fossil fuel industry got the media to think climate change was debatable," *Washington Post*. Available at https://www.washingtonpost.com/outlook/2019/01/10/how-fossil-fuel-industry-got-media-think-climate-change-was-debatable/ [https://perma.cc/8AUS-NN68].

West Virginia v. Environmental Protection Agency [EPA] (2022) 597 US ___. Available at https://www.law.cornell.edu/supremecourt/text/20-1530.Whitehouse. (2022) *The Scheme: How the Right Wing Used Dark Money to Capture the Supreme Court*, New York: The New Press.

Whitehouse, S., Blumenthal, R., Sanders, B. and Warren, E. (2022) *No. 20-1530, Brief as Amici Curiae in Support of Respondents, West Virginia v. EPA*. Available at https://www.whitehouse.senate.gov/imo/media/doc/WV%20v.%20EPA%20Amicus.pdf.

Wilson, P. (ed.) (1992) *Open Letters: Selective Writings 1965-1990 by Václav Havel*, New York: Vintage Books.

PART IV

Ethical Issues Raised by Various Policy Responses to Climate Change

23

ETHICAL ISSUES RAISED BY CAP-AND-TRADE REGIMES

Nicholas S. Bryner

Introduction

The first wave of modern environmental law was built in the 1970s around the use of prescriptive regulatory controls to limit environmental pollution. In the United States, these 'command-and-control' policies include, for example, centrally mandated pollution control technology in the Clean Air Act and Clean Water Act, as well as specific regulatory requirements for the handling and disposal of hazardous waste (Lazarus 2004).

In the 1980s, as a response to these early laws, economists began advocating for 'market-based' environmental regulation that would, in theory, provide greater flexibility to regulated industries and promote greater economic efficiency in achieving environmental policy goals (Anderson & Leal 2001). Such market-based tools encompass both the use of taxes or economic subsidies and the creation of commodity-like, regulated markets for environmental goods or pollution emissions. One widespread market-based policy tool is the formation of 'cap-and-trade' systems. In such a system, a governmental authority sets a quantitative cap on emissions of a particular pollutant. So long as the system-wide emissions remain below the cap, individual firms may trade allowances or credits that correspond to quantities of pollutant emissions. Beginning in the early 2000s, policymakers at subnational, national, and international levels have adapted the cap-and-trade concept for use in climate mitigation by regulating the emission of greenhouse gases (GHGs) – in particular, carbon dioxide (CO_2) (Bryner 2008).

This chapter identifies and addresses three dimensions of ethical issues raised by the design and use of cap-and-trade regimes in the climate context. First, cap-and-trade policies create an environmental ethics concern by establishing 'allowances' for pollution. Giving actors a license or right to emit GHGs that can be earned, purchased, or sold as a commodity presents the issue of whether and under what circumstances polluting activity can or ought to be legitimized and legally recognized. Second, the design of cap-and-trade programs raises equity issues based on where and how the policies apply. Trading pollution credits has led to distributional justice concerns about the creation of windfall benefits from cap-and-trade systems and the placement of inequitable burdens on communities located near emissions sources that choose to purchase credits or allowances rather than limiting their local emissions. Third, cap-and-trade regimes raise ethical concerns due to the separation of 'climate' impacts from other environmental and public health externalities associated with fossil fuel combustion and from other ecological values and non-climate ecosystem services.

DOI: 10.4324/9781003039860-28

The chapter proceeds with a brief explanation of cap-and-trade regimes and the history of their use for conventional air pollution regulation and climate mitigation. It then turns to address the three types of ethical issues mentioned above, illustrated through discussion of the state of California's cap-and-trade program, which was established in 2006 and, more recently, reauthorized in 2017 despite severe criticism from environmental justice advocates. The chapter then provides a critique of cap-and-trade on its own terms, questioning the efficiency of these policies due to the increased transaction costs and complexity of establishing and regulating cap-setting, allocation, and trading regimes. Contextualization of cap-and-trade policies, among other climate mitigation tools, can suggest solutions to respond to these significant ethical issues.

A Brief Explanation and History of Cap-and-Trade Regimes

Cap-and-Trade Economics

A cap-and-trade regime for controlling pollution emissions, as the name suggests, includes two components: (1) the setting of a market-wide 'cap' on pollution; and (2) the creation of a trading market in which regulated entities can exchange pollution emission allowances or credits. Proponents of cap-and-trade regimes point to advantages in both of these processes.

First, the idea of setting a cap on pollutant emissions can convey a clear and certain emissions target in a way that other pollution control policies, such as taxes or technology-based mandates, do not. Regulators can phase in cap-and-trade programs by beginning with a high cap, based on historical emissions, with gradual reductions coming into place each year toward a long-term goal.[1] If scientific modeling suggests a desirable pollution target (for example, the number of tons of sulfur dioxide emissions in a region that is acceptable to minimize environmental damage or health impacts), setting a cap prioritizes that quantitative target. In an effective cap-and-trade program, compliance with that target will ultimately dictate the practical impact of other policy elements. The trading price of emissions allowances and the types of pollution control measures or technologies employed by regulated entities will be dependent on the stringency of the cap.

Second, cap-and-trade proponents view the trading component of the program as a tool for maximizing economic efficiency. When pollution allowances are tradeable, regulated parties may choose among various compliance options: paying for additional allowances, changing behavior to reduce emissions, developing and employing new control technologies, or some combination of these actions. Some actors will be able to more cheaply change their behavior to reduce emissions than others. Tradability of credits incentivizes these actors to expend resources in competition and reduce emissions more than other policies would require, in order to sell credits or allowances to less efficient entities in the market. The effect, in theory, is to reduce the overall system-wide compliance costs in reducing pollution emissions while setting up a market that distributes those costs efficiently (Driesen 2016).

Figure 23.1 illustrates a simplified example of how a cap-and-trade policy can affect the distribution of pollution emissions. Suppose that Factory A and Factory B each historically emit 150 tons of SO_2 per year, for a system-wide total of 300 tons. A regulator imposes a new system-wide cap of 200 tons per year. Under a command-and-control policy, each entity would be required to cut emissions to 100 tons per year. Suppose that it costs Factory A $10,000 to cut emissions by each 10-ton increment, while for Factory B, due to technical or other differences, each 10-ton reduction costs $20,000. Under the command-and-control regime, A spends $50,000 to comply and B spends $100,000, for a total regulatory cost of $150,000.

Under a cap-and-trade regime, the regulator sets an overall cap of 200 tons and allocates 100 allowances (each allowance corresponding to one ton of permitted SO_2 emissions) to Factory A and to Factory B. Factory A can spend $100,000 to cut its emissions by 100 tons below its

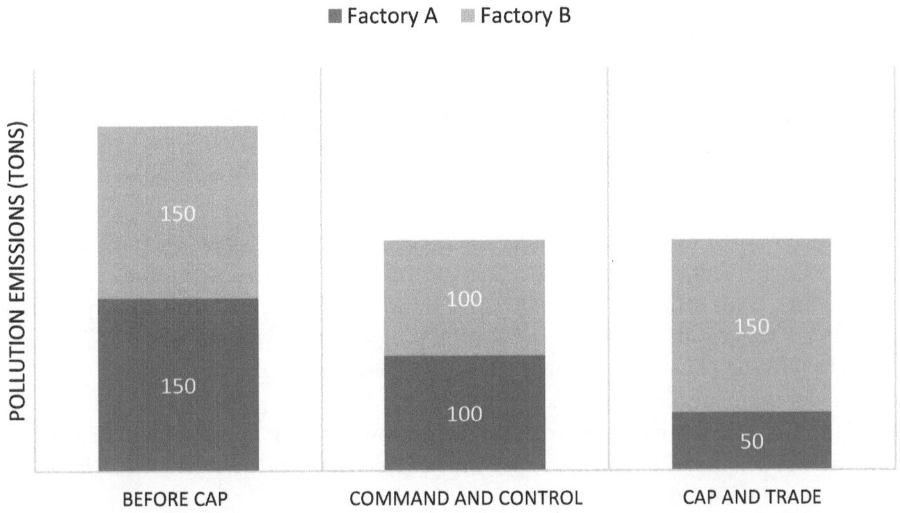

Figure 23.1 Simplified example of pollution emissions under a command-and-control policy vs. cap-and-trade system

historical level of 150, down to a total of 50 tons. Factory A now has 50 extra, unused 'allowances' that it can sell to B. Factory B purchases those allowances or credits from A for some negotiated price, allowing B to continue to emit 150 tons. The total regulatory cost is now $100,000, split between the two parties as the market for allowances dictates.

The regulatory body must make several decisions in addition to setting the cap level in order to run a cap-and-trade system (World Bank 2016). One of the key issues in establishing a functioning program is the initial allocation of allowances under the cap. In a market-oriented system, the government regulator conducts a competitive auction, in which regulated entities bid on the price they are willing to pay for an emissions allowance. In most existing regimes, the government grants at least some allowances for free based on historical emissions – an anti-competitive measure that eases the burden on existing industries in the transition to a new regulatory regime but that disadvantages potential new entrants to the market and those industries that have already taken efficiency measures to reduce their emissions in the past. Establishment of the rules, regulations, and policies for allowances, as well as the terms, policies, and guidelines for trading those allowances, unavoidably involve political decisions that affect industrial and environmental constituencies.

Under this cap-and-trade system, Factory A can reduce its emissions more efficiently and can then sell allowances or credits to Factory B, resulting in lower system-wide compliance costs. Over time, the regulator may tighten (lower) the cap to reach long-term policy goals.

The Development of Cap-and-Trade: The Acid Rain Program

Cap-and-trade programs today are frequently associated with policy discussions about climate change mitigation. However, the idea for these types of policy regimes predates climate regulation. As a response to what some regulated industries and stakeholders saw as inefficiently high costs associated with traditional environmental regulation, environmental policymakers began drawing on economists' advocacy for market-oriented pollution control policies.

In the United States, Congress was in the process of updating the Clean Air Act in the late 1980s in order to address (among other things) the problem of acid rain pollution in the northeast portion of the country (Bryner 1993). The most significant driver of acid rain in the area is the emission of SO_2 from coal-fired power plants in the Midwest and northeast United States. Electric utilities advocated for greater flexibility in any new statute that would impose restrictions on SO_2 emissions. As part of a compromise approach, the new statute required power plants to install continuous emissions monitoring systems – ensuring accurate reporting and compliance – while also providing for flexible compliance under a cap-and-trade system (Clean Air Act, 42 U.S. Code § 7651k).

The result, the 1990 Clean Air Act Amendments, created the first regional cap-and-trade program for environmental pollution. The statute called for implementing the program in two phases: in the first, beginning in 1995, only a small number of power plants were required to comply with the cap; other sources could opt into the program on a voluntary basis (Clean Air Act, 42 U.S. Code § 7651c). The second phase expanded the number of sources in the cap and began a gradual drawdown of available allowances toward the statute's identified goal by 2010 (42 U.S. Code § 7651d). In general, existing sources received allowances for free based on historical energy output and an expected rate of emissions per unit of electricity generated. A small proportion of allowances under the cap are sold via auction.

Many economists, legal scholars, and other commentators have hailed the Clean Air Act Acid Rain Program as a significant policy success. From 1980 to 2019, the United States reduced nationwide SO_2 emissions by 92 percent (U.S. EPA 2020), with most of the reductions coming after implementation of the trading program – and at a fraction of the projected cost. Compliance has been nearly universal, due to accurate reporting and high penalties for violations (initially $2,000 per ton of SO_2 emissions, now indexed for inflation) (40 Code of Federal Regulations § 77.6). Others have questioned the extent to which the cap-and-trade program actually reduced costs; rather than install control equipment at many locations, utilities opted to switch to lower sulfur coal, changing the economics (and the social impacts in mining communities) of power plant operations.

Cap-and-Trade as Climate Policy

The concept of emissions trading in climate policy dates to early days of international climate negotiation in the 1990s. Early on the negotiation of the UN Framework Convention of Climate Change, several countries discussed the possibility of creating an international climate regime with tradeable emissions rights (Bodansky 1993: 522–523). The Convention itself, though, lacked quantitative commitments or targets, leaving eventual questions about implementation for future meetings.

The Kyoto Protocol to the UNFCCC, adopted in 1997, established the first binding GHG emissions targets in international law. The Protocol set targets only for the developed countries and 'economies in transition,' listed together in Annex I of the UNFCCC, at 5 percent below those countries' aggregate 1990 GHG emissions levels during a commitment period of 2008–2012 (Kyoto Protocol Art. 3, Annex B). The Kyoto Protocol does not explicitly use the language 'cap-and-trade' and does not create a centralized trading marketplace for allowances or credits. However, the Protocol explicitly endorses trading among countries with binding targets and lays out the framework for the so-called 'flexible mechanisms,' including the Clean Development Mechanism (CDM) for cooperation with developing countries (Kyoto Protocol Arts. 12, 17).

At the global level, subsequent discussions on emission trading regimes have been hindered both by the failure to create further overall binding targets for GHG emissions and by the political difficulties of setting up programs for facilitating transfers of resources from one group of countries to another. However, several regional, national, and subnational governments began setting up cap-and-trade programs for compliance with their own climate mitigation policies.

Building on the success of the United States' acid rain program, European policymakers turned toward cap-and-trade as an option for regional climate action. In 2005, the European Union set up the EU Emission Trading System (ETS), which has become the continent's major policy piece in implementing European nations' Kyoto Protocol commitments and a significant part of the EU's contributions under the 2015 Paris Agreement (European Union 2015).

In the United States, nationwide efforts to mandate GHG emissions reductions have floundered. In 2009, the U.S. House of Representatives passed a nationwide cap-and-trade bill to regulate CO_2, but the bill was never voted on in the Senate. Republican Party officials have opposed any binding action of any kind to reduce CO_2, and among Democrats, there has been no serious push for traditional command-and-control regulation. The new Democratic presidential administration has taken an ambitious approach to executive-branch action on climate (Grandoni & Eilperin 2021), but legislative action remains a much more difficult task.

Instead, innovation on U.S. climate policy has primarily taken place at the regional, state, and local levels. Two major cap-and-trade initiatives have developed – the Regional Greenhouse Gas Initiative (RGGI) in the northeastern United States and the State of California's cap-and-trade program under the Global Warming Solutions Act of 2006, commonly known as AB 32. California's statute created a target for the state to return to 1990 GHG emissions levels by 2020, delegating authority to the California Air Resources Board (CARB) to implement a cap-and-trade policy (CARB 2018).

Commodifying the Climate: Cap-and-Trade Ethics

The use of a cap-and-trade system to regulate GHG emissions raises a host of ethical considerations. The central ethical critique of market-based environmental regulations is that they, by design, equate environmental damage with economic value. Economic theory of environmental regulation – with a goal of optimizing welfare – is primarily concerned with maximizing efficiency. In contrast, an ethics-based approach to environmental regulation is centered on a duty to avoid harm to others (Brown 2008). Here, we can explore at least three ways in which cap-and-trade systems in the climate context must grapple with these ethical issues.

The Ethics of a 'Right to Pollute'

Cap-and-trade systems operate by creating a market for pollution 'allowances.' In a climate context, one allowance typically represents one ton of carbon dioxide or equivalent measure of global warming. At the end of a compliance period, each regulated entity must hold an equivalent number of allowances to the tons of carbon dioxide emitted. In the cap-and-trade marketplace, allowances may be granted by the regulator, purchased in an auction, and bought or sold between private actors. The allowance therefore takes on some of the characteristics of a property right: a 'right' to emit a certain amount of pollution.

By establishing an allowance-based system, the cap-and-trade policy converts a public good – the stability of the climate and the equilibrium of climate-forcing gases into the atmosphere – into a commodity. This can change societal perceptions about the value of environmental protection and the culpability of environmentally harmful activities. According to Sandel, "turning pollution into a commodity to be bought and sold removes the moral stigma that is properly associated with it" (Sandel 1997).

In a tradeable market, a large facility that emits GHGs at a high rate can comply with climate regulation without undertaking any changes on site; by simply purchasing allowances, polluting entities maintain the legal ability to continue with business as usual. From the perspective of atmospheric science, there is no functional difference between one molecule of CO_2 emitted in

Los Angeles and another in São Paulo. Any curtailment of GHGs is beneficial, and economic efficiency is maximized when the cap incentivizes least-cost reductions. However, the ethics of paying a penance in allowances for pollution are not neutral. In an attempt to maintain a social license to operate, enterprises can tout their green credentials as climate-conscious or carbon-neutral companies for a monetary price without making any on-the-ground changes. In this way, cap-and-trade "provides a profit-making means for complying with a preexisting ethical duty not to pollute" (Burkett 2008).

To be sure, traditional (command-and-control) environmental regulation can also confer a gloss of legitimacy on polluting activity. A permit-based system also bestows a right-like license or authorization to pollute at a certain level deemed legally acceptable. However, traditional regulation in the form of standards and mandates applied to all sources implicitly conveys a sense of shared responsibility for addressing the problem of climate change. An ethical approach to climate change mitigation recognizes that GHG emissions are undesirable because of the harm they cause to others, "requir[ing] those who are responsible for human-induced climate change harms to comply with their duty to not harm others without regard to the economic value of costs and benefits of climate change policy responses" (Brown 2008).

Further, some climate harms cannot be easily quantified. One reason for this is uncertainty: i.e., scientific uncertainty in the degree of warming that will occur and in the degree of sea level rise. Second is the difficulty of quantifying non-economic impacts, including, for example, the loss of biodiversity and the socio-cultural impacts on societies from forced displacement or climate-induced conflicts. Market trading of GHG emissions implies that these harms can be reduced to economic value, obscuring other negative impacts from a changing climate.

Distributional Justice and Environmental 'Hot Spots'

A second major ethical concern in cap-and-trade systems is the issue of distributional justice in experiencing environmental harms. The environmental justice movement has drawn attention to the discriminatory impacts of pollution. In the United States, polluting activities are disproportionately located in African-American and other minority communities (Bullard 1997; Cole & Foster 2000). In other places around the world, sources of pollution are also concentrated in less desirable locations, where poor or marginalized communities live.

Most of the largest sources of GHG emissions – power plants, oil refineries, automobiles, etc. – also emit conventional air pollutants that negatively impact public health, such as nitrous oxide and particulate matter. This fact means that mitigating GHG emissions has considerable 'co-benefits': that is, reducing climate change by driving less and burning less coal also has the simultaneous benefit of reducing nearby communities' exposure to harmful air pollution. Depending on how the costs of climate change are calculated, the co-benefits from closing a power plant for climate purposes may be even more significant in terms of premature deaths and respiratory diseases avoided than the expected climate benefits. For example, the US EPA's Clean Power Plan, put in place by the Obama Administration (but subsequently challenged in court and repealed by the Trump Administration) to regulated coal-fired power plants, estimated a range for health-related co-benefits from $12 to $28 billion per year by 2030, on the same order of magnitude as the regulation's estimated climate benefits of $20 billion per year (US EPA 2015: 64681).

The notion that climate regulation can also help alleviate other longstanding environmental problems is positive; it incentivizes collaboration between community advocates and those concerned about climate change from a broader perspective. However, when market-based regulation specifically targets climate change and GHG emissions, the de-coupling of these related pollution concerns can cause devastating distributional impacts.

A climate-focused cap-and-trade system can create a misconception that climate is the only harm or is of a greater priority than other pollution problems. The economic incentives of trading under a cap have the effect of privileging climate concerns over public health harms from conventional air pollution. As Kaswan notes,

> Market-based systems are designed to achieve aggregate reductions; they are distribution neutral. The efficiency of pollution reduction is key; spatial distribution is not. In the world of environmental justice, in contrast, distributive justice is key: the issue is who suffers the impacts of pollution. Economic efficiency is less important than distributional fairness.
>
> *(Kaswan 2008: 10294)*

The ethical consideration in whether and how to use a cap-and-trade system is one of fairness and distributive justice. An ethical approach will focus on avoiding or minimizing harm. Although the harm of climate change is of significant weight, the fact that it is intertwined with deep-rooted social, cultural, and racial inequities means that all climate action might not necessarily be equal. While market-based regulation takes a neutral approach –seeking the lowest-cost GHG reductions, regardless of location – the result may be to reinforce existing systemic inequities.

Environmental justice advocates contend that when GHG sources pay for compliance, rather than make operational changes, this can lead to the creation of conventional pollution 'hot spots,' although there is debate among legal and economic scholars about the extent to which this phenomenon occurs in practice. Firms' decisions about how to generate, buy, or sell GHG emission allowances are presumably economic. The same factors that lead one facility operator to purchase allowances might not extend to another in a nearby geographic area.

At the beginning of California's cap-and-trade experiment, Kaswan concluded that "the hot spot risk is generally caused by the existing regulatory system, [rather than the] GHG cap-and-trade system" (Kaswan 2008: 10299). Thus, it is important to consider the ongoing role of the existing regulatory structure for conventional air pollutants and the degree to which it might be adjusted to respond to environmental justice concerns. If a facility purchases GHG allowances and *increases* emissions, the cap-and-trade system will provide no protection; however, other technology-based and command-and-control-type regulations of conventional co-pollutants will still apply. In this manner, the cap might not be solely responsible for hot spots but would allow them to continue and would provide no constraint if complementary measures on conventional air pollution are ineffective.

The political momentum of establishing a cap-and-trade presents an unusual opportunity for reform; if designed well, a cap-and-trade system can help to alleviate hot spots from climate co-pollutants, rather than exacerbate them (for example, this can be done by redistributing funds raised by auctioning allowances to improve environmental conditions or by establishing stricter standards of enforcement of both climate and conventional pollution standards). However, without input and pressure from environmental justice communities, this distributionally conscious policy changes may be unlikely to occur.

Missing the Forest for the Carbon in the Trees

A third, and related, ethical issue for cap-and-trade programs in the climate context is the use of outside-the-cap 'offsets.' A simple cap-and-trade regime allows only for trading of emissions allowance among regulated parties. An offset, on the other hand, is the generation of a credit by undertaking some emissions-reducing activity *outside* the cap, which is then used toward compliance with the cap requirements.

The use of offset mechanisms is a logical extension of the economic rationale for cap-and-trade systems. If one ton of GHG emissions has the same planetary impact regardless of where it is emitted, the actual site of emissions reductions should not matter. Offsets, then, may increase possibilities for economic efficiency.

Supporters of the use of offsets in climate cap-and-trade systems point to both the economic justification and also to potential sociocultural and ecological benefits. In the international climate treaty system, the Kyoto Protocol created a CDM that incentivized a transfer of wealth and technology toward poorer, developing countries (see above). Regulated parties in developed countries can, by financing projects that avoid GHGs (such as clean energy generation or forest preservation), acquire credits used toward Kyoto Protocol compliance (Kyoto Protocol: Art. 12).

From an equity and ethics standpoint, there are positive consequences from allowing and encouraging these kinds of investments. It can represent the fulfillment of a duty to use resources in abundance to mitigate climate change and avoid other environmental harms in vulnerable countries. However, there are also serious practical and ethical concerns about using these mechanisms as they convey a notion that 'wealthy' actors can absolve themselves from the responsibility to make changes to their harmful behavior. Purchasing carbon offsets, while done in the name of preserving the climate and maximizing co-benefits, can extend colonialist systems by undermining local control and self-determination over natural resources and by extending an ethically problematic economic model in which 'carbon' is commodified while biodiversity and other ecosystem services are undervalued.

As a practical matter, ensuring the effectiveness of an offset regime requires extensive consideration to monitoring and verification. In order to ensure that offsets represent *real* reductions in GHG emissions, the action outside the cap that gives rise to the carbon credit must be *additional*; that is, it must be some action that would not have otherwise occurred. Actions must also lead to *permanent* emissions reductions in order to have real value as offsets. Ascertaining the additionality of a climate action is notoriously difficult; the actors involved may have subjective biases about what would be likely to occur without intervention (Gillenwater 2012). Permanence imposes an obligation on those undertaking the offset activity to maintain certain conditions and requires follow-up.

Because of this difficulty, cap-and-trade regulators – as well as advocates – demand thorough verification processes and monitoring to assess both the 'baseline' before carbon credits are generated as well as the on-the-ground implementation and durability of offset projects. This need for verification and monitoring presents a dilemma for governance in developing countries and among vulnerable or marginalized communities. Carbon offsets are intended, in part, to give such communities an alternative to outside pressures from extractive industries (timber, mining, etc.) seeking to make economic use of their land. However, generating the offset and establishing a credit creates a vested interest in the land.

Countries that would like to receive investment for carbon offsets must balance that desire with concerns about the colonialist nature of the bargain. If a corporation pays a community not to cut a forest, in order to generate carbon credits, what ownership or property-like interest does the corporation now hold as a means of ensuring the permanence of the offset? Foreign-owned interests and externally-imposed monitoring or enforcement, while perhaps useful from an environmental perspective, threaten sovereignty.

For extractive industries, international law requires the free, prior, and informed consent (FPIC) of communities that own the land or resources to be exploited (ILO Convention 169). While cap-and-trade offset programs offer a valuable alternative to extractive economic development, they must also be subjected to the same consultation and consent-based standards. Otherwise, offsets fail an ethical standard to avoid doing harm to marginalized communities. Carbon offset funding can provide a meaningful counterweight against the real, violent pressures of extractive operations. However, unregulated carbon offsets undermine self-determination and autonomy, exchanging the mercantilism of resource expropriation for a carbon-sequestered colonialism.

In the context of land-use offsets (e.g., forest management), there is also an ethical concern in placing value on the carbon stocks of the forest for climate purposes in isolation from the value of biodiversity and non-climate ecosystem services. Again, as with the discussion of pollution co-benefits above, the synergy of climate mitigation and biodiversity conservation can be extremely valuable and can be self-reinforcing. However, prioritizing climate change mitigation in a vacuum can distort ethical priorities.

Case Study: Ethical Issues in the Reauthorization of California's Cap-and-Trade Program

When California's legislature enacted AB 32 in 2006, the state created a cap-and-trade system with a target of reducing statewide emissions to 1990 levels by the year 2020. Six years passed while the California Air Resources Board (CARB) established the regulatory framework for the program; CARB held the first auction of GHG emissions allowances in November 2012 and began running the trading market under a binding cap in 2013 (CARB 2018). Originally, in 2013, the cap applied only to the industrial and electricity generation sectors in the state. The cap expanded in 2015 to include transportation emissions – the largest source of GHGs (EDF n.d.).

By 2016, the state legislature and CARB began considering how to plan for the program beyond the initial 2020 target. California law required a two-thirds majority of state legislators to expand and reauthorize the state's suite of climate statutes; this required broad political support throughout the state's Democratic Party lawmakers for the cap-and-trade program design beyond 2020.[2]

In 2017, the state legislature enacted two bills as part of a joint package to extend the cap-and-trade system to 2030 and to address air quality issues in the state (AB 398 and AB 617). Throughout the process, environmental justice advocates in the state opposed many of the provisions that were ultimately enacted. Advocates had been frustrated with CARB's lack of attention to justice and equity issues in the previous decade of creating and implementing the state's cap-and-trade program. Despite progress in peaking and reducing GHG emissions, as of 2017, "[m]ore than half of the facilities covered under cap-and-trade, including 15 of refineries in the state, [were still] located in or within one-half mile of an environmental justice community" (Vanderwarker & Cuajunco 2017).

Ethics of climate change regulation were at the forefront in the reauthorization debate with regard to pollution hotspots as well as the use of offsets in the program. Legislators allied with the oil industry in the state successfully lobbied for the state law to preempt local air districts' efforts to regulate carbon dioxide (Mernit 2017). While the local authorities retain the ability to regulate conventional pollutants within their jurisdiction, advocates expressed concern that this preemption authority could lead to legal uncertainty for the local regulation of co-pollutants and could therefore undermine efforts to reduce pollution hot spots.

In addition, environmental justice advocates argued for an elimination of the state's offset program. In order to reduce local pollution, advocates emphasize that GHG emitters in the state should be obligated to take action within the state, within the cap, to address climate change – bringing along the co-benefits for local communities (Horowitz 2017). The final legislation reduced, but did not eliminate, offsets (from a maximum of 8 percent of compliance obligations to 4–6 percent (Biber 2017).

Although the cap-and-trade reauthorization is seen as an environmental success – in carries forward the program with an ambitious new target of reducing emissions by 40 percent by 2030 – the concerns raised by the environmental justice community loom large. From their perspective, the legislative debate resulted in a missed opportunity to rectify the ethical problems associated with the cap-and-trade program that still allow disproportionate impacts to poor and minority communities to continue. Advocates have continued to make their frustration with CARB and the state's implementation of climate programs known. In December 2020, longtime CARB Chair Mary Nichols was rumored to be considered by incoming President Biden for the position of EPA Administrator. A coalition of over 70 organizations in California, led by environmental justice advocates, sent a letter opposing Nichols' nomination, declaring that the cap-and-trade program had 'turned working class communities of color into environmental sacrifice zones' (California Environmental Justice Alliance 2020). Biden ultimately chose a different person for the job.

Critique of Cap-and-Trade as an Efficiency Mechanism

Environmental economists' support for market-based approaches to pollution mitigation (including climate mitigation) is based on theory that these approaches will lead to cost savings and greater economic efficiency that other forms of regulation. If cap-and-trade programs can overcome the ethical difficulties outlined in this chapter, there remains the question of whether cap-and-trade is an efficient means for achieving economic goals. If not, then an ethical approach to climate change ought to consider different regulatory systems.

The cap has one major benefit to other market-based approaches: it allows for a firm, quantitative target for pollution emissions. This helps in articulating and achieving a clear goal that can be communicated to the public and that can be tied to scientific recommendations about the likely climate effects of a given concentration of GHGs.

However, the major economic and political drawback to setting up a cap is that it creates opaque layers of governance and administration that can alienate the public. Overseeing the newly created carbon market requires establishing rules and guidelines for emissions trading, as well as regulation and oversight of prices. Many cap-and-trade regimes, including both the EU ETS and the California scheme, have struggled with fluctuating and low prices that have failed to provide an adequate incentive for firms to change behavior. Oversight and market administration add transaction costs that can undermine the efficiency of the cap regime, clawing back the economic benefits of allowing trades. Furthermore, the complexity of accounting for emissions allowances, auctions, and rules or limitations for offsets (including rules for offset verification and monitoring) can be difficult for market participants to understand, let alone the interested public who may express concern about environmental justice issues.

Economists have also described how the giving of emissions allowances represents a 'windfall' for regulated parties. Because the allowances have monetary value, the opportunity cost of selling or generating credits can cause a cost increase that is passed on in the form of higher prices for consumers.

In reality, governments that use cap-and-trade systems include them as simply a part of a network of laws and policies that regulate GHGs. Other existing laws that govern pollution emissions from major stationary sources or automobiles, as well as those that regulate agriculture, forestry, and other GHG-emitting activities, can complement one another and can incentivize reductions beyond the strictures of a set cap. How much 'work' the cap does in setting the price of GHG emissions will vary from case to case. The challenge for regulators is that this interaction does not match up with public perception that the cap is the cornerstone of the climate regulatory framework. Together with the ethical considerations about cap-and-trade, decision makers should weigh the advantages of a cap with other policy tools. Though they may not be able to easily express a clear quantifiable target, both command and control and subsidy/tax-based market tools are simpler for a government to administer and can lead to greater pricing predictability for regulated parties and for the public.

Conclusions

Economic approaches to climate change regulation have become dominant in many parts of the world. The potential benefit of market-based climate regulation is economic efficiency, lower costs, and ultimately a greater magnitude of GHG emissions reductions. However, cap-and-trade systems raise serious ethical concerns by legitimizing commodity-like 'rights' to pollute at a certain level. Ethical climate mitigation is focused on minimizing harm. A cap-and-trade system, on the other hand, can focus only on aggregate actions, with the potential to exacerbate existing inequities in the distribution of environmental benefits and burdens.

When making policy decisions about how to address climate change, governments and regulators ought to consider lessons from how cap-and-trade systems have been applied in climate change mitigation and in other environmental contexts. Ethical use of these programs can be possible if done with particular attention to and consultation with marginalized communities. Efforts must also be made to ensure that climate benefits are not considered in the abstract, but as an interlinked part of ecosystem management, along with the conservation of biodiversity and essential ecosystem services.

Notes

1 The Clean Air Act of 1990s Acid Rain cap-and-trade program included two phases of implementation, beginning with application to small set of sources, and then to a wider group of regulated parties. Other existing cap-and-trade systems have operated similarly. See 42 U.S. Code §§ 7651c–7651d.

2 California law requires any law that levies a 'tax' to be enacted with a two-thirds majority. A court had previously held that the original cap-and-trade program was valid under an earlier definition of 'tax,' but lawmakers wanted to avoid any possibility of legal challenge on these grounds, pushed for supermajority approval of the program's reauthorization in 2017.

References

Anderson, T. & Leal, D. (2001) *Free Market Environmentalism: Revised Edition*, New York: Palgrave.

Biber, E. (2017) 'Thoughts on AB 398,' *Legal Planet*, https://legal-planet.org/2017/07/14/thoughts-on-sb-398/ (last visited July 14, 2017).

Bodansky, D. (1993) 'The United Nations Framework Convention on Climate Change: A Commentary,' *Yale Journal International Law,* 18(2), 451.

Brown, D. (2008) 'Ethical Issues with Relying on Pricing Carbon as a Policy Response to Climate Change,' *Ethics and Climate*, https://ethicsandclimate.org/category/ethics-of-carbon-pricing/ (last visited January 29, 2021).

Bryner, G. (1993) *Blue Skies, Green Politics*. Washington, DC: Congressional Quarterly Press.

Bryner, G. (2008) 'Reducing Greenhouse Gases Through Carbon Market,' *Denver University Law Review*, 85(4), 961–982.

Bullard, R. (ed.) (1997) *Unequal Protection: Environmental Justice & Communities of Color*. San Francisco: Sierra Club Books.

Burkett, M. (2008) 'Just Solutions to Climate Change: A Climate Justice Proposal for a Domestic Clean Development Mechanism,' *Buffalo Law Review*, 56(1), 169

California Air Resources Board. (2018) *AB 32 Global Warming Solutions Act of 2006*, https://ww2.arb.ca.gov/resources/fact-sheets/ab-32-global-warming-solutions-act-2006 (last visited January 15, 2021).

California Environmental Justice Alliance. (2020) *70+ Organizations Oppose CARB Chair Mary Nichols for EPA* https://caleja.org/2020/12/press-release/ (last visited December 3, 2020)

Cole, L. & Foster, S. (2000) *From the Ground Up: Environmental Racism and the Rise of the Environmental Justice Movement*, New York: NYU Press.

Driesen, D. (2016) 'Decisions About Emissions Trading Design,' in *Decision Making in Environmental Law* Glicksman, R., Paddock, L. & Bryner, N. (eds.), 187–198. Edward Elgar Publishing.

Environmental Defense Fund. (n.d.) *California's Cap-and-Trade Program Step by Step*, https://www.edf.org/sites/default/files/californias-cap-and-trade-program-step-by-step.pdf (last visited January 16, 2021).

European Union. (2015) *EU ETS Handbook*, https://ec.europa.eu/clima/sites/clima/files/docs/ets_handbook_en.pdf.

Gillenwater, M. (2012) 'What Is Additionality? Part 1: A Long Standing Problem,' GHG Management Institute, Discussion Paper No. 001, https://ghginstitute.org/wp-content/uploads/2015/04/AdditionalityPaper_Part-1ver3FINAL.pdf.

Grandoni, D. & Eilperin, J. (2021) 'Biden Swells the Ranks of His White House Climate Team' *Washington Post*, https://www.washingtonpost.com/climate-environment/2021/01/14/biden-climate-staff/ (last visited January 24, 2021).

Horowitz, C. (2017) 'California Extends its Cap-and-Trade Program Through 2030,' *Legal Planet*, https://legal-planet.org/2017/07/17/california-extends-its-cap-and-trade-program-through-2030/ (last visited July 17, 2017).

Kaswan, A. (2008) 'Environmental Justice and Domestic Climate Change Policy,' *Environmental Law Report News and Analysis*, 38, 10287–10315.

Lazarus, R. (2004) *The Making of Environmental Law*, Chicago: University of Chicago Press.

Mernit, J. (2017) 'Cap-and-Trade Showdown: Will Big Oil Have the Last Word?' *Capital & Main*, https://capitalandmain.com/cap-and-trade-showdown-will-big-oil-have-the-last-word-0712 (last visited July 12, 2017).

Sandel, M. (1997) 'It's Immoral to Buy the Right to Pollute,' *New York Times*, https://www.nytimes.com/1997/12/15/opinion/it-s-immoral-to-buy-the-right-to-pollute.html (last visited December 15, 1997).

U.S. Environmental Protection Agency. (2015) 'Carbon Pollution Emission Guidelines for Existing Stationary Sources: Electric Utility Generating Units,' *Federal Register*, 80(205), 64661–65120 (last visited October 23, 2005).

U.S. Environmental Protection Agency. (2020) *Sulfur Dioxide Trends*, https://www.epa.gov/air-trends/sulfur-dioxide-trends (last visited October 9, 2020).

Vanderwarker, A. & Cuajunco, K. (2017) 'Equity at the Center: SB 775 and AB 378 Create New Path towards More Equitable, Effective Climate Policy,' *Legal Planet*, https://legal-planet.org/2017/05/12/guest-bloggers-amy-vanderwarker-and-kay-cuajunco-equity-at-the-center-sb-775-and-ab-378-create-new-path/ (last visited May 12, 2017).

World Bank. (2016) *Emissions Trading in Practice: A Handbook on Design and Implementation*, https://openknowledge.worldbank.org/handle/10986/23874 (last visited January 16, 2021).

24

ETHICAL ISSUES RAISED BY CARBON TAXING REGIMES

Cristiane Derani

Der Zauberlehrling

(Goethe 1797)

(...)
Wärst du doch der alte Besen!
Immer neue Güsse
Bringt er schnell herein.
Ach! und hundert Flüsse
Stürzen auf mich ein.
Nein, nicht länger
Kann ich's lassen,
Will ihn fassen.
Das ist Tücke!
Ach! nun wird mir immer bänger!
Welche Miene! welche Blicke!
O, du Ausgeburt der Hölle!
Soll das ganze Haus ersaufen?
Seh' ich über jede Schwelle
Doch schon Wasserströme laufen (...)[1]

The road to hell is paved with good intentions.

Introduction

Despite being one of the biggest challenges of humanity, climate change is communicated with counter-information, thus weakening the social concern about its reality. And despite the impacts of climate change being widespread and consequential, vested interests and ignorance minimize the collapse we are producing with greenhouse gas (GHG) emissions.

Climate change could become a galvanizing force for humanity, leaving us all not just safer from extreme weather, but with societies that are safer and more equitable, as well (Klein 2014) – a catalyzing force for positive change, for individual, for societies, and for species, from inside, out. With its complex, but widespread, issues, it has the potential to transform our thinking and acting, our worldviews that brought us to this crisis, through a true ethical revolution. But instead, what seems to be happening is the last ball before the Titanic sinks – even with us fully aware of the iceberg ahead.

DOI: 10.4324/9781003039860-29

An ethical revolution requires knowing the roots to what placed us as protagonists in an era of global warming and the consequent changes in species and biomes, and this chapter advocates that ethics is at the root of fighting the causes and the effects of the harms created by Anthropocene (Crutzen 2002).[2] Yet even the word 'Anthropocene,' applied to explain the magnitude of the effects caused by human beings on this planet, seems inappropriate from a historical and ethical point of view. It leads to the idea that the human-caused destruction of ecosystems is simply a matter of destiny. In reality, the destruction of our environment and the solid threat to our existence result from historical choices and not an intrinsic human characteristic.

When in reality, there are those in control of our future. That power, however, of knowledge and for change, does not lie with the ecstatic pawns, mere zombies vampirized by the illusions of consumerist selfishness, but with oligopolies and political leaders. These are the masters of the capitalocene, or carbonocene (Campagne 2017) – lords of a necro-politics, which uses race, gender, geographical origin, philosophy, and global financial power to seduce any legal-political organization with instruments that are inefficient but profitable. And all under the guise of 'reducing emissions.'

The State and politics are today an arena of different expressions of power, where the political power of the people is confronted with the power of scientific knowledge and the power of money, therefore reconfiguring the State's entire constitution and action. Political action through State action should combine these three powers, for the sake of the general interest that it represents and in the people's name for which it was created. Yet, these powers are not balanced and do not address the same goals. They fight each other to dominate the discourse and build their own reality.

Taxing carbon emissions is a practice that is both political and economic, and it also requires scientific knowledge. Therefore, these three powers would need to be composed in a way that would favor the achievement of the goals for which the taxation instrument was created. The values and intentions that nourish this instrument and the environmental implications, as well as the balance of collective and private gains, are considerations that draw on ethical assumptions about human society and its relations with each other and with other living beings and global structures.

The Ethics of Scientific Knowledge

In his book *Collapse* (2005), Jared Diamond asks, "What did the Easter Islander who cut down the last palm tree say while he was doing it? Like modern loggers, did he shout 'Jobs, not trees!'" or, 'Technology will solve our problems, never fear, we'll find a substitute for wood!' The ecological collapse of the island, like other regions whose nature was exhausted in ancient times, may be compared with the collapsing reality that we are experiencing today. Two differences, however, are worth being noted; first, an objective one – the collapsed area is the entire globe; and second, a subjective one – we are fully aware of what is happening and the causes and consequences thereof.

Since the rise of *Homo sapiens*, much has been destroyed. The Australian and American megafaunas and the Neanderthals themselves are examples of the destructive power of our species. Along with the ability to represent the world with drawings and symbols came the capacity to change it, and therefore also, the capacity to destroy it (Kolbert 2014: 258). However, the destructive actions of human beings today have a global dimension and make the already known migratory movements in search of new lands unfeasible. Additionally, science provides us with sufficient data, models, evidence, and technological alternatives to nourish our knowledge and produce changes in conduct.

We are heading toward a point of no return, where the terrestrial ecosystem begins to collapse and cannot be healed, its capacity for resilience part of the destruction. And for some, we might be reaching that point already (Cattano et al. 2020; Lovejoy and Nobre 2020). Such a self-destructive process, starting with the destruction of our natural bases, forces the recognition that we are also Nature. At the same time that we cause the extinction of non-human beings, the threats to human existence are increased. It seems that we are paradoxically directing our creative capacity to a process of self-destruction in proportion to a war against all living and non-living elements on earth.

Modernity was built upon the scientific/technological revolution, the strengthening of States, and the concentration of capital brought in from large naval expeditions directed toward industrial expansion. This process dissociates itself from social values, and both science and the State end up being colonized by the ethics of monetary accumulation.[3] From the moment when money becomes the reason for economic production rather than just the means of its attainment, we lose track of the real under the illusion of what wealth could be.

The submissiveness of science and technology to capital production removes from knowledge its connection to the common good. Science is brought into disrepute by society and, despite all scientific knowledge built over three centuries, there is still ignorance and contempt for it – weakened in the public sphere as it has been subjected to an ethics different from the common good. Without democratic and clear channels of communication, without the State's political strength to detach itself from the colonization of money – and to properly communicate scientific evidence and take effective measures – society is at the mercy of social media, popularity, and tribalism, which overrides scientific information and disdains the ethics of knowledge, throwing the population into a well of blindness and illusion.

The Ethics of Politics

> No ethics in the world can dodge the fact that in numerous instances the attainment of 'good' ends is bound to the fact that one must be willing to pay the price of using morally dubious means or at least dangerous ones – and facing the possibility or even the probability of evil ramifications. (…) As a matter of fact, logically it has only the possibility of rejecting all action that employs morally dangerous means.
>
> *(Weber et al. 1958: 121–122)*

The desire to be efficient at any cost, to be unencumbered by complexity, to minimize neighborliness, respect, reverence, responsibility, accountability, and self-subordination – this is the culture of which our present leaders and heroes are the spoiled children (Berry 2018).

In this vein, the creation of the State is, according to Thomas Hobbes in *Leviathan* (1651), a condition of restraint necessary for humans, who "naturally love liberty and dominion over others," and for the benefit of "their own preservation and of a more contented life" (Hobbes: 112). Hobbes did not have a positive outlook on human nature. According to him, good feelings such as justice, equity, modesty, mercy could only be expressed out of fear of some power. This Hobbesian ethical assumption has nourished and hypnotized the moral imagination of our civilization. Although prejudiced and not empirically verifiable, this adversarial perception of human nature has been resorted to by the State to justify political structures and the monopoly of violence and norms.

In his *Essays*, French jurist and philosopher Michel de Montaigne (1533–1592) had a much more complex and realistic perception of the human being, in part validated by reading accounts from travelers who came to America (1993). For him, feelings of friendship and generosity are a part of human nature, who finds joy in expressing them. In Montaigne's pursuit of human nature, he dared to think beyond his immediate space, challenging Eurocentrism at the dawn of modernity and the age of discovery.

Not even Hobbes' contemporary René Descartes (1596–1650) understood human nature and the idea of freedom so miserably. For Descartes, real freedom is essential to the spirit, and the human being is endowed with innumerable passions that act positively and negatively in the world. In his work *The Passions of the Soul* (1649), he does not discuss the much talked about selfishness, as does his contemporary, but human generosity (Guinsburg et al. 2020). In Descartes' view, generosity serves as a remedy for anger's excesses and the disorders of the passions, including the accumulation of goods.

Despite being contested, the ideology that deems human beings as selfish by nature and as individuals who need the State to express their positive feelings prevails and travels forward toward the theory of economic liberalism. *Homo economicus* is therefore the one who seeks the greatest advantage, with a selfish and competitive behavior. It is under the auspices of free choice that economic instruments, aimed at the best advantage, are used to discourage investment and consumption activities, as is the case of taxes imposed to push behavioral changes.

The Ethics of Taxation

Taxes, in general, are part of the private wealth destined to maintain the functioning of the State. Taxation finds its first theoretical rationality in David Ricardo's, *On the Principles of Political Economy* and Taxation (1817). In this book, he seeks to rationalize taxation, its utility and necessity, and differentiates it based on the objects upon which it falls. He devotes a chapter to taxes on raw produce. He does not use oil as an example, of course, but takes another fundamental energy source in the industrial process of his time: corn. Corn is the source of food for all the human workforce, which, in Ricardo's time, was the great driving force of production. His reasoning is as follows: a tax on corn would fall on the consumers of corn, and at the first moment will decrease the affordability (access) of working-class families.

Transposing Ricardo's reasoning on corn into GHG-emitting sources of energy, basically oil and coal, it can be seen that taxation on these GHG-emitting sources will increase the cost of all products that employ such energy sources in their production chain. In our fossil energy-based civilization, production in general will suffer an increase in cost, which therefore will be transferred onto those with less economic power and less exchange option. When posing the problem, Ricardo expresses a concern with the working class, which is the most affected because of their limited income (1817: 121).[4] Ricardo argues that taxes on the raw material cause a rise in the price of all products that use it and remove international competition, in addition to leading the working class to misery.

One possible counterpoint to this logic would be, as regards to the carbon emission tax, the existence of an option for another source of energy. The thought, with reference to the *Homo economicus*, i.e., the selfish and free pursuit of individual advantage, is that the individual would go on a quest for energy alternatives or reduce the consumption of products that rely on GHG-emitting sources of energy (namely, almost everything). Thus, the destiny of all humanity relies on the individual while the State, the center of political composition and governance development, assists these movements while filling its own coffers with tax payments.

When it comes to taxing essential raw materials with very low substitutability, the price will only cause market distortion, in addition to impoverishment or inflation. The neoliberal assumption that there would be an incentive to innovation for the high cost involved is again a wish in no way supported by evidence or science. The history of capitalism shows that innovations that substitute profitable products require a vigorous action from the State, through imposition or on its own initiative.

Oddly, politics – the realm of composition, negotiation, and the pursuit of the best alternatives for society – co-exists poorly with the selfishness of individual advantage. Moreover, the political adoption of a carbon tax affects politics by depriving it of its power over the results, at the same time that it hands the destiny of human life to the sum of individual decisions. The carbon tax is an economic instrument, within the same economic matrix of liberal production, operated by selfish and consumerist subjects who are always ambitious and dissatisfied. As such, the future does not look promising.

The Ethics of Individual Choice

In addition to Ricardo's warning in the early 19th century, Arthur Cecil Pigou (1877–1959), presents taxation as a manner of correcting a market failure based on what he calls 'negative externalities' The carbon tax is the implementation of a Pigouvian tax, to correct the greatest externality in the market since the Industrial Revolution, namely global warming. As previously mentioned, the guiding ethics of taxation is concerned with controlling the pursuit of individual advantage; it is not an effective ethics of building a new energy matrix or of protecting the human species, even less of stimulating solidarity or generosity.

Pigou recognizes that the economic process brings private benefits, at the same time that it produces negative externalities, which are social costs. His solution is to internalize the social cost in the form of taxes (Pigou 2013). Regarding the burning of fossil fuels, for instance, anyone who burns these fuels must pay the State an amount that would compensate for the damage that they cause to society. The first question that arises is whether it would really be possible to individualize and financially translate the value of the damage caused by global warming into a monetary expression. If the answer is yes, the second question is whether it would be possible for the production chain to bear this cost. If the answer is affirmative, the denial of which Ricardo argued long ago, then this third question will apply: to what extent would this rate lead to a change in the energy matrix and prevent society from building its own collapse?

Based on the freedom-selfishness dyad, tax on products or activities that produce negative social outcomes would reach the private entities' willingness to pay and, theoretically, this would then reduce their willingness to pay and make them seek alternatives that are less harmful to society.

The ethics underlying the maximum willingness to pay carries the necessary material attribute of the ability to pay. A relationship based on this subjective interest does not guarantee the effective protection of the environment. Additionally, as it depends on an individual economic situation, it conforms to other criteria, such as the possibility of product substitutability, a competitive game that leads to alternatives free of GHG emissions, a stable macroeconomic situation, balanced consumption, and inflation, among others. In this context, immediately there would be an increase in the cost of production, tending to expel people from the market and increase the concentration of income and capital.

In this way, the market becomes even more distorted and less balanced. Instead of moving away from pollution, privileges are granted in the form of rights to pollute. It is the reverse of the 'polluter-pays' principle. Obliged to pay tax, the ones with the greatest economic power will do so and will not be concerned with reducing emissions, but will take the opportunity to concentrate market power and weaken smaller competitors with less capital reserves. Considering that the use of fossil fuels is the basis of the world's economic production, all production chains are affected by the cost increase, which then is unevenly distributed in society due to the income-to-expense ratio.

A simple and clear example is the incidence of the carbon tax on food. The lower a family's income, the greater the proportion of their expenditure on food. This sector (agriculture, forestry, and other land use) is the second largest GHG emitter, second only to energy production (FAO 2017). The sector is estimated to contribute 30 percent of total GHG emissions, with the livestock sector responsible for 14.5 percent of the total emissions (Wellesley et al. 2015).

In this regard, Caillavet, Fadhuile, and Nichèle (2019) conducted a thorough literature review, demonstrating that any policy increasing the cost of energy will disproportionately impact low-income households, making equity a major concern when taxes are discussed. They found that, "The key issue of introducing compensating mechanisms with a combination of taxes and subsidies has been used in certain carbon scenarios designed for food consumption; however, they did not target distributional neutrality or frequently measure its effect" (Caillavet et al. 2019).

An Ethical Radical Change

The carbon tax as an instrument for changing individual decisions is not only fallible in the expectation of the individual's behavior, but also, at best, would result in no more than the reduction – not the elimination – of GHG emissions in the atmosphere. The idea is that, with the carbon tax and the subject's interest to lessen its use and seek new technological alternatives, there would be a reduction such that it would balance the maximum possible pollution with the maximum possible protection, thus reaching Pareto optimality (1848–1923).[5] From the Paretian perspective, the market economy reaches its optimum degree when it establishes a satisfactory relationship between the use of a natural resource and its conservation, i.e., by determining a price that allows using the good while at the same time it is conserved (Pareto 1984).

According to the 2022 IPCC report, there is a need for urgent action to preserve the planet from further adverse effects of climate change.[6] At its current rate of change, the temperature is expected to increase by 2.7°C before the end of this century, far exceeding the temperature specified in the 2015 Paris Agreement. The challenge facing us today is the elimination of emissions as soon as possible, toward a zero-carbon economy, thereby ensuring that the rising global temperature reaches a point that still allows human life on Earth. This requires more than minimizing emissions. Our civilization is 90 percent based on energy from fossil hydrocarbons. Therefore, the issue calls for a revolution of civilization, not a simple adjustment from within the same structures (Ayres 1989: 107).

The energy market concentrates all the market failures pointed out by economic theories. It is a market that offends against all liberal theoretical assumptions: it is not competitive, its product has no immediate substitutability, and the consumer has little information and little chance to stop buying what the market imposes. Moreover, energy is a public good, because it is found in nature and lies at the root of all economic activity.

Specifying in other words, the GHG-emitting energy market limits the consumers' choices because it forms the basis of every economic activity, so it is present in the goods regardless of individual preferences. The damages caused by GHG-emitting energy are dissociated from its immediate use, making it difficult to understand the cause-effect relationship around it, in an inexorable asymmetry of information. The energy market is part of a productive oligopoly, which is not competitive and with a very low degree of substitutability, where agents unite in practices that go against 'free trade,' combining price adjustments (aided by governments) and associated economic practices (trusts and cartels).

In fact, global warming represents the largest and most comprehensive result of the market failure caused by the accumulation of failures (negative externalities + public good + information distortion + non-competitive market + almost total stagnation in terms of finding alternatives for products), affecting all economic activities, whereas the main causers of pollution are simply not paying the fair share of this bill (Aydos 2017).

Even if it were argued that the internalization of negative externalities could be achieved by the carbon tax, all other failures would work so that the final outcome of substitutability, innovation, and reduced use thereof would not occur. It is no wonder that, although there is abundant literature on the purpose and use for the carbon, there is not a single positive result of significant emission reductions.

There is a profusion of articles devoted to attempting to resolve the fallacies arising from the carbon tax. The main idea is to spur cost-effective mitigation and cost-reducing innovation, but above all, ensure that emissions are held to the target level.[7] As if said unsatisfactory results were not enough, the articles advocating carbon taxation ignore reality and remain immersed in liberal economic theory. For instance, according to Boyce,

> In the short run, a carbon price provides an incentive for households, firms, and governments to reduce emissions cost-effectively. In the long run, the prospect of continuing and rising carbon prices also provides an incentive for innovations to lower the cost of cutting emissions. These static and dynamic efficiency effects are independent of the policy's design.
>
> *(2018: 52)*

All of these arguments are deprived of evidence yet stay as natural assumptions for many economists, reproducing the maxim stated by Kapp and Vilmar (1972: 78): the market flows well if you dehumanize it.[8]

Taxation on carbon co-exists with strong subsidies. As stressed by Coady et al.

> Most actually existing carbon prices are well below the levels recommended by climate policy analysts, whether on the basis of the efficiency criterion of neoclassical economics or the safety criterion embodied in the Paris Agreement. Also, the political influence of groups with vested interests in continued use of fossil fuels are probably responsible for these precarious results. Another indicator of this influence is the fact that many countries actively subsidize use of fossil fuels by means of policies that are tantamount to a negative carbon price. Direct fossil fuel subsidies by governments to consumers and producers worldwide amounted to \$333 billion/yr in 2015, according to a study by IMF researchers; by a broader measure that includes unpriced externalities, the study estimated the worldwide subsidy at \$5.3 trillion/yr. By the narrower definition, the average global subsidy was equivalent to about \$10/mt CO_2, roughly five times more than the average global carbon price of \$2/mt CO_2.
>
> *(2017)*

According to Boyce (2018), who regards carbon taxation as a solution, prescriptions for the 'right' carbon price necessarily rest on an ethical foundation. In neoclassical economics, where the reigning normative criterion is efficiency, the optimal price – termed the 'social cost of carbon' (SCC) – is one that maximizes the net present value of the benefits of emission reduction minus its cost. Apart from the technical difficulties involved in this application of cost-benefit analysis to climate change, it is important to note that different normative criteria can lead to quite different prescriptions.

The difficulties in measuring the monetary benefits of emission reductions have been discussed extensively in the literature.[9] However, the proposed economic models fail to bring concreteness to the application of carbon taxes to significantly reduce emissions. To exemplify the abundant literature that defends the carbon tax but is unable to bring concrete data of success, the present study resorts to two other articles that will demonstrate the failure of the tax in Australia and the United States. The first one dates back to 2013 and is relevant because it offers a field report analysis on dividend return resulting from the application of the carbon tax under a 2011 Australian law. In the article, the field research conducted by Perry, Rosewarne, and White reveals the impossibility of combatting so many market failures with carbon tax alone. The authors argue that the behavioral assumptions overstate the consumer and producer substitution possibilities, failing to consider the possibility of technical re-switching, and ignore the impact that oligopolistic market structures would have on price increases and infrastructure investment. The full ramifications of compensation for overall government expenditure and therefore the capacity of the government to continue to fund a range of elements of the social wage, the potential for unemployment, and transitioning workers to less carbon-intensive industries, are also overlooked (Perry et al. 2013).

A paper published in 2019 analyzes the situation in the United States and comes to a similar conclusion: in addition to the inability to change the amount of carbon emissions, there is an increase in the already unfair distribution of wealth in American society. In their article, Anders Fremstad and Mark Paul (2019) estimate households' carbon footprints and examine the impact of a revenue-neutral tax of $50 per ton of CO_2 on multiple forms of inequality. It demonstrates that Americans in the richest decile emit over five times as much CO_2 as Americans in the poorest decile. Nevertheless, a carbon tax would cost poor households a higher percentage of their expenditures (or incomes) than the rich, making it a regressive tax (Fremstad 2019: 96). Taking into account that society owns the atmosphere in common and equal measure, there is a strong moral case for paying every person an equal amount for the use of this natural asset. As carbon pricing initiatives spread around the world, it is imperative to understand the impact of carbon tax policies on inequality.

Even though carbon taxation was proposed thirty years ago as an economic instrument to reduce emissions, there is no single evidence in the world of its successful application. This, *per se*, would be enough to warn about its inefficiency, which in turn could be countered by the fact that one cannot refute an instrument that has not yet been put into practice. Now, the ongoing difficulty of putting it into practice in the face of such urgency would be enough to show its political fragility. This also demonstrates that anchoring political outcomes in individual selfish decisions of investment and production. reinforces Hobbes' thesis and endangers the human species. Paradoxically, the lack of success of the carbon tax is not reflected, to the same extent, on the coffers of the Hobbesian State, which obtains yet another source of income and therefore ends up protecting the exercise of the human predator over other living beings, a contradiction to the reasons for its creation.[10]

In conclusion, if from an economic and ecological perspective the carbon tax has not been successful, rooted in an ethics that praises selfishness and individualism, it is not efficient for the achievement of collective goals (the very purpose of a State). The carbon tax does not reflect the cost of the environmental damage caused by emissions thereof and distorts even the inspiring idea inherent to it, which is to internalize the negative costs inflicted on society. Therefore, taxation represents neither environmentally, economically, nor ethically the solution to the issue of fossilization in our society.[11]

The Ethics of Joint-and-Several Responsibility and the Meaning of Freedom

The economic response to the ecological issue consists mainly of handling situations of uncertainty and conflict the best way possible, through an approach that takes into consideration the interconnected socioeconomic structures. If the dynamics of the prevailing direction of the market economy are not clear, the chances are small that any implemented change will envisage the durability and sustainability of natural resources and, consequently, the sustainability of economic relations. In economics, there is not only monetary illusion; there is also a conceptual illusion. Schumpeter (1976) even spoke of 'the word-mindedness' of economists, losing sight of their meanings in pertinent discussions (Maier-Rigaud 1992: 24).

A new economic-environmental paradigm should first draw on the image of the Earth as a spaceship, that is to say, the earth is limited in its resources and in its carrying capacity. Second, the idea of temporal continuity and infinity should be questioned. We need to give up on the idea that we have to take care of the next four or ten generations only. The consequence of this is that we need to adapt to a process of zeroing fossil fuels and reject nonrenewable energy. It is then a question of building a sustainable economy for the environment and society (Maier-Rigaud 1992: 42). Respect for ecological limits is built upon social limits. An ethics of relationships and joint-and-several responsibility that values healthy human passions – and that neutralizes the passions nourished by fear and the weakness of sociopathy – is the basis for policies to drastically reduce emissions.

To sum up, this is not about combating the carbon tax instrument. The purpose of this chapter is to demonstrate that the ethical foundation of an instrument is sufficient to show the direction toward its success or failure. A collective positive outcome for the protection of a common good is not possible with the use of private pricing instruments grounded in individual production and consumption decisions. In this regard, supporters of the idea of freedom should know that those who are bound by their own customs and rules are not free.

The concept of freedom based on the idea of freedom of choice is very restricted, as pointed out by Amartya Sen (1993: 176). The economist reminds us that society cannot be organized in such a way that each person controls all aspects that affect his or her personal life. What a person *would* choose if they had the power to decide is an important consideration when judging that person's freedom. However, this possible choice depends on the 'de-ignorantization' of society. Which would the individual choose: preserve life or walk to death? The answer seems clear to us. What Sen calls indirect freedom is a choice in the name of an outcome, which would equate to an individual's free choice if he or she truly had the conditions to choose.

In short, it is paramount to build an ethics targeted to combatting ignorance, so that freedom can be truly exercised toward unequivocally socially desired outcomes. Even so, the choice of paying the tax as a form of mediation of 'free' choices on the decision to play a role or not in GHG emissions does not lead to indirect freedom because the outcome thereof would not reflect the free choice of society, driven by the wandering pathways of ignorance, purchasing power, and uncertain outcomes from the set of decisions. This is closer to the expression of random than freedom.

According to Sen, the idea that freedom is in the power of choice hides the real possibility of building freedom as a social choice, since individual decision can lead to unwanted roads different from those freely chosen. That is why he maintains that it is fundamentally inappropriate to describe freedom in terms of the real domain that the individual enjoys regardless of the nature of the outcomes. This 'indirect freedom' evinces the fallacy of individual options as the exercise of freedom. There is a conflict between individual choices and the final outcome when these choices multiply. Therefore, stating that there is a social direction for the carbon tax to lead to reduced emissions and that society, in deciding to increase the cost of products that cause emissions is protecting itself from the effects of global warming, is not true (as proven in practice), nor does it reflect the conscious freedom of choice of those who want to prevent the collapse of the planet.

On true freedom, which mirrors a fair social construction, the French philosopher Edgard Morin speaks of an anthropo-ethics or planetary ethics (2004), and the German philosopher Karl-Otto Apel treats it as a macro-ethics (1990) that comes out of a privatist and de-politicized morality. For Apel, the achievement of a socially just world society would consist of actions that save the eco-sphere (Apel 1990: 39).

The big ethical question now must be answered by everyone collectively, as for the first time in history, every human species is under threat. For that purpose, consciousness at different levels is necessary (Morin 2004: 186), namely, human beings need to become aware of their common human identity and their common destiny. They need to understand the finiteness of their existence and thereby increase their spiritual awareness and relativize their material attachment. They need to become aware of their identity as nature, their vital relationship with the biosphere, and the respect they should have for all living and non-living beings that support them. Their social consciousness should expand toward a civic and community idea, as well as solidarity with future generations. These teachings of Morin are ethical commandments for a mindset change, for the development of ecological humanism and human ecologization so that human beings can recover the value of their existence and their commitment to the whole planet.

Conclusion

"A thing is right when it tends to preserve the integrity, stability, and beauty of the biotic community. It is wrong when it tends otherwise" (Leopold 1949).

In a situation in which the economy has already crossed ecological limits,[12] the ethics of existence and co-existence rejects any attitude that leaves selfishness the free decision about what and how much to consume of the Earth's resources. We are facing the most important ethical cause in the history of humankind, which should not be left to individual decision, substantiated by imperfect information and guided by individual selfish passions.

Thirty years of international environmental policy to reduce carbon emissions, with exhaustive studies and scientific models warning about life's extinction on Earth by humans, did not provoke the necessary reaction from global society. In reality, it was the COVID-19 pandemic that set the record for reducing emissions.[13] The pandemic forced collective and associated practices. The countries that best fought the coronavirus were those that acted collectively. The realm of liberalism and selfishness were record setters for deaths and infections. We *Homo sapiens* have only survived because of our spirit of solidarity, one which the pandemic is re-teaching us. In other words, the practices that value collective action, as well as norms that demand responsible commitment in each act that may impact society, constitute the basis for a law that envisage the balance of our home, namely ecological law.

We are Nature, our production is natural and social, and it reflects our social idea of a nature-society relationship. The present shows that we have taken a wrong route. The sixth mass extinction and global warming are social constructions of a particular choice of *Homo sapiens* to relate to the environment and one another. The industrial process evolves upon an ideology antagonistic to the ethics of joint-and-several responsibility. It is individualistic and competitive; it is hedonistic and immediate; it seeks wealth by generating scarcity; and it is founded on the future time that despises the past and is not satisfied with the present. The ecological foundation of the law consists of rescuing circular time, a solidary consciousness that recognizes abundance within the limits of the Earth. Perhaps it will be a global ecological order, with instruments that respond to this ethical foundation.

This chapter sought to demonstrate that the carbon tax, in addition to not achieving the purposes for which it was created, lulls us into a false feeling that we are taking care of the planet when in fact we are exchanging our last hope for survival for coins. Quoting Morin,

> For the first time in human history, the universal has become a concrete reality: it is the objective inter-solidarity of humankind, in which the global fate of the planet decides the individual destinies of nations and in which the individual destinies of nations upset or change the global destiny.
>
> *(2004: 183)*

The reflection upon the unsuccessful attempt to transform the destruction of the planet into monetary value to be paid by those who are causing it demands an urgent correction, of course. It calls for an ethical shift that will lead to a turning point in terms of instruments ethically compatible with the society we want. There is no solution outside of humanistic values and virtues of solidarity that builds freedom in the complex relationship between different beings, and that finally encourages the awareness of who we really are on this planet as individuals, society, and species.

Notes

1 *The Sorcerer's Apprentice*

(...)
Now your efforts, spell-begotten!
Still his work abhorrent
Does the wretch resume;
Where I look a torrent
Threatens me with doom.
No, no longer
Shall I suffer
You to offer
Bold defiance.
I have brains, I am the stronger
And I shall enforce compliance
You, hell's miscreate abortion,
Is this house doomed to perdition?
Signs I see in every portion
Of impending demolition. (...)

Translation by Paul Dyrsen, 1878. Available at: https://germanstories.vcu.edu/goethe/zauber_e4.html. Accessed on: 28 December 2020.

2 The Anthropocene could be said to have started in the latter part of the 18th century, when analyses of air trapped in polar ice showed the beginning of growing global concentrations of carbon dioxide and methane. This date also happens to coincide with James Watt's design of the steam engine in 1784 (Crutzen 2002: 22).

3 Adventure, ambition, curiosity created the new society, which forgot its limits. In the words of Harari "Self-made gods with only the laws of physics to keep us company, we are accountable to no one. We are consequently wreaking havoc on our fellow animals and on the surrounding ecosystem, seeking little more than our own comfort and amusement, yet never ending satisfaction." Yuval Noah Harari, Sapiens. A brief history of humankind 2015.

4 Tax would be equitable to the extent to which it affects consumers, but it would be partial in its incidence on utilities. The prices of necessity goods would be distorted due to this taxation on the raw produce. Even more brutal is Ricardo's revelation when he states that the capitalists, whose income stems from the capital gains, will not be affected because they do not live off an activity that depends on the raw produce.

5 **Vilfredo Pareto** was born July 15, 1848, in Paris, France and died August 19, 1923, in Geneva, Switzerland. He was an Italian economist and sociologist who is known for his theory on mass and elite interaction as well as for his application of mathematics to economic analysis.

6 2022 IPCC Mitigation of Climate Change Report declares that human-induced climate change is a consequence of more than a century of net GHG emissions from unsustainable energy use, land-use and land use change, lifestyle and patterns of consumption and production. Without urgent, effective and equitable mitigation actions, climate change increasingly threatens the health and livelihoods of people around the globe, ecosystem health and biodiversity. (https://www.ipcc.ch/report/ar6/wg3/downloads/report/IPCC_AR6_WGIII_SPM.pdf) access in 12.12.2022.

7 The carbon prices needed to meet this constraint are likely to be considerably higher, however, than existing prices and conventional measures of the social cost of carbon. This poses issues of distributional equity and political sustainability that can be addressed by universal dividends funded by carbon revenues. (Boyce 2018: 52). The fallacy of this mechanism is proved by Aydos 2017.

8 The authors point to the need for a substantive-human inquiry capable of placing human beings and the material content of their needs as a starting point. (...) Only with the concomitant observance of the conditions of well-being outside the acquisition of material goods could the loss of humanity be recovered and link scientific knowledge with practical action, capable of meeting the requirements of the future.

9 See, for example, Azar (1998); Ackerman et al. (2009); Pindyck (2013, 2017); Van den Bergh and Botzen (2014); Howard and Sterner (2017).

10 In 2014, the World Bank estimated that in order to limit global warming to 2.5°C above pre-industrial levels at the least cost, a global CO_2 price should be set at approximately US$30 per ton (in 2014 dollars) in 2020, and then rise at around 5 percent per year. Furthermore, the World Bank's State and Trends

of Carbon Pricing 2020 report demonstrates that although practically ineffective in terms of reducing emissions, the practice of pricing has expanded countries' coffers, with thirty countries adopting the carbon tax. Governments raised more than $45 billion from carbon pricing in 2019. https://openknowledge.worldbank.org/bitstream/handle/10986/33809/9781464815867.pdf?sequence=4&isAllowed=y

11 A national carbon tax is currently implemented in 25 countries around the world, including various countries in the EU, Canada, Singapore, Japan, Ukraine and Argentina. However, according to the 2019 OECD Tax Energy Use report, current tax structures are not adequately aligned with the pollution profile of energy sources. For example, the OECD suggests that carbon taxes are not harsh enough on coal production, although it has proved to be effective for the electricity industry. A carbon tax has been effectively implemented in Sweden; the carbon tax is USD $127 per tonne and has reduced emissions by 25 percent since 1995, while its economy has expanded 75 percent in the same time period. https://earth.org/carbon-tax-a-shared-global-responsibility-for-carbon-emissions/#:~:text=A%20national%20carbon%20tax%20is,%2C%20Japan%2C%20Ukraine%20and%20Argentina; https://taxfoundation.org/carbon-taxes-in-europe-2020/

12 According to an international team of researchers, four of nine planetary boundaries have now been crossed as a result of human activity. These are: climate change, loss of biosphere integrity, land-system change, altered biogeochemical cycles (phosphorus and nitrogen). Publishing in the in the journal Science, the team of 18 researchers explain that climate change and biosphere integrity are "core boundaries." https://unfccc.int/news/scientists-say-planetary-boundaries-crossed. https://www.footprintnetwork.org/our-work/ecological-footprint/.

13 https://www.usatoday.com/story/news/nation/2020/12/11/covid-climate-change-record-drop-worlds-carbon-dioxide-emissions/3896686001/ (accessed in 12.12.2022)

References

Apel, K-O. (1990) *Diskurs und Verantwortung: Das Problem des* Übergangs *zur Postkonventionellen Moral*. Frankfurt am Main: Suhrkamkp.

Aydos, E. (2017) *Paying the Carbon Price – The subsidization of Heavy Polluters under Emissions Trading Schemes*. Cheltenham: Edward Elgar Publishing Limited.

Ayres, R.U. (1989) "Externalities: Economics & thermodynamics." In: F. Archibugi and P. Nijkamp (eds) *Economy and Ecology: Towards Sustainable Development*. The Netherlands: Kluwer Academic Publishers.

Berry, W. (2018) "Faustian economics." *Happer's Magazine*, May 2018. 35–42.

Boyce, J.K. (2018) "Carbon pricing: Effectiveness and equity." *Ecological Economics*, 150, 52–61.

Caillavet, F. et al. (2019) "Assessing the distributional effects of carbon taxes on food: Inequalities and nutritional insights in France." *Ecological Economics*, 163, 20–31.

Campagne, A. (2017) *Le Capitalocène: Aux Racines Historiques du Dérèglement Climatique*. Paris: Éditions divergences.

Cattano, C., Agostini, S., and Harvey, B.P., et al. (2020) "Changes in fish communities due to benthic habitat shifts under ocean acidification conditions." *The Science of the Total Environment*, 725, 138501.

Coady et al. (2017) "How large are fossil fuel subsidies?" *World Development*, 91, 11–27.

Crutzen, P. (2002) "Geology of mankind." *Nature*, 415, 23.

Diamond, J. (2005) *Collapse - How Societies Choose to Fail or Succee*d. New York. Penguin Books.

Fremstad, A., and Paul, M. (2019) "The impact of a carbon tax on inequality." *Ecological Economics*, 163, 88–97.

Guinsburg, J., Romano R., and Cunha, N. (2020) Descartes: Obras escolhidas. (Textos). *Editora Perspectiva S/A*. Kindle edition.

Hobbes, T. (1651) *Leviathan*. London: Oxford University Press.

Kapp, K.W., and Vilmar, F. (1972) "Die enthumanisierung der wirtschaftswissenschaft und der gesellschaftlichen realität." In: *Sozialisierung der Verlust? Die Sozialen Kosten eines Privatwirtschaftlichen Systems*. München: Hanser.

Klein, N. (2014) *This Changes Everything: Capitalism vs the Climate*. New York: Simon & Schuster Paperbacks.

Kolbert, E. (2014) *The Sixth Extinction: An Unatural History*. UK. Bloomsbury.

Leopold, A. (1949) *The Land Ethic*. https://www.cooperative-individualism.org/leopold-aldo_land-ethic-1949.pdf. Accessed in 12 December 2022.

Lovejoy, T.E., and Nobre, C. (2020) *When Will the Amazon Hit a Tipping Point? Scientists Say Climate Change, Deforestation and Fires Could Cause the World's Largest Rainforest to Dry Out. The Big Question Is How Soon That Might Happen*. https://www.nature.com/articles/d41586-020-00508-4. Accessed 25 February 2020.

Maier-Rigaud, G. (1992) "Die herausbildung der umweltökonomie: Zwischen axiomatischem Modell und normativer Theorie." In: F. Beckembach (ed) *Die Herausforderung für Die Ökonomische Theorie*. Marburg: Metropolis.

Montaigne, Michel de. (1993) *The Complete Essays* (1st edition 1752). Translated by Charles Cotton. London: Penguin Classics.

Morin, E. (2004) *La Méthode 6 – Ethique*. Paris: Seuil.

Pareto, V. (1984) *Manual de Economia Política* (1st ed. 1909). São Paulo: Abril Cultural.

Perry, N. et al. (2013) "Clean energy policy: Taxing carbon and the illusion of the equity objective." *Ecological Economics,* 90, 104–113.

Pigou, A.C. (2013) The Economics of Welfare (1st ed. 1920). London: Palgrave-Macmillan.

Ricardo, D. (1817) *On the Principles of Political Economy and Taxation* (1st ed.). London: John Murray.

Schumpeter, J.A. (1976) *History of Economic Analysis* (1st ed. 1954). London. Oxford University Press.

Sen, A. (1993) *Ethique et Economie*. Paris: Presses Universitaires de France.

Weber, M., Gerth, H., and Wright, C. 1916–1962 Mills (1958) *From Max Weber: Essays in Sociology*. New York: Oxford University Press.

Wellesley, L., Happer, C., and Froggatt, A. (2015) Changing climate, changing diets. In: *Chatham House Report FAO, 2017. The Future of Food and Agriculture – Trends and Challenges*. London: Chatham House.

25

AN ETHICAL APPROACH TO ASSESSMENTS OF HARM FROM CLIMATE CHANGE AND GLOBAL HEATING

The Problem with Integrated Assessment Models and Other Value-Laden Tools

Geoffrey Garver

Introduction

The increasingly recognized urgency to act collectively around the world to mitigate climate change derives from the growing awareness of the catastrophic harm that will result if adequate mitigation is not achieved. The ethical case is simple and clear: acting to prevent horrific harm, despite some inconvenience, is the right thing to do if you can do it. Yet, the question of how much to mitigate greenhouse gas emissions, and how fast, persists (Gupta 2020). Resistance to acting arises because the systems dynamics of climate change and global change are extraordinarily complex, such that the causes and effects of climate change and global heating are spatially and temporally removed from each other and shrouded in varying levels of uncertainty. For example, much of the most dire harm to which past and present emissions contribute—the harm that threatens to cause broad civilizational collapse—will occur far in the future. And, island states like Vanuatu and the Maldives that have contributed little to climate change would be among the first places to suffer catastrophic harm. Further, much of this future harm will be due to complex reinforcing feedbacks that are both difficult to predict with precision and virtually impossible to reverse once they take hold, such as the replacement of Arctic ice by dark land surface that will absorb instead of reflect solar radiation.

In addition, the inconveniences of taking sufficient action in the short term to prevent future harms from climate change and global heating have immediacy that those futures harms, the precise nature and timing of which are uncertain, lack. Imposing easily conceivable near-term costs or inconveniences to address uncertain and hard to conceive future harms gives rise to arguments against urgent and strong action to mitigate climate change that many people find compelling. In *Merchants of Doubt*, Oreskes and Conway (2010) explain in detail how powerful interests, especially the fossil fuel industry, have swayed public opinion, especially in North America, against strong climate mitigation action by exaggerating and mischaracterizing uncertainty about future costs and harms of climate change and casting aggressive efforts to control greenhouse gases as an assault on economic growth, freedom, preferred lifestyles and individual wealth and opportunity.

DOI: 10.4324/9781003039860-30

Scientifically and ethically sound assessments of likely future harms from climate change and global heating are essential for countering the skewed framing and factual distortions of the well-funded and politically connected 'merchants of doubt' that have largely succeeded to date in preventing adequate government-led action to mitigate climate change. The five assessments to date of the Intergovernmental Panel on Climate Change (IPCC), with the sixth in progress, are an important foundation of this work. They make a compelling case for action. The tricky part is translating the scientific assessments from the IPCC and other rigorous and reliable climate researchers for policymakers and decision makers who must weigh them against other considerations—such as those put forth by the climate skeptics and doubters. A common way to do this kind of translation is to estimate the value of future harms in monetary terms (in essence, the future costs of not mitigating now), so that, with discounting, the estimated present monetary value of those future harms—the social cost of carbon (Snyder 2019)—can be compared to the present costs of imposing restrictions on greenhouse gas emissions or of other measures to mitigate climate change.

The integrated assessment models (IAMs) that merge climate and economic dynamics so as to estimate the social cost of carbon in this way have proliferated in recent years. Typically, IAMs allow integrated assessment of the long-term effects of different mitigation policies on the economy and on the climate system (Nikas et al. 2019). Many IAMs raise serious questions, including ethical ones, regarding what kind of future harms of climate change and global heating should be considered; how to address uncertainty and catastrophic risks; whose views regarding the nature and extent of those harms is represented; the expression of those harms in monetary terms; the use of discounting, pure rate of time preference (which estimates the preference for consumption today as opposed to in the future) and other techniques that affect the present monetary value of those harms; and, fundamentally, the worldviews and numerous assumptions that are built into the models.

This chapter presents a review of the ethical appropriateness of relying on IAMs to make policy decisions regarding whether and to what extent action is warranted now to mitigate climate change and global heating. This analysis, consistent with several other reviews of IAMs, reveals that flaws in adequately presenting future climate harms in present day monetary terms are nearly impossible to overcome. The Dynamic Integrated Climate-Economy (DICE) model of William Nordhaus, who won the 2018 Sveriges Riksbank Prize in Economic Sciences in Memory of Alfred Nobel (a.k.a. the Nobel Prize in Economics), illustrates these flaws well. Yet, DICE is likely the best known and most influential IAM to date, and because it tends to disfavor strong action now to mitigate future climate harms (Nordhaus 2008), climate skeptics and conservative politicians have relied on it to oppose climate regulations (Inhofe 2007).

Because of the many problems with expressing future climate harms in monetary terms using IAMs, policymakers and decision makers should rely first and foremost on the consensus assessments of climate scientists regarding the nature and likelihood of future impacts from climate change and global heating for a range of greenhouse gas emissions scenarios. In other words, the work of the IPCC and other rigorous climate assessments should guide policy on questions of future climate damages with little or no regard to estimates of the monetary value of those impacts. Policy and decision makers should then favor multi-criteria assessment, participatory dialogue and tribunals focused on justice and fairness over monetary cost-benefit assessments for making value-based decisions about climate change policy. These alternative approaches to cost-benefit analysis would better account for the perspectives of the most vulnerable victims of climate change and for intangible and immeasurable harms to people and ecosystems, many of which are compounded, interconnected and virtually impossible to express in economic terms.

A Systems View of Future Climate Harms

The fifth report of the IPCC is clear about the risks of damage due to climate change and global heating under a range of emissions scenarios, "Rising rates and magnitudes of warming and other changes in the climate system, accompanied by ocean acidification, increase the risk of severe, pervasive and in some cases irreversible detrimental impacts" (IPCC 2014: 13). Human-induced climate change will lead to rising sea levels; regional increases in both drought and extreme precipitation events and associated wildfires; loss of polar ice; and ocean acidification and other changes to ocean chemistry. The severity and risk of catastrophic crossing of systemic climate tipping points increases with emissions and the associated rise in atmospheric greenhouse gas concentrations and global surface temperature (IPCC 2014). For example, the IPCC has documented significantly more severe climate impacts for a post-industrial temperature rise of 2°C than for a temperature rise of 1.5°C (IPCC 2018: 7–11). Conversely, not only can timely mitigation actions prevent sea level rise from becoming catastrophic, but also, slowing the rate of sea level rise through aggressive mitigation likely "enables greater opportunities for adaptation in the human and ecological systems of small islands, low-lying coastal areas and deltas" (IPCC 2018: 7).

These climate impacts will in turn harm other features of the global ecosystem (IPCC 2014; Steffen et al. 2015; Segan et al. 2016). Where several different thresholds are used to define safe operating space for human and other life, as with planetary boundaries (Rockström et al. 2009; Steffen et al. 2015), the boundaries' interrelatedness and interactions must be considered. Because "[i]nteractions among planetary boundaries may shift the safe level of one or several boundaries" (Rockström et al. 2009: 24), as pressure on one boundary, such as climate change, increases, other boundaries will likely have to be adjusted so as to shrink humanity's safe operating space (Rockström et al. 2009). Thus, the interactions among different planetary boundaries have important implications. For example, ocean acidification is directly related to increased levels of atmospheric carbon dioxide, for which the overarching consequence of concern is climate change (Steffen et al. 2015). Impacts of climate change and global heating on biodiversity are particularly noteworthy. Climate change is likely to significantly exacerbate habitat loss, the greatest global threat to biodiversity (Segan et al. 2016). Many plant and animal species face increased risk of extinction as climate change and global heating increase, due to severe impacts on habitat, the inability to shift geographical ranges fast enough, and broad-scale impacts such as lower oxygen levels and acidification in marine ecosystems (IPCC 2014).

Human victims of climate change include anyone whose basic human needs of survival (air, water, food, shelter, and procreation), safety (security stability and peace), social engagement (family and community), esteem (status and recognition) and self-fulfillment are diminished or threatened by the impacts of climate change and global heating (Popovski & Mundy 2012: 8–9). Among these needs, the cascading impacts of climate change and global heating are projected to pose perhaps the clearest threats to human safety and survival. The interrelated impacts of climate change and declining biodiversity will affect food security, as marine fisheries decline and terrestrial plant and animal species face pressures from high temperatures, drought and reduced surface and groundwater for irrigation (IPCC 2014). Climate change and global heating will also pose clear risks to human health and the ability of people to work outdoors in some regions (IPCC 2014). The IPCC (2014: 15) states with very high confidence that in "urban areas climate change is projected to increase risks for people, assets, economies and ecosystems, including risks from heat stress, storms and extreme precipitation, inland and coastal flooding, landslides, air pollution, drought, water scarcity, sea level rise, and storm surges." Meanwhile, shocks to rural areas will include "major impacts on water availability and supply, food security, infrastructure and agricultural incomes, including shifts in the production areas of food and non-food crops around the world" (IPCC 2014: 15). The consequent increase in poverty and in mass migrations and climate refugees will increase the risk of violent conflict and social strife in many parts of the world (IPCC 2014).

Although these impacts of climate change and global heating, and the resulting damages and harms, can be described in broad terms, the exact nature, timing and magnitude of the impacts are impossible to predict for any of the IPCC's possible future emissions scenarios. Not only are the specific details of known types of impacts and potential human responses uncertain, but also, some future impacts and possible responses may currently be unknown altogether. For example, the risk of war and violent conflict may increase as the climate changes, but questions as to what kind of war, with what weapons, affecting what geographic area, and with what ancillary impacts on ecosystems and human security remain virtually unanswerable. All that can be said is that those impacts could well be catastrophic in and of themselves.

The Problems with DICE and Other IAMs

Linking the global climate system with economic systems involves two basic causal relationships. In one direction, economic activity and economic growth over time drives impacts on the climate. For example, in a fossil fuel economy, economic growth typically correlates with an increase in greenhouse gas emissions that exacerbate climate change and global heating (Steffen et al. 2011; Bai et al. 2016). In the other direction, climate change and global heating have many adverse effects on the economy, for example, by contributing to lower crop yields or more intense weather events and sea level rise (Nikas et al. 2019). These adverse impacts of climate change can be modeled with complex climate models, such as those on which the IPCC reports rely. With IAMs, the goal is to translate these climate impacts "into monetary terms, or processed through a model of the economy, to give a single measure of the economic cost of climate change" (Nikas et al. 2019: 2). Ultimately, IAMs assess climate policies first and foremost on the basis of their impact on economic growth in gross domestic product (GDP) terms (Snyder 2019).

Nordhaus's DICE model is an 'optimal growth' model that seeks to determine the balance of consuming or investing capital (including the climate system, or more precisely atmospheric concentration of greenhouse gases, as a form of natural capital) now and in the future that will produce the maximum social welfare, expressed through utilitarian metrics, through time (Nordhaus 2008; Nikas et al. 2019; Gupta 2020). The model

> Links together the factors affecting economic growth, CO2 emissions, the carbon cycle, climate change, climatic damages, and climate-change policies. The equations of the model are taken from the different disciplines – economics, ecology, and the earth sciences. They are then run using mathematical optimization software so that the economic and environmental outcomes can be projected.
>
> *(Nordhaus 2007: 15)*

Policies that protect natural capital, such as a well-functioning climate, for the future (that is, that involve investments in climate change mitigation) are assessed in regard to how much they (1) reduce present consumption, (2) displace investments that would generate other kinds of future consumption or economic benefits and (3) allow for future consumption that unmitigated climate change would render impossible.

A prominent criticism of the DICE model (Stanton et al. 2011: 12–13), and IAMs more generally (Pezzey 2018), is of the assumptions underlying the discount rate and the pure rate of time preference it uses, both of which factor heavily in Nordhaus's conclusion that aggressive action to mitigate climate change is not economically justified (Nordhaus 2007). Discount rates reflect ethical choice, not scientific reasoning (Snyder 2019). Discount rates can be based on either the expected return on a risk-free investment, or on the estimated "pure rate of time preference (e.g., the value of future generation's welfare compared the present generation's welfare), the elasticity of marginal utility, and the growth rate of consumption" (Snyder 2019: 1571). Not surprisingly, the range and form of discount rates, in IAMs or more generally, varies considerably. Lower discount rates tend to increase the present value of future costs, and it is possible to use graduated discount rates that decrease over time so as to increase the present value of future generations, or even to use negative discount rates if the future is valued more than the present. Discount rates, above all, account for the sharply divergent conclusions of Nordhaus (2007), who used a discount rate of 4.5% and advocated against strong climate mitigation, and Stern (Stern & Taylor 2007), who used a discount rate of 1.4% and advocated for strong climate mitigation action, in a high-profile public clash in 2007 (Nordhaus 2007a).

Another set of criticisms of DICE and other IAMs targets the way they model climate-economy interactions (Pezzey 2018). For example, the climate damage function in the DICE model and others, which typically derives an aggregated economic cost associated with a given global temperature increase, which is in turn derived from economically-driven emissions increases, obviously has a significant effect on model results (Gupta 2020: 17). As it turns out, the most important simplification in the DICE model's climate subsystem is reflected in the relationship of global surface temperature to climate damages. Climate damages include "estimated damages to major sectors such as agriculture, sea-level rise, health, nonmarket damages, as well as estimates of catastrophic damages" (Nordhaus 2007: 47). They are proportional to world output and related to temperature by an equation that "is extremely conjectural given the thin base of empirical studies on which it rests" (Nordhaus 2007: 47). The DICE model also

> Assumes there are no genuinely catastrophic outcomes that would wipe out the human species or destroy the fabric of human civilizations [because e]stimating the likelihood of and dealing with catastrophic outcomes is beyond the frontier of current research in this area.
>
> *(Nordhaus 2007: 33)*

Concerns about climate damage functions go well beyond DICE, such that Pezzey (2018: 4) concludes that, especially in estimating damages associated with global heating of 3°C and higher, "[climate damage functions] should be regarded as 'highly unknowable' … [and] alternatives to using [social costs of carbon] should be considered."

The DICE model's weakness in relating climate change to economic impact on human and natural systems is well acknowledged (Costanza et al. 2007; Gupta 2020). Gupta (2020) focuses on two critical flaws in the way DICE handles climate damage: first, it does not use a sufficiently precautionary estimate of the sensitivity of the climate to rising emissions, and second, it uses global temperature rise as the only measure of climate damage. Specifically, the DICE model uses a climate sensitivity of 3°C (that is, with a doubling of pre-industrial greenhouse gas emissions, the global temperature will stabilize in the long term at 3°C above the pre-industrial average), whereas the likely range is 1.5°C–4.5°C and a 10 percent chance exists that it will be 6°C or more. In other words, the DICE model ignores a "non-negligible probability of [climate sensitivity] being very high" (Gupta 2020: 13n5). As well, the DICE model predicts a relatively slow rise in climate damages as global temperature rises, with no real attempt to account for catastrophic non-linear shifts that have some probability of occurring sooner than the most likely estimates, for example through accelerated reinforcing feedbacks from loss of Arctic and Antarctic snow and ice (Gupta 2020). A parallel set of critiques could be applied to the adequacy with which the DICE model and other IAMs model the cost of adaptation to climate change and global warming—for example, protection against sea level rise, measures to protect against or avoid floods and forest fires, increased air conditioning and so on—if mitigation efforts are minimized or shifted to the future.

Gupta (2020) suggests that the DICE model could be improved by changing the climate damage equations, but leaving the overall structure intact. Tweaking the model in this way, however, fails to address more fundamental problems with DICE and other IAMs that are founded on a growth-insistent, anthropocentric view of the economy and of the human-Earth relationship. A major assumption in DICE and other IAMs, inherent in neo-classical economic growth theory, is that the best measure of social welfare is utility expressed in terms of per capita consumption (Costanza et al. 2007). Further, IAMs are typically built on the shaky foundations of the weak sustainability version of capitalism, by which all forms of capital—economic, social or natural, for example—can be expressed in monetary terms and substituted for one another without limit (Gowdy & McDaniel 1999; Speth 2008). This approach "is in contradiction to the assumptions of environmental ethics and most major ecotheologies which share the assumption that non-human life has intrinsic value" (Snyder 2019: 1570, citations omitted).

Even without the homogenization of capital involved in weak sustainability, the mere conceptualization of Earth's communities of life, its ecosystems, as a form of capital excludes many perspectives and value systems. Seeing ecosystems as capital is necessarily anthropocentric. It means objectifying, economizing, and commodifying ecosystems and instilling them with instrumental value to humans above all else, as with the notion of ecosystem services. This economist perspective is even more perverse when, as in the DICE model, the economy, the climate and climate impacts are highly aggregated, with the economy expressed in a single monetary metric and the climate and impacts on it expressed in a single temperature value. This aggregation masks inequalities in wealth, dilutes the moral force of climate injustices and erases the wide diversity with which different people and peoples around the world relate to and value each other and the ecosystems that sustain them (Snyder 2019). Aggregation also masks issues regarding historical greenhouse gas emissions, which open up important ethical and justice-based discussions regarding who is responsible for addressing the harms from those past emissions and who is entitled to emit the limited remaining amount of emissions that humanity as a whole can afford to emit if safe operating space for human and other life on Earth is to be maintained. Although some note the relative simplicity of the DICE model (Costanza et al. 2007), the price of that homogenization is a distortion or minimalization of questions of inequality, fairness and justice that are highly relevant to climate policy.

Beyond the convoluted nature of most methods for expressing ecosystem harms and other non-market consequences of climate change in monetary terms, a more fundamental problem is that the conversion of climate damages into monetary terms makes it much easier to force comparisons of incommensurable values in order to make the math work. The tendency to incorporate the language of money into policy domains is rooted ultimately in the conception of humans as apart from nature, and as rational actors free to accommodate and own elements of nature in the quest to maximize personal wealth and well-being (Nadeau 2006: 81–123). As with monetary valuation of ecosystem services, monetary valuation of climate damages—even assuming all the relevant categories are included—requires assigning values less than infinity in order to derive values that can be used in policy assessments. Thus, IAMs are intolerant to viewing certain things, such as generations-old attachment to place, or the role of pollination in maintaining ecosystems and food supplies, as infinitely valuable. In fact, "climate change is expected to have profound humanitarian impacts which cannot (or should not) be monetized including increases in war and conflict, droughts and malnutrition, migration and disease" (Snyder 2019: 1570, citations omitted), yet those impacts are implicitly accounted for in the assumptions used in IAMs.

Alternative Approaches: Science-Led Determinations of the Need for Climate Mitigation

When one steps backs from the details of IAMs and overly economic arguments on how to address threats of severe and ultimately catastrophic consequences of climate change and global heating, the problematic moral essence of such approaches becomes clear. The fundamental ethical flaw with the monetary valuation of climate damages in IAMs is akin to the ethical flaw in the famous Ford Pinto case. In a rush to compete with foreign auto manufacturers, Ford callously weighed the benefits of cheaper car production against the cost of damages it would have to pay for human injury or death because of its negligence in not addressing a known risk of Pintos catching fire when rear-ended (De George 1981; Bazerman & Tenbrunsel 2011). The moral depravity of the cost-benefit calculations used to justify the rush to produce the Pinto without adequate safety technology, which in effect put a price on human life, hit home when real victims, such as the three young girls who died after the motor of the Pinto they were in exploded after it was rear-ended (De George 1981), became known. The stark lesson of the Ford Pinto case and others like it is that monetary damages are not sufficient to pay for certain kinds of harm or injury. Instead, measures are needed to prevent the harm from occurring in the first place, using a precautionary approach, because allowing the injury to occur crosses a moral boundary that should not be crossed.

The IPCC has high confidence that "[i]mpacts on natural and human systems from global warming have already been observed" (IPCC 2018: 5). Thus, as with victims of the Ford Pinto, already the real identities behind the economic abstractions that represent victims of the failure to mitigation climate change in IAMs are becoming known. A remote community on the Carteret Islands in Papua New Guinea has been recognized as "the first recorded people on Earth, in the context of human-induced climate change in the twenty-first century, to have lost their ancestral lands to rising sea levels" (Popovski & Mundy 2012: 7). The residents of coastal cities such as Maputo (Mozambique), Dar Es Salaam (Tanzania), Mombasa (Kenya), Knulna (Bangladesh) and Cotonou (Benin) are known to be particularly vulnerable to sea level rise in the near term, due to human-induced climate change (Popovski & Mundy 2012: 7). These are some of the most obvious victims. Taking into account already existing diminution and threats from climate change to the full range basic human needs—survival, safety, social engagement, esteem and self-fulfillment—the extent of present and future victimization from climate change is much broader. And, humans are but one species. Already, many other species are significantly harmed and threatened by climate change and global heating, including coral reefs and the species that depend on them; Arctic species like caribou, lemmings and tundra breeding shorebirds (Gilg et al. 2012), and mountaintop species like the Giant mountain lobelia of Ethiopia (Chala et al. 2016) and the Sierra Nevada blue butterfly of Spain (Aragon et al. 2019). Although it will never be fully possible to know what part of a severe drought, storm or other climate-related event is due to human-induced climate change, or, consequently, to identify all of the victims of those impacts with precision, those impacts and victims are becoming increasingly clearer, putting the ethical dimensions of climate harms and injuries in ever sharper focus.

In light of the impossibility of accurately or ethically expressing future climate damages and harms in monetary terms, Pezzey (2018) recommends abandoning altogether the use of climate damage functions that assign monetary value to those harms. Such an approach would preclude the approach in DICE, which reduces the expression of harm to a single global temperature value that is then linked to a monetary damage function. IAMs could still be used, but only to model the climate mitigation associated with different carbon prices, with the ultimate carbon emissions reductions policy targets determined using methods that avoid giving climate damages a monetary value (Pezzey 2018).

Fortunately, cost-benefit calculations like those used in DICE and other IAMs are not necessary to derive timetables and emissions reductions needed to adequately mitigate climate threats. Normative limits on greenhouse gas emissions based on scientific and ethical considerations and systems thinking are already being determined. The planetary boundaries research that began in 2009 supports a normative governance approach by which global policy should focus ultimately on staying within global ecological limits determined using a precautionary approach (Rockström et al. 2009). For climate change, the updated precautionary boundary is still an atmospheric CO_2 concentration of 350 ppm, with the zone of uncertainty 350–450 ppm. This climate change boundary correlates relatively well to the Paris Agreement ambition to keep the post-industrial increase in global temperatures to well below 2°C and preferably no more than 1.5°C (Hansen et al. 2008). The IPCC has indicated that to keep post-industrial global heating below 1.5°C, net CO_2 emissions must decline by about 45 percent from 2010 levels by 2030 and reach zero around 2050, and to stay under 2°C, net CO_2 emissions would need to decline by about 25 percent by 2030 and reach zero around 2070 (IPCC 2018: 12). Thus, for both outcomes, a downward total global emissions pathway leading to zero net emissions can be derived, with the area under the curve from now until the end point representing the total net amount of greenhouse gas emissions that likely will allow the 1.5°C or 2°C goal to be met—the remaining carbon budget (IPCC 2018: 104). The IPCC's most precautionary estimate of the remaining (post-2017) carbon budget, which has a 66 percent probability of keeping post-industrial temperature rise to 1.5°C or less, is 420 GtCO2, not counting the possibility that carbon and methane releases from melting permafrost and wetlands would require the budget to be significantly smaller (IPCC 2018: 12).

In its assessment of how to keep post-industrial global temperature rise under 1.5°C or 2°C, the IPCC confirms planetary boundaries researchers' affirmation that "myriad pathways" exist for respecting the boundaries (Rockström et al. 2009), which for climate change means emissions pathways that stay within the remaining carbon budget (IPCC 2018). For climate change, selecting policies that respect this budget first involves broad consideration of the countless economic drivers of greenhouse emissions and destruction or impairment of carbon sinks, such as forests. These economic drivers and their impacts can be further broken down in several ways, for example geographically, into sectors and through analysis of material and energy flows. The resulting pictures of the distribution of economic drivers of climate change and their impacts then provide the basis for development societal responses, aggregated to the global level from broadly distributed political and governance contexts. The best that can be said about the Paris Agreement is that it represents the first efforts to pull together aggregated national responses toward the global goal. Although the IPCC (2019) has noted that national commitments under the Paris Agreement collectively fall short of the 1.5°C target, the agreement contains provisions for continuous improvement. Any hope that climate damages and harm can be limited to those associated with a 1.5°C post-industrial rise in global temperature is strongly tied to whether sufficient tightening of emissions reduction targets and mandates takes place through those provisions.

Conclusion

IAMs and the cost-benefit analysis they support purvey a veneer of sophistication and objectivity that masks often dubious assumptions about the nature of the social welfare and the economy, about future impacts of climate change and global heating on human societies and the ecosystems that support them and, ultimately, about the human-Earth relationship. Reducing the consideration of the harmful and ultimately catastrophic future impacts of climate change and global heating to assessments grounded in cold hard cash strips climate policy of humanity. Thus, it is morally imperative that IAMs not be relied on to establish target levels of atmospheric carbon dioxide or post-industrial temperature rise based on estimates of the monetary value of future climate damages and harm. It is best to leave that work to the IPCC and similar scientific bodies. Once those targets are established, IAMs might play a limited role in assessing the most economical pathways for staying within the allowable remaining carbon budget, as with cap-and-trade regulatory schemes. For example, the IAMs that the IPCC used in assessing socio-economic pathways for attaining the goal of limiting post-industrial temperature rise to 1.5°C were "typically… GHG pricing mechanisms to induce emissions reductions and associated changes in energy and land uses consistent with the imposed climate goal" (IPCC 2018: 100).

Meanwhile, some climate damage has already occurred, and much more will occur in the future, even staying within a temperature rise of 1.5°C. The victims of those negative climate impacts, human and otherwise, are very often geographically and temporally removed from those who cause them. Assigning responsibility and accountability for those damages in the quest for climate justice will involve complex questions of causation and liability that will pose additional challenges for policymakers, judges and human rights tribunals for many years to come.

References

Aragon, P., Sánchez-Fernández, D. and Hernando, C. (2019) "Use of satellite images to characterize the spatio-temporal dynamics of primary productivity in hotspots of endemic Iberian butterflies," *Ecological Indicators* 106: 105449.

Bai, X., van der Leeuw, S., O'Brien, K., Berkhout, F., Biermann, F., Brondizio, E.S., Cudennec, C., Dearing, J., Duraiappah, A., Glaser, M., Revkin, A., Steffen, W. and Syvitski, J. (2016) "Plausible and desirable futures in the Anthropocene: A new research agenda," *Global Envtl Change* 39: 351–362.

Bazerman, M.H. and Tenbrunsel, A.E. (2011) "Ethical breakdowns," *Harvard Business Review,* April 2011.

Chala, D., Brochmann, C., Psomas, A., Ehrich, D., Gizaw, A., Masao C.A., Bakkestuen, V. and Zimmermann, N.E. (2016) "Good-bye to tropical alpine plant giants under warmer climates? Loss of range and genetic diversity in *Lobelia rhynchopetalum,*" *Ecology and Evolution* 6: 8931–8941.

Costanza, R., Leemans, R., Boumans, R. and Gaddis, E. (2007) "Integrated global models," in Costanza, R., Graumlich, L.J. and Steffen, W. (eds.), *Sustainability or Collapse: An Integrated History and Future Of People on Earth,* Dahlem Workshop Report 96, Cambridge, MA: MIT Press.

De George, R.T. (1981) "Ethical responsibilities of engineers in large organizations: The Pinto case," *Business & Professional Ethics Journal* 1(1): 1–14.

Gilg, O., Kovacs, K.M., Aars, Fort, J., Gauthier, G., Grémillet, D., Ims, R.A., Meltofte, H., Moreau, J., Post, E., Schmidt, N.M., Yannic, G. and Bollache, L. (2012) "Climate change and the ecology and evolution of Arctic vertebrates," *Annals of the New York Academy Science* 1249: 166–190.

Gowdy, J.M. and McDaniel, C.N. (1999) "The physical destruction of naura: An example of weak sustainability," *Land Economics* 75(2): 333–338.

Gupta, S. (2020) "Has economics caught up with climate science?" *Ecology, Economy and Society – The INSEE Journal* 3(1): 11–30.

Hansen, J., Sato, M., Kharecha, P., Beerling, D., Berner, R., Masson-Delmotte, V., Pagani, M., Raymo, M., Royer, D.L. and Zachos. J.C. (2008) "Target atmospheric CO2: Where should humanity aim?" *Open Atmospheric Science Journal* 2: 217–231.

Inhofe, Sen J. (2007) "Floor speech: Inhofe, boxer debate global warming on senate floor," *Press Release* (29 October 2007), Washington, DC: U.S. Senate online:<http://epw.senate.gov/public/index.cfm?-FuseAction=Minority.PressReleases&ContentRecord_id=eddb00e1-802a-23ad-4b39-b70900f-52b98&Region_id=&Issue_id=>.

IPCC. (2014) *Climate Change 2014: Synthesis Report, Summary for Policymakers,* Geneva: Intergovernmental Panel on Climate Change.

IPCC. (2018) *Global Warming of 1.5°C. An IPCC Special Report on the Impacts of Global Warming of 1.5°C Above Pre-Industrial Levels and Related Global Greenhouse Gas Emission Pathways, in the Context of Strengthening the Global Response to the Threat Of Climate Change, Sustainable Development, and Efforts to Eradicate Poverty,* Masson-Delmotte, V., Zhai, P., Pörtner, H.-O., Roberts, D., Skea, J., Shukla, P.R., Pirani, A., Moufouma-Okia, W., Péan, C., Pidcock, R., Connors, S., Matthews, J.B.R., Chen, Y., Zhou, X., Gomis, M.I., Lonnoy, E., Maycock, T., Tignor, M. and Waterfield, T. (eds.), Geneva: Intergovernmental Panel on Climate Change.

Nadeau, R.L. (2006) *The Environmental Endgame: Mainstream Economics, Ecological Disaster, and Human Survival*, New Brunswick, NJ: Rutgers University Press.

Nikas, A., Doukas, H. and Papandreou, A. (2019) "A detailed overview and consistent classification of climate-economy models," in Doukas, H., Flamos, A. and Lieu, J. (eds.), *Understanding Risks and Uncertainties in Energy and Climate Policy* (pp. 1–54), Cham, Switzerland: Springer.

Nordhaus, W. (2007) *Accompanying Notes and Documentation on Development of DICE-2007 Model: Notes on DICE-2007. Delta. v8 as of September 21, 2007.* Miscellaneous publication, New Haven CT: Yale University

Nordhaus, W. (2007a) "Critical assumptions in the stern review on climate change," *Science* 317: 201–202.

Nordhaus, W. (2008) *A Question of Balance: Weighing the Options on Global Warming Policies,* New Haven, CT: Yale University Press.

Oreskes, N. and Conway, E. M. (2010) *Merchants of Doubt,* New York: Bloomsbury Press.

Pezzey, J.C.V. (2018) "High unknowability of climate damage valuation means the social cost of carbon will always be disputed," *CCEP Working Paper 1802,* Crawford School of Public Policy, The Australian National University.

Popovski, V. and Mundy, K.G. (2012) "Defining climate change victims," *Sustainability Science* 7: 5–16.

Rockström, J., Steffen, W., Noone, K., Persson, Å., Chapin, III, F.S., Lambin, E., Lenton, T. M., Scheffer, M., Folke, C., Schellnhuber, H., Nykvist, B., De Wit, C.A., Hughes, T., van der Leeuw, S., Rodhe, H., Sörlin, S., Snyder, P.K., Costanza, R., Svedin, U., Falkenmark, M., Karlberg, L., Corell, R.W., Fabry, V.J., Hansen, J., Walker, B., Liverman, D., Richardson, K., Crutzen, P. and Foley, J. (2009) "Planetary boundaries: Exploring the safe operating space for humanity," *Ecology and Society* 14(2): 32.

Segan, D.B., Murray, K.A. and Watson, J.E.M. (2016) "A global assessment of current and future biodiversity vulnerability to habitat loss—climate change interactions," *Global Ecology and Conservation* 5: 12–21.

Snyder, B.F. (2019) "Beyond the social cost of carbon: Negative emission technologies as a means for biophysically setting the price of carbon," *Ambio* 49: 1567–1580.

Speth, J.G. (2008) *The Bridge at the Edge of the World*, New Haven, CT: Yale University Press.

Stanton, E.A., Ackerman, F. and Kartha, S. (2011) *Inside the Integrated Assessment Models: Four Issues in Climate Economics,* Working Paper WP-US-0801, Stockholm: Stockholm Environment Institute.

Steffen, W., Grinevald, J., Crutzen, P. and McNeill, J. (2011) "The anthropocene: Conceptual and historical perspectives," *Philosophical Transactions of the Royal Society A* 369: 842–867.

Steffen, W., Richardson, K., Rockström, J., Cornell, S.E., Fetzer, I., Bennett, E.M., Biggs, R., Carpenter, S.R., de Vries, W., de Wit, C. A., Folke, C., Gerten, D., Heinke, J., Mace, G.M., Persson, L.M., Ramanathan, V., Reyers, B. and Sörlin, S. (2015) "Planetary boundaries: Guiding human development on a changing planet," *Science* 347(6223): 1259885-1–1259885-10.

Stern, N. and Taylor, C. (2007) "Climate change: Risk, ethics, and the stern review," *Science* 317(5835): 203–204.

26

ETHICAL ISSUES RAISED BY GEOENGINEERING TECHNOLOGIES TO REDUCE THE THREAT OF CLIMATE CHANGE

Shi Jun

Introduction

The international community is not taking the action necessary to avert dangerous increases in atmospheric concentrations of greenhouse gases (GHGs). Facing a potentially bleak future, humanity is confronting the question of whether the best of bad alternatives may be to counter global warming through human-engineered climate interventions. It is widely accepted in the scientific community that climate change poses a severe threat to current and future generations, as well as to the rest of nature. Nevertheless, the countries of the world are not currently on track to meet their stated goal of avoiding dangerous anthropogenic interference with the climate system, understood in terms of the internationally agreed targets of limiting average global temperature rise to 2°C and pursuing efforts to achieve 1.5°C. Moreover, the political prospects for further robust action remain questionable. Consequently, many scientists are concerned that in practice deliberate large scale technological interventions into the climate system ('geoengineering') are already, or may soon become, unavoidable if the 1.5°C or 2°C targets are to be met.

Geoengineering is an idea that has been in circulation for decades and perhaps centuries, though it is difficult to trace the origins of the term (Fleming 2010). The idea of geoengineering to modify the climate has been around for decades (Buck 2012). A 2006 editorial in Climatic Change by Paul Crutzen made geoengineering a respectable topic of discussion in the scientific community and it is now beginning to be discussed more widely. The growing interest in geoengineering is partly a reaction to the steady increase in GHG atmospheric concentrations which has continued despite scientific consensus, international treaties, and solemn declarations. In the background is the fact that many people find Promethean solutions to urgent problems almost irresistibly attractive. The Paris Agreement, by setting the aim of limiting the temperature increase to 2°C or even 1.5°C in this century compared to pre-industrial levels, may have unwittingly given an impetus to the carbon removal option; that is to say the 'negative emission' technologies. The vast majority of mitigation scenarios now include massive deployment of negative emission technologies, notably bio-energy combined with carbon capture and storage (BECCS) (Clark & Herzog 2014; Mander et al. 2017; Rogelj et al. 2015; Schleussner et al. 2016). As Anderson (2015) has convincingly demonstrated, the only plausible assumption behind the overwhelming majority of 2°C mitigation scenarios depicted by the Intergovernmental Panel on Climate Change (IPCC) is, to Anderson's dismay, the reliance on large-scale removal of CO_2 from the atmosphere.

 DOI: 10.4324/9781003039860-31

Combined with the insufficient mitigation commitments by countries (Rogelj et al. 2016), this makes the use of some form of geoengineering seem inevitable.

The Royal Society defines geoengineering as "...deliberate large-scale intervention in the Earth's climate system, in order to moderate global warming" (Royal Society 2009). Geoengineering methods or technologies are divided into two overarching types: carbon dioxide removal (CDR), which aims to reduce atmospheric carbon dioxide levels, and solar radiation management (SRM), which aims to reflect some of the sun's heat away from the earth.

CDR geoengineering techniques are intended to draw carbon dioxide from the atmosphere, but do not affect GHG emissions. Examples include large-scale reafforestation, BECCS, and machines that 'suck' carbon from the air. To have a long-term effect, many of these techniques rely on geological storage of the extracted carbon. Other examples include processes to enhance the ocean's biological capacity to draw down and sequester carbon in the deeper ocean. These include ocean fertilization using iron or other nutrients, mechanical promotion of circulation of deeper nutrient-rich waters to the surface, and ocean liming to reduce ocean acidity and increase carbon removal (Lenton 2014). There is no consensus in the geoengineering literature whether large-scale biochar, which is plant-based charcoal used to increase soil carbon retention, or large-scale afforestation should also be included in the definition of CDR. There is serious disagreement about exactly what technologies count as geoengineering (US GAO 2010). The Royal Society counts afforestation as geoengineering (presumably only if it is undertaken with the right intention), while the IPCC considers afforestation to be a form of mitigation.

SRM techniques are intended to limit warming by intercepting solar radiation before it reaches the earth's surface. Proposed methods involve atmospheric measures such as injecting particles into the stratosphere to deflect sunlight or spraying salt particles into low-lying marine clouds to make them whiter and more reflective; terrestrial measures such as covering the ocean or deserts with reflective material; and space-based approaches such as placing sunshades or mirrors in orbit (Vaughan & Lenton 2011). Modeling suggests that SRM could reduce the magnitude of climate change impacts caused by elevated GHG concentrations (IPCC 2013), although with regional variations (Kravitz et al. 2014). Should atmospheric GHG concentrations reach dangerous levels, some SRM methods could, in theory, be deployed quickly to bring temperatures down – an important consideration for those communities and species most vulnerable to rising temperatures. CDR offers the capacity to achieve the practical effect of negative emissions, which climate models suggest will be necessary to hold global average temperatures within 2°C in the absence of effective SRM (IPCC 2013).

Is Geoengineering the Last Resort?

Global warming is a manifestation of the imbalance of the earth's ecosystem caused by human activities, which can be compared to a 'climate disease' suffered by the earth. The main reason for supporting geoengineering is that the earth's 'climate disease' has reached an advanced stage, and the 'prescriptions' such as mitigation and adaptation have thus far failed. We can only justify the use of extreme technical means constituted by geoengineering to fight the likely catastrophic impacts of climate change. In other words, the emergence of serious consideration of geoengineering options has arisen due to the current failure of global climate politics and governance, and therefore sometimes viewed as a necessary last resort.

According to this logic, geoengineering is a kind of 'Plan B' – it will not be put into operation until 'Plan A' fails. Deploying geoengineering is like using a 'life-saving straw' experimental drug for dying patients. What's more, geoengineering promises to be a kind of 'morning-after pill' for societies that do not practice safe consumption (Goodell 2010), or a kind of 'painless therapy' (Barrett 2008; Nordhaus 1992) – geoengineering is typically treated as nearly cost-free (especially SRM).

Is geoengineering a last resort? We can examine this problem from three related questions: The first question: has the earth's climate system become 'terminally ill' so that we have reached the moment when the last resort must be used? Of course not. Although the threat of climate change is growing, it is not yet the last moment, although there is growing evidence that some of the positive feedback 'tipping points' are showing signs of destabilization. Once a feedback threshold is tripped, it will not only speed up global warming, it runs the risk of tripping other feedbacks, further increasing warming rates. Yet, the earth's climate system has certain self-healing ability to remove atmospheric GHG concentrations although even after global GHG emissions are reduced to net zero, the climate will not immediately return to the temperature regime that existed before human-induced warming began because although 80 percent of GHG emissions are removed naturally from the atmosphere by carbon sinks in the first two decades, 20 percent remain for thousands of years (Brown 2020). Climate change is a medium and long-term process. The most serious impact of climate change will occur in the future, and its negative impacts are not evenly distributed. Poor countries and poor people in the future will suffer the most serious harms. This is also a selfish reason for the low enthusiasm of contemporary people to deal with climate change – the unwillingness to sacrifice personal immediate interests for the benefit of others, especially residents of poor countries and those who are not yet born. In fact, in response to the climate crisis, human beings are far from reaching the time when they need to use the 'last resort.' The reason for resorting to climate geoengineering is that human beings are not willing to give up the huge benefits brought by high emission production mode and hope to continue to enjoy the unsustainable high emission lifestyle.

If people mistakenly believe that medical technology can save people's lives at any time, they will rely on medical technology as a 'life-saving straw,' thus reducing the use of preventive measures such as physical exercise. This is not only a wrong understanding of medical technology, but also a wrong understanding of ourselves. Climate geoengineering reflects man's wrong understanding of himself, nature, and society. The wrong understanding of self leads to the 'moral hazard' of human arrogance, anthropocentrism, consumerism, and geoengineering; the wrong understanding of nature leads to the 'ecological risk' of mechanical view of nature, technology omnipotence, and geoengineering; the wrong understanding of society leads to social injustice at the level of intrageneration and intergeneration, and the 'social risk' of relying on geoengineering.

The second question: is there no alternative to geoengineering – a risky approach to climate change? Obviously not. In fact, both mitigation and adaptation are 'special medicine' to treat 'climate diseases.' Prevention is at the heart of climate change action, and GHG emission reduction is the 'best climate vaccine' to prevent the threat of climate change. It can be said that, even if geoengineering is effective, it is only the 'temporary medicine' to delay the crisis caused by 'climate disease,' while mitigation and adaptation are the 'radical medicine' to solve the disease. Scientists have diagnosed the earth's 'climate disease': if the current warming trend continues, the earth's ecosystem and human civilization system will soon collapse. But the sad thing is that people know how to prevent climate disaster, but they are still reluctant to take effective action. In essence, it's not that people are reluctant to deal with climate change, but that they are reluctant to give up their short-term benefits.

The third question: even if geoengineering is a 'last resort,' should we use it, or when? There is also a close relationship between geoengineering and risk society: the natural environmental changes caused by global warming increase the risks we are facing, and the geoengineering technologies strengthen these risks. Medical technology may not only delay the deaths of patients, but also may bring new risks and accelerate the death of patients. The same is true for climate geoengineering, which may not only reduce global climate impacts, but also bring new and greater risks. Climate warming is a high risk, so is the climate geoengineering to prevent global warming harms. Facing the high risk of the use of climate geoengineering, how should we choose? Is it worth the risk? Nonetheless, the science of many geoengineering methods is complex and uncertain. How should we choose? Is it worth the risk? The science of geoengineering method is extremely complex and full of uncertainty. Current geoengineering options, including the most mentioned carbon capture and storage technologies, are unproven technologies. Due to the limitation of human knowledge, the chaos of climate system, and the uncertainty of climate change and its impact, we cannot accurately predict the complex interaction and its possible impact to the earth system (and social system). Therefore, geoengineering options may open the 'Pandora's box' and produce unexpected 'unintended consequences' to the earth's ecosystem. And human social systems bring new dilemmas and disasters. The science behind many of the more controversial geoengineering proposals is highly uncertain and some geoengineering techniques such as ocean fertilization are likely to be inefficient and risky. The deployment of global geoengineering methods, or the application of local geoengineering methods at a scale needed to have global effect, are likely to have regionally disparate impacts potentially adverse climatic and/or environmental effects (National Research Council 2015a, b). The associated potential environmental and human impacts, and their possible uneven regional distribution, support the conclusion suggest that any prospect of geoengineering warrants measured consideration and debate at multiple jurisdictional scales. In some cases, the use of a 'last resort' with high uncertainty and serious side effects is better than tolerating climate change, because its side effects and uncertainty may be broadly catastrophic. Yet, blind use of 'last resort' is a gambler's strategy – the bet is the future of the earth and mankind. 'Last resort' should not be used until the 'final moment,' but who is obliged and have the ability to judge that climate change has put human society in the 'last moment.' A full risk assessment of each possible geoengineering option should be required prior to use. However, due to a large number of uncertainties in the science, technology, ecology, and social impacts that any risk assessment would seek to describe, this kind of risk assessment would face challenging issues in generating political confidence. This results in state actors being either disengaged or adopting a 'wait and see' response while supporting further academic research.

Proponents of geoengineering should respond to the following questions that warrant further discussion:

1 Is 'Plan A' such as the deployment of one geoengineering technology enough to deal with climate change?

   Most environmentalists reject geoengineering because they rightly believe that it is better to prevent climate change than to cover up its symptoms – only a substantial reduction in GHG emissions can truly address climate change. As long as all countries take immediate action to reduce GHG emissions and turn to green economy, human society will have a safe and stable climate and sustainable society. However, in the current global political, economic and value system with egoism as the 'universal value,' the 'free rider' strategy is often preferred by people, which makes it impossible for global cooperation to slow down and adapt. A zero emission, moral, low-carbon society may be a utopia, because capitalism and its desire to encourage economic growth will continue until the earth collapses. Therefore, the dubious strategy of geoengineering is realistic. Unfortunately, man-made warming is now a reality, so we have to choose not between mitigation of climate change and ambitious geoengineering, but between hazards.

2 Is 'Plan A' and 'Plan B' mutually exclusive, and so that only one of them could be chosen from them?

   Some argue that climate policy should be a toolkit and should not be seen as an absolute competition between geoengineering and other approaches to climate change, because "the real problem we face is not in choosing the right class of responses to climate change but rather in selecting and implementing a portfolio of specific strategies" (Jamieson 2013). Therefore, geoengineering research should be supported as part of climate change options to compare with other possible climate change policy options. In other words, in order to provide backup and comparative options, geoengineering research should be supported, but not necessarily deployed. Most geoengineering advocates say that what they want is not deployment, but a dedicated research program. The program would start small, but if appropriate, could gear up to the scale of the Manhattan Project. They claim that there is a bright line between research and deployment. The tension between the two can be resolved if research and deployment can be cleanly separated. Deployment will take place only after rigorous research to ensure that the potential risk of a geoengineering option is sufficiently small. Bright lines, however, have a way of fading, particularly when it comes to dedicated research programs. Yet research and deployment cannot be clearly divorced from each other for some geoengineering technologies, either physically or from a governance perspective. Many geoengineering methods require years of large-scale experimental studies in natural ecosystems (such as the ocean or atmosphere), and these studies are actually deployment. Therefore, often there is no clear line between meaningful field-testing of geoengineering and deployment (Keith et al. 2010). There is also the potential for a financial or technological slippery slope from research to deployment (Oldham et al. 2014).

3 Is geoengineering a global public good?

In early policy work, geoengineering is often described as a global public good. What is the significance of identifying geoengineering as public goods? If geoengineering is a public good, its research and deployment should be supported. Oxford Principles declares that geoengineering is "to be regulated as a public good" for all humanity (Rayner et al. 2009, 2013). It is commonly assumed that genuine public goods benefit everyone, and a pure public good is a good that satisfies two conditions: nonrival and nonexcludable. But there is no reason to assume that any geoengineering will have a general net benefit – some geoengineering interventions will have an adverse impact on some people, others will be detrimental to all (such as those that could lead to climate runaway). Therefore, Gardiner (2013) argued that the public good framing is seriously misleading, in part because it arbitrarily marginalizes ethical concerns. David R. Morrow (2014), contrary to Gardiner's claim, holds that geoengineering meets the two standards of public goods, and that geoengineering is a kind of public good even if it is not fully deployed. 'Public goods' is not a value neutral conception; it must be beneficial to but not harmful to public interests. In addition to the two conditions of nonrival and nonexcludable, public goods also need to have a value premise: it must be a 'good' thing. Public goods can produce public benefits such as the benefits to the public like a lighthouse. Although geoengineering may bring benefits, its high uncertainty and high-risk characteristics can also bring disaster to much of all mankind or some regions. Therefore, geoengineering may also be a kind of 'public evil.'

Ethical Issues Raised by Geoengineering

Many geoengineering techniques pose issues common to other transboundary environmental problems, including challenges of achieving harmony among competing political institutions, worldviews, and values that frequently confront attempts to govern the global commons. It is generally recognized that many geoengineering project proposals would generate serious social conflicts, thus mandating that 'ethical considerations are central to decision-making in this field,' and that 'analysis of ethical and social issues associated with research and deployment' should be a central research priority (Shepherd et al. 2009).

Moral Hazard

A central debate in combatting climate change: whether we should focus on reducing CO_2 emissions or on removing the emitted CO_2 from the atmosphere. The perception of geoengineering as a viable option reduces the willingness to reduce CO_2 emissions. If the Paris Agreement targets are to be met, pursuit of a reduction in emissions seems all the more rational (Turaj et al. 2018).

'Moral hazard' is a unique ethical problem faced by geoengineering: these technologies will distract people's attention and effort on global mitigation and adaptation (Reynolds 2014). However, reducing carbon emissions is the most basic challenge we face (Keith et al. 2010).

Even talking about geoengineering has, already to some extent, weakened our willingness to reduce emissions. Such as William Nordhaus (1992), wrote: geoengineering, would introduce a hypothetical technology that provides costless mitigation of climate change. This could occur, for example, if one of the geoengineering options proved technically feasible and environmentally benign. Two interesting proposals include shooting smart mirrors into space with 16-in. naval rifles or seeding the oceans with iron to accelerate carbon sequestration.

This 'low-cost geoengineering dream' has caused moral hazard to GHG emission reduction. The possibility of geoengineering provides an excuse for the continuation of unsustainable production and lifestyle, potentially encouraging emissions. Research and development of geoengineering may also lead to the transfer and loss of funds for climate change mitigation and adaptation spending. Governments will need to decide whether they should provide huge amounts of money for geoengineering research and development, or for emission control, to prevent the adverse consequences of climate change, or to compensate the victims.

The very idea of intentionally changing climate strikes many people as arrogant and may result in greater disrespect for nature, an attitude that is responsible for climate change. Many attempts to manipulate and 'improve' nature have caused environmental problems. Although it is impossible for human beings not to intervene in nature at all, there is a growing sense that modern societies have erred on the side of excessive intervention so that they have become arrogant and intrusive in attempting to manage all elements of nature. Human beings have a long history of controlling and destroying the ecology, but people who have a strong desire to manipulate nature can never escape potential adverse impacts of nature. Many human behaviors that have attempted to manipulate nature have produced unintended consequences – and attempts to control climate change through various biotechnology methods risks numerous unintended consequences.

'Jevons Paradox' and Capital Logic

Technological progress itself is also an important factor that intensifies the development and utilization of the earth's resources and the release of a huge amount of GHGs. Technological criticism has become a common target of ecological movement. Ecologists believe that their most basic work is to eliminate the modern society's uncritical confidence in technology, and to eliminate the modern society's belief in unlimited economic growth. The dangerous energy consumption pattern of modern society has been created despite 50 years of warnings from the mainstream scientific community about the harms of human-induced warming because of a widespread belief that our science and technology will protect us. Many developed nation citizens believe that technological progress will eventually solve the problems caused by its energy systems including climate change. Proponents of green developmentalism often regard it as a belief that technological breakthroughs will enable society to substantially reduce resource use, make production 'dematerialized,' and decouple growth from GHG emissions, so that production can grow forever while resource consumption declines. However, the development and application of green technology is always subject to the availability of capital.

Historically, higher energy efficiency has always been accompanied by an increase in energy consumption. As early as 1865, William Jevons, one of the founders of the economic marginalism school, pointed out that the use of increasingly efficient steam locomotives would lead to a large increase in coal consumption. In other words, lower unit energy prices due to higher energy efficiency will encourage people to consume more energy. Every progress of energy development and utilization technology is accompanied by the geometric exponential growth of energy consumption, and the growth of consumption far exceeds the speed of technological progress. In the face of the unlimited expansion of desire for consumption, technology seems to be unable to meet its needs.

Technological progress is a double-edged sword of creation and destruction. Science and technology endow human beings with the power to use a large amount of energy, and thereby constantly cause great impact on the earth's ecosystems. It was soon discovered that technology might be more of a threat than an antidote for some problems. For example, in the context of climate change and energy crisis, new energy technologies seem to open up 'unlimited' possibilities for the expansion and growth of capital. New energy is regarded as a promising cheap, low-carbon, sustainable power source to replace the traditional energy, and becomes the hope of realizing the sustainable growth of capitalism. However, can new energy technology and its new energy really enable human society to avoid the situation of energy resource depletion and the environmental crisis? New energy is part of the solution to many environmental problems, but not the vast majority. More and more evidence shows that new technology and new energy cannot solve many of the problems that capitalism causes, but may worsen the problems of energy and environment. Modern science and technology have achieved economic success, but it may be an ecological failure. If new energy is to really help solve the climate crisis, energy crisis and associated social problems, it needs to be sustainable not only in energy, ecology, and economy, but also in society as a moral matter.

Ethical Principle for Geoengineering Governance

Geoengineering could have significant global environmental and social impacts. Since any geoengineering has great potential to create natural and social risks, moral considerations should be an important focus of attention in this field. So far, there is no particular set of norms that specifically apply to proposals that have been introduced in discussions of geoengineering projects. The governance of geoengineering needs to consider moral issues raised by the technologies.

Several scholars from Oxford University have put forward five 'Oxford principles' for geoengineering governance: geoengineering to be regulated as a public good; public participation in geoengineering decision-making; disclosure of geoengineering research and open publication of results; independent assessment of impacts; and governance questions before deployment (Rayner et al. 2009). However, these principles emphasize issues that arise in governance, and they do not sufficiently lay the groundwork for the more substantive ethical debate that is needed, especially around values such as justice, respect, and procedural legitimacy (Gardiner & Fragniere 2018). The ethical issues need to be examined before governance decisions are made.

As a large-scale intervention in the Earth's climate system to deal with climate change, geoengineering decisions must be guided by ethical considerations. In addition to other ethical considerations, terms of ethics, geoengineering governance should follow the principles of emission reduction priority, prudent application, and risk prevention.

Emission Reduction First

Geoengineering may be worth trying, but it cannot by itself solve the problem of climate change. In any case, there is no reason to think that we can delay mitigation and adaptation and hope that geoengineering will eventually save us. It will eventually prove ineffective to rely on geoengineering to solve global warming without active emission reduction actions. Geoengineering only diminishes the carbon budget that must constrain global carbon emissions to prevent dangerous climate change, but cannot eliminate the need to reduce global GHG emissions. All CO_2 equivalent emissions will raise atmospheric CO_2 concentrations unless they are removed by natural sinks or biotechnology and about 20 percent remain in the atmosphere for thousands of years (Brown 2020), To achieve the Paris Agreement's warming limit goals of 1.5°C and 2.0°C the international community must reduce global CO_2 emissions to net zero by 2045 to achieve the 1.5°C goal and net zero by 2070 to achieve the 2.0°C goal in absence of carbon removal technologies (Brown 2020). The unlimited emission demand will consume the remaining carbon budget that must constrain global emissions to achieve the Paris Agreement's warming limit goals, and thus threaten climate catastrophe. If we do not significantly change the status quo on human activities causing climate change, the risk of destruction for all mankind will be greatly increased.

We still have enough time to reduce emissions to avert rapid non-linear climate damages. We still have enough time to reduce emissions and transform (economic, social, and conceptual transformation) but the international community will have to converge on net zero carbon emissions in several decades. In short, geoengineering can be used as a supplement to mitigation and adaptation measures, but it cannot replace mitigation and adaptation measures. Thus, the indispensable core of dealing with climate change is to implement GHG emissions reductions. Any plan that cannot reduce carbon dioxide emissions cannot solve the problem.

Prudential Deployment

Although geoengineering research should proceed, before geoengineering technology is deployed, it will be necessary to consider several ethical issues and procedural matters that ethical issues raise. As we will see, the deployment of geoengineering technologies raises a host of ethical issues because geoengineering has the potential to disproportionally harm people around the world and daunting uncertainty issues must be resolved before geoengineering technologies can be deployed.

In the study of geoengineering, scientific research and practical application should be distinguished. On the one hand, research and peer review should be allowed, as it helps to eliminate bad proposals and encourage good ones, and knowledge itself is valuable for determining the efficacy and danger of geoengineering technologies.

A Precautionary Approach to Geoengineering Science

Because of the complexity of the global climate system, it may be difficult to prove the harms that could be caused by geoengineering technologies particularly to some local and regional areas. We expect science to not only prove and explain causation but to warn people of risks before proof is in – particularly if the risks are significant and civil society can't wait until proof is demonstrated before the harm occurs. Some local and regional harms from some geoengineering technologies may be particularly difficult to predict. For example, some SRM technologies may reduce harmful warming but increase ocean acidification in some parts of the world.

Science often discovers sound scientifically based reasons for concern about potential harm from some human activities, but for practical and theoretical reasons can't reach ideal high levels of proof before the harm occurs. Because of this, scientists are sometimes expected by the law or social norms to warn people of potential harms. For instance, sometimes scientists are expected to make reports based upon the 'balance of the evidence' or regulatory agencies are expected by law to take preventative action as long as the scientific reasons for taking preventative action are not 'arbitrary and capricious.' Because of the difficulty in determining with high degrees of certainty all the adverse impacts of geoengineering technologies, scientists should follow a precautionary approach in determining the harms that geoengineering could cause.

In examining human-environment interactions as well as human health-environment questions, scientists often uncover scientifically sound reasons for significant concern about potential harms, but for practical or theoretical reasons are unable to prove cause and effect. In such cases, ethics requires that scientists apply a precautionary approach.

There are many uses of science in its 'public policy' role about which there appears to be widespread social agreement that it would be imprudent or otherwise unethical to wait until all the proof is in before taking appropriate action. This is so if: (a) waiting guarantees that the harm will occur if the risk turns out to be real, (b) the potential harms are grave to some people or ecological systems, and (c) those being put at risk have not consented to be put at risk.

The precautionary principle here serves as a norm requiring inaction on deploying geoengineering until the consequences of this technology are better understood. Because of the uncertainty of human-induced warming, we should take active preventive measures instead of relying on blind technological optimism. 'Prevention first, treatment later' is a basic principle of environmental risk management. Governments should invest a smaller amount of money to prevent climate harms, rather than suffer the enormous losses and damages of adverse climate impacts.

Due Process

The harms and benefits of various geoengineering technologies will be differentially experienced in different parts of the world. Those who may be harmed have a right to participate in any decision on geotechnology that may adversely affect them. For this reason, which governments may participate in any geoengineering deployment decision raises serious due process ethical questions. A strong argument can be made that an international institution that has a wide international governance structure should approve deployment of geoengineering technologies that have potential adverse international impacts.

Distributive Justice

Within some countries, some subpopulations may be more vulnerable to adverse impacts of geoengineering technologies than others. For instance, SRM may reduce excessive heat in one part of a nation while affecting rainfall in other parts of a nation such that farming communities that are sensitive to rainfall amounts. Thus, geoengineering deployment decisions should consider the interests and views of all who may be affected by geoengineering deployment.

Conclusion

There are four possible options for geoengineering. The first is prohibition. Due to the huge risks and uncertainties of some geoengineering impacts, we could completely prohibit research or application that has any potential to significantly modify the climate system. But what if the risk of climate warming is far greater than the potential risk of climate geoengineering? If human-induced climate warming threatens to bring human beings to the brink of destruction, and geoengineering is the only plausible option, then geoengineering may be viewed as the last straw for dying human beings. No matter how risky it is, the global community may need to try it. Even if the result of deploying geoengineering may be to destroy human lives, governments may need to run the risks because climate change is an existential threat to much of life on earth.

The second is to encourage geoengineering research. If the international community can't effectively deal with climate change by reducing emissions, it may be necessary to deploy some kind of geoengineering? To answer this question, the international community will need to consider whether climate change is an existential threat to much of life on earth. Is geoengineering really the last option for mankind? At present, even if climate warming may destroy human being, the international community most likely has several decades to put global carbon emissions on a pathway to net zero greenhouse emissions. But because use of several decades of research may be necessary before any geoengineering technology can be deployed with confidence, an argument can be made that research on geoengineering technologies should be encouraged by the international community.

The third option is priority research questions. Because of the potential need to deploy geoengineering as a last resort in the decades ahead and because of the need to answer many research questions before geotechnologies are deployed safely, research priorities should be set and funded. Meanwhile, efforts should be made to reduce emissions and reduce associated costs.

And the fourth is that research funding amounts should be secured from the international community. Because the development of any new capabilities, including geoengineering, requires research resources, a global research strategy on geoengineering deployment should be created by the international scientific community.

In the end, geoengineering research is urgently needed to find a 'technological antidote' to the 'climate disease,' but the international community also needs a 'social antidote.' Ethicists have found the 'root cause' of 'climate disease': unsustainable social production and consumption mode. Climate change is a wake-up call to the unsustainability of society, which reminds human society to make fundamental adjustments in politics, economy, and social system. There is no way to build a sustainable future within an unsustainable social relationship. There is no technological solution to climate change on a planet dominated by unfettered free markets and consumerism.

References

Anderson, K. (2015). Duality in climate science. *Nature Geoscience*, 8(12), 898–900.

Barrett, S. (2008). The incredible economics of geoengineering. *Environmental Resource Economics,* 39, 45–54.

Brown, D. (2020) *Seven features of climate change that citizens and the media need to understand to critically evaluate a government's response to this existential threat and the arguments of opponents of climate policies*, https://ethicsandclimate.org (accessed 8 January 2021).

Buck, H. J. (2012). Geoengineering: Re-making climate for profit or humanitarian intervention? *Development and Change*, 43(1), 253–270.

Clark, V. R., & Herzog, H. J. (2014). Can "stranded" fossil fuel reserves drive CCS deployment? *Energy Procedia*, 63, 7261–7271.

Fleming. (2010). *Fixing the Sky*. New York: Columbia University Press.

Gardiner, S. M. (2013). Why geoengineering is not a 'global public good,' and why it is ethically misleading to frame it as one. *Climatic Change*, 121(3), 513–525.

Goodell, J. (2010). *How to cool the planet*. New York: Houghton, Mifflin, Harcourt.

IPCC. (2013). Intergovernmental panel on climate change. In T. Stocker, D. Qin, G. Plattner, M. Tignor, S. Allen, J. Boschung, et al. (Eds.), *Climate change 2013: The physical science basis. Contribution of working group I to the 5th assessment report of the intergovernmental panel on climate change*. Cambridge: Cambridge University Press.

Jamieson, D. (2013). Some whats, whys and worries of geoengineering. *Climate Change*, 121, 527–537.

Keith, D., Parson, E., & Morgan, M. (2010). Research on global sun block needed now. *Nature,* 463, 426–427.

Kravitz, B., MacMartin, D. G., Robock, A., Rasch, P. J., Ricke, K. L., & Cole, J. N. S., et al. (2014). A multimodel assessment of regional climate disparities caused by solar geoengineering. *Environmental Research Letters*, 9(7), 074013.

Lenton, T. M. (2014). The global potential for carbon dioxide removal. In R. M. Harrison, R. E. Hester & Royal Society of Chemistry (Great Britain) (Eds.), *Geoengineering of the climate system* (pp. 52–79). Cambridge, MA: Royal Society of Chemistry.

Mander, S., Anderson, K., Larkin, A., Gough, C., & Vaughan, N. (2017). The role of bio-energy with carbon capture and storage in meeting the climate mitigation challenge: A whole system perspective. *Energy Procedia*, 114, 6036–6043.

Morrow, D. R. (2014). Why geoengineering is a public good, even if it is bad. *Climatic Change*, 123, 95–100.

National Research Council. (2015a). *Climate intervention: Carbon dioxide removal and reliable sequestration*. Washington, DC: The National Academies Press. http://www.nap.edu/catalog/18805/climateintervention-carbon-dioxide-removal-and-reliable-sequestration.

National Research Council. (2015b). *Climate intervention: Reflecting sunlight to cool earth*. Washington, DC: The National Academies Press. http://www.nap.edu/catalog/18988/climate-interventionreflflecting-sunlight-to-cool-earth.

Nordhaus, W. (1992). An optimal transition path for controlling greenhouse gases. *Science*, 258, 1315–1319.

Oldham, P., Szerszynski, B., Stilgoe, J., Brown, C., Eacott, B., & Yuille, A. (2014). Mapping the landscape of climate engineering. *Philosophical Transactions of the Royal Society A*, 372, 20140065.

Rayner, S., Heyward, C., Kruger, T., Pidgeon, N., Redgwell, C., & Savulescu, J. (2013). The Oxford principles. *Climatic Change*, 121(3), 499–512.

Rayner, S., Redgwell, C., Savulescu, J., Pidgeon, N., & Kruger, T. (2009). *Memorandum on draft principles for the conduct of geoengineering research*. London: House of Commons Science and Technology Committee.

Reynolds, J. L. (2014). A critical examination of the climate engineering moral hazard and risk compensation concern. *The Anthropocene Review*, 2(2), 174–191.

Rogelj, J., den Elzen, M., Höhne, N., Fransen, T., Fekete, H., & Winkler, H., et al. (2016). Paris Agreement climate proposals need a boost to keep warming well below 2 C. *Nature*, 534(7609), 631–639.

Rogelj, J., Luderer, G., Pietzcker, R. C., Kriegler, E., Schaeffer, M., & Krey, V., et al. (2015). Energy system transformations for limiting end-of-century warming to below 1.5 [deg] C. *Nature Climate Change*, 5(6), 519–527.

Schleussner, C.-F., Rogelj, J., Schaeffer, M., Lissner, T., Licker, R., & Fischer, E. M., et al. (2016). Science and policy characteristics of the Paris agreement temperature goal. *Nature Climate Change*, 6(9), 827–835.

Shepherd, J., Caldeira, K., Cox, P., Haigh, J., Keith, D., Launder, B., & Watson, A. (2009). *Geoengineering the climate: Science, governance, and uncertainty* (pp. 1–98). London: The Royal Society.

Stephen M. Gardiner, & Augustin Fragnière. (2018). The tollgate principles for the governance of geoengineering: Moving beyond the Oxford principles to an ethically more robust approach. *Ethics, Policy & Environment*, 21, 2, 1

Turaj, S. Faran, & Lennart Olsson. (2018). Geoengineering: neither economical, nor ethical--a risk-reward nexus analysis of carbon dioxide removal. *International Environmental Agreements*, 18, 63–77.

US GAO. (2010). *Climate change: A coordinated strategy could focus federal geoengineering research and inform governance efforts*. GAO-10–903.http://www.gao.gov/products/GAO-10-903

Vaughan, N. E., & Lenton, T. M. (2011). A review of climate geoengineering proposals. *Climatic Change*, 109(3–4), 745–790.

27

THE ETHICS OF CARBON OFFSETS AND THEIR AFTERLIVES IN TROPICAL FORESTS

Maron Greenleaf

Introduction

Climate change raises numerous complex ethical issues. So too do many of the efforts to mitigate it – even if they fail or are implemented differently than planned. Those mitigation efforts include carbon offsets, a commodity invented in the effort to mitigate climate change. Offsets represent a reduction or removal of greenhouse gas (GHG) emissions. In carbon markets, polluters may be able to buy them to 'offset' their own emissions based on the premise that, from a climate change perspective, it does not matter where emissions or emissions reductions occur. Individuals, organizations, and corporations can buy carbon offsets voluntarily, for reputational or personal reasons or as part of corporate 'carbon neutrality' programs. Polluting entities, such as power plants, might also be able to buy offsets to compensate for some of their own emissions in compliance emissions trading or cap-and-trade programs. This chapter traces some of the ethical issues raised by carbon offsets. It pays particular attention to their climatic impacts, the capitalist practices primarily responsible for climate change, and local communities where offsets are developed.

Offsets have been highly controversial, dividing environmentalist, Indigenous, and other communities. In its attention to the veracity of offsets and their global and local effects, much critical scholarship, as well as nongovernmental and activist research, is ethically grounded. The case for offsets has ethical components as well, found within the contentions that offsets are a cost-effective way to combat climate change in the short term and to channel needed resources into underfunded climate mitigation. Offsets also raise ethical issues even when they fail to develop as expected, as they have in some sectors. Like other developmental and conservation policies and programs (e.g., Ferguson 1990; Escobar 1995), offsets can have long-term, unintended, and sometimes negative effects, even when they do not achieve their stated goals. Depending on whether one is supportive or critical of offsets, these effects can be understood as evidence that offsets are ethically problematic or as evidence that they are needed. These effects are part of what this chapter explores as offsets' ethical afterlife.

The focus in this chapter is on carbon offsets created through efforts to reduce GHG emissions from deforestation and forest degradation and conservation, enhancement, and sustainable management of forest carbon stocks in tropical forests (REDD+). Tropical forests are both an essential and climate-stabilizing carbon sink and a significant source of GHG emissions (Baccini et al. 2017; Mitchard 2018). What came to be known as REDD+ was developed in the 2000s through the United Nations Framework Convention on Climate Change (UNFCCC) negotiations as a way to mitigate climate change. For a time, it seemed like REDD+ would be funded through carbon

DOI: 10.4324/9781003039860-32

offset sales within the UNFCCC system. Yet, while some REDD+ carbon offsets are sold and private funding has recently increased, REDD+ funding has primarily come from multilateral and bilateral payments. In this, REDD+ funding has been characterized as a form of climate-linked conditional aid (Seymour & Angelsen 2012; Angelsen 2017), rather than the market transactions that were projected to bring significant private money into forest protection. At the same time, REDD+'s scope has widened, taking on an increasingly broad array of developmental, conservation, and land use-governance priorities. These changes to REDD+'s funding structure and scope are part of offsets' ethical afterlife.

The chapter proceeds as follows. First, it offers a brief history of carbon offsets and REDD+, focusing on how they have changed over the past two decades. Next, it examines the ethical issues and arguments surrounding REDD+ carbon offsets, first those connected with climate change, then those related to the global economy, and finally those focused on offsets' local impacts. The chapter examines the ethical case for and against offsets and then offsets' ethical afterlife across these three dimensions.

Carbon offsets

Take, for example, a company that was planning to build a coal-burning powerplant, but instead built a wind farm. In doing so, it could potentially produce not only renewable energy, but also a strange kind of commodity called carbon offsets. The project's offsets would likely be the equivalent of the GHGs the coal-fired powerplant would have emitted minus the GHGs produced by the wind farm. In other words, the project could earn offsets for 'avoided emissions,' also called emissions reductions. Offsets here are produced by reducing GHG emissions from a counterfactual 'business-as-usual' or other baseline emissions rate. They might also be created by removing GHGs from the atmosphere, as with tree planting. All sorts of entities and actors are involved in producing carbon offsets, including landowners, carbon offset developers, auditors, and carbon offset certifiers.

Carbon offsets are part of carbon markets, which are based on the premise that the location of GHG emissions does not matter climatically because emissions mix in the atmosphere, impacting the global climate system. Based on this understanding of GHG fungibility, carbon markets attempt to harness 'the market' to combat climate change. The aim is to reduce emissions where it is most efficient (i.e., cheapest) to do so effectively. Carbon markets are not the first instance of this kind of market-based approach; a similar kind of pollution trading was used to address acid rain under the 1990 amendments to the US Clean Air Act (Calel 2013).

There are two basic kinds of carbon markets, voluntary and compliance, and carbon offsets can be a component of both. Both kinds of markets use metric tons of carbon dioxide equivalent ($MtCO_2e$) as a common metric into which all GHGs can be translated (MacKenzie 2009). In the voluntary carbon market, businesses, organizations, and individuals who want to compensate for their own GHG emissions can buy carbon offsets from those who have reduced emissions elsewhere as part of carbon neutrality and related programs. In 2020, for example, voluntary offset trading reached a record high 188.2 MtCO2e, worth US$473 million (Donofrio et al. 2021: 4, 3).

Compliance carbon markets are created by governments seeking to address climate change. In these markets, also known as cap-and-trade or emissions trading systems, the government sets an overall emissions cap and then issues permits (also known as allowances) to regulated polluters to emit up to the cap. This issuance can happen via allowances, auction, or a secondary market. The cap is ramped down over time to the desired emissions level and regulated polluters, such as power plants, need to have such permits to emit GHGs. These entities can reduce their own emissions, but they can also buy and sell the permits, effectively putting a price on carbon emissions and creating an incentive for regulated entities to further reduce their emissions so they can sell permits.

Cap-and-trade systems may also admit carbon offsets into the market – emissions reductions or removals from non-regulated entities.

Carbon markets and offsetting were adopted as part of the UNFCCC's 1997 Kyoto Protocol. Under the Kyoto Protocol, the European Union developed its EU-ETS trading system and carbon offsets were bought and sold through the Clean Development Mechanism (CDM) and the Joint Implementation (JI) mechanism. Over 7,000 CDM projects in 140 countries were developed in subsequent decades, with almost two billion tons of CO_2e reduced (UNFCCC 2018: 2–3). The stalling of international climate negotiations in the beginning of the 21st century put the future of a global carbon market in doubt (Angelsen et al. 2017). But in the meantime, carbon trading was adopted at various lower jurisdictional scales, including at the subnational level in China, Canada, and the United States (International Climate Action Partnership 2021). These programs vary significantly in the number of offsets they accept. For example, New Zealand firms have been able to compensate for all emissions with purchased offsets, while regulated entities in the northeastern United States are only able to offset 3.3 percent of their emissions with locally produced offsets (Anderson & Bernauer 2016: 388). Thousands of voluntary offset initiatives have also been developed as the voluntary carbon market continues to grow in the post-Paris Accord era.

In the meantime, carbon markets and offsets may have a resurgence at the international level through Article 6 of the 2015 Paris Agreement, the subsequent Marrakech Partnership for Global Climate Action (Blum & Lövbrand 2019), and in the wake of the 2021 Glasgow Climate Pact. The voluntary market has also continued to grow in recent years (Berkeley Carbon Trading Project n.d.; Donofrio et al. 2021), with thousands of new voluntary offset initiatives developed and support for funding in tropical forests facilitated through initiatives like the LEAF Coalition. Offsets also remain central to aviation and some other corporate mitigation strategies (Lang et al. 2019). As of 2021, for example, almost 500 large companies (with annual revenues of more than US$16 trillion) had adopted carbon neutrality commitments, most of which allowed for offsets purchases (Kreibich & Hermwille 2021: 941, 942). Voluntary offset transactions have now exceeded US$6.7 billion in value and have grown substantially in recent years (Donofrio et al. 2021: 3).

REDD+ and Offsets

There are many different kinds of carbon offsets, often with their own specific dynamics. The subset that is the focus of this chapter are those related to efforts to reduce emissions from deforestation and forest degradation and forest conservation, known now as REDD+. REDD+ was developed through and around the UNFCCC in the 2000s, though offsets from avoided deforestation were excluded from the Clean Development Mechanism. Eventually, REDD+ was incorporated into UNFCCC agreements, including the 2015 Paris Agreement. In the meantime, hundreds of private and subnational REDD+ initiatives were developed in tropical forests around the world as part of the voluntary carbon market and, aspirationally, compliance carbon markets (Simonet et al. 2020).

During this time, what constituted REDD+ changed significantly. What was initially conceptualized as a way to compensate people and institutions for forgoing climate changing deforestation became something broader (Angelsen & McNiell 2012; Turnhout et al. 2017; Wunder et al. 2020). It moved from centering on deforestation to including forest degradation and then to also include forest conservation. Environmental and social co-benefits – biodiversity conservation and poverty alleviation, for example – then also became part of it as well. Finally, in recent years, REDD+ has often been advanced as a form of landscape-centered rural governance that links forest protection with agricultural and other land uses, increasingly through jurisdiction-wide programs (Nepstad et al. 2013; Boyd et al. 2018). Through these changes, REDD+ has, in some ways, come to look more like other conservation and development programs, often with a conditional payment component.

REDD+ also became less centered on carbon offsets as a funding mechanism. The majority of international REDD+ funding has come from multi- and bi-lateral sources (Norman & Nakhooda 2015) and REDD+ offsets have not been admitted into compliance cap-and-trade markets. Moreover, carbon markets themselves did not develop as was once expected. "A global carbon market," as Angelsen et al. (2017: 718) state, "has not materialized and is unlikely to emerge." They point out that the amount pledged to fund REDD+ – $10 billion – is much less than the $10–15 billion per year estimated to be needed to reduce deforestation (by half Angelsen et al. 2017: 718). In this sense, Angelsen and some other scholars have argued that REDD+ became more like a form of conditional foreign aid than like the market mechanism it was envisioned to be (Seymour & Angelsen 2012; Angelsen 2017), though increasing voluntary market transactions in recent years may counter this argument to some extent.

The Ethics of Offsets

Below, some of the ethical issues surrounding REDD+ carbon offsets are traced, categorized into three groups: those related to climate change, those related to capitalism (the economic system linked to climate change), and those related to offsets' local effects in the places where emissions are reduced. While somewhat specific to REDD+, many of these ethical issues are also raised by other types of carbon offsets.

Ethical Arguments for Offsets

The case for carbon offsets has been presented as a way to facilitate climate change mitigation by reducing its cost (e.g., Anderson & Bernauer 2016). According to this logic, since GHGs have a global effect on the climate system, they should be reduced wherever it is cheapest to do so. Offsets can reduce the cost of carbon within a cap-and-trade system and offer arguably a low-cost way to mitigate emissions voluntarily, garner corporate and popular support for regulating emissions (see Newell & Paterson 2010).

This kind of efficiency argument contains an ethical component: since climate change itself, with its unequal impacts, raises many ethical issues, addressing it is ethically necessary. In as much as offsets are an effective form of climate change mitigation – a contested proposition, as explored below – both in and of themselves and as a way to gain corporate support for mitigation (or at least acquiescence to it), there is an ethical argument for them. Efficiency arguments contain this kind of ethical proposition. As Newell and Paterson (2010: 95), for example, write about the US's acid rain legislation that was a precursor to carbon markets: "By dealing with the need to cut emissions efficiently, the scheme does so fairly."

There is also an ethical argument for carbon offsets as a way to change how capitalist economies work. Climate change is sometimes depicted as a 'market failure' due to polluters' ability to emit GHGs without paying for them (Stern 2006). Pricing pollution is therefore a way to 'internalize' these 'externalities' so that pollution and its effects are included in corporate profit-based decision-making. Such efforts to price climate costs might be seen, as Michel Callon (2009) characterizes it, as an experiment in 'civilizing markets.' Moreover, in monetizing standing forests and other forms of carbon sequestration, carbon markets and their offsets can give monetary value to what some conceptualize as 'ecosystem services' – the workings of ecosystems on which humans rely but for which we generally do not pay. By giving monetary value to forms of emissions reductions and removals outside of cap-and-trade markets, offsets might be seen as part of this 'civilizing' project and, in this, as an effort to reshape capitalism – the system that many fault with engendering climate change (e.g., Moore 2017).

Finally, there is an ethical argument for offsets focused on the places they are produced. Offsets have been seen as a way to channel climate mitigation funding into places and sectors, such as tropical forests, that are not included in carbon markets and where climate mitigation (among many other things) is underfunded. More generally, they have been seen as a way for the global North to funnel mitigation funding to the global South. Since tropical forests are an important source of emissions (when burned), as well as an important carbon sink, there is a climate component to this argument. But it goes beyond this: from quite early in REDD+'s development, social and environmental co-benefits became a focus (Turnhout et al. 2017). Including tropical forests in carbon markets was thus meant, in part, to channel significant amounts of money into tropical forest protection in ways that could benefit local people and ecosystems. Through requirements in various offset certification initiatives, offset initiatives' social and environmental 'safeguards,' and REDD+ readiness programs (such as the World Bank's Forest Carbon Partnership Facility), offsets seemed like they could be a way to financially support forest conservation, poverty alleviation, rural development, and rights' recognition. For example, land tenure rights that have been a goal of Indigenous and other forest communities have been described as a prerequisite to equitable and effective REDD+ (see Larson et al. 2013; Duchelle et al. 2013; Sunderlin et al. 2014).

Ethical Critiques of Offsets

Often seen as part of an ethically-critiqued neoliberalism that aims to make markets the primary mode of governance, carbon offsets have been controversial from the start. Most fundamentally, there is an overall critique of offsets as a way for polluters to avoid a duty to reduce emissions themselves – essentially buying a right to pollute (Dhanda & Hartman 2011; Hyams & Fawcett 2013; Page 2013). This has been compared to the medieval church's sale of indulgences to sinners in a monetization of absolution (Monbiot 2006; Smith 2007).

There are a number of other related critiques, some of them empirically-based, which fall into different categories. Some center on the issue of climate change. Since climate change requires extensive reductions in GHGs in the very near term, reductions should be pursued in all locations, rather than reductions in one place offsetting continued emissions elsewhere. In other words, location of emissions does matter and, while forest protection and other forms of mitigation should be happening, emissions reductions need to happen now in the industries and places that are responsible for most emissions – fossil fuel use, historically in the global North (Lohmann 2008; FOE 2009).

Moreover, if carbon offsets do not actually reduce GHG emissions, then they are not an effective form of climate change mitigation and only allow it to continue. In as much as climate change itself, with its widespread but highly unequal impacts, is an ethical issue, carbon offsets' veracity is an ethical issue as well. Moreover, if emissions reductions are not real, then offset purchases may amount to 'green washing' or fraud, with polluters obtaining positive publicity and sometimes the right to pollute through false claims of emissions reductions.

There are multiple ways in which offsets might not represent real emissions reductions. Most prominently, there are the issues of 'additionality' – the question of whether a reduction in emissions would have occurred without a mitigation intervention (Newell & Paterson 2010; Erickson et al. 2014). There is also the question of permanence (Galik & Jackson 2009) – what if forests that are protected today as the basis for carbon offsets are burned tomorrow? 'Leakage' is another issue. Perhaps a forest within the borders of one particular REDD+ project is not converted into a soy field, resulting in the issuance of carbon offsets, but then another forested area outside of the project is turned into a soy field. Many strategies have been developed to address these and other seemingly 'technical' issues, as well as other issues with early CDM offsets. Yet questions persist as to what extent offsets represent 'real' emissions reductions, which remain essential because their sale to polluters enables continued emissions.

Offsets are also critiqued for enabling and extending forms of global capitalism that have driven climate change and created other extensive inequalities and injustices. For some, offsets obscure the real causes of climate change and thus enable the continuation of extraction and pollution, with their unequal impacts, climatic and otherwise (e.g., Lohmann 2008; Ervine 2013; Gilbertson 2017). Carbon offsets have been analyzed as a form of 'accumulation by decarbonization' (Bumpus & Liverman 2008, building on the conception of "accumulation by dispossession" (Harvey 2003; Luxemburg 2015; West 2016)). This can be seen as "a form of unequal exchange in which commodities that are produced in the global South are priced at less than those in the North" (Bumpus & Liverman 2008: 142), allowing offset buyers in the global North to keep profiting by polluting (Bryant et al. 2015). With the global South as a source of cheap emissions reductions, offsets can reinforce global inequalities, including via the dispossession of land and labor (Harvey 2006; Kosoy & Corbera 2010; McAfee 2012) and might be seen as a form of 'carbon colonialism' (Bumpus & Liverman 2011) and 'green grabbing' (Fairhead et al. 2012; Leach et al 2012). In this critique, offsets enable wealthy polluters and places to continue to benefit from GHG emissions-fueled economic growth whilst undermining Indigenous and other locally-based forms of life in the global South. Moreover, the expertise that offsetting requires privileges more powerful actors, such as companies and well-resourced NGOs (Bryant et al. 2015; Wang & Corson 2015).

Offsets have also been critiqued as contributing to environmental injustice by enabling the continued use of and emissions from GHGs. There are multiple issues here. For one, GHGs are often emitted with co-pollutants, such as nitrogen oxides (NO_x) and sulfur oxides (SO_x), with significant negative impacts on local communities and ecologies. Since marginalized communities tend to live near such polluting facilities because of dynamics like environmental racism (see Pulido 1996), they can be disproportionately harmed by offset-enabled GHG emissions (Bullard 2000). Moreover, fossil fuel extraction, processing, and transport can have negative local impacts (e.g., Watts 2004; Whyte 2017), so fossil fuel's perpetuation via offsets is implicated in these local harms.

A third set of ethical issues is raised by the impact of carbon offsets in communities and landscapes where emissions are meant to be reduced. Here, the issue is that even if carbon offsets do effectively and efficiently reduce GHG emissions, any negative impacts on people or ecosystems would be ethically problematic. There are a number of types of impacts to highlight here. First, the creation or discovery of new resources can exacerbate and/or engender new struggles over rights and access in 'frontier spaces' (Rasmussen & Lund 2018). Researchers have examined how offset and other carbon forestry programs have led to or exacerbated dispossession or the criminalization of land uses (Bond 2012; Cavanagh & Benjaminsen 2014; Leach & Scoones 2015; Fischer et al. 2016) or undermined common property relations (Osborne 2015). This has sometimes been paired with the (re)centralization of forest control (Phelps et al. 2010; Osborne 2015; Milne et al. 2019) and/or profits accruing to outside actors (Fairhead et al. 2012). Additionally, offset programs and REDD+ initiatives may provide insufficient or uneven benefits (Milne et al. 2019) and conflict with local forms of knowledge and land use practice (Fairhead et al. 2012).

Moreover, obtaining consent – whether to fulfill the International Labor Organization's Convention 169 or other consent requirements – can be difficult to do ethically (e.g., Hendrickson & Corbera 2015; Dawson et al. 2018; Boer 2019). Relevant factors that can undermine consent include unequal power relations, insufficiently narrow conceptions of consent, inadequate information about project impacts, and community disagreements about offsets. Insufficient consent processes are themselves unethical, and they can also further divide communities and enable ineffective and/or harmful offset projects.

There is also concern that the commodification of carbon can have negative impacts through elevating and isolating carbon sequestration from the social and ecological relations that enable it (e.g., Castree 2003). Through such 'carbonization' (Gupta et al. 2012), the concern is that a landscape managed to maximize carbon can negatively impact other aspects of social and environmental relations, such as biodiversity and a landscape's cultural value and local meanings, knowledge, and land use practices (Fairhead et al. 2012).

Such critiques and related forms of activism have had some effect on offsets and their governance. For example, offset certifications such as the Gold Standard were developed as correctives to identified problems with CDM offsets through the adoption of social and environmental safeguards, project eligibility criteria, additionality and transparency rules, and the required demonstration of at least two co-benefits (Blum & Lövbrand 2019: 3, 5). There are also various strategies that have been developed for dealing with risks and uncertainty surrounding REDD+ emissions reductions (see Angelsen 2017: 249–250, 251). And some Indigenous and other forest communities have successfully used REDD+ to influence national policies and to raise their concerns in international negotiations (Joidon 2017). Yet offsets remain ethically controversial since their adequacy, effects, and efficacy are disputed, as is the offsets' basic premise that emissions reductions in one place can compensate for continued emissions somewhere else.

Offsets' Ethical Afterlife

Carbon offsets have not developed in the form or to the level anticipated, as detailed above, though their role in the Paris Agreement's implementation remains unclear. This may be due in part to critiques of the sort laid out above, as well as the stalled effort to create a global cap-and-trade system and REDD+ offsets' continued exclusion from those national and subnational systems that do exist. Both the anticipation that carbon offsets would fund forest protection and their failure to do so have had impacts that might be considered from an ethical perspective. In other words, the ethical dimensions of offsets extend beyond the arguments for and against them to include their afterlife.

One aspect of offsets' afterlife is the inadequacy of the funding for tropical forest protection that they provide in the absence of another robust form of climate mitigation funding. Much of the REDD+ bilateral funding was framed as 'bridge funding' to prepare for a more lucrative carbon market (Guizol & Atmadja 2008) that has failed to materialize within the UNFCCC structure. At least during the 2010s, non-market and voluntary REDD+ funding did not replace the level of funding promised by carbon markets, the price of carbon in voluntary and compliance markets has been volatile and often low, and pledged REDD+ funding is vastly insufficient to compensate for emissions reductions (Angelsen 2017). As Turnhout et al. (2017: 3) argue:

> [W]hile countries are getting ready for REDD+ and are preparing to supply carbon credits, it is unclear what exactly they are getting ready for, in terms of the financial mechanisms to be put in place, the conditions attached to it, and whether there will be sufficient finance available.

Corbera and Martin (2015: 2024) reflect on how the declining price of carbon, oversupply of carbon offsets, and uncertainty around the future of carbon markets creates a "potentially difficult situation for those poor, agrarian communities who have been, and continue to be, encouraged to enter into the carbon offset business." This is particularly a problem given the performance-based payment structure of REDD+ and other mitigation. The prospects of payment need to be credible. REDD+ may now be aid-like not only in terms of its structure and source, but also in terms of its inadequacy and the unequal power relations that it engenders.

The impact of this inadequacy is exacerbated by the fact that REDD+ offsets seemed like they would bring a lot of new private money to tropical forests. This created a kind of 'economy of expectations' (Fletcher et al. 2016) recognized by both offset proponents and critics. The prospect of compliance market funding was 'a powerful motivating factor' (Angelsen & McNeill 2012: 35) for governments at different levels, NGOs, businesses, and landholders to invest time, money, and political capital in costly initiatives for REDD+ funding. Much of the funding has not materialized. This should be considered when interpreting continued and, in some places, increasing deforestation (e.g., INPE 2020). For example, some government officials in the Amazonian state of Acre, Brazil, where I conducted research, felt that they had done a lot of work to create a system to protect forests financed by offset sales, only to have offset buyers never show up. Such unmet expectations can undermine future climate mitigation efforts.

Even as REDD+ offset funding failed to materialize, REDD+ initiatives and some other offsets came to encompass and be made responsible for an increasing array of rural development priorities. As outlined above, while REDD+ started with a focus on carbon, it soon came to include social and environmental co-benefits. As part of this, institutions like the Gold Standard offset certification have sought to make sustainable development co-benefits central to emissions reductions by quantifying social and environmental impacts in relationship to the Sustainable Development Goals (Blom & Lövbrand 2019: 3–4). In recent years, landscape approaches that bring together forestry and agriculture have also been advanced as best practice. Meanwhile, the rise of jurisdictional REDD+ programs has meant increasingly large and institutionally complex REDD+ programs that are meant to be integrated into governmental priorities, policies, and programs.

All of this is costly. And yet, even with its expanded environmental and social scope, as Turnhout et al. (2017) point out, REDD+ programs are still generally funded (and at a low price) based on the carbon emissions avoided, if they are funded at all. The argument for attaching payments to emissions reductions is clear: "Reduced emissions are the primary goal of REDD+ and performance should be measured as directly as possible" (Angelsen 2017: 245). Yet this focus on emissions does not match the wide variety of developmental and environmental goals REDD+ now encompasses.

Not only do payments for emissions reductions not fully reflect the broad array of low-emissions development for which REDD+ and other mitigation have become a stand-in. Scholars have also pointed out the tension between carbon sequestration and other environmental, social, and governance priorities increasingly attached to REDD+. There may be a trade-off, for example, between goals of poverty alleviation, sustainable development, and community benefits on the one hand and market-based forest conservation on the other (Chhatre & Agrawal 2009; McAfee 2012; Osborne & Shapiro-Garza 2018). Where REDD+ represents low emissions rural development in tropical forest areas writ large, the continued focus on carbon as the rubric for payments means that tropical forest jurisdictions are being asked to use relatively meager REDD+ funding to support low-emissions rural development overall. This is an ethical issue of offsets' afterlife because it tends to be marginalized communities and ecologies that suffer the most.

Conclusion

The ethical issues raised by climate change include those connected to efforts to mitigate it, whether or not those efforts are successful. This chapter has traced the ethical dimensions of carbon offsets using REDD+ offsets as an example. Offsets include ethical issues related to their efficacy as a form of climate change mitigation, their impact on global capitalism, and their local effects in places where they are created. Moreover, offsets' overall contention that emissions are fungible itself is ethically debatable.

Arguments for and against carbon offsets contain ethical dimensions. Where some see carbon offsets as a form of effective mitigation that also enables other mitigation via cost containment, others see dubious emissions reductions that salve the conscience but do not address the climate crisis. Where some see an effort to bring consideration of environmental harm into corporate profit-seeking, others see an extension and retrenchment of global capitalism that causes not only climate change but also many unequally borne harms. Where some see REDD+ as an opportunity to channel funding into poverty alleviation, land rights protection, and biodiversity protection, others see it as a threat to humans and nonhumans.

It may be unlikely that a global carbon market or its envisioned offsets will ever be created, though offsets continue through a number of subnational cap-and-trade efforts and the voluntary market. But offsets also have an ethical afterlife, regardless of their future. They have cast a long and ethically complex shadow in the form of heightened expectations, limited funding, and extensive environmental and developmental responsibilities that are now linked to climate mitigation. Proponents and critics of offsets may disagree as to whether this ethical afterlife is a result of too few offsets or too many.

Regardless, those with different views on the ethics of offsets may agree that countries and companies in the global North should fund tropical forest protection, as well as other forms of climate change mitigation, and that what has been promised and actually given to date has been inadequate. The case for doing so is about both ethics and efficacy, which are, in the case of climate change at least, inextricably linked: climate change cannot be effectively combated without forest protection and other mitigation measures in the global South, even though those areas are not its primary cause. Whatever form this global mitigation funding may take, it will raise difficult ethical issues. And some of these will likely be quite similar to those raised by carbon offsets. Hopefully, whether or not offsets remain a prominent mitigation strategy, the experience of trying to make them can inform more equitable forms of mitigation in tropical forests and beyond.

References

Anderson, B., and Thomas, B. (2016). 'How Much Carbon Offsetting and Where? Implications of Efficiency, Effectiveness, and Ethicality Considerations for Public Opinion Formation.' *Energy Policy* 94: 387–395.

Angelsen, A. (2017). 'REDD+ as Result-Based Aid: General Lessons and Bilateral Agreements of Norway.' *Review of Development Economics* 21 (2): 237–264. https://doi.org/10.1111/rode.12271.

Angelsen, A., Brockhaus M., Duchelle, A.E., Larson, A., Martius, C., Sunderlin, W.D., Verchot, L., Wong, G., and Wunder, S. (2017). 'Learning from REDD+: A Response to Fletcher et al.' *Conservation Biology* 31 (3): 718 720. https://doi.org/10.1111/cobi.12933.

Baccini, A., Walker, W., Carvalho, L., Farina, M., Sulla-Menashe, D., and Houghton, R.A. (2017). 'Tropical Forests Are a Net Carbon Source Based on Aboveground Measurements of Gain and Loss.' *Science* 358 (6360): 230–234. https://doi.org/10.1126/science.aam5962.

Berkeley Carbon Trading Project. n.d. *Voluntary Registry Offsets Database*. Available at: https://gspp.berkeley.edu/faculty-and-impact/centers/cepp/projects/berkeley-carbon-trading-project/offsets-database. Accessed April 28, 2021.

Blum, M., and Lövbrand, E.. (2019). 'The Return of Carbon Offsetting? The Discursive Legitimation of New Market Arrangements in the Paris Climate Regime.' *Earth System Governance* 2: 100028.

Boer, H.J. (2019). 'Deliberative Engagement and REDD+ in Indonesia.' *Geoforum* 104 (August): 170–180. https://doi.org/10.1016/j.geoforum.2019.04.025.

Bond, P. (2012). 'Emissions Trading, New Enclosures and Eco-Social Contestation.' *Antipode* 44 (3): 684–701. https://doi.org/10.1111/j.1467-8330.2011.00890.x.

Boyd, W., Stickler, C., Duchelle, A.E., Seymour, F., Nepstad, D., Bahar, N.H.A., and Rodriguez-Ward, D. (2018). 'Jurisdictional Approaches to REDD+ and Low Emissions Development: Progress and Prospects.' *World Resources Institute*. Available at: https://www.cifor.org/knowledge/publication/6933/. Accessed on April 15, 2021.

Bryant G., Dabhi, S., and Böhm, S. (2015). ''Fixing' the Climate Crisis: Capital, States, and Carbon Offsetting in India.' *Environment and Planning A* 47: 2047–2063.

Bullard, R.D. (2000). *Dumping in Dixie: Race, Class, and Environmental Quality*. 3rd ed. Boulder, CO: Westview Press.

Bumpus, A.G., and Liverman, D.M. (2008). 'Accumulation by Decarbonization and the Governance of Carbon Offsets.' *Economic Geography* 84 (2): 127–155. https://doi.org/10.1111/j.1944-8287.2008.tb00401.x.

Bumpus, A.G., and Liverman, D.M.. (2011). 'Carbon Colonialism? Offsets, Greenhouse Gas Reductions, and Sustainable Development.' In *Global Political Ecology,* edited by R. Peet, P. Robbins and M. Watts. London: Routledge, 203–224.

Calel, R. (2013). 'Carbon Markets: A Historical Overview.' *WIREs Climate Change* 4 (2): 107–119. https://doi.org/10.1002/wcc.208.

Callon, M. (2009). 'Civilizing Markets: Carbon Trading between in Vitro and in Vivo Experiments.' *Accounting, Organizations and Society* 34 (3–4): 535–548. https://doi.org/10.1016/j.aos.2008.04.003.

Castree, N. (2003). 'Commodifying What Nature?' *Progress in Human Geography* 27 (3): 273–297. https://doi.org/10.1191/0309132503ph428oa.

Cavanagh, C., and Benjaminsen, T.A. (2014). 'Virtual Nature, Violent Accumulation: The 'Spectacular Failure' of Carbon Offsetting at a Ugandan National Park.' *Geoforum* 56 (September): 55–65. https://doi.org/10.1016/j.geoforum.2014.06.013.

Chhatre, A., and Agrawal, A. (2009). 'Trade-Offs and Synergies between Carbon Storage and Livelihood Benefits from Forest Commons.' *Proceedings of the National Academy of Sciences* 106 (42): 17667–17670. https://doi.org/10.1073/pnas.0905308106.

Corbera, E., and Martin, A. (2015). 'Carbon Offsets: Accommodation or Resistance?' *Environment and Planning A: Economy and Space* 47 (10): 2023–2030. https://doi.org/10.1177/0308518X15611666.

Dawson, N.M., Mason, M., Mujasi Mwayafu, D., Dhungana, H., Satyal, P., Fisher, J.A., Zeitoun, M., and Schroeder, H. (2018). 'Barriers to Equity in REDD+: Deficiencies in National Interpretation Processes Constrain Adaptation to Context.' *Environmental Science & Policy* 88 (October): 1–9. https://doi.org/10.1016/j.envsci.2018.06.009.

Dhanda, K.K., and Hartman, L.P.. (2011). 'The Ethics of Carbon Neutrality: A Critical Ex- Amination of Voluntary Carbon Offset Providers.' *Journal f Business Ethics* 100: 119–149.

Donofrio, S., Maguire, P., Myers, K., Daley, C., and Lin, K. (2021). 'Markets in Motion: State of the Voluntary Carbon Markets 2021, Installment 1.' *Ecosystems Marketplace.* https://www.forest-trends.org/publications/state-of-the-voluntary-carbon-markets-2021/. Accessed May 16, 2022.

Duchelle, A. E., Cromberg, M., Fernanda Gebara, M., Guerra, R., Melo, T., Larson, A., and Cronkleton, P., et al. (2013). 'Linking Forest Tenure Reform, Environmental Compliance, and Incentives: Lessons from REDD+ Initiatives in the Brazilian Amazon.' *World Development* 55 (March): 53–67. https://doi.org/10.1016/j.worlddev.2013.01.014.

Erickson, P., Lazarus, M., and Spalding-Fecher, R. (2014). 'Net Climate Change Mitigation of the Clean Development Mechanism.' *Energy Policy* 72:146–154. https://doi.org/10.1016/j.enpol.2014.04.038.

Ervine, K. (2013). 'Carbon Markets, Debt and Uneven Development.' *Third World Quarterly* 34 (4): 653–670. https://doi.org/10.1080/01436597.2013.786288.

Escobar, A. (1995). *Encountering Development: The Making and Unmaking of the Third World.* Princeton: Princeton University Press.

Fairhead, J., Leach, M., and Scoones, I.. (2012). 'Green Grabbing: A New Appropriation of Nature?' *The Journal of Peasant Studies* 39 (2): 237–61. https://doi.org/10.1080/03066150.2012.671770.

Ferguson, J. (1990). *The Anti-Politics Machine: 'Development,' Depoliticization, and Bureaucratic Power in Lesotho.* Cambridge: Cambridge University Press.

Fischer, R., Hargita, Y., and Günter, S. (2016). 'Insights from the Ground Level? A Content Analysis Review of Multi-National REDD+ Studies Since 2010.' *Forest Policy and Economics* 66 (May): 47–58. https://doi.org/10.1016/j.forpol.2015.11.003.

Fletcher, R., Dressler, W., Büscher, B., and Anderson, Z.R. (2016). Questioning REDD+ and the future of market-based conservation. *Conservation Biology* 30: 673– 675. https://doi.org/10.1111/cobi.12680

Friends of the Earth (FOE). (2009). *A Dangerous Distraction. Why Offsetting Is Failing the Climate and People: The Evidence.* Available at: https://policy.friendsoftheearth.uk/sites/files/policy/documents/2019-02/dangerous_distraction_June_2009.pdf. Accessed April 1, 2021.

Galik, C. S., and Jackson, R.B. (2009). 'Risks to Forest Car- Bon Offset Projects in a Changing Climate.' *Forest Ecology and Management* 257 (11): 2209–2216. https://doi.org/ 10. 1016/j.foreco.2009.03.017.

Gilbertson, T. (2017). 'Carbon Pricing A Critical Perspective for Community Resistance.' *Climate Justice Alliance and the Indigenous Environmental Network.* Available at: https://www.ienearth.org/wp-content/uploads/2017/11/Carbon-Pricing-A-Critical-Perspective-for-Community-Resistance-Online-Version.pdf. Accessed on April 4, 2021.

Guizol, P., and Atmadja, S. (2008). 'Overview of REDD proposal submitted to UNFCCC'. In *Moving Ahead with REDD,* edited by A. Angelsen. Bogor: CIFOR.

Gupta, A., Lövbrand, E., Turnhout, E., and Vijge, M.J. (2012). 'In Pursuit of Carbon Accountability: The Politics of REDD+ Measuring, Reporting and Verification Systems.' *Current Opinion in Environmental Sustainability* 4 (6): 726–731. https://doi.org/10.1016/j.cosust.2012.10.004.

Harvey, D. (2003). *The New Imperialism.* Oxford: Oxford University Press.

Harvey, D. (2006). *Spaces of Global Capitalism.* London: Verso.

Hendrickson, C. Y., and Corbera, E. (2015). 'Participation Dynamics and Institutional Change in the Scolel Té Carbon Forestry Project, Chiapas, Mexico.' *Geoforum* 59 (February): 63–72. https://doi.org/10.1016/j.geoforum.2014.11.022.

Hyams, K., and Fawcett, T. (2013). 'The Ethics of Carbon Offsetting.' *WIREs Climate Change*, 4: 91–98.

INPE. (2020). *PRODES-Amazônia.* http://www.obt.inpe.br/OBT/assuntos/programas/amazonia/prodes. Accessed April 30, 2021.

International Climate Action Partnership. (2021). *ETS Map.* https://icapcarbo naction.com/en/ets-map. Accessed April 27, 2021.

Jodoin, S. (2017). *Forest Preservation in a Changing Climate: REDD+ and Indigenous and Community Rights in Indonesia and Tanzania.* Cambridge: Cambridge University Press.

Kosoy, N., and Corbera, E. (2010). 'Payments for Ecosystem Services as Commodity Fetishism.' *Ecological Economics* 69 (6): 1228–1236. https://doi.org/10.1016/j.ecolecon.2009.11.002.

Kreibich, N., and Hermwille, L. (2021). 'Caught in between: Credibility and Feasibility of the Voluntary Carbon Market Post-2020.' *Climate Policy* 21 (7): 939–957. https://doi.org/10.1080/14693062.2021.1948384.

Lang, S., Blum, M., and Leipold, S. (2019). 'What Future for the Voluntary Carbon Offset Market after Paris? An Explorative Study Based on the Discursive Agency Approach.' *Climate Policy* 19 (4): 414–426. https://doi.org/10.1080/14693062.2018.1556152.

Larson, A.M., M. Brockhaus, W.D. Sunderlin, A. Duchelle, A. Babon, and T. Dokken, et al. 2013. 'Land Tenure and REDD+: The Good, the Bad and the Ugly.' *Global Environmental Change*, 23(3), 678–689. https://doi.org/10.1016/j.gloenvcha. 2013.02.014.

Leach, M., J. Fairhead, and Fraser, J. (2012). 'Green Grabs and Biochar: Revaluing African Soils and Farming in the New Carbon Economy.' *Journal of Peasant Studies* 39 (2): 285–307. https://doi.org/10.1080/03066150.2012.658042.

Leach, M., and Scoones, I. (2015). *Carbon Conflicts and Forest Landscapes in Africa.* 1st ed. New York: Routledge. https://doi.org/10.4324/9781315740416.

Lohmann, L. (2008). 'Carbon Trading, Climate Justice and the Production of Ignorance: Ten Examples.' *Development* 51 (3): 359–365. https://doi.org/10.1057/dev.2008.27.

Luxemburg, R. (2015 [2013]). *The Accumulation of Capital.* Translated by Agnes Schwarzschild. Mansfield Center, CT: Martino Fine Books.

MacKenzie, D. (2009). 'Making Things the Same: Gases, Emission Rights and the Politics of Carbon Markets.' *Accounting, Organizations and Society* 34 (3–4): 440–455. https://doi.org/10.1016/j.aos.2008.02.004.

McAfee, K. (2012). 'The Contradictory Logic of Global Ecosystem Services Markets.' *Development and Change* 43 (1): 105–131. https://doi.org/10.1111/j.1467-7660.2011.01745.x.

Milne, S., Mahanty, S., To, P., Dressler, W., Kanowski, P., and Thavat, M. (2019). 'Learning From 'Actually Existing' REDD+: A Synthesis of Ethnographic Findings.' *Conservation & Society* 17 (1): 84–95.

Mitchard, E.T.A. (2018). 'The Tropical Forest Carbon Cycle and Climate Change.' *Nature* 559 (7715): 527–534. https://doi.org/10.1038/s41586-018-0300-2.

Moore, J.W. (2017). 'The Capitalocene, Part I: On the Nature and Origins of Our Ecological Crisis.' *The Journal of Peasant Studies* 44 (3): 594–630. https://doi.org/10.1080/03066150.2016.1235036.

Monbiot, G. (2006). *Selling Indulgences* [online]. Available from: http://www.monbiot.com/2006/10/19/selling-indulgences/. Accessed April 20, 2021.

Nepstad, D. C., Boyd, W., Stickler, C.M., Bezerra, T., and Azevedo, A.A. (2013). 'Responding to Climate Change and the Global Land Crisis: REDD+, Market Transformation and Low-Emissions Rural Development.' *Philosophical Transactions of the Royal Society B: Biological Sciences* 368 (1619): 20120167. https://doi.org/10.1098/rstb.2012.0167.

Newell, P., and Paterson, M. (2010). *Climate Capitalism: Global Warming and the Transformation of the Global Economy.* Cambridge: Cambridge University Press. https://doi.org/10.1017/CBO9780511761850.

Norman, M., and Nakhooda, S. (2015). 'The State of REDD+ Finance.' *Center for Global Development Working Paper No. 38.* https://doi.org/10.2139/ssrn.2622743.

Osborne, T. (2015). 'Tradeoffs in Carbon Commodification: A Political Ecology of Common Property Forest Governance.' *Geoforum* 67 (December): 64–77. https://doi.org/10.1016/j.geoforum.2015.10.007.

Osborne, T., and Shapiro-Garza, E. (2018). 'Embedding Carbon Markets: Complicating Commodification of Ecosystem Services in Mexico's Forests.' *Annals of the American Association of Geographers* 108 (1): 88–105. https://doi.org/10.1080/24694452.2017.1343657.

Page, E.A. (2013). 'The Ethics of Emissions Trading.' *Wiley Interdisciplinary Reviews: Climate Change* 4: 233–243.

Phelps, J., Webb, E.L., and Agrawal, A. (2010). 'Does REDD+ Threaten to Recentralize Forest Governance?' *Science* 328 (5976): 312–313. https://doi.org/10.1126/science.1187774.

Pulido, L. (1996). 'A Critical Review of the Methodology of Environmental Racism Research*.' *Antipode* 28 (2): 142–159. https://doi.org/10.1111/j.1467-8330.1996.tb00519.x.

Rasmussen, M.B., and Lund, C. (2018). 'Reconfiguring Frontier Spaces: The Territorialization of Resource Control.' *World Development* 101 (January): 388–399. https://doi.org/10.1016/j.worlddev.2017.01.018.

Seymour, F., and Angelsen, A. (2012). 'Summary and Conclusions: REDD+ Without Regrets.' In *Analysing REDD+: Challenges and Choices,* edited by A. Angelsen, M. Brockhaus, W.D. Sunderlin, and L.V. Verchot. Bogor, Indonesia: Center for International Forestry Research, 317–334.

Simonet G., Atmadja, S., Agrawal A., Bénédet F., Cromberg M., de Perthuis C., Haggard D., et al. (2020). *ID-RECCO, International Database on REDD+ Projects and Programs, Linking Economic, Carbon and Communities Data. Version 3.0.* Available online at: http://www.reddprojectsdatabase.org

Smith, K. (2007). *The Carbon Neutral Myth: Offset Indulgences for Your Climate Sins.* Amsterdam: Transnational Institute.

Stern, N. (2006) 'Stern Review on the Economics of Climate Change.' *Government of the United Kingdom.* Accessed April 30, 2021. Available at: https://webarchive.nationalarchives.gov.uk/20100407172811/; http://www.hm-treasury.gov.uk/stern_review_report.htm. Accessed April 12, 2021.

Sunderlin, W.D., Larson, A.M., Duchelle, A.E., Resosudarmo, I.A.P., Huynh, T.B., and Awono, A., et al. (2014). 'How are REDD+ proponents addressing tenure problems? Evidence from Brazil, Cameroon, Tanzania, Indonesia, and Vietnam.' *World Development* 55: 37–52.

Turnhout, E., Gupta, A., Weatherley-Singh, J., Vijge, M.J., de Koning, J., Visseren-Hamakers, I.J., Herold, M., and Lederer, M. (2017). 'Envisioning REDD+ in a Post-Paris Era: Between Evolving Expectations and Current Practice.' *WIREs Climate Change* 8 (1): e425. https://doi.org/10.1002/wcc.425.

Wang, Y., and Corson, C. (2015). 'The Making of a 'Charismatic' Carbon Credit: Clean Cookstoves and 'Uncooperative' Women in Western Kenya.' *Environment and Planning A: Economy and Space* 47 (10): 2064–2079. https://doi.org/10.1068/a130233p.

Watts, M. (2004). 'Resource Curse? Governmentality, Oil and Power in the Niger Delta, Nigeria.' *Geopolitics* 9 (1): 50–80. https://doi.org/10.1080/14650040412331307832.

West, P. (2016). *Dispossession and the Environment: Rhetoric and Inequality in Papua, New Guinea.* New York: Columbia University Press.

Whyte, K.P. (2017). 'The Dakota Access Pipeline, Environmental Injustice, and U.S. Colonialism.' *Red Ink: An International Journal of Indigenous Literature, Arts, & Humanities* 19.1: 154–169.

Wunder, S, Duchelle, A.E., de Sassi, C., Sills, E.O., Simonet, G., and Sunderlin, W.D. (2020). 'REDD+ in Theory and Practice: How Lessons From Local Projects Can Inform Jurisdictional Approaches.' *Frontiers in Forests and Global Change* 3: 1–17. https://doi.org/10.3389/ffgc.2020.00011.

28

KEY ETHICAL ISSUES FOR CLIMATE-FOREST POLICIES

Brendan Mackey and Nicole Rogers

Introduction

The world has lost around one-third of its pre-human forests, yet around 4 billion hectares of forest remain standing, with about 1.11 billion hectares still in a natural condition, largely untouched by modern development. In addition to their intrinsic value, the world's remaining natural forests provide critical ecosystem services that benefit humans and nature. Of particular importance here is their role in the global carbon cycle, as forest has a critical role to play in stabilizing atmospheric concentrations of CO_2 (Mackey et al. 2013). Yet, despite their well-documented benefits for climate mitigation (Mackey et al. 2020), they continue to be cleared and degraded (Fao 2020a). What is perhaps more surprising is that this ongoing loss is occurring despite a massive increase in attention given by the international community to forests, their recognition as a matter of global concern as evidenced by the New York Declaration on Forests (Summit 2014), and the fact that forest policy is now germane to the implementation of the Paris Agreement on Climate Change (Grassi et al. 2017).

A major reason for the lack of progress in protecting and restoring forests is that they lie at the nexus of multiple global crises and challenges and especially the biodiversity extinction crisis, pandemics and sustainable development (Barber et al. 2020; Pörtner et al. 2021). Furthermore, forests are exploited as a source of raw materials for the production of manufactured goods and fuel, or they are destroyed for other land uses. The role of forests in climate change policy, and the ethical issues they entail, therefore cannot be untangled from their other functions and roles. So, while there is a plethora of international forest policy, there is no agreed, coherent policy framework for their use and protection. Rather, coherent governance is impeded by a clutch bag of uncoordinated and often contradictory policies, fragmented across multiple treaties, agreements, and jurisdictions. To date, these have failed to stem the tide of forest loss and degradation, even in the world's most well-known forests such as the Amazon (Qin et al. 2021).

Here we address key ethical issues in forest policy that are typically ignored or dismissed, sometimes out of ignorance, other times wilfully. We focus on three forest policy problems. First, issues arising from forests being understood as subject to sovereign owned resources for exploitation, rather than being recognized as part of the global environmental commons. Second, we examine the trade-offs between the various forest beneficiaries and between the multiple forest ecosystem services. Third, we consider the ethics of developed countries offsetting their fossil fuel emissions through forest-based carbon offsets in developing countries.

DOI: 10.4324/9781003039860-33

In discussing these forest-climate related ethical issues, we draw upon a set of justice principles that have proven useful in evaluating climate justice issues more generally (Mackey & Rogers 2015): consequential justice; procedural justice; intergenerational justice; Earth justice; deontological justice; and distributive justice. By consequential justice, we are referring here to the harmful consequences of climate change for humans, especially for those communities who are most vulnerable and at risk. Consequential justice requires climate change harm to be avoided or minimized and when harm occurs for due compensation be forthcoming to the injured. Procedural justice is concerned with the fairness and the transparency of the processes by which decisions are made, that is, procedural fairness or due process. Intergenerational justice derives from the fact that it is our children and future generations who will bear the burden of the climatic disruption which we are now causing. Earth justice is related to the rights of nature as it is concerned with the wellbeing of the greater community of life and the integrity of the Earth system. The philosopher Hans Jonas argued that if we accept that life is better than non-life, it follows that we must be concerned for the quality of life lived. Earth justice encompasses our responsibilities to the community of life with whom we share Earth as home. Deontological justice is what many people understand to mean by what is 'right,' according to ethical principles framed as commandments, imperatives, or moral laws. Examples of these are the Earth Charter (Dower 2019) and the United Nations Declaration on the Rights of Indigenous Peoples. Finally, distributive justice concerns the way in which collective goods and responsibilities are fairly apportioned among people who stand in a social community (Dutton 2022).

Forests as Sovereign Owned or Global Environmental Commons

Forests are understood to be subject to state sovereignty, in the sense that national governments have control over the land and use of the natural resources within their territories.[1] One consequence is that forests are owned and managed by different actors with varying degrees of agency including governments (i.e., publicly owned), individuals, local, tribal, and Indigenous communities, as well as business entities and institutions (Table 28.1). It is not surprising that forest serve a multitude of uses and that about 1.15 billion ha of forest is managed for the production of wood and non-wood forest products. In addition, 749 million hectares is designated for multiple use, which often includes wood production. The production and multiple use forests produced annually around 2,028 million tonnes of industrial roundwood for wood and paper manufactured

Table 28.1 Summary statistics of who owns the world's forests

Category of forest owner	
All forests	*Percentage of total forest*
Publicly owned	73
Privately owned	22
Unaccounted	4
Privately owned public forest	*Percentage of privately owned forest*
Individuals	51
Local, tribal, and indigenous communities	29
Business entities and institutions	20
Holders of management rights in publicly owned forest	*Percentage of publicly owned forest*
Public administration	83
Business entities and institutions	13
Local, tribal, and Indigenous communities	2

Source: FAO (2020b).

Table 28.2 Global aggregate wood production from production forests and total product value (exports)

Product	*Unit*	*2,019*
Roundwood	million m^3	3,966
Wood fuel	million m^3	1,945
Industrial roundwood	million m^3	2,021
Wood pellets and other agglomerates	million tonnes	46
Sawn wood	million m^3	488
Wood-based panels	million m^3	357
Plywood	million m^3	107
Particle board, OSB and fibreboard	million m^3	250
Wood pulp	million tonnes	190
Pulp from fibres other than wood	million tonnes	12
Recovered paper	million tonnes	228
Paper and paperboard	million tonnes	404
Forest products value (exports)	US$ billion	244

product and 1,943 million tonnes of wood fuel (3,971 in total), with an export value of US$240m billion (Table 28.2) (Secretary-General IGS 2019). Most of the world's forests, therefore, are exploited for extractive uses based on logging trees and harvesting woody biomass. These are used as raw materials for manufactured goods and energy generation for electricity, heating, and cooking. Governments earn revenue and businesses substantial profits from these activities, that have a total global export value of US$240 billion (Table 28.2).

At the same time as they are valued as a source of raw materials for manufactured products and energy, forests have been recognized by the international community as a common concern to humankind. They provide services, such as climate regulation, that make them an integral component of Earth's life support systems (UN 1992). From this perspective, forests belong to the global environmental commons. The 'global commons' are four areas that fall outside of any national jurisdiction: the high seas; the atmosphere; Antarctica, and outer space. It has been proposed that this concept be expanded to other commons which may lie within well-defined national or regional jurisdictions, but whose continuing existence confers benefits beyond them, including tropical rain forests, land, biodiversity, and climate. The rationale here is that these comprise various large-scale biomes and systems that contribute directly or indirectly to the functioning of the Earth system, and hence to supporting life. These global environmental commons can be understood as a stock of natural capital from which flows benefits often shared across humanity (Secretary-General 2019). From a scientific perspective, all forest biomes – boreal and temperate as well as tropical – are part of the global environmental commons (Keith et al. 2009).

An estimated 700 million hectares of forest is in legally established protected areas (FAO 2020b). To the extent they are 'no-go' areas for extractive industries, forests in these protected areas can be considered a national level commitment to the global environmental commons. However, in many jurisdictions, categories of protected areas allow logging, mining, and dam construction (Mackey et al. 2017). A key ethical issue therefore concerns the conflict between the short-term profits to some people[2] that arise from their exploitation of forests, versus the wider public good benefits to the world community and future generations from keeping them protected as part of the global environmental commons. There is no international treaty governing the world's natural forests as part of the global environmental commons. In its absence, and in light of the imposition of sovereignty claims by national governments on the use of forests within their territories, what causes of legal action are open to citizens?

Legal Action Open to Protect Forests as Part of Global Environmental Commons

A ground-breaking precedent in this context is the *Future Generations* case: a successful tutela, or legal action, launched by 25 young Colombians against their government, Colombian municipalities, and various corporations. The plaintiffs invoked their 'supra-legal rights,' encompassing rights to a healthy environment and rights to life and health, in seeking a halt to deforestation in the Colombian Amazon. The Colombian Supreme Court acknowledged not only the fundamental importance of the Amazon rainforest for the health and wellbeing of future generations, but was prepared to acknowledge that the forest itself had rights. It ordered the government to create an "intergenerational pact for the life of the Colombian Amazon" (Dejusticia 2018).

This ruling is significant for a number of reasons. First, in recognising that forests have rights, the case constitutes part of an emerging global movement to confer rights upon rivers, forests, and other, more-than-human entities. Contemporaneous developments in, *inter alia*, New Zealand, India, Bangladesh and Ecuador demonstrate a growing willingness to, as Alessandro Pelizzon has put it, "construe nature as a legal *subject* within human legal systems" (Pelizzon 2020). In the context of forests, literary works have facilitated such developments. The underlying premise of Richard Powers' novel *The Overstory*, shortlisted for the Booker Prize, is that we co-exist with "slow deliberate creatures with … elaborate vocabularies" which are "the most wondrous products of four billion years of creation": trees (Powers 2018). To wantonly destroy or devalue these "wondrous products" is profoundly ignorant and short-sighted. A popular appreciation of forests as communities of cooperative, sentient living entities has been enhanced by non-fiction accounts from forester Peter Wohlleben (Wohlleben 2016) and ecologist Suzanne Simard (Simard 2021).

Secondly, the Court highlighted that the wellbeing of today's youth, and the unborn, is intrinsically connected to halting deforestation and, thereby, contributing to the stabilisation of the world's climate. Forest conservation is critical to the implementation of the principle of intergenerational equity, framed by leading legal philosopher Edith Brown Weiss thus: "[e]very generation needs to pass the Earth and our natural and cultural resources on in at least as good condition as we received them" (Weiss 2008). As the Colombian Supreme Court put it, "[i]n terms of intergenerational equity, the transgression is obvious" (Dejusticia 2018).

Finally, the case constitutes part of a growing suite of youth climate lawsuits. In these actions, young people are seeking to extrapolate from established applications of legal concepts, including the public trust doctrine, human rights, and duty of care, to compel governments to respond to the existential challenge of climate change. Notably, in an Australian Federal Court ruling in *Sharma v Minister for the Environment*, Justice Mordecai Bromberg identified a specific constraint upon the exercise of statutory decision-making powers linked to greenhouse gas emitting projects: a common law duty of care to avoid causing death or personal injury to young people. The judge described climate change as "the greatest inter-generational justice ever inflicted by one generation of humans upon the next" (FCA 2021). Applied in the *Sharma* case to a pending decision on a proposed coalmine extension, the duty of care could well have much broader application. Arguably, it fetters the exercise of statutory powers in relation to logging operations in old-growth forests; deforestation, along with the burning of coal from coalmines, contributes to the steep rise in greenhouse gas emissions. Significantly, Justice Bromberg, in reaching his decision, considered the incineration of vast amounts of Australian east coast forest in the 2019/2020 Black Summer megafires, and scientific predictions of more frequent and severe bushfires and ensuing forest destruction in a climate changed future; young people, he pointed out, will inhabit a "harsher, far more extreme and devastatingly brutal" physical environment (FCA 2021).

Trade-Offs Between Forest Beneficiaries and Between Forest Ecosystem Services

We must also consider issues of distributive justice related to who wins and who loses from forests being allocated for commons, private or customary benefit. In most countries, though with varying legal status, forests are the customary lands of Indigenous Peoples and fundamental to their culture and sustainable livelihoods (Garnett et al. 2018). In addition, non-Indigenous local communities depend on forests for food and livelihoods (Newton et al. 2016). Many Indigenous and local communities do seek to, or are put in a situation where they need to exploit forests for commodity production and conventional land development purposes (Curtis et al. 2018). Also, many local communities in developing countries rely on forest biomass for cooking and clear forest for subsistence farming. However, there are also many who seek an accommodation with modernity and a market economy to find sustainable development pathways that preserve their forest's ecosystem integrity (He et al. 2019; Mackey & Claudie 2015). For these communities, forests are much more than simply a supply of raw materials or vacant land for development. A confounding factor therefore is that forests provide multiple services which benefit different people, communities, organisations, and governments, spanning scales from local to global.

The many benefits provided by forests can be recognized and accounted for using the concept of ecosystem services and related framework. Well-regarded approaches are those developed by the Millennium Ecosystem Assessment, the Intergovernmental Platform on Biodiversity and Ecosystem Services (IPBES), and UN System of Environmental Economic Accounting (Keith et al. 2021). What is particularly useful here is valuation of the aggregate flows of ecosystems from a particular forest, and indeed from all forests, known as Total Ecosystem Service Value (TESV) (Buckwell et al. 2020). Forest services include extractive uses such as logging but also so-called 'non-use values', including the cultural significance of forests for customary Indigenous peoples for whom these landscapes are their ancestral territory. Non-use values also include 'existence value' where we value being aware of the simple fact that a forest remains in a healthy condition and is supporting a rich abundance and diversity of wildlife.

The role of forests in climate mitigation can also be understood as an ecosystem service. Forest ecosystems sequester and store large quantities of carbon. There is more carbon stored in terrestrial ecosystems than in known fossil fuel reserves. The majority of biomass carbon is stored in forests, and with equally vast stocks stored below ground in the boreal biome (Keith et al. 2009). There is sufficient carbon stored in tropical forests alone which, if emitted, would breach the remaining global carbon budget for 1.5°C of warming since pre-industrial levels (Mackey et al. 2020). Previously logged forest has significant sequestration potential and even old growth continues to sequester and store carbon for many centuries (Moomaw et al. 2019).

As most of the biomass carbon in a forest is stored in big old trees, primary forests – those natural forests that have not been subject to degradation and clearing from modern land use such as commercial logging or clearing for ranching and commercial cropping – store the most carbon compared to forests used for wood production. They also produce the cleanest water, having the least sediment and nutrients compared to logged and re-growth forest (Mackey et al. 2015). Forests are also reservoirs of biodiversity at all levels of ecological complexity and organisation – from the genetic diversity with the populations of a species; to the intricate food webs that arise, the complex patterns of co-evolution, the essential pollination of plants by wildlife, especially insects and birds; through to the emergent system level properties that give the whole ecosystem its stability, resilience, and adaptive capacity. It is, in fact, these evolutionary, ecological, and biological processes that comprise, generate, and sustain the forest itself.

In addition to the benefits provided to people from a forest's ecosystem services, including their significance to the spirituality, cultural practices, and ways of living of their Indigenous Peoples, the rich biodiversity they comprise also has intrinsic value. As articulated by the Earth Charter, intrinsic value recognizes that all beings are interdependent, and every form of life has value regardless of its worth to human beings. A practical application of intrinsic value is that people should respect and care for the community of life, recognizing that all beings are interdependent, and every form of life has value regardless of its worth to human beings.

As the time of writing, the COVID-19 pandemic continues unabated, with chronic and acute impacts – a deadly reminder of the reality and power of biological evolution and how humans remain part of the planetary web of life. Viruses reproduce in the cells of other organisms, with natural selection working on the genetic diversity in an ongoing interplay with the host organism's immune system. The loss and fragmentation of tropical forests results in increased spill-over of pathogens from wildlife, to livestock and humans (Faust et al. 2018). Protecting the ecosystem integrity of forests is, therefore, a key component of a wholistic approach to preventing future pathogen spill-overs and pandemics. Protecting forests resides at the nexus of the climate, biodiversity and pandemic crises: the nexus representing both the shared underlying causes of these problems, as well as the solution space for addressing them wholistically (Barber et al. 2020).

Understanding the Trade-Offs

If we are to advance policies that promote the climate mitigation value of forests in ways that achieve a just and equitable distribution of forest services and the benefits they provide people, it is important to understand who benefits and who suffers from their protection or exploitation. As well as benefits arising at different scales, some locally, others globally, and for different beneficiaries, there are trade-offs between ecosystem services. Figure 28.1 illustrates examples of the trade-offs and synergies between different kinds of forest ecosystem services. These comparisons are based on detailed assessments of forest ecosystem services and the impacts on forests of logging and biomass harvesting (Booth 2018; Mackey et al. 2020; Taye et al. 2021).

Ecosystem service	Description							
	Biodiversity conservation	⊘	✓	✓	✕	✓	✓	✓
	Climate mitigation	✓	⊘	✓	✕	✕	✓	✓
	Fresh water	✓	✓	⊘	✕	✕	✓	✓
	Commercial wood supply	✕	✕	✕	⊘	✓	✕	✕
	Industrial biomass energy feed stock	✕	✕	✕	✓	⊘	✕	✕
	Ecotourism	✓	✓	✓	✕	✕	⊘	✓
	Bio-prospecting	✓	✓	✓	✕	✕	✕	⊘

Figure 28.1 Indicative tradeoffs and synergies between different kinds of forest ecosystem analyses. A forest managed for commercial wood supply and biomass energy feedstocks will degrade other ecosystem services. The main reasons relate to the destructive impacts of these land uses in depleting biomass carbons stocks, opening up of the canopy and increasing vegetation dryness thereby increasing flammability, causing soil erosion and increasing sediment and nutrient runoff, loss of species and genetic diversity, and disruption of food webs and other ecological dependences. Biodiversity conservation refers here to *in situ* conservation measures.

Sources: Taye et al. 2021; Mackey et al. 2020; Booth 2018.

Prioritizing forest management for some services will degrade others, with consequences for who benefits. Prioritizing logging degrades many other services including climate mitigation, freshwater supplies, and biodiversity conservation. The financial benefits from the ecosystem services of commercial wood supply and industrial biomass energy feedstock, which both rely upon logging and removing forest trees, flow to for-profit private companies and corporations, with royalties and taxes flowing to national and sub-national governments. A private commercial company may benefit from the logging, but the world community will lose the climate mitigation benefit the forest provides, and the local community is left with degraded non-wood forest resources. Critically, aggregating the TESV of all non-extractive ecosystem services gives a higher value than any one category ecosystem service, including commercial wood supply (Taye et al. 2021). Yet, it is the economic value of wood and land for development that dominates policy debates (Ajani 2007). Despite increasing attention being paid by experts to the TESV of forests (Dasgupta 2021), the loss of TESV from those ecosystem services that are not compatible with commercial wood supply is usually ignored and the costs of the degraded forest are carried by the Indigenous and local communities and the world community.

Fossil Fuel Offsets

A related ethical issue concerns the use of forests as offsets for fossil fuel emissions. About 1/3 of the anthropogenic emissions accumulated in the atmosphere that are driving global warming and human influenced climate change, and about 12 percent of annual emissions, are from deforestation and degradation (Mackey et al. 2013). Furthermore, ongoing forest growth in primary forests, as well as allowing previously logged and degraded forests to regrow, results in significant removals of atmospheric CO_2 (Moomaw et al. 2019). Avoiding emissions from deforestation and allowing forests to regrowth without further land use impacts, especially commercial logging and biomass harvesting, is therefore an important mitigation strategy (Mackey et al. 2020).

Many developed countries now seek to meet their mitigation targets under the Paris Agreement through 'carbon offsets' in developing countries (Mcafee 2021). The underlying idea is that they can 'cancel out' the impact of some of their emissions by paying developing countries to change forest management or invest in projects that reduce or store the same amount of carbon. A carbon offset is formally defined as occurring when an individual, company, NGO, or state invests in a project elsewhere that results in a reduction of GHG emissions that would not have occurred in the absence of the project (WBCSD 2005). The use of offsets to meet "net zero emissions" targets under the Paris Agreement is a widely used approach with the majority of countries and many of the world's largest companies intending to use offsets to meet their "net zero" commitments (Rogelj et al. 2021).

Offsetting fossil fuel emissions is ethically problematic in at least three ways. First, they do not lower the concentration of CO_2 in the atmosphere and they send the misleading message that we can avoid or post-pone the rapid phase-out of fossil fuels (Mcafee 2021). Second, 'off-shoring' developed countries' emissions has the perverse effects of placing the mitigation burden on developing countries who have been least responsible for the problem and have the greatest development needs and least adaptive capacity. Third, the benefits to developing countries from changing their forest management to make use of the mitigation benefits from forest protection and natural regrowth are lost to the developed country which records them as having 'netted out an equivalent amount of fossil fuel emission in their national greenhouse gas inventories (Becken & Mackey 2017).

Achieving the deep cuts in greenhouse gas emissions to stabilize atmospheric concentrations of CO_2 at levels that will limit global warming to 1.5°C–2.0°C above pre-industrial levels, in line with the Paris Agreement long-term temperature goal, will require: (i) preventing further emissions by ending the use of fossil fuel (oil, coal, gas) and protecting forest; as well as (ii) removing atmospheric carbon by enabling forest restoration and continued growth. Using the removal of atmospheric carbon by forest restoration and growth to offset fossil fuel emissions does not solve the problem. At best we could get an 'accounting net zero' outcome when what we need is for the atmospheric concentrations of CO_2 to be reduced in absolute terms.

Offsets are ethically problematic because climate change is a 'non-net-zero sum' game. In a zero-sum game, for someone to gain, someone else has to lose by the same amount. In a non-net-zero sum game, we all win or we all lose. Offsets are fundamentally unjust because they do not help solve the problem faster than we are causing it to worsen, with the result that, ultimately, we all lose.

Conclusion

From one perspective, the climate change problem is the result of the unintended consequences of actions, which are in themselves seemingly morally neutral or positive. Most greenhouse gas emissions result from burning energy that underpins everyday economic and social activity in the modern world. As such, we are all complicit in the climate crisis when we light, heat, and cool our homes and offices, drive our cars, catch planes, and consume goods, using energy generated from burning fossil fuel. As Timothy Morton puts it, climate change "make[s] hypocrites of us all"; "we are always in the wrong" (Morton 2013). However, the problem cannot be solved by individual actions alone as the vast majority of people are locked into systems of power generation and transmission, and forms of transportation, that do not provide options for emission-free ways of living. Rather, transformative changes are needed in all sectors, including energy generation, transportation, manufacturing and land management (Masson-Delmotte et al. 2018). Such large-scale, systemic changes require government leadership and the cooperation of all sectors, along with a strong social licence for action.

Where such leadership is lacking, and sectors with a vested interest play a blocking or 'go slow' role, the issue of culpability arises as the human influence on climate change is now a scientifically established fact (Masson-Delmotte et al. in press). Given the impacts and risks from current and projected climate change, to deny or ignore the science leaves these decision makers open to the interrogation of consequential justice, as they will have deliberately taken a course of action knowing the harm it would cause to people in all countries, to future generations and, as well, to other species.

National government policy remains dominated by conventional notions of security and development, which inhibit the progressive policies needed for forest protection and restoration. The non-zero-sum game that is the climate change problem, however, is expanding our understanding of what constitutes 'security' and what is needed for sustainable development. We face the ultimate post-normal (or wicked) problem where stakes are high, values contested and there is no single correct solution (Funtowicz & Ravetz 1993). Ethical reflection as to the rights and wrongs of alternative solutions is critical to arriving at the necessary societal consensus in these circumstances.

The ethical dimension of climate-forest policy and management decisions is often buried under layers of narrowly framed short-term national interest and blinkered economic rationales. The 'economic growth' paradigm dominates the realpolitik of public policy discourse everywhere with often destructive consequences for a forest's Total Ecosystem Service Value. As part of the global environmental commons, forests are irreplaceable for their role in climate mitigation and maintaining other planetary life support systems for people and the greater community life with whom we share Earth as home.

Notes

1 Absolute national government control can be subject to internal constitutional arrangements regarding sub-national responsibilities (e.g., The Australian Constitution allocates powers between the national and state governments and any pooled international sovereignty arrangements (such as the European Union)).

2 The "some" typically includes corporations, other private businesses, and national and sub-national governments who dominate large scale exploitation. However, small scale exploitation can provide benefits to small groups and families.

References

Ajani, J. (2007). *The Forest Wars.* Melbourne University Press.

Barber, C. V., Petersen, R., Young, V., Mackey, B. & Kormos, C. (2020). *The Nexus Report: Nature Based Solutions to the Biodiversity and Climate Crisis.* F20 Foundations, Campaign for Nature and SEE Foundation.

Becken, S. & Mackey, B. (2017). What role for offsetting aviation greenhouse gas emissions in a deep-cut carbon world? *Journal of Air Transport Management,* 63, 71–83.

Booth, M. S. (2018). Not carbon neutral: Assessing the net emissions impact of residues burned for bioenergy. *Environmental Research Letters,* 13, 035001.

Buckwell, A. J., Fleming, C., Smart, J. C., Ware, D. & Mackey, B. (2020). Challenges and sensitivities in assessing total ecosystem service values: Lessons from Vanuatu for the Pacific. *The Journal of Environment & Development,* 29, 329–365.

Curtis, P. G., Slay, C. M., Harris, N. L., Tyukavina, A. & Hansen, M. C. (2018). Classifying drivers of global forest loss. *Science,* 361, 1108–1111.

Dasgupta, P. (2021). *The Economics of Biodiversity: The Dasgupta Review.* HM Treasury.

Dejusticia. (2018). *Climate Change and Future Generations Lawsuit in Colombia: Key Excerpts from the Supreme Court Decision.* DeJusticia.

Dower, N. (2019). Chapter 10: Global ethics and the Earth Charter. *In:* Burdon, P., Bosselmann, K. & Engel, K. (eds.) *The Crisis in Global Ethics and the Future of Global Governance: Fulfilling the Promise of the Earth Charter.* Elgaronline, 156–167.

Dutton, B. D. (2022). Benedict De Spinoza (1632–1677). *In:* J. Fieser & Dowden B. (ed.) *Internet Encyclopedia of Philosophy,* Internet Encyclopedia of Philosophy; https://iep.utm.edu/spinoza/.

FAO. (2020a). *Global Forest Resources Assessment 2020 – Key Findings: Food And Agricultural Organisation of the United Nations.*

FAO. (2020b). *Global Forest Resources Assessment 2020 Main Report.* Rome: Food and Agriculture Organization of the United Nations.

Faust, C. L., Mccallum, H. I., Bloomfield, L. S., Gottdenker, N. L., Gillespie, T. R., Torney, C. J., Dobson, A. P. & Plowright, R. K. (2018). Pathogen spillover during land conversion. *Ecology Letters,* 21, 471–483.

FCA. (2021). *Sharma v Minister for the Environment [2021] FCA 560 [293].* Federal Court of Australia.

Funtowicz, S. O. & Ravetz, J. R. (1993). Science for the post-normal age. *Futures,* 25, 739–755.

Garnett, S. T., Burgess, N. D., Fa, J. E., Fernández-Llamazares, Á., Molnár, Z., Robinson, C. J., Watson, J. E., Zander, K. K., Austin, B. & Brondizio, E. S. (2018). A spatial overview of the global importance of Indigenous lands for conservation. *Nature Sustainability,* 1, 369–374.

Grassi, G., House, J., Dentener, F., Federici, S., Den Elzen, M. & Penman, J. (2017). The key role of forests in meeting climate targets requires science for credible mitigation. *Nature Climate Change,* 7, 220–226.

He, Y., Baldiviezo, J. P., Agrawal, A., Candaguira, V. & Perfecto, I. (2019). Guardians of the forests: How should an indigenous community in eastern Bolivia Defend their land and forests under increasing political and economic pressures? *Case Studies in the Environment,* 3(1), 1–14.

Keith, H., Mackey, B. G. & Lindenmayer, D. B. (2009). Re-evaluation of forest biomass carbon stocks and lessons from the world's most carbon-dense forests. *Proceedings of the National Academy of Sciences,* 106, 11635–11640.

Keith, H., Vardon, M., Obst, C., Young, V., Houghton, R. A. & Mackey, B. (2021). Evaluating nature-based solutions for climate mitigation and conservation requires comprehensive carbon accounting. *Science of the Total Environment,* 769, 144341.

Mackey, B., Cadman, S., Rogers, N. & Hugh, S. (2017). Assessing the risk to the conservation status of temperate rainforest from exposure to mining, commercial logging, and climate change: A Tasmanian case study. *Biological Conservation,* 215, 19–29.

Mackey, B. & Claudie, D. (2015). Points of contact: integrating traditional and scientific knowledge for biocultural conservation. *Environmental Ethics*, 37, 341–357.

Mackey, B., Dellasala, D. A., Kormos, C., Lindenmayer, D., Kumpel, N., Zimmerman, B., Hugh, S., Young, V., Foley, S. & Arsenis, K. (2015). Policy options for the world's primary forests in multilateral environmental agreements. *Conservation Letters*, 8, 139–147.

Mackey, B., Kormos, C. F., Keith, H., Moomaw, W. R., Houghton, R. A., Mittermeier, R. A., Hole, D. & Hugh, S. (2020). Understanding the importance of primary tropical forest protection as a mitigation strategy. *Mitigation and Adaptation Strategies for Global Change*, 25, 763–787.

Mackey, B., Prentice, I. C., Steffen, W., House, J. I., Lindenmayer, D., Keith, H. & Berry, S. (2013). Untangling the confusion around land carbon science and climate change mitigation policy. *Nature Climate Change,* 3, 552–557.

Mackey, B. & Rogers, N. (2015). *Climate Justice and the Distribution of Rights to Emit Carbon.* Routledge.

Masson-Delmotte, V., Z. P., Pirani, A., Connors, S. L., Péan, C., Berger, S., Caud, N., Chen, Y., Goldfarb, L., Gomis, M. I., Huang, M., Leitzell, K., Lonnoy, E., Matthews, J.B.R., Maycock, T. K., Waterfield, T., Yelekçi, O., Yu, R. & Zhou, B. (eds.) (in press). IPCC, 2021: Summary for Policymakers. In: *Climate Change 2021: The Physical Science Basis. Contribution of Working Group I to the Sixth Assessment Report of the Intergovernmental Panel on Climate Change.* Cambridge University Press.

Masson-Delmotte, V., Zhai, P., Pörtner, H.-O., Roberts, D., Skea, J., Shukla, P.R., Pirani, A., Moufouma-Okia, W., Péan, C., Pidcock, R., Connors, S., Matthews, J. B. R., Chen, Y., Zhou, X., Gomis, M. I., Lonnoy, E., Maycock, T., Tignor, M. & T., W. (2018). *Global Warming of 1.5°C. An IPCC Special Report on the Impacts of Global Warming of 1.5°C Above Pre-Industrial Levels and Related Global Greenhouse Gas Emission Pathways, in the Context of Strengthening the Global Response to the Threat of Climate Change, Sustainable Development, and Efforts to Eradicate Poverty.* IPCC.

Mcafee, K. (2021). Shall the American association of geographers endorse carbon offsets? Absolutely not! *The Professional Geographer,* 74: 1–7.

Moomaw, W. R., Masino, S. A. & Faison, E. K. (2019). Intact forests in the United States: Proforestation mitigates climate change and serves the greatest good. *Frontiers in Forests and Global Change,* 2, 27.

Morton, T. 2013. *Hyperobjects. Philosophy and Ecology After the End of the World.* University of Minnesota Press.

Newton, P., Miller, D. C., Byenkya, M. a. A. & Agrawal, A. (2016). Who are forest-dependent people? A taxonomy to aid livelihood and land use decision-making in forested regions. *Land Use Policy,* 57, 388–395.

Pelizzon, A. (2020). An intergenerational ecological jurisprudence: The supreme court of Colombia and the rights of the amazon rainforest. *Law, Technology & Humans,* 2, 33.

Pörtner, H.O., Scholes, R.J., Agard, J., Archer, E., Arneth, A., Bai, X., Barnes, D., Burrows, M., Chan, L., Cheung, W.L., Diamond, S., Donatti, C., Duarte, C., Eisenhauer, N., Foden, W., Gasalla, M. A., Handa, C., Hickler, T., Hoegh-Guldberg, O., Ichii, K., Jacob, U., Insarov, G., Kiessling, W., Leadley, P., Leemans, R., Levin, L., Lim, M., Maharaj, S., Managi, S., Marquet, P. A., McElwee, P., Midgley, G., Oberdorff, T., Obura, D., Osman, E., Pandit, R., Pascual, U., Pires, A. P. F., Popp, A., ReyesGarcía, V., Sankaran, M., Settele, J., Shin, Y. J., Sintayehu, D. W., Smith, P., Steiner, N., Strassburg, B., Sukumar, R., Trisos, C., Val, A.L., Wu, J., Aldrian, E., Parmesan, C., Pichs-Madruga, R., Roberts, D.C., Rogers, A.D., Díaz, S., Fischer, M., Hashimoto, S., Lavorel, S., Wu, N., Ngo, H.T. (2021). *IPBES-IPCC Co-Sponsored Workshop Report on Biodiversity and Climate Change.*

Powers, R. (2018). *The Overstory.* William Heinemann.

Qin, Y., Xiao, X., Wigneron, J.-P., Ciais, P., Brandt, M., Fan, L., Li, X., Crowell, S., Wu, X. & Doughty, R. (2021). Carbon loss from forest degradation exceeds that from deforestation in the Brazilian Amazon. *Nature Climate Change*, 11, 442–448.

Rogelj, J., Geden, O., Cowie, A. & Reisinger, A. (2021). Three ways to improve net-zero emissions targets. *Nature*, 591, 365–368.

Independent Group of Scientists appointed by the Secretary-General (Secretary General IGS) (2019). *Global Sustainable Development Report 2019: The Future Is Now – Science for Achieving Sustainable Development.* New York: United Nations.

Simard, S. (2021). *Finding the Mother Tree. Discovering the Wisdom of the Forest.* Alfred A Knopf.

Summit, U. C. (2014). New *York Declaration on Forests.* New York: United Nations.

Taye, F. A., Folkersen, M. V., Fleming, C. M., Buckwell, A., Mackey, B., Diwakar, K., Le, D., Hasan, S. & Saint Ange, C. (2021). The economic values of global forest ecosystem services: A meta-analysis. *Ecological Economics,* 189, 107145.

UN. (1992). Convention on Biological Diversity. Available at https://www.cbd.int/convention/text/.

WBCSD. (2005). GHG protocol for project accounting. *WRI WBCSD*, www. ghgprotocol. org/files/ghg_ project_ protocol. pdf.

Weiss, E. B. (2008). Climate change, intergenerational equity, and international law. *Vermont Journal of Environmental Law*, 9, 615–627.

Wohlleben, P. (2016). *The Hidden Life of Trees. What They Feel, How They Communicate. Discoveries From a Secret World*. Black Inc.

29

PARIS' ARTICLE 6 DEADLOCK

Climate Change, from Globalization to Nationalism

José Domingos Gonzalez Miguez and Thiago de Araújo Mendes

Introduction

Global climate change is one of the most pressing problems for humanity and our planet, particularly when considering the link of sustainable development and international cooperation agendas. The United Nations Framework Convention on Climate Change (UNFCCC) can be considered one of the international regime's more developed and complex international agreements and has evolved over the past 30 years.

However, recently, there has been a gradual decrease in the UNFCCC's influence on the international agenda. The most significant events include:

1 The canceling movement to reduce the political importance of the Kyoto Protocol (an international cooperation and common rules-based system) to favor the Paris Agreement (a low comparability and nationalist rules-based system) and its emergent implementation of a vague 'Nationally Determined Contributions' (NDC)[1] institutional arrangement;
2 The transfer of the 2019 COP 25 host country from Brazil to Chile, and from there to Spain, and the consequent de facto presidency of a European country (dominating what was supposed to be a Conference led by Latin America and the Caribbean Group), which continued a pattern that privileged a European vision of the process, considering that from the last seven COPs in a row, six were hosted by European countries or led by European de facto presidencies;
3 The lack of leadership by the European Union that generated Paris Agreement Article 6s deadlock by blocking the transition from the Kyoto Clean Development Mechanism (CDM) to the Paris Agreement's NDC and proposing corresponding adjustments to paragraph 4 of Article 6, including methodologies and baselines requisite approval by countries (therefore decentralizing the operation of the mechanism and jeopardizing environmental integrity);
4 The attempt to change the Paris Agreement, creating a lack of confidence in the market mechanisms and the reliability of the UN treaties as a stable regulatory backing among developing and developed nations;
5 Brexit and the approval to host COP 26 by the United Kingdom;
6 The postponement of COP 26 from 2020 to 2021 due to the COVID-19 pandemic and consequential postponement of Kyoto Protocol's CMP 16;
7 The rise of nationalism among world leaders, including the United States, Brazil, Italy, Hungary, Poland, the Philippines, and others, that challenges international cooperation and multilateralism; and

DOI: 10.4324/9781003039860-34

8 The recent blockade of the CDM operation by European members[2] on the 108th meeting of CDM Executive Board of the Kyoto Protocol, even with the entry into force of the Doha Amendment in October 2020, trying to render the CDM inoperative and ineffective in contradiction with the Convention from Vienna. All of these events generated a decrease in ambition for the implementation of climate change mitigation and undermined the impetus for international cooperation. They also created a political environment that created perverse incentives for developed countries to cooperate less and to avoid presenting more ambitious short-term mitigation targets under the Kyoto Protocol.[3]

It is noteworthy that the cumulative effect of greenhouse gas (GHG) emissions from developed Parties since the Industrial Revolution is overlooked in the international debate. By ignoring such concrete data, developed nations – those that are most responsible for climate change and its impacts – aim to shift the burden of mitigation to developing countries and make them responsible for this mitigation. These developed nations' narratives emphasize the growing emissions from developing countries, with a view to making them responsible for global inaction in combating climate change. This emphasis on developing countries is occurring at a time when they are not only overwhelmed with recovering from the pandemic but have urgent problems to prioritize, such as poverty, health, and education, with financial resources and support from developed countries lacking.

This chapter highlights the scientific and political relevance of these problems today. Its scientific value is associated with dimensions of theoretical application, as well as the extensive practical knowledge of the authors in this area. More specifically on the theoretical dimension, the application, dissemination, and research of the methodological frameworks of neoliberal-institutionalist and neorealist approaches will be highlighted to better understand the rational action of states in cooperating to combat climate change in the face of the dilemma assessment of absolute gains and relative gains in an anarchic international environment.

The political value of this chapter is to produce scientific data that assists in the formulation of international policies and agreements favoring sustainable development approaches. Currently, there is a preponderance of the international discussion on climate change and an increase in international pressure for the progressive strengthening of the GHG emission reduction targets for large economies located in developing countries. Such conditions have territorial and political implications on the costs of real access to energy reserves, uses of natural resources, patterns of technological evolution, and trade, significantly affecting the current symmetries and asymmetries present in the global economy.

From a social perspective, this chapter aims to clarify the implications of interactions between recent nationalist movements in the evolution of the international climate change regime and publicize the importance of the concept of common but differentiated responsibilities in the global warming debate and the implications of applying this concept in the formation of national strategies in external relations engagement. Therefore, it seeks to inform society about the importance of studies on the transfer of responsibilities from developed countries to developing countries and the impact of the lack of international cooperation in deepening the problem of global warming instead of mitigating it.

The multiplicity of variables of the proposed problem and the different perspectives indicates the need to delimit effective scientific instruments for the implementation of this analysis. To this end, the following guiding questions are taken as possible explanatory ways to illuminate this problem: From which political-philosophical assumption was the distribution of responsibilities in the international climate change regime based before and after the approval of the Paris Agreement? What are the consequences of the emergence of recent nationalism in the process of reducing international cooperation reflected in the negotiating text of the Paris Agreement? What are the impacts of the slowdown in the globalization process on the evolution of the UNFCCC regime? Does the principle of differentiation adopted by the Paris Agreement reflect the ethical conditions and material support needed to avoid irreversible adverse impacts of global climate change?

Based on such questions, the central purpose of this chapter is to contribute to the debates that deal with the role of ethics and the dynamics of globalization and nationalisms in the formation and evolution of international regimes, especially about the regime on climate change.

The Climate Change Regime

The Intergovernmental Panel on Climate Change (IPCC) developed from an international scientific body, the Consultative Group on Greenhouse Gases, created in 1985 by the International Council of Scientific Unions, the United Nations Environment Program (UNEP), and the World Meteorological Organization (WMO), to provide research-based recommendations. At the time, the temperature anomaly was starting to increase from the historical average.[4] The United States government was the main political force in forming the IPCC as an autonomous intergovernmental body in which scientists participated, both as science experts and as official representatives of their governments, to produce reports that were strongly supported by the leading scientists of the world on the topic, and those reports then had to obtain the consensus of each of the participating governments. In this way, it was formed as a hybrid panel between a scientific body and an intergovernmental political organization. The IPCC has been tasked with reviewing peer-reviewed scientific literature and other relevant publications to provide information on the state of knowledge about climate change.

The origin of the UNFCCC dates back to 1988 in response to the concern arising from the establishment of the IPCC in 1988. The concern was the strong influence of science and scientists on the topic, which was not under the command of governments and, especially, their veto power. Therefore, there was a call by the IPCC to establish a treaty on combating global warming, led by states but with the participation of scientists from all member countries of the UN.

The UN General Assembly included the topic, 'Protection of the global climate for present and future generations of humanity,' on the agenda for its forty-third session at the initiative of the Government of Malta. The General Assembly recognized, inter alia, that "climate change is a common concern of humanity" and determined that "necessary and timely actions must be taken to deal with climate change in a global framework" (UNGA 1988).

In December 1990, the General Assembly decided to establish a single intergovernmental negotiation process under the auspices of the General Assembly, supported by UNEP and the WMO. These arrangements were organized with the goal to prepare by an Intergovernmental Committee for the Negotiation of an effective framework convention on climate change and any related instruments that may be agreed by UN Parties. It also set the objective that these negotiations should be concluded before the United Nations Conference on Environment and Development (UNCED) in December 1989 and be open for signature during the conference in 1992.

The Intergovernmental Negotiating Committee, open to all members of the UN and its specialized agencies, met in four sessions between February and December 1991. On May 9, 1992, the Committee agreed and adopted the text of the UNFCCC and recommended it for signature during the UNCED. The UNCED took place in Rio de Janeiro from 4 to 14 June 1992. The UNFCCC was then opened for signature at the conference and, by the end of the conference on June 14, 1992, it had been signed by 154 states and a regional economic integration organization (European Commission). The UNFCCC entered into force on March 21, 1994.

The history of globalization, according to Thomas L. Friedman (2005), is divided into three periods: Globalization 1 (1492–1800), Globalization 2 (1800–2000), and Globalization 3 (2000–present). He states that Globalization 1 involved the globalization of countries, Globalization 2 involved the globalization of companies, and Globalization 3 involved the globalization of individuals. The establishment of the UNFCCC, in this context, corresponds to the transition from the globalization of companies to the globalization of individuals and corresponds to a possible shift of power from governments to companies and the growing concern of individuals[5] with the greenhouse effect, anthropogenic GHG emissions and removal by sinks, global warming, and its adverse impacts in communities and ecosystems.

Globalization is a social, cultural, political, and legal phenomenon. Socially, it leads to greater interaction between various populations. Culturally, globalization represents the exchange of ideas, values, and artistic expression between cultures. Globalization could also represent a trend toward the development and emergency of a common world culture that can reflect the diversity of local cultures into a global citizenship. Politically, globalization has shifted attention to intergovernmental organizations such as the UN and the World Trade Organization (WTO). Legally, globalization has changed the way international law is created and applied, enhancing complexity and the number of stakeholders engaged. Both the IPCC scientific literature review process and the negotiation of commitments to mitigate anthropogenic GHG emissions and reduce global warming and mitigate its impacts are clearly associated with this process of globalization along these four axes (social, cultural, political, and legal).

Regarding the advantages of globalization, its supporters believe that it allows developing countries to become industrialized nations through increased manufacturing, diversification, economic expansion, and improvements in living standards. Expansion and integration of global supply chains, as well as outsourcing by companies could generate new jobs and the transfer of technology to developing countries, supporting their economies' escalation. Trade initiatives also increase cross-border trade, removing supply-side and trade-related restrictions. Globalization has also created room for advancing social justice and human rights, including by enhancing working conditions and workers' protections on an international scale.

Regarding the disadvantages of globalization, an economic slowdown in a country can create a domino effect through its trading partners. Globalization detractors argue that it has created a concentration of wealth and power in the hands of a small corporate elite that can swallow smaller competitors around the globe. Globalization has become a polarizing issue in developed countries such as Europe, the United States, and Japan, with the transfer of entire industries to new locations in developing countries, an important factor in the economic squeeze of the developed countries' middle classes.

The advantages of globalization led to the establishment of the IPCC and the UNFCCC. The disadvantages of the globalization process will be the focus of the negotiating process, with its most consequential changes perceived as starting in 2012. In the context, the underlying assumptions that guided developed nations' narratives are that the advantages of globalization are highlighted as benefiting developing countries, while the disadvantages highlight the difficulties imposed on developed countries. The assessment of such diverse readings of the implications of globalization helps to understand the changes in the process of negotiating measures to mitigate climate change.

Three characteristics of the UNFCCC are important for its understanding: its objective and commitments, the negotiating groups between the different Parties, and the principles embedded in Article 3. The ultimate objective of the UNFCCC is to achieve a stabilization of GHG concentrations in the atmosphere at a level that would prevent dangerous anthropogenic interference in the climate system. This level must be reached in a sufficient period of time that allows ecosystems to adapt naturally to climate change, to ensure that food production is not threatened, and to allow economic development to proceed in a sustainable manner. It should be noted that the text mentions "at a level that prevents dangerous anthropogenic interference in the climate system," without determining what that level would be (UNFCCC 1992). This complicates its implementation, given the existing uncertainties about the modeling of the climate system, as well as the large number of variables that influence the climate system and the complexity of its influences.

In addition, the statement, "This level must be reached in a sufficient period that allows ecosystems to adapt naturally to climate change," does not define this period or its detailed features (UNFCCC 1992).

The UNFCCC has two types of commitments for developed countries: general and specific. The general commitments require (1) to prepare, periodically update, publish, and make available national inventories of anthropogenic emissions by sources and removals by sinks of all greenhouse gases, which aims to make a diagnosis of how countries emit greenhouse gases; and (2) formulate, implement, publish, and regularly update national and, as the case may be, regional programs that include measures to mitigate climate change, addressing anthropogenic emissions by sources and removals by sinks of all greenhouse gases not controlled by the Montreal Protocol, as well as measures to allow adequate adaptation to climate change.

Among the specific commitments for developed countries, within the principle of Common but Differentiated Responsibilities, the UNFCCC established "that these anthropogenic emissions of carbon dioxide and other greenhouse gases not controlled by the Montreal Protocol return, individually or jointly, to their 1990 levels" (1992). It is important to note that the text of the UNFCCC is almost universally ratified by the UN Parties, including the Republican Administration of George Bush and US Congress.

The negotiating groups allow a certain order of negotiations and discussions through the dialogue of a few representatives of these groups during the meetings of the COPs and subsidiary bodies. The developed countries (countries listed in Annex I of the UNFCCC) represent a group formed by the countries that were originally members of the Organization for Economic Co-operation and Development (OECD) in 1990 and negotiate in different groups with different views, of which the European Union group and the other group of developed countries stand out. Non-Europeans gathered in the so-called 'Umbrella,' also known as JUSCANZ, originally composed of Japan, the United States, Canada, Australia, and New Zealand, and to which the Russian Federation, Ukraine, Kazakhstan, Belarus (CIS and Annex I), Iceland, Norway (Europeans not from the European Union), and Israel (independent) have joined.

Developing countries come together under 'the G77 and China,' currently composed of 134 countries.[6] However, despite being developing economies, these countries are quite different, whether in their stage of development such as small island states, least developed countries, group of large emerging BASIC countries (Brazil, South Africa, India and China), or whether by common economic, regional interests, cultural or ideological perspectives (OPEC, coal producers, African group, AILAC – Independent Alliance of Countries in Latin America and the Caribbean, ALBA – The Bolivarian Alliance for the Peoples of Our America, Arab group, Coalition for Rainforest Nations (CfRN), Like-Minded Developing *Countries* on Climate Change (*LMDC*), and the recently formed group of Argentina, Brazil and Uruguay. In negotiation, these groups are the main speakers; however, depending on the agenda item, there may be changes in alliances ('shift alliances') for specific issues and different national circumstances.

It should be noted that the commitments of developed countries – that their anthropogenic emissions of carbon dioxide and other greenhouse gases not controlled by the Montreal Protocol return to their 1990 levels – were not being fulfilled when it came into force. This situation led to the establishment of a negotiating mandate, the Berlin Mandate, at the first Conference of the Parties to the UNFCCC, held in Berlin in 1995. This established a process that allows the COP to take appropriate measures for the subsequent period to 2000, through the adoption of a protocol or other legal instrument.

Brazil, within the scope of the Berlin Mandate, in May 1997, presented a proposal[7] for the elaboration of a protocol. This proposal played an important role in the agreement that led to the adoption of the Kyoto Protocol. It contained two elements: (1) an objective measure to assign historical responsibility to different countries for causing global warming; and (2) for the creation of a clean development fund that would allow greater cooperation between countries, involving developing countries in mitigation measures of anthropogenic emissions of GHG and, also, the participation of developing countries that had no commitment to mitigate emissions, therefore inserting them in the global effort.

The objective feature for estimating the responsibility of each country stemmed from the physics of global warming, where anthropogenic emissions from different countries increase the concentration of greenhouse gases in the atmosphere, in particular the three main gases, carbon dioxide (CO_2), methane (CH_4), and nitrous oxide (N_2O), and this increase in concentration increases the deposition of infrared radiation on the Earth's surface, causing an increase in the average temperature of the Earth's surface. In relation to the possible impacts resulting from this global warming, the increase in the average sea levels would be an additional accumulation resulting from the increase in the average temperature, as well as the rate of variation in the time of this temperature increase would lead to a greater or lesser capacity of the biodiversity of adapt to this temperature rise.

The Clean Development Fund was proposed as a way to reduce the global cost of implementing the Kyoto Protocol, since reducing emissions in developing countries would in many cases have a lower cost of implementation than in developed countries with the same effect. The funds for the development of these actions would come from penalties imposed on developed countries that did not meet their goals.

The Brazilian proposal for a Clean Development Fund, modified in the negotiations for a mechanism, was adopted as Article 12 of the Kyoto Protocol, the so-called CDM. The strict liability criterion of the countries that did not meet their mitigation targets was not adopted. The rationale behind promoted incentives was for cooperative early action to mitigate the risks of potential non-compliance Parties.[8]

The Kyoto negotiations in 1997 led to the adoption of the Kyoto Protocol. The Kyoto Protocol, adopted almost universally, represents the highest point of the globalization perspective within the climate change regime. Under this process, all developed countries have comparable and well-defined emission reduction targets. Based on such targets, objective cooperation with developing countries could be implemented through the CDM and with the Economy in Transition parties through the joint implementation mechanism, similar to the CDM, but between parties that all had comparable and quantified commitments in the Protocol (all were Annex I Parties). In addition, the targets are translated into emissions allowances and a system of units is built that would allow the fungibility of all these units. Comparability was the basis of the rational action of countries to cooperate with a common goal to reduce GHG emissions. Hence, to create stability and predictability in the system it was essential to make sure that one ton of CO_2 was calculated and verified as one ton of CO_2, regardless of where it was reduced.

Therefore, the explanatory driver of the rational action of Parties to justify cooperation was the absolute gains reached in terms of tons of GHG emissions. The assessment was based on the climate change outcomes and in the identification of lower costs possible to reach this absolute gain – mitigate emissions. In short, all the Parties involved in the cooperation would get higher absolute gains by engaging in the cooperative action. For the 'buying' side, it would be 'cheaper' to invest in reducing emission 'abroad,' and for the 'selling' side it would represent a potential source of financing that would not occur in the absence of the cooperation. For the purpose of atmosphere protection, the localization of the emission reductions is much less relevant than the gain possibility to anticipate this mitigation earlier.

This process would be interrupted with the announcement of the Byrd and Hagel resolution, which would create an obstacle to the ratification of the Kyoto Protocol by the United States

(Byrd-Hagel Resolution 1998). This was followed by an announcement by President George W. Bush that the Protocol was flawed, and that the United States would leave the Protocol, starting a period of conflict in the relationship between countries and weakening both globalization and international cooperation on climate change. The rationale behind these actions was to only co-operate based on relative gain calculation. Therefore, even if the cooperation may result in a better situation to the Parties involved, if one of the partners considers that the relative gain with the cooperation outcome is smaller than the counterpart, this is enough to justify a no-cooperation result.

The UNFCCC's original targets were to stabilize emissions at 1990 levels. On the other hand, the Kyoto Protocol's reduction target, in the first commitment period, was supposed to reduce 5 percent from 1990 levels during the period centered on 2010 (2008–2012), representing a small additional mitigation effort and a ten-year deferral of mitigation efforts. This postponement represents in practice a further ten-year delay in implementing actions to combat climate change.

The departure from the United States was accompanied by Canada and New Zealand, which further weakened the Kyoto Protocol. The negotiation of the rules of the Kyoto Protocol is quite conflicting, starting with the announcement of the departure from the United States, and leading to the deadlock at the COP in The Hague, Netherlands in 2000, which resumed in March 2001 with the official announcement by the United States of its departure from the Protocol.

After the withdrawals from the United States, the largest global issuer, along with Canada and New Zealand, the remaining developed countries with targets represented only 18 percent of world emissions. The growth in emissions from developing countries with no reductions commitments led to the negotiation of a new route with two tracks. A first track, within the scope of the UNFCCC involved not only developed countries that had no commitment in Kyoto, but also developing countries seeking to renegotiate their commitments; and a second track that, determined by the Kyoto Protocol, negotiated the second commitment period. The Bali Conference ended with the establishment of these two negotiating trails.

Globalization versus Nationalism

The negotiation of the two tracks approach proved to be conflictive and led to a deadlock in Copenhagen at COP 15 in 2009. In 2012, corresponding to the end of the first commitment period of the Kyoto Protocol, COP 18 adopted the Doha Amendment, which established additional emission reduction commitments of 18 percent in 2020 for developed countries in relation to the levels from 1990. Again, the goal of the Doha Amendment was a small additional reduction compared to the first commitment period and, again, implementation was delayed for another ten years.

The announcement that the Russian Federation and Japan would have no targets in the second Kyoto commitment period deepened the perception that this second protocol period would not come into effect. This was because the ratification process for the Doha Amendment required ratification by three-fourths of the Parties to the Kyoto Protocol, and the ratification process in developing countries was cumbersome and taking too long. Therefore, it would be innocuous if not entered into force. The Doha Amendment clearly represents the division between developed countries that want to put an end to globalization and the developing countries, beneficiaries of the process, that defend its continuation.

With the departure of Japan and the Russian Federation (as they no longer have commitments), the goals of developed countries are, in the vast majority, reduced to the goals of European Union countries and Australia. These remaining countries did not want the second period of the Kyoto Protocol to come into force and practically determined the end of globalization and international cooperation within the framework of the UNFCCC.

The Paris Agreement was adopted at the Paris COP in 2015 and embodies the nationalist and anti-globalization vision and, therefore, was an act against international cooperation. The Paris Agreement replaced[9] the comparable and quantified targets, or 'commitments,' of the Kyoto Protocol with the vague and non-defined 'Nationally Determined Contribution'. This clearly shows the intention of developed countries to start a movement against globalization, accepted by developing countries which are led by the extreme right-wing political ideology, who, paradoxically, are opposed to the Paris Agreement and either do not to ratify it or wish to leave it after ratification.

The characteristic of NDCs means that the postponement of the deadline to combat climate change becomes variable in different countries. Countries with greater concern with solving the problem opt for shorter deadlines, such as 2025, and countries less concerned with combatting climate change and more concerned with social issues inside the country opt for longer deadlines of 2030. These longer deadlines again delay ten years for a global mitigation effort. Since the perception of relative gains drives the rational action of the Parties, and that it is not possible to have a clear comparable outcome to be able to assess the absolute gains, no cooperation becomes the standard of the regime and the Paris Agreement 'early' implementation.

The same occurs with NDCs that are of a quite different nature, ranging from quantifiable commitments in relation to actual emissions in the past, to commitments that will depend on future trajectories and, therefore, not quantifiable, and uncertain because they are based on future growth and development. Others relate their NDCs to mitigation efforts only in one sector or one greenhouse gas and others only in adaptation measures with no 'national determined contribution' to mitigation.

Once assessing the Paris Agreement, potential international cooperation opportunities were nearly always restricted to an Article 6 legal text which encompassed completely different views within the same article; for example, in its paragraph 2, the European Union's regional vision of cooperation is only within its borders (anti-globalization); in paragraph 4, Brazil's vision of international cooperation and international assistance is to achieve national goals, reduction of the global cost, and joint global effort (globalization); and in paragraph 8, "non-market methods" encompass the view of extreme left-wing countries (against capitalism and international trading systems).

During COP 21 in France, these three paragraphs were originally presented in draft texts of the Paris Agreement as three different articles. A concern of Venezuela that the non-market mechanism would be taken away from the agreement led the French Presidency to put together all three articles about mechanisms under Article 6. This led to an extension of the idea of corresponding adjustments to be demanded by the developed Parties to apply to Article 6.4, despite the fact that this paragraph has no mention to corresponding adjustments. The result of this demand from developed countries was that if the market mechanism is used by developing countries, then developing countries would have to increase their NDC ambition through corresponding adjustments.

This would cause several consequences. First, the NDC would no longer be 'nationally determined' once the buying of units by developed countries would imply increased ambition in the NDC through corresponding adjustments (and therefore increases difficulty to achievement of the contribution by the developing country). Second, the profit of the private sector project developer would correspond to a burden to the developing country government to enhance its NDC ambition and would lead to a backstop of the market mechanism by developing governments and consequential political pressure by the private sector on their governments. Third, projects under Article 6.4 would be additional to what would have been implemented in the baseline (in the absence of the existence of the mechanism) and therefore would not be part of the NDC. It would be beyond an NDC, and any corresponding adjustments would not make sense; on the contrary, it would only prevent action as it obliges an extra mitigation by governments because of the mitigation by the private sector – or a double mitigation caused by 'corresponding adjustments'.

The attempt to enhance the application of corresponding adjustments to Article 6.4, together with preventing CDM to be operated, are in line with a nationalist view of the European Union to stop international cooperation under the market mechanisms. This is because the EU is diverting European investments by multinationals to China, India, and Brazil,[10] clearly representing an anti-globalization view. By adopting such an approach, the EU becomes the front runner of a strategy that will enhance the speed of terminating the historical remnants of the Kyoto Protocol's rules-based approach that could lead to potential cooperation based on absolute gain calculation through comparable targets. In addition, it facilitates the potential erasure of developed countries footprints that reflected historical and consistent trajectory of non-compliance of past mitigation commitments since the beginning of the UNFCCC regime. During subsequent COPs from 2016 to 2019, these conflicts led to Article 6 negotiations deadlock. As a consequence, it was not possible to regulate the Paris Agreement, not only in this Article, but also in the definition of future commitments and future periodicity, deepening the division of countries in relation to the objective of the Agreement.

The Paris Agreement also presents a major change in the objective of reducing GHG emissions (the cause) to limit the increase in the average temperature of the Earth's surface. The increase in the average temperature of the Earth's surface (the effect) appears as a central element of the objective of the Paris Agreement, although the shock of the different views leads, once again, to the lack of consensus on the level of increase in the average temperature that prevents dangerous interference in the climate system. The Paris Agreement's objective is defined as a range between 2°C, defended by developed countries, and 1.5°C, defended by small island countries, least developed countries, and the African group.

Present and Future Generations

The election of President Donald Trump in the US in November 2016 and his nationalist rhetoric marked an important point for international action on climate change, most clearly evidenced by the US withdrawal of the Paris Agreement. And although President Joseph Biden defeated Trump in 2020 and re-joined the Paris Agreement, nationalism continues to permeate the country, now normalized and embraced by the US Republican Party. This, together with the rise of other nationalist leaders around the world, and the withdrawal of the United Kingdom from the European Union within the framework of the 'Brexit' process, reinforces the idea that nationalism interrupts the process of globalization. Hence, opportunities for international cooperation are clearly reduced compared to the initial stage of the regime. The political relevance of the UNFCCC weakens, and therefore the climate change regime as a whole leads to uncertainty about the ability of countries to combat global warming.

The European Union, in turn, aims to stop the Kyoto Protocol and the Doha Amendment by seeking to stop the CDM, with the argument that there is no third commitment period after 2020. In addition, other minor technical details like the lack of Global Warming Potential (GWP) metrics decisions and the lack of indicators of the commitment period in the Certified Emissions Reductions (CERs) bar code also support arguments to this position. However, that can be easily fixed by the CDM Executive Board and confirmed by a simple CMP decision. Therefore, this suggests that the root cause and the intention of the EU is to unilaterally stop the Kyoto Protocol once there is no willingness to have commitments in a third commitment period, making the CDM inoperative. Hence, stopping mitigation actions in developing countries and halting the CDM in CMP 16 for lack of consensus (blocked by all of the member states of the EU).

Following this rationale, the Kyoto Protocol post-2020, despite not having an expiration clause, will be extinguished in practice and, as a result, so will the CDM. This argument was facilitated by the strategy of postponing the COP later held in Glasgow, Scotland (which opposed Brexit) from 2020 to 2021 due to the pandemic. This is also supported with the absence of the possibility of negotiating a possible third period of the Kyoto Protocol's commitment and confirming the adjustment for the minor details that were proposed by the CDM EB. The European Members on the CDM EB are still trying to stop even voluntary cancellation of CERs highlighting the attempt to stop the Kyoto Protocol by means of turning CDM ineffective.

This conflict that can be generalized as between globalization and nationalism (understood as anti-globalization) and leads to the possibility of two future scenarios of global warming. The first scenario is a positive vision, privileging mitigation, more concerned with the growing impacts of global warming (in addition to growing impacts of the pandemic), which seeks a greater commitment to reduce emissions[11] with international cooperation. The second scenario could be characterized as more pessimistic, nationalist, skeptical, and negationist, denying global climate change and adopting a narrative of greater concern with internal social and economic conditions, and moreover, underestimating or denying the impacts of global climate change in the short term.

The pandemic raised further difficulties to future negotiations, such as the logistics of in-person COPs. Moreover, attempts to have on-line COPs creates advantages for developed Parties, and especially for the EU in terms of time schedules, but jeopardizes even more the climate change regime. And although the pandemic is becoming more manageable, it has normalized online and hybrid gatherings, exacerbating the difficulties of participation by developing nations. These facts can lead to enhancing the inequality of the negotiation process among developed and developing Parties and delaying even further the fight against climate change and its adverse impacts.

Notes

1 It is worth mentioning that the Paris Agreement text is silent on presenting a definition of "Nationally Determined Contribution." Therefore, comparability of efforts is a mirage once actual NDCs are carefully analysed.

2 Following a letter of the European Commission to the CDM Executive Board (EB), the European members blocked any attempts to have a consensus of deciding these issues by voting because the voting process of the EB requires ¾ majority of ten members. These issues could be easily solved by the EB and recommended to the CMP of the Kyoto Protocol. One of the authors was a member during EB 108.

3 Under the Paris Agreement rationale, the GHG mitigation ambition debate on developed nations focuses on 2050–2060 net zero emissions but is almost always silent to the period from 2020 to 2030, featuring a clear backslide compared to the subsequent commitment period approach adopted under the Kyoto Protocol.

4 See Figure 3.20 of the Second IPCC Assessment Report (IPCC 1995: 175).

5 As it was presented above, the scientists and IPCC authors participated in the processes under their individual capacity, not necessarily representing their respective nationalities.

6 See https://www.g77.org/doc/.

7 See https://unfccc.int/resource/docs/cop1/07a01.pdf.

8 The feeding of penalties resources to build the Fund and only afterwards implement mitigation is not coherent with the time dimension of the physics of global climate change discussed before. Hence, the impact of the compensation of a reduced tonne of CO_2 is also a function of when this reduction happened. That's why the delay of action that may reduce the same tonne of CO_2 today or in 2060 maybe be a great backslide on mitigation ambition, despite having the same numerical target.

9 This was proposed by a lawyer from the United States, according to sources inside of the process.

10 This is a direct statement by the EU director in a bilateral meeting with Brazil, in which one of the authors was participating.

11 See UN Climate Change News (2020).

References

Byrd-Hagel Resolution. (1998) *US Senate Res. 98 – A Resolution Expressing the Sense of the Senate Regarding the Conditions for the United States Becoming a Signatory to Any International Agreement on Greenhouse Gas Emissions Under the United Nations Framework Convention on Climate Change*. Available at https://www.congress.gov/bill/105th-congress/senate-resolution/98

Friedman, T. (2005) It's a Flat World, After All. *New York Times Magazine*. Available at https://www.nytimes.com/2005/04/03/magazine/its-a-flat-world-after-all.html.

Intergovernmental Panel on Climate Change (IPCC). (1995) *Second Assessment Report*. Available at https://www.ipcc.ch/site/assets/uploads/2018/05/2nd-assessment-en-1.pdf.

Kyoto Protocol. (1997) Available at https://unfccc.int/resource/docs/cop3/07a01.pdf.

UN Climate Change News (2020) 2020 Is a Pivotal Year for Climate – UN Chief and COP26 President. Available at https://unfccc.int/news/2020-is-a-pivotal-year-for-climate-un-chief-and-cop26-president.

UN Framework Convention on Climate Change. (UNFCCC) (1992) Available at https://unfccc.int/resource/docs/convkp/conveng.pdf.

UNGA. (1988) "Protection of global climate for present and future generations of mankind." 43rd UN *General Assembly*. Available at https://digitallibrary.un.org/record/54234?ln=en.

Additional Reading

Mendes, T. (2015) "Desenvolvimento sustentável – A construção do conceito no discurso do relatório "nosso futuro comum." In M. Souza et al. (eds.) *Meio Ambiente: Temática Política Central do Século XXI*. Conselho Editorial da Editora PUC Minas em 2013. Minas: Editora PUC.

Mendes, T. and Rodrigues Filho, S. (2012) Antes do pré-sal: Emissões de gases de efeito estufa do setor de petróleo e gás no Brasil. *Estudos Avançados,* [S.l.], 26(74), 201–218. Available at: http://www.revistas.usp.br/eav/article/view/10633/12375.

Miguez, J. D. G. (2002) "Equity, responsibility and climate change". In L. Pinguelli-Rosa and M. Munasinghe (eds.) *Ethics, Equity and International Negotiations on Climate Change* (pp. 7–35). Cheltenham: Edward Elgar.

Miguez, J. D. G. and Oliveira, A. (2011) "The importance of historical responsibility in the context of the international regime on climate change." In BASIC experts, 2011. *Equitable Access to Sustainable Development: Contribution to the Body of Scientific Knowledge*. Beijing, Brasilia, Cape Town, and Mumbai: BASIC Expert Group.

Moss, R. H. et al. (2008) *Towards New Scenarios for Analysis of Emissions, Climate Change, Impacts, and Response Strategies*. Geneva: IPCC.

Moss, R. H., et al. (2010) "The next generation of scenarios for climate change research and assessment." *Nature*, 463, 747–756.

Teitelbaum, B. (2021) *War for Eternity: The Return of Traditionalism and the Rise of the Populist Right*. New York: Penguin.

30

NEW ZEALAND'S NET ZERO CARBON LEGISLATION

Obfuscation and Missed Opportunities to Move beyond the Dominance of National Self-Interest

Prue Taylor

Introduction

In November 2019, the New Zealand Parliament passed the Climate Change Response (Zero Carbon) Amendment Act 2019 (CZAA or Act). Initially intended as stand-alone legislation, the Act was eventually passed as an amendment to existing legislation, the Climate Change Response Act 2002 (the principal Act), in order to create an emissions trading scheme. The CZAA was largely the initiative of a youth organization, Generation Zero. They worked hard to advocate for and popularize new legislation based on the United Kingdom's Climate Change Act 2008 (CCA) ahead of the 2017 New Zealand (NZ) general election. The basic architecture of the CZAA is very similar to that of the CCA. The CZAA creates a binding legislative emissions reduction target, net zero by 2050, and a process for achieving it via five-yearly budgets. In addition, an independent expert Climate Change Commission (CCC) is established to advise the New Zealand Government (Government) on emissions budgets, reduction, and adaptation plans, and to monitor progress. The Act intends to create a stable and credible policy environment, enabling long-term planning, decision-making, and investment, as well as address both mitigation and adaptation, with the latter requiring the development of a national risk assessment and national adaptation plan.

The CZAA also embodies some elements unique to New Zealand. For example, the Government has legal obligations under *Te Tiriti o Waitangi* (Treaty of 1840) to actively protect Māori (the indigenous peoples of Aotearoa New Zealand) in the use of their land and resources. As a result, some provisions explicitly address the special interests and needs of Māori. For example, under sections 5ZI and 5ZG, emissions reduction plans must be developed in consultation with Māori and include a strategy to recognize and reduce the impacts of mitigation action on Māori and *iwi* (or tribes). Another unique element is the separate gross reduction target for biogenic methane emissions (defined in CZAA Sections 6 and 5Q as methane produced from agriculture and waste sectors). This partly reflects NZ's emissions profile; around 48.1 percent of emissions are from agriculture, of which around 35.2 percent is methane. This high methane emissions rate reflects a massive growth in the dairy industry between 1994 and 2017 (Ministry for the Environment 2019a).

The Government engaged the public in two rounds of consultation prior to finalizing the Act, and submission records demonstrate a significant level of civil society engagement on many ethical issues. Generation Zero, academics, Māori, the NZ Parliamentary Commissioner for the

DOI: 10.4324/9781003039860-35

Environment, and some non-governmental organizations demonstrated considerable attention to a wide range of ethical issues.[1] However, the CZAA only partially reflects this civil society input. For example, it refers to the need to consider justice issues without using the phrase "just transitions" (CZAA 2019: sections 5M and 5ZC). Emissions reduction and adaptation plans, together with risk assessment, must take account of vulnerable groups and sectors in NZ (CZAA 2019: sections 5ZG and 5ZS) and distributive justice, inter- and intra-generational equity, and other unspecified equity principles are also relevant to review of the target and to budget setting (CZAA 2019: sections 5ZC and 5T). However, as the analysis in this chapter demonstrates, the treatment of ethical issues is generally muddled and superficial. The precautionary principle is not specifically mentioned (the target can be reviewed if there is a significant change in "scientific understanding," but this is not an implementation of the precautionary principle or approach) and human rights, including indigenous peoples' rights, are omitted, as is interspecies justice. The omission of human and indigenous rights, together with the limited enforceability of the target and budgets, has even led to calls for an environmental human right to be added to NZ's Bill of Rights Act 1990 (LCANZ n.d.). There is also no specific provision acknowledging the role and vulnerability of women, and the Act does little to raise public awareness and facilitate active participation in decision-making and action through, for example, knowledge provision, financial support, and access to resources per the UNESCO Declaration of Ethical Principles in Relation to Climate Change (UNESCO 2017: Article 4).

While it would be churlish to deny the importance of the Act's specific references to ethical issues and relevant principles, it is also true that in critical respects, the Act reflects both obfuscation and missed opportunities. This chapter outlines selected ethical issues that are both important and relevant to climate change mitigation, and while the Act covers both mitigation and adaptation, the latter is not the focus of this analysis. This analysis intends to inform further policy and legal work in order to strengthen the Act and provide fairer and more effective outcomes nationally, regionally, and globally. This analysis also seeks to inform the design and content of climate change legislation in other jurisdictions (see also Averchenkova et al. 2017), as a growing number of nations are adopting a "net zero" target and utilizing the CCA as legislative precedent (see Taylor 2021). Ultimately, the whole is greater than the sum of the parts in the context of both domestic legislation and collective global action. Policy and law genuinely framed and implemented around an ethic of collective human responsibility, and which ensures open and informed public debate/critique about the justifications for decisions, including the role of ethical principles, economic rationality, and science, is much more likely to result in fair and just outcomes that motivate and sustain the high levels of public and individual commitment needed to urgently address climate change (e.g., Dunlap & Brulle 2015; Wetts 2020).

It should be noted that the CZAA only creates a legal framework for mitigation and adaptation planning. It has not yet: (a) resulted in any new policies for mitigation; nor (b) repealed existing laws which explicitly prevent authorities (e.g., local government) from directly regulating emissions (Harker et al. 2017). It is not yet clear when new policies will be released and whether pre-existing laws will be repealed. Both will be an important test of the extent to which explicit ethical principles (e.g., inter- and intra-generational equity) are implemented. It is also important to note that the information in this chapter was accurate as of the date of original completion, 2020. A new NDC was released in November 2021 stating an intention to reduce net GHG emissions to 50% below gross 2005 levels by 2030.

The Warming Limit

The stated purpose of the CZAA is to "*contribute to* the global effort under the Paris Agreement to limit the global average temperature increase to 1.5°C above pre-industrial levels" (2019: Section

4, emphasis added). It is appropriate that the low end of the Agreement's temperature range has been selected given the connection between temperature, NZ's status as a developed nation, and harms caused, including loss of life and livelihood. However, NZ's target and level of 'contribution' to achieving this target must be read in light of: (a) target setting and review; (b) the separate methane target; and (c) budgeting. These elements are considered below.

Target Setting and Review

The Act sets a statutory target for reducing net accounting emissions of greenhouse gases (excluding biogenic methane) to zero by 2050 (CZAA 2019: section 5Q). This target is said to be based on scenarios developed by the Intergovernmental Panel on Climate Change's (IPCC) 2018 report aimed at limiting global atmospheric warming to 1.5°C (Explanatory note to the Climate Change Response (Zero Carbon) Amendment Bill 2019: 4). Adopting a net zero approach is common in legislation of this kind (e.g., the United Kingdom recently announced a new target of net zero by 2050). However, as the independent Parliamentary Commissioner for the Environment (PCE) points out, this target is not well understood and was selected by the Government without following a transparent independent decision-making process (PCE 2019: 3–4, Annex A). To this end, and prior to enactment, the PCE recommended that the independent CCC advise the Government on the initial target following a statutory framework based on both what science requires to achieve the warming limit and a list of specific factors, including equity principles. These factors and principles should include international and inter-generational equity, historical responsibility, future emissions per capita and development rights. This list includes New Zealand's economic and fiscal circumstances, but it is not the only factor (PCE 2019: 3–4, Annex A). The point is that a transparent process such as this would enable the public to understand and debate the justifications for decisions, including the application of relevant ethical principles, the rationale for and acceptability of any trade-offs, and the ethical implications of those decisions. Interestingly, the PCE's submission included the comment that most national legislation omits this kind of procedural and substantive analysis prior to target setting (PCE 2019: 4). Unfortunately, the Government did not accept the PCE's submission.

A similar point can be made about the processes and considerations relevant to reviewing the target and to budget setting (CZAA 2019: Sections 5T and 5ZC). At present, the statutory criteria are a 'grab bag' of factors that heavily prioritize economic and fiscal matters. They are also openly drafted, enabling, for example, NZ to scale back its mitigation ambition if global action on climate change reduces (CZAA 2019: Section 5T). There is no requirement to otherwise account for emissions reductions avoided. As currently written, this provision raises real risks that economic rationality will dominate, creating a legal justification for mitigation measures that will contribute to an excess of 1.5°C of warming and resultant harms, including loss of life and livelihoods. The Government can use this provision to avoid the need to offer morally justifiable reasons for prioritizing national economic interests.

The Government has explicitly chosen not to pursue a global carbon budget approach. As a consequence, it has avoided, and without any supporting justification, consideration of the nation's 'fair share' of that budget, including how to determine 'fair share.' This is a critical point of departure for statutory target setting and supporting budgetary processes. Using a global carbon budget approach is not without difficulties and calculating this figure is highly complex (for some of the more recent research, *see* Rogelj and others (2019)). However, there is considerable support for its use as a means to address some of the equity requirements of the Paris Agreement, including equality, or equal rights to emit, using a per capita calculation as a start point (see, e.g., Brown et al. 2018).

The absence of any real justification for avoiding a global carbon budget approach leaves real uncertainty about the motives of the Government. Generation Zero's CZAA submission outlines how a global carbon budget approach could be used. In their evaluation, it requires: (a) determination of New Zealand's 'fair share' of the remaining global carbon budget (for keeping within 1.5°C) using both science and equity principles (including NZ's historical emissions); (b) alignment of NZ's target and all budgets with both the 1.5°C temperature goal and keeping within NZ's 'fair share' of the remaining global carbon budget; (c) setting a 2030 target for 50 percent net reductions of non-methane emissions from 2010 levels (consistent with the IPCC's 2030 advice) (IPCC 2018: C.1); and (d) setting a specific methane reduction target for 2050 using IPCC science for achievement of the 1.5°C temperature goal (Generation Zero 2019: 9–11).

The CZAA's failure to adopt a target and budget process based on NZ's 'fair share' of the remaining global carbon budget is particularly egregious given a cold hard fact: decades of high-emitting economic growth and consumption has resulted in New Zealanders having the fifth-highest per capita emissions profile in the OECD (OECD 2017). New Zealand's gross and net emissions continue to grow under a business-as-usual scenario. Gross emissions increased 2.2 percent between 2016 and 2017, accounting for around 80.9 million tonnes of carbon dioxide equivalent. This brought the increase in gross emissions from 1990 to 2017 to 23.1 percent Net emissions (taking into account carbon dioxide absorbed by forestry) increased by 65 percent since 1990. The two largest sectors for gross emissions in 2017 were agriculture (48.1 percent and energy (40.7 percent Emissions from electricity generation and transport continue to grow significantly and are responsible for the increases in carbon emissions (Statistics NZ 2019). Clearly, NZ has benefited from the wealth generation that sits behind this emissions profile but not accounted for the costs to others. This has not escaped the attention of domestic litigants. Legal action against the Government was initiated in 2019 arguing that neither the approach of the CZAA (long-term net target) nor NZ's nationally determined contribution (NDC) under the 2015 Paris Agreement (to reduce emissions by 11 percent below 1990 levels by 2030) (NEW ZEALAND: Submission n.d.)[2] are sufficient to meet NZ's "fair share" of the emissions reductions required to keep within 1.5°C (Neilson 2019). Recent Carbon Tracker research supports this, noting NZ's current targets and policies are "not in line with any interpretation of a 'fair' approach to the former 2°C goal, let alone the Paris Agreement's 1.5°C limit" (Fair Share – Climate Action Tracker New Zealand).

Rejection of the global carbon budget approach also distracts from the ethical considerations inherent in choosing a *net* emissions approach for carbon emissions. There is growing discomfort with net zero approaches to climate change mitigation. In principle, it enables wealthy economies, such as NZ, to continue to benefit from increasing greenhouse gas emissions levels for another 30 years, provided these are at some distant point, i.e., by 2050, offset by removals. The net approach has the potential to enable certain economic sectors to benefit from delays in making the profound socio-economic transformations required, and as a result, impose higher risks and unfair burdens on others, including vulnerable communities, future generations, and nature. For example, research suggests that the use of net zero targets that combine emissions reductions and offsets may well provide excuses for continued delays in reductions and put misplaced trust in the achievement of significant negative emissions (offsets) in the future (McLaren 2019). This is partly because negative emissions can appear cheaper, and therefore more palatable, than accelerated mitigation. In short, negative emissions (offsets) can be a substitution for emissions reductions rather than a supplement (McLaren 2019).

There are also considerable uncertainties and therefore risks about the efficacy of many carbon offset mechanisms. Reflecting this uncertainty, the PCE has warned of significant risks inherent in using forestry for offsetting. These risks include: the difficulty of maintaining forests due to the effects of climate change (e.g., water scarcity, fires, disease, soil change); effects of forestry on rural communities; and impacts of land-use change and forests being a default solution used instead of gross emissions reductions at source (PCE 2019: 6–7). Given these uncertainties and the risks posed for increasing emissions and therefore resultant harms, the PCE and others recommended a precautionary approach – namely, that forestry offsets, the primary offset mechanism favored in NZ, be prohibited or strictly limited (PCE 2019: 6–7; Wise Response 2019: 15). Neither of these suggestions were adopted, but the Government did agree to restrict use of offshore mitigation measures as a means to offset (CZAA 2019: Sections 5Z, 5ZA, 6).

From a policy design perspective, some research suggests the merits of "formal separation of negative emissions targets and accounting for emissions reduction, rather than combining them in a single 'net-zero' goal" (McLaren 2019), together with the use of explicit targets and timetables. This is not the approach taken by the CZAA. The CCC can advise, but not direct, the Minister for Climate Change on

> The proportions of an emissions budget that will be met by *domestic emissions reductions and domestic removals*, and the amount by which emissions of each greenhouse gas should be reduced to meet the relevant emissions budget and the 2050 target.
>
> *(CZAA 2019: section 5ZA, emphasis added)*

Both the Minister and the CCC must have regard to "risks and uncertainties associated with emissions reductions and removals," and the "distribution of those impacts [of actions to achieve the budget] across the regions and communities of New Zealand, and from generation to generation" (CZAA 2019: Section 5ZC). While this language is better than nothing, it falls well short of providing the strict separation that risks associated with the net zero approach merits (removals is defined in CZAA 2019: Section 6).

There is a significant difference between a low or zero emissions economy and a *net zero* emissions economy. Clear separation between reductions and offsets helps expose interests, politics, benefits, and burdens between nations, sectors, communities, generations, and between humanity and nature. The intra- and inter-generational equity implications were recently targeted by Greta Thunberg. Speaking to the 2020 Davos World Economic Forum, she urged nations to "forget net zero" and achieve "real zero" (BBC News 2020). The potential for obfuscation inherent in the net zero approach has the potential to place a much greater burden on others without their full knowledge or consent.

New Zealand's use of a net approach is deeply rooted in an early interpretation of the 1992 UN Framework Convention on Climate Change as enabling a policy approach of 80 percent of reductions through offsetting and 20 percent emissions reduction at source. Related to this is an intention to use the existing emissions trading scheme (ETS) as the primary policy tool to achieve the CZAA target and budgets (Ministry for the Environment 2018: 14).[3] However, the Government acknowledges that significant changes must be made to the ETS for it to be effective (Ministry for the Environment 2018: 14). To date, the ETS has failed to adequately drive emissions reductions and offsets. Changes may include a specified reducing emissions cap, together with an effective carbon price (currently around NZ$25 per unit) and, most controversially, bringing agricultural (methane) emissions into the scheme at some future point.[4] There is no suggestion that carbon credits for domestic offsets will be dropped from the ETS as a means of creating greater clarity between reductions and offsets (McLaren 2019).

The Split-Gases Approach

The CZAA creates separate targets for biogenic methane, defined as produced by the agricultural and waste sectors. The targets are: (a) reducing gross emissions within the range of 24–27 percent below 2017 levels by 2050 and (b) including to 10 percent below 2017 levels by 2030. This approach was highly controversial and is not supported on scientific or ethical grounds (PCE 2019: 7). The premise used by Government was that methane is a short-lived gas and a stable rate of emissions will not lead to increases in atmospheric concentrations (Explanatory note to the Climate Change Response (Zero Carbon) Amendment Bill 2019: 4). However, what is a 'fair target' for methane is not solely a question of stabilizing the temperature impact of a gas by a given year. Given methane's potency (80 percent more potent than other GHGs over a 20-year time scale: National Geographic 2019), short-lived nature, and the urgency of climate change, significantly reducing methane can contribute to rapid cooling of the global atmosphere. The PCE notes, "[g]iven the need to head off peak global temperatures by *whatever means we can, contributions are needed from wherever they can be found*" (PCE 2019: 8, emphasis added). He opines that, "[s]imply attempting to 'grandfather' our existing contribution to warming from methane [using 2017 baseline] could be seen to be self-serving …" (PCE 2019: 8). The Wise Response submission is more straightforward: given the considerable scientific uncertainties around global methane emissions, it is essential that New Zealand reduces them immediately (Wise Response 2019: 14). This is a clear reference to protection of NZ's economically significant agricultural sector, which is responsible for around 35.2 percent of the nation's emissions. A Climate Analytics report notes that mere stabilization of methane is not sufficient. A global reduction of 30–50 percent below 2010 levels by 2030 and around 50 percent below 2010 levels by 2050 is required by climate science (Climate Analytics 2018: 5). The Generation Zero submission cites a report of the Biological Emissions Reference Group, which found that NZ's methane emissions could feasibly be reduced by 22 percent to 48 percent by 2050 (Generation Zero 2019: 11). Accordingly, New Zealand's approach will "not set the comprehensive policy targets required and will not be consistent with achieving the Paris Agreement goals" (Climate Analytics 2018: 2). In other words, this approach places an additional burden on other nations to compensate for the lack of action by NZ to reduce its non-CO_2 gases. NZ has transferred its responsibility to reduce emissions elsewhere, without adequate transparency, justification, or acknowledgment.

The Act's treatment of methane is a clear reflection of political compromise. The Government's objective was bipartisan support for the CZAA. To achieve this, it needed the agreement of the major opposition party, which is the traditional "farmer's party." The split-gases approach, with two methane targets, was sufficient to achieve bipartisan support for the enactment of the CZAA. However, the opposition party has clearly stated an intention to amend the CZAA should it form a government in the future. It wants delayed introduction of a biogenic methane target based on advice from the CCC and enhanced consideration of economic impacts (The Spinoff 2019). It seems that the much-vaunted bipartisan support has a potential expiry date, at least for the methane targets, as national elections are scheduled in September 2020. To reiterate, this is significant because 48.1 percent of NZ's emissions come from agriculture, of which around 35.2 percent is methane. NZ will not achieve its target unless it takes ambitious action to reduce methane emissions at source, in a timely manner. The Government has not discussed if, and if so, how, this deficit will be accounted for through, for example, greater cuts in carbon emissions. This would, of course, transfer an extra burden to other sectors of NZ society and the economy.

Accounting – Banking, Borrowing, and Offshore Mitigation Units

The CZAA allows banking and borrowing between budgets with the latter being subject to some restrictions (2019: section 5ZF).[5] Banking and borrowing are justified on the basis of creating accounting flexibility, but in the absence of very robust restrictions and transparency requirements there is potential for delayed mitigation action creating a liability for future governments, future generations, and greater burdens for communities and nature. As regards to offshore mitigation units, the legislation puts limits on their use as a means to prioritize domestic reductions and removals (CZAA 2019: section 5Z). It also defines "offshore mitigation" units in an effort to ensure robust accounting, including avoidance of double accounting and actual additional, measurable, and verifiable reduction or removals (CZAA 2019: section 6(1)). While these elements are welcome and will limit the use of bogus units, the CZAA makes no mention of controlling for units generated in circumstances that harm or otherwise disadvantage communities and/or nature. A whole host of ethical issues has been raised around, for example, carbon-forestry policies such as REDD (Reducing Emissions from Deforestation and Degradation) (Friends of the Earth International 2014). The Act does not acknowledge these ethical concerns as a criterion for rejecting offshore mitigation units. As a consequence, these will likely be dismissed as not being legally relevant to decision-making.

Role and Expertise of Climate Change Commission

The functions of the CCC include advising on any review of the target, providing advice on budgets and emissions reduction plans, and monitoring/reporting on progress towards the target and budgets. It also monitors implementation of the national adaptation plan. The statutory "matters" relevant to performance of their functions and duties include scientific knowledge, but not indigenous or traditional knowledge; technology; economic effects; social, cultural, environmental, and ecological circumstances; distribution of benefits; costs and risks between generations; relationship between the Crown and Māori; and responses taken by Paris Agreement parties (CZAA 2019: section 5M). The listed expertise of CCC is designed to support proper consideration of these diverse "matters." Expertise includes professional skills in "the environmental, ecological, social, economic, and distributional effects of climate change and climate change policy interventions" (CZAA, section 5H). The CCC must also collectively have an understanding of the Treaty of Waitangi (*Te Tiriti o Waitangi*) (CZAA 2019: section 5H) and a specific list of Māori cultural concepts, including the Māori worldview and traditional knowledge (CZAA 2019: section 5H).

At first glance, there are some welcome references within the statutory 'matters and 'expertise' which could support the CCC's engagement with Government on a range of ethical issues. However, a few qualifiers are merited. First, it is highly likely that the CCC's initial focus will be primarily technocratic and economically focused given (a) use of the ETS as the primary policy tool and (b) political/economic sensitivity over methane targets. This was reflected in the expertise and focus of the Interim Climate Change Commission (ICCC 2019). Second, all references to distributional effects and costs/risks between generations will be limited to the New Zealand domestic context and therefore omit important ethical issues concerning vulnerable peoples/communities and ecological systems beyond NZ's jurisdiction. Third, while the CCC can provide advice on emissions reduction plans, its ability to evaluate the policy settings chosen by Government is limited. For example, the Government is not required to explain the reasons for any discrepancy or gap between the CCC's advice and Government policy decisions on emissions reduction plans. This potential lack of transparency will potentially obscure when and why policy trade-offs have been made between what science requires to meet the target and budgets and what Government has decided to do or not do, having regard to "economic feasibility" and "economic circumstances" (the Minister must "consider" the Commission's advice; CZAA 2019: section 5ZI). Given the tendency for economic rationality and political compromises to consistently trump ethical principles, this is a glaring omission. It also greatly weakens the ability of the CCC to use its expansive range of statutory matters and expertise to influence critical policy outcomes – that is, emissions reduction plans – before they are made operative and to hold Government to account.

Indigenous Peoples

The UNESCO Declaration of Ethical Principles in Relation to Climate Change articulates a core ethical concept – the need to give priority to vulnerable groups, including but not limited to indigenous peoples (UNESCO 2017: Article 10). This language in mirrored in the Paris Decision and Agreement, and in multiple international documents on sustainability. In NZ, Māori are uniquely vulnerable to climate change impacts due to their intimate relationship with the natural world and to high levels of poverty within their communities. For example, coastal communities are reliant on marine ecosystems for food and perform a traditional role as *kaitiaki* (or guardians) of these ecosystems. The CZAA's statutory references to matters and expertise specific to the needs and interests of Māori and to the Crown–Māori relationship are welcome. However, there is a significant critique from the Māori community that the provisions are inadequate. They do not fully integrate an indigenous worldview, nor do they fully acknowledge Māori as a Treaty of Waitangi (*Te Tiriti o Waitangi*) partner. One commentator noted that full recognition of Māori as a partner, not just a stakeholder, would merit the *co-development* of reduction and adaptation plans to acknowledge Treaty rights, *mātauranga* Māori (traditional knowledge) and "the fact that we are the indigenous people here and have a key role to play …" (for further analysis on strengthening of the Act, see Bargh 2019). In the absence of a true partnership relationship between Māori and the Crown, there is a tendency for statutory consultation and participation processes to become exclusionary (Bargh & Van Wagner 2019). This compounds the fundamental lack of justice for Māori as evidenced in multiple reports by the Waitangi Tribunal (Waitangi Tribunal 2011).

While the CZAA makes an effort to be inclusive of the needs and interests of Māori, consistent with its obligations under Treaty of Waitangi (*Te Tiriti o Waitangi*), unlike other environmental legislation it does not explicitly require the Crown to observe "the principles of the Treaty." Courts have found that the Treaty places a moral and legal duty of "active protection" upon the Government (*New Zealand Maori Council v Attorney General* [1987] 1 NZLR 641). The potential of this moral and legal duty to drive increased climate change action is currently being tested. In 2016, a claim was lodged by Māori before the Waitangi Tribunal arguing the Government's inadequate efforts to mitigate climate change are a breach of this moral and legal duty of 'active protection' (Paul et al. 2016). The claim is still before the Tribunal awaiting hearing.

Legal Enforceability

The CZAA explicitly excludes any legal remedy or relief for failure to meet the target and budgets, beyond a declaratory judgment and associated award of costs (Climate Change Response (Zero Carbon) Amendment Bill 2019: clause 5ZJ).[6] While it is understandable that Government departments should not be exposed to damages claims for failing to meet the target or budgets, the legislation goes too far "because it ousts the courts from applying or creating remedies in appropriate cases in a proportionate and reasonable manner" (LCANZ 2019: 5). In short, Government is accepting legal responsibility through the CZAA but actively preventing New Zealanders from holding it legally accountable for a failure to fulfill that responsibility (UNESCO 2017: Article 4). This is at odds with the willingness of the NZ courts to conduct judicial review proceedings that cautiously test the lawfulness of decisions but do not substitute their own policy views for those of Government Ministers (*Thomson v The Minister for Climate Change Issues* [2017] NZHC 733).

From a legal perspective, the CZAA's privative clause goes too far, potentially offending against environmental justice concepts, including developing climate justice litigation; international law principles; the right to natural justice; and constitutional principles (Daya-Winterbottom 2019; Generation Zero 2019: 18–19). In general terms, given the urgency and existential risks of climate change, resort to media scrutiny, political accountability, and parliamentary oversight are insufficient (Daya-Winterbottom 2019). A declaratory judgment and costs are a very poor substitute for judicial review and holding the Executive to account. More consideration should be given to timely scrutiny, through judicial processes if needed, of emissions reduction plans (Generation Zero 2019: 18), including their capability to meet the target and budget before any failure to meet a budget occurs. Directing the Government to put in place measures to achieve a numerical target or budget does not encroach on Executive prerogative if it does not extend to directing Government to act in a specific way. In addition, the existing Executive discretion to ignore, or pay lip service to, the Paris Agreement's equity requirements, and in favor of other measures of national self-interest, in setting the NZ NDC could be constrained by relevant provisions in the CZAA. These could then form the basis for domestic judicial review proceedings.

Conclusion

The very existence of the CZAA reflects some progress in New Zealand's understanding of its ethical responsibilities to respond to climate change. Until recently, the nation's failure to adequately address climate change was justified on the basis that its 0.17 percent contribution (to global net emissions) was insignificant, deliberately overlooking growing gross, net, and per capita emissions (Hadley 2014). The change in position is partly attributable to a combination of Labour, Green, and centralist political parties forming a coalition to govern in late 2017.

New Zealand has also begun to more openly acknowledge its responsibilities towards the Pacific and vulnerable small island developing states. However, these responsibilities will be met primarily through financial and technical development assistance and not by NZ raising the level of its ambition. Political rhetoric focuses on NZ's leadership in adopting a legal framework designed to "*contribute* to limit[ing] the global average temperature increases to 1.5° Celsius" (CZAA 2019: Section 4, emphasis added). This target is, in part, a reflection of the existential risk posed to Pacific Island states by temperatures in excess of 1.5°C. However, referring to a temperature target alone is not sufficient. Doing so easily distracts from the reality that NZ's current NDC (11 percent below 1990 levels by 2030) is a wholly inadequate 'contribution' to meeting that target. Carbon Tracker research indicates that NZ's targets and policies are consistent with warming of between 2 and 3°C. Critically, the Paris Agreement's 1.5°C target could only be met if "other countries make much deeper reductions and comparatively greater effort" (Fair Share – Climate Action Tracker New Zealand). As has been argued elsewhere (Brown et al. 2018), inadequate contributions of this nature raise ethical obligations to provide reasoned justifications for inaction and to compensate through, for example, enhanced adaptation and loss and damage obligations. These issues are not addressed by the CZAA. Like the UK CCA, for legislation designed to address a global problem, it has an almost exclusively domestic focus.

The NZ Government has not yet signaled its intention to increase the level of its NDC. A key opportunity for this is in the lead-up to the 26th Conference of the Parties in 2020. Official documents suggest that NZ may update or recommunicate its current NDC, delaying any further change until 2023 (OIA 2019). In the event that it does increase its NDC, this will not necessarily have a direct impact on CZAA targets and budgets. This is because the NDC may be taken into account in reviewing the target and setting budgets but is not a mandatory requirement that drives either element (CZAA 2019: sections 5T and 5ZC). Despite the link to the Paris Agreement's 1.5°C temperature limit, the CZAA is not a domestic tool for the direct implementation of the NDC.[7]

The Government has not yet responded to calls for a climate emergency declaration (MacManus 2019), but it has previously refused to develop policies and plans providing for climate migrants/refugees from the Pacific (Rive 2015). This latter policy omission may change in the near future, especially in light of the United Nations Human Rights Committee's recent finding that NZ (and other nations) may not return to their country of origin people fleeing from imminent climate change risks if their essential human rights are at risk (UNHCR 2020).

A reading of political statements made before the NZ House of Representatives ahead of the final vote on the CZAA reveals fine rhetoric about responsibility to current and future generations and to vulnerable nations in the Pacific. However, these statements also reveal that both major political parties have, in reality, framed their policies primarily around national economic interests, particularly those of the economically and politically powerful agricultural sector. The opposition political party justifies continued high per capita emissions around NZ's self-appointed role as the "world's most efficient food producer." The Prime Minister referred extensively to climate action being necessary to protect NZ exporters who trade on a reputation of *sustainability* and environmental responsibility (The Spinoff 2019). The Prime Minister's statement highlights yet another curious omission from the CZAA – it makes no reference at all to sustainability or sustainable development (Taylor & Scanlen 2018).[8] This omission deprives NZ of the opportunity to draw upon a rich international and domestic discourse on the alternative ethical paradigm inherent in the mode of 'strong' as opposed to 'weak' or 'triple bottom line' sustainability. The former recognizes the finite limits of ecological systems and human dependency. The latter justifies false 'balancing' of interests, enabling economic benefits to outweigh environmental goods (Bosselmann 2017; Sustainable Aotearoa NZ 2009). The outcome is a legally justifiable cumulative degradation of ecological systems.

Overall, with the exception of a few nods toward ethical principles, domestic economic and political interests coupled with climate science remain the dominant considerations under the Act, setting both ambition and pace. This reflects a continuance of the state-centric focus of international climate change law that allows states to 'pass the buck' and creates a trajectory for catastrophic global warming. The CZAA deprives NZ and New Zealanders of the opportunity to engage with and critique climate change policy and law from the perspectives of human responsibility, justice, and equity. These concepts speak to both our hearts and our minds and have to power to inspire, motivate and sustain the level of collective action needed for the common good. They are also fundamental to lifting the level of public critique and discourse needed to challenge an underlying assumption that climate change (Otero 2020; Wetts 2020) and related global issues can be addressed without challenge to ecologically and socially destructive paradigms of economic growth and wealth production.

Notes

1 Copies of public submissions are available at: https://www.parliament.nz/en/pb/sc/scl/environment/.

2 This target is also expressed as 30 percent below 2005 levels by 2030.

3 France-Hudson argues the NZ ETS avoids creating private property (private rights) to pollute the global atmosphere. See France-Hudson 2015: 105. Ethical analysis of emissions trading schemes is beyond the scope of this chapter.

4 At the time of writing, indications were for limited changes, including delays in bringing agriculture into the ETS (not until around 2025) and a low carbon price of NZ$50 per unit (recommendations are for between NZ$150 and $250). See NZ Productivity Commission 2018: 41; Ministry for the Environment 2019b.

5 If the total quantity of emissions in a budget period are less than provided for in that budget, the excess may be carried forward to the next budget period (banked). If the total quantity of emissions in a budget period are greater than provided for in that budget, up to 1 percent may be carried back (borrowed) from the next emissions budget.

6 It is not clear whether this privative clause extends beyond substantive obligations (to meet the target and budgets) to include procedural obligations, including those in respect of the reduction plans.

7 The Explanatory note to the Climate Change Response (Zero Carbon) Amendment Bill 2019: 6 provides that the Act does not impact commitments to communicate and achieve NDCs. Further, limiting use of offshore mitigation measures to meet the Act's 2050 target *does not preclude New Zealand's ability to use them to achieve its NDC.*

8 See also NZ's Voluntary National Review on progress towards achievement of the SDGs including SDG13 Climate Action: MFAT 2019.

References

Averchenkova, A., Fankhauser, S., and Nachmany, M.. (eds.) (2017) *Trends in Climate Change Legislation*, Cheltenham: Edward Elgar.

Bargh, M. (2019) "A Tika Transition," in David Hall (ed.) *A Careful Revolution: Towards a Low-Emissions Future*, Wellington: Bridget Williams Books.

Bargh, M., and Van Wagner, E.. (2019) "Participation as Exclusion: Māori Engagement with the Crown Minerals Act 1991 Block Offer Process," *Journal of Human Rights and the Environment* 10: 118–139.

BBC NEWS. (2020) *Davos: 'Forget About Net Zero, We Need Real Zero' – Greta Thunberg*, 21 January <https://www.bbc.com/news/av/world-51193460/davos-forget-about-net-zero-we-need-real-zero-greta-thunberg>.

Bosselmann, K. (2017) *The Principle of Sustainability: Transforming Law and Governance*, 2nd ed., Abingdon: Routledge.

Brown, D., et al. (2018) "A Four-Step Process for Formulating and Evaluating Legal Commitments Under the Paris Agreement," *Carbon & Climate Law Review* 12: 98–108.

Climate Analytics. (2018) *New Zealand's Zero Carbon Bill – Getting the Paris Agreement Right*, 19 July <https://climate-analytics.org/publications/2018/new-zealands-zero-carbon-bill-getting-the-paris-agreement-right/>.

Daya-Winterbottom, T. (2019) "Zero Carbon, Climate Justice and Privative Clauses," *Resource Management Bulletin* 12: 173–174.

Dunlap, R. E. and Brulle, R. J. (eds) (2015) *Climate Change and Society: Sociological Perspectives*, New York & Oxford: Oxford University Press.

Explanatory note to the Climate Change Response (Zero Carbon) Amendment Bill (2019) <http://www.legislation.govt.nz/bill/government/2019/0136/11.0/LMS183736.html>.

Fair Share – Climate Action Tracker New Zealand <https://climateactiontracker.org/countries/new-zealand/fair-share/>.

France-Hudson, B. (2015) "No Private Property Rights in the Atmosphere," in P. Martin and others (eds.) *The Search for Environmental Justice*, Cheltenham: Edward Elgar.

Friends of the Earth International. (2014) *The Great REDD Gamble*, The Netherlands: Friends of the Earth International <https://1bps6437gg8c169i0y1drtgz-wpengine.netdna-ssl.com/wp-content/uploads/2018/10/The-great-REDD-gamble.pdf>.

Generation Zero. (2019) *Submission to the Environment Select Committee on the Climate Change Response (Zero Carbon) Amendment Bill 2019* <https://www.parliament.nz/en/pb/sc/scl/environment/>.

Hadley, C. (2014) "New Zealand's Consideration of Ethics and Justice Issues in Formulating Climate Change Policies," in D. A. Brown and P. Taylor (eds.) *Ethics and Climate Change: A Study of National Commitments*, IUCN Environmental Policy and Law Paper No. 86: 95–108.Bonn: IUCN Environmental Law Center.

Harker, J., Taylor, P., and Knight-Lenihan, S,. (2017) "Multi-Level Governance and Climate Change Mitigation in New Zealand: Lost Opportunities," *Climate Policy* 17: 485–500.

Intergovernmental Panel on Climate Change (IPCC) (2018) "Summary for Policymakers," in V. Masson-Delmotte and others (eds.) *Global Warming of 1.5°C. An IPCC Special Report on the Impacts of Global Warming of 1.5°C Above Pre-Industrial Levels and Related Global Greenhouse Gas Emission Pathways, in the Context of Strengthening the Global Response to the Threat of Climate Change, Sustainable Development, and Efforts to Eradicate Poverty*, Geneva, Switzerland: World Meteorological Organization.

Interim Climate Change Commission (ICCC). (2019) *Our Work*, 26 September <https://www.iccc.mfe.govt.nz/what-we-do/our-work/>.

McLaren, D. (2019) *Guest Post: The Problem With Net-Zero Emissions Targets*, 30 September <https://www.carbonbrief.org/guest-post-the-problem-with-net-zero-emissions-targets>.

Lawyers for Climate Action NZ (LCANZ). (n.d.) *NZBORA + The Right to a Sustainable Environment* <https://www.lawyersforclimateaction.nz/nzbora>.

——— (2019) *Submission to the Environment Select Committee on the Climate Change Response (Zero Carbon) Amendment Bill 2019* <https://www.parliament.nz/en/pb/sc/scl/environment/>.

MacManus, J. (2019) *PM Jacinda Ardern: Government 'Not Opposed to' Idea of Declaring a Climate Emergency*, 3 July <https://www.stuff.co.nz/environment/climate-news/113946213/more-than-50-of-new-zealands-top-scientists-call-on-government-to-declare-climate-emergency>.

Ministry for the Environment. (2018) *Zero Carbon Bill Consultation: Summary of Submissions*, Wellington: Ministry for the Environment.

——— (2019a) *Environment Aotearoa 2019 Summary*, Issue 4, Wellington: Ministry for the Environment <www.mfe.govt.nz/Environment-Aotearoa-2019-Summary>.

——— (2019b) *Reforming the New Zealand Emissions Trading Scheme: Proposed Settings*, consultation document ME 1484, Wellington: Ministry for the Environment.

Ministry of Foreign Affairs and Trade (MFAT). (2019) *He Waka Eke Noa: Towards a Better Future, Together – New Zealand's Progress towards the SDGs – 2019* <https://www.mfat.govt.nz/assets/Uploads/New-Zealand-Voluntary-National-Review-2019-Final.pdf>.

National Geographic. (2019) *Methane Facts and Information*, 23 January <https://www.nationalgeographic.com/environment/global-warming/methane/>.

Neilson, M. (2019) *Iwi Leader Suing Government Over 'Failure' to Protect Māori from Climate Change*, 16 July <https://www.nzherald.co.nz/nz/news/article.cfm?c_id=1&objectid=12250045>.

NEW ZEALAND: Submission under the Paris Agreement: New Zealand's Nationally Determined Contribution. (n.d.) <https://www4.unfccc.int/sites/ndcstaging/PublishedDocuments/New%20Zealand%20First/New%20Zealand%20first%20NDC.pdf>.

NZ Productivity Commission. (2018) *Low-Emissions Economy: Final Report*, Wellington: NZ Productivity Commission <www.productivity.govt.nz/low-emissions>.

Organisation for Economic Co-operation and Development (OECD). (2017) *Environmental Performance Review: New Zealand* <https://www.oecd.org/newzealand/oecd-environmental-performance-reviews-new-zealand-2017-9789264268203-en.htm>.

Official Information Act (OIA). (2019) *Documents Relating to NZ's Nationally Determined Contributions (NDCs) per Article 4, paragraph 10 of the Paris Agreement Released to the Author.*

Otero, I., et al. (2020) "Biodiversity Policy Beyond Economic Growth," *Conservation Letters* 13(4): 1–18. DOI: 10.1111/conl.12713.

Parliamentary Commissioner for the Environment (PCE). (2019) *Submission to the Environment Select Committee on the Climate Change Response (Zero Carbon) Amendment Bill 2019* <https://www.parliament.nz/en/pb/sc/scl/environment/>.

Paul, C. M., Potter, D., and Paterson, A. (2016) *Statement of Claim WAI 2607,* dated 30 May, received 27 September, <https://forms.justice.govt.nz/search/Documents/WT/wt_DOC_106784185/Wai%202607%2C%201.1.001.pdf>.

Rive, V. (2015) "Safe Harbours, Closed Borders? New Zealand Legal and Policy Responses to Climate Displacement in the South Pacific," in P. Martin and others (eds.) *The Search for Environmental Justice* (pp. 221–238), Cheltenham: Edward Elgar.

Rogelj, J., et al. (2019) "Estimating and Tracking the Remaining Carbon Budget for Stringent Climate Targets," *Nature* 571: 335–342.

Statistics New Zealand. (2019) *New Zealand's Greenhouse Gas Emissions*, 18 April <https://www.stats.govt.nz/indicators/new-zealands-greenhouse-gas-emissions>.

Sustainable Aotearoa New Zealand (SANZ). (2009) *Strong Sustainability for New Zealand: Principles and Scenarios*, Auckland: SANZ.

Taylor, P. E. (2021) "The New Zealand Legislation: Pursuing the 1.5 ° C Target Using a Net Zero Approach," in Thomas L Muinzer (ed.) *National Climate Change Acts: The Emergence, Form and Nature of National Framework Climate Legislation*, Oxford: Hart Publishing.

Taylor, P. and Scanlen, K. (2018) "The UK Climate Change Act: An Act to Follow?" *Policy Quarterly* 14(3): 66–73.

The Spinoff. (2019) *119 of 120 MPs Just Voted to Pass NZ's Historic Zero Carbon Bill into Law. Here's the Best of What They Said*, 7 November <https://thespinoff.co.nz/politics/07-11-2019/119-of-120-mps-just-voted-to-pass-nzs-historic-zero-carbon-bill-into-law-heres-the-best-of-what-they-said/>.

United Nations Educational, Scientific and Cultural Organization (UNESCO). (2017) *Declaration of Ethical Principles in Relation to Climate Change* <https://unesdoc.unesco.org/ark:/48223/pf0000260889.page=127>.

United Nations High Commissioner for Refugees (UNHCR). (2020) *UN Human Rights Committee Decision on Climate Change Is a Wake-Up Call, According to UNHCR*, 24 January <https://www.unhcr.org/news/briefing/2020/1/5e2ab8ae4/un-human-rights-committee-decision-climate-change-wake-up-call-unhcr.html>.

Waitangi Tribunal. (2011) *Ko Aotearoa Tēnei: A Report into Claims Concerning New Zealand Law and Policy Affecting Māori Culture and Identity*, Wai 262, Wellington: Legislation Direct <https://waitangitribunal.govt.nz/news/ko-aotearoa-tenei-report-on-the-wai-262-claim-released/>.

Wetts, R. (2020) "Models and Morals: Elite Orientated and Value-Neutral Discourse Dominates American Organizations' Framing of Climate Change," *Social Forces* 98(3): 1339–1369. DOI: 10.1093/sf/soz027.

Wise Response. (2019) *Submission to the Environment Select Committee on the Climate Change Response (Zero Carbon) Amendment Bill 2019* <https://www.parliament.nz/en/pb/sc/scl/environment/>.

31

HUMAN RIGHTS-BASED APPROACHES TO CLIMATE LITIGATION

A Critical Perspective of the Athabaskan Petition

Giada Giacomini

Introduction

Human rights-based climate litigation is an expanding trend. The success reported by the case *State of the Netherlands v. Urgenda Foundation*[1] demonstrated that human rights arguments can provide a viable way to seek injunction to compel a government to respect its emission reduction obligations (Setzer & Birnes 2020). This chapter focuses on a climate litigation case, the *Petition to the Inter-American Commission on Human Rights Seeking Relief from Violations of the Rights of Arctic Athabaskan Peoples Resulting from Rapid Arctic Warming and Melting Caused by Emissions of Black Carbon by Canada* (hereinafter *Athabaskan Petition*), in order to provide a critical legal reflection on the use of human rights-based approaches to climate litigation as it relates to Indigenous peoples. The *Athabaskan Petition* was filed in 2013 but has not yet been decided by the Inter-American Commission (IACHR). The Athabaskans are seeking redress from the wrongful conduct of Canada concerning black carbon emissions.

The application of a critical legal perspective to the *Athabaskan Petition* deals with a two-fold issue concerning the international human rights law realm. On the one hand, a crucial reflection is provided on the construction of the abstract legal subject entitled to the protection of human rights. The chapter aims at evidencing how international law, throughout the history of the recognition of human rights, has needed to expand the scope of recognition of full human beings to women, slaves, Indigenous peoples, and other embodiments of the 'other.' In fact, even the supposed universal personhood of the United Nations Declaration on the Rights of Indigenous Peoples (UNDRIP) features an *exclusive universality*, where a multitude of other instruments needed to emerge to protect "specific categories of human beings" such as Indigenous peoples, who had to undertake a 20-year negotiation process to have their rights recognized in the UNDRIP. This type of theoretical framework is the foundational basis of a renewed paradigmatic order in international human rights law. It shows that ethical considerations in law, politics, and governance, are needed to unravel the deep interconnectedness between colonial and neo-colonial practices and the construction of the liberal legal subject.

On the other hand, it is argued that the abstract human being entitled to rights under the current legal realm is a disembodied construct, personified by a flawless rational capitalistic mind which tends to maximize its condition. From this consideration derives the exclusion of Indigenous' and

DOI: 10.4324/9781003039860-36

other minorities' views and cosmologies in international human rights law-making, and a substantial colonization of capitalistic ideologies in legal and juridical processes.

From such a critical perspective, the struggle and impairments Indigenous peoples face when trying to access a court by using human rights-based approaches in climate litigation fully emerges. This chapter seeks to analyze how the Athabaskans framed their case through such an approach, considering the objective difficulties in the translation of their own understanding of the environment and its deep interconnection with their cultural aspects into a Westernized type of juridical claim. It will aim to put into evidence the limits of human rights-based approaches to climate litigation from the perspective of critical legal theory and argue that a decolonization of law and policy-making through an ecological justice perspective is necessary, where Indigenous views and cosmologies are an integral part of the legal tools aimed at proving redress to climate and environmental injustices.

Critical Legal Perspectives in International Human Rights Law and Climate Litigation

It is of paramount importance to take into account the legacy of colonialism on issues related to Indigenous peoples' rights in the climate change context. Such legacy is not only deeply embedded in the governance of extractivism, political exclusion of Indigenous peoples' in international and national decision-making, and environmental racism, but also in certain aspects of international law that often are not considered by legal scholars when analyzing the different aspects of the international human rights law realm. Thus, this section aims at providing a critical reflection on the legal regime aimed at protecting Indigenous peoples' rights, going beyond the surface of law's prescriptions and normative aspects, and unpacking its doctrinal underpinnings in a critical legal perspective. This is needed to sincerely understand Indigenous struggles when availing international human rights law to bring their claims before human rights courts, such as in the *Athabaskan Petition* presented in this chapter.

The first critical reflection revolves around the origins of international law and the subsequent construction of the Westernized liberal legal subject as the recipient of rights in international human rights law. In order to better understand the challenges and the complexities of Indigenous peoples' rights and titles to territory, it is necessary to consider the legal impositions and injustices of colonization that are ultimately reflected in the provision of corporate legal personhood. Upon encountering native people in the Americas, new legal doctrines followed that shaped the interrelations between the colonizers and the colonized. Theories of subjugation served as the theoretical basis permitting the conquest of territories, enslavement of the people inhabiting them, and physical and cultural genocide. This history imparts meaning upon current laws which, on their face, provide legal protection for Indigenous peoples (Acuña 2015). At the time of colonization, public international law was an instrument utilized to deny Indigenous peoples their rights. Political doctrine justified centuries of occupying ancestral lands and legitimized the inequality of power, use of force, coercion, and legal subterfuges. This model of non-consensual occupation and exploitation of Indigenous lands and territories, which began in the XVI century, is still resonant today, since it exists in new forms which are embodied, for example, in modern resource extractivism and land grabbing (Gilbert 2016).

The encounter between the so-called *Indios* and the Europeans raised a series of questions: could it be considered legitimate to wage war against them even though they never attacked Europe? Was colonization and the establishment of international trade to be considered legitimate? Was it fair to enslave those people? The answers to these questions were essential in shaping international law. The conquest of America has been described as a "legal enterprise" (Foster 1985) and a "civilizing mission" (Baxi 2006). Legitimization through law was essential for Europeans to assert sovereignty and land claims over Indigenous territories. Medieval legal doctrine was relied upon for justification of the colonial initiative. In 1492, Columbus proceeded to officially declare the Bahamian islands as territories of the Spanish Crown given the barbaric nature of the peoples he encountered (Foster 1985). The first treaties dividing spheres of legitimate conquest were signed by Portugal and Spain in 1494.[2] Colonial practices such as forced labor, shipment of slaves and the *encomienda,*[3] caused the death of most of the islands' populations within less than two decades (Williams 1990) For the first time, this led to critique of Spanish colonial practices, particularly those of Francisco de Vitoria and of Bartolome de las Casas (Marks 1992).

It is not odd that in this context the first multinational enterprises, such as the East Indian company, began to have recognition in law long before human beings had the same (Baxi 2006). The predatory patterns of colonialism and their ideological justifications continued to be resonant throughout the 20th century and they are still resounding today as the juridical order often favors the biases of corporations and industries over human and environmental interests, the interests of the Global North over those of the Global South (Gallas et al. 2016).

International law has long supported the subjugation, enslavement, killing and marginalization of Indigenous peoples as well as the occupation of their lands.[4] However, in the XX century, the decolonization process of the UN, the rise of new global powers, and the US-Russia polarization caused Eurocentrism to collapse. Even though the rhetoric of international law remained state-centered, new non-state actors began to appear on the international stage. Individuals, international organizations, NGOs, labor unions, and transnational corporations started to participate in international legal debates, contributing to the shaping of international law. Anaya, who has taught and written extensively on international human rights and issues concerning Indigenous peoples, argues that legal discourse witnessed a re-emergence of the classical-era naturalism philosophy, but free from the Eurocentric perspective (Anaya 2004). The new era of universal, fundamental human rights law had begun.

However, for the purposes of this chapter, it is pivotal to consider the inner features of the abstract legal subject that is entitled under international law to withhold such fundamental human rights. In other words, who is the *Anthropos* of this new legal era, considering the premises made in the previous paragraphs?

According to Grear, an important contemporary scholar expert critical legal theory, the European construction of legal and political subjectivity greatly influenced the shaping of the liberal legal subject with which such rights are awarded. The construction of legal subjectivity has been forged through a "powerful exclusionary structure," which purportedly excluded and dispossessed Indigenous peoples under colonialism (Grear 2015). The liberal legal subject is a disembodied structure, a fully capitalistic legal mind that resembles the characteristics of the corporation: the corporate *Anthropos* is characterized by being rational, white, man, with the capability of enjoying his rights because they are deeply entrenched to his capacity of reason and to the biological determinism of a 'civilized' maturity (Grear 2011). The corporation, and not the human being, is thus the perfect realization of the liberal legal subject under Westernized human rights law, a disembodied, abstract entity that pursues the maximization of its well-being.

This corporate, abstract legal embodiment has made its way through the international governance system and through the judicial means of resolution of dispute. Capital already pervades all aspects of international law, from market-based solutions to climate change, to social, economic, and cultural rights, to development policies (Guerrero 2018). Corporations have assumed the role of victim and they are able to defend themselves in tribunal and court proceedings, as is being demonstrated in new climate litigation trends: no 'first-wave' climate litigation trials have been decided against a corporation (Ganguli et al. 2018).

From the perspective of climate justice, this last point is particularly significant. In fact, the so-called 'Carbon Majors,' the 90 largest fossil fuel and cement producers, have emitted nearly two-thirds of all carbon dioxide since the 1750s. Most of these producers still operate today, such as ExxonMobil and Chevron (CIEL 2017). The political meta-power that connects liberal states and corporations, masked under the 'best interest of the nation,' has led to the violation of human rights connected to air pollution, climate change, and the destruction of biodiversity. States have played a pivotal role in the rising power of corporations and their recognition as legal subjects within the global justice realm, whilst the worst impacts of climate change – that such companies have created – are falling disproportionately on the most emarginated communities.

In this paradoxical global order, Indigenous peoples, alongside other vulnerable people such as women, the disabled, and children, have had to fight to emerge as legal subjects and have their rights recognized. This is precisely because the fully capitalistic, liberal legal *Anthropos* was created as a disembodied entity, far from being a human with comprehensive characteristics that included such heterogenous embodiments of the human (Federman 2003). This consideration is paramount for a critical understanding of the difficulties of Indigenous peoples in gaining legal recognition in the Western – and Westernized – systems of law. This burdensome process, as evidenced by the twenty years of negotiations that led to the approval of UNDRIP, is still ongoing, and it will be insofar as human hierarchies are foundational in the global legal order (Grear 2015).

Bearing such considerations in mind, an attentive reading of the *Athabaskan* case unravels the many challenges Indigenous peoples face when filing a case against a state entity. For Indigenous peoples to be accepted at the table of international governance, they need to translate their systems of knowledge into viable means of communication to have their claims recognized, always adapting themselves to legal requirements based on Westernized legal subjectivity. This process was evident when the Athabaskans filed their claim against Canada. The next section focuses on this critical interpretation of the text of the *Athabaskan Petition* to the Inter-American Commission on Human Rights.

The *Athabaskan Petition*: framing a legal claim

The Athabaskan peoples have inhabited the states now known as Alaska and Canada since before the arrival of European colonizers, 6000–9000 years ago (Langdon 1992). They are organized into different cultural groups which include Gwich'in, Kuskokwim, and Tanana Han, among others. Their ancestors were part of the second wave of three major migrations from Asia that crossed the Bering land bridge (Gary et al. 2010). Their region was first colonized by Canadian fur traders in the 19th century, and other colonizers, including Americans, started to settle in the area in the 20th century. The colonization process led to a decline of the Athabaskan population because of the illnesses carried by the colonizers, whilst the assimilationist policies enacted in the 20th century provoked a generalized cultural loss (Gary et al. 2010).

Athabaskan history and traditions have long been neglected by traditional historiography, and the first studies on Athabaskan history began in the 1970s (Heaton 2012). Such studies have contributed in re-discovering the important role of Athabaskan people in shaping history and the modern nations they now live. The intimate relationship between Athabaskan people with their ancestral lands and territories constitutes a special cultural trait by which their social processes are interlinked to the ecology of their territories. The harsh environment has profoundly molded Athabaskan culture and traditions; in fact, there is little difference in the ways Athabaskan people adapted, whether they lived in either Canada or Alaska (Van Stone 1974). Because of the harsh environment they have inhabited since time immemorial, their culture, traditions and livelihoods are highly adaptable, resilient, and flexible. Traditionally, they lived in small family groups that used to migrate seasonally according to where survival resources were available (Natcher 2004). After colonization in the 20th century, however, they tended to establish their residence permanently in villages due to the enacted assimilation policies – such as mandatory schooling – and because of the competition in land claims with non-Indigenous settlers (Gary et al. 2010).

Notwithstanding such crucial challenges faced by Athabaskan people, in the last decades climate change has exacerbated critical social and economic issues. In fact, according to the Intergovernmental Panel on Climate Change (IPCC),

> in both polar regions there is strong evidence of the ongoing impacts of climate change on terrestrial and freshwater species, communities and ecosystems (very high confidence). [...] These changes will have cascading effects on key regional bio-physical systems and cause global climatic feedbacks, and in the north will affect socio-economic systems (high confidence).
>
> *(Anisimov et al. 2007)*

For Arctic Indigenous peoples, this means an unprecedented challenge in adaptation and cultural survival, and, in many cases, a violation of their fundamental human rights.

The Arctic is indeed facing disproportionate rates of warming, with huge negative consequences for Athabaskan traditional livelihoods. In 2013, Canadian Athabaskan, together with the nonprofit environmental law firm Earthjustice, filed a petition to the IACHR against the government of Canada. The petition, not yet decided, it is an important means of ascertaining the impacts of climate change in the Athabaskan territories and their consequences for the survival of Indigenous culture. It is also a crucial example of how Indigenous people must necessarily frame their claims within the only possible instrument available for them in climate change litigation: a human rights-based approach.

Climate litigation perhaps presents a viable opportunity to obtain redress for past and current climate injustices. The Indigenous peoples of the Arctic, Athabaskan, Inuit, and Eskimos are the least contributors to global emissions, yet they are facing the negative impacts of climate change which are impairing the patterns of social and economic exclusion already present in their societies.[5] According to the petition, Canada's deficiency in the effective regulation of black carbon emissions is the beginning of a causal chain that culminates in the violation of the Athabaskans' rights to enjoy the benefits of their culture, to property, to the preservation of health, and to their own means of subsistence as established by the American Declaration on the Rights and Duties of Man (hereinafter the Declaration). In addition, this failure produces a violation of Canada's duties to avoid transboundary harm and to protect the environment consistent with the precautionary principle as established in international environmental law.[6]

The petition is framed using international human rights law as a tool to help reduce the complexity of climate litigation, linking impacts of global warming, severe weather events, and environmental degradation to immediate danger and violations of fundamental rights. The petition relies on the rights enshrined in the Declaration that, although is a non-binding instrument, is vested with considerable moral power, and is the only enforceable human rights document against the United States and Canada, which are not parties to the American Convention on Human Rights.[7] In 2010, the IACHR stated that "the IACHR has competence, in accordance with the Organization of the American States (OAS) Charter and with the Commission's Statute, to consider alleged violations of the American Declaration by OAS member States that are not yet parties to the American Convention."[8] This approach was confirmed in two cases, *Marie and Carrie Dann v. US*[9] and *Grand Chief Michael Mitchell v. Canada*,[10] where the IACHR considered the States subject to its jurisdiction.[11]

The Petition outlines the characteristics of Athabaskan culture and its deep embedment with the environment.[12] It could succeed if the Athabaskans unequivocally establish a specific link between Canada's acts and the violation of human rights and if the IACHR expands its interpretation of the law, as explained later on in the section (McCrimmon 2016).

The Athabaskan people alleged violations of their right to property, culture, health, and subsistence (namely, Articles XIII, XXIII, and XXI of the Declaration) due to Canada's acts and omissions in limiting the emissions of black carbon. The plaintiffs specified that such omissions have caused the depletion of food resources due to changes in the distribution and availability of animals, and loss of traditional culture and hunting methods due to the increase in the frequency of extreme weather events and the degradation of the Arctic environment.[13]

In relation to the right to culture, the *Athabaskan Petition* relies on previous case-law such as *Moiwana v Suriname, Yakye Axa v. Paraguay, Sawhoyamaxa v. Paraguay and Saramaka v. Suriname.* In these cases, the plaintiffs attempted to establish the direct correlation between damage to the environment and violation of the right to culture. However, the IACHR, although recognizing the important holistic relationship between Indigenous culture and the environment, has never established a *tout court* violations of the right to culture connected to environmental damage. It has, however, recognized violations of the right to property, life, preservation of health, and well-being (Antkowiak 2013; Fuentes 2017). Thus, the possibilities that such an allegation would be recognized by the IACHR appears quite improbable.

In relation to the right to property, the Athabaskans contend that Arctic warming, caused by black carbon emissions, has made their lands unfamiliar and less valuable.[14] They argue that the changes to regional topography of cultural meaning have compromised their ability to make sense of their environment, for purposes ranging from subsistence to maintenance of cultural traditions. Despite the legal deficiencies of the right to culture argument, the allegation of the violation of the right to property remains strongly framed within the right to culture, which might prevent the claim from being successful due to the inherent difficulties of linking black carbon emissions with the violation of something intangible, such as cultural aspects. This, however, shows the clear difference in interpretation of human rights in Indigenous views and in Westernized legal systems. For the Athabaskans, environmental elements and their ancestral culture are deeply entrenched within each other, as the Petition reads,

> [A]ccelerated Arctic warming also interferes with Arctic Athabaskan peoples' right to intellectual property. Arctic Athabaskan peoples possess intangible property in the form of traditional knowledge, [...] Athabaskan peoples' educational system of passing on and building upon knowledge from one generation to the next has tremendous value to their cultural survival. Arctic Athabaskan peoples have spent millennia developing knowledge about their physical surroundings. Rapid regional warming has altered weather patterns and therefore has made Athabaskan peoples' traditional knowledge less valuable.[15]

Thus, for the Athabaskans, their culture is not only represented by language, beliefs, and knowledge, but also by their complex relationships and interactions with the environment, and with the possibility to transmit this knowledge to future generations.[16]

Affirmation of such a violation of culture would require the IACHR to establish a causal link between impacts deriving from black carbon emissions, alterations of the environment, and the violation of intellectual property rights. In *Awas Tigni v. Nicaragua*, the IACHR included "intangible object capable of having value" in the definition of property. Yet such definitions do not find an equivalent in the Declaration, which narrowly defines the right to culture.[17] In order to recognize the violation of such a right, the IACHR would have to expand its current interpretation of the law. In the cases cited in the *Athabaskan Petition, Moiwana v. Suriname, Yakye Axa v. Paraguay,*

Sawhoyamaxa v. Paraguay, and *Saramaka v. Suriname*, while a reference is made to the impacts of environmental harm on Indigenous culture, once again these impacts never amount to a violation of the right to culture (McCrimmon 2016).

On another note, and also in relation to the right to property, the plaintiffs argue that climate impacts cause two different phenomena that affect their right: increased rainfall and reduction of permafrost.[18] Permafrost provides the foundation upon which Athabaskan communities live, and when it thaws it can damage roads, buildings, and infrastructure. Melting permafrost also releases huge quantities of water, increasing the risk of flooding. Increased rainfall patterns also cause extensive flooding: the Athabaskan cite a case of a village that was completely destroyed by a combination of heavy snowfall and warm record temperatures.[19] Therefore, climate impacts undermine the right of the Athabaskan to enjoy their traditional lands and the right to use their personal property. Whilst this aspect would not require a major reinterpretation of the Declaration, it relies on an extensive interpretation of the law in relation to the contents of the American Convention – to which Canada is not part.[20]

Finally, the Athabaskans argue for the violation of the right to health and subsistence due to a consistent change in their diet, by the incrementation of the potential of physical harm and by a profound alteration of the environment that affects their mental health. At the same time, they argue that their right to subsistence impedes the practice of traditional hunting and gathering.[21] However, this claim does not appear to rely on a consistent legal basis, as the Declaration does not include a right to subsistence. The IACHR has previously referred to the subsistence of Indigenous peoples in relation to their traditional livelihoods, hunting, and gathering methods, and the interlinkage between subsistence and environment. However, the violation found by the Court was connected to the right to property, not to a right of subsistence.[22] Perhaps this claim could be considered by the IACHR as being part of the right to health, since the Athabaskans have alleged that their health is threatened by the overall increase in average temperature associated with climate change. They claim that it allows new organisms to move into Athabaskan territory bringing new pathogens, and it changes the traditional presence of game, forcing them to switch to a different diet and to buy unhealthy processed food to compensate.[23]

The *Athabaskan Petition* represents an important means of understanding the magnitude of climate change impacts in ancestral lands because it is based on the Indigenous traditional knowledge of their territory. In addition to its legal significance, the *Athabaskan Petition* also constitutes an important way to address an alternative narrative of the territory, focusing on how climate change is re-shaping the traditional life of Indigenous peoples. It shows the clear contrast between Indigenous peoples' views and lived experience of their lands and what the law should do to preserve the environment to which Indigenous livelihoods are deeply embedded. As such, it highlights the inadequacy of Canadian law to address the breadth and depth of climate change issues.[24] This inadequacy is directly correlated with the difficulties of Athabaskans to access the Canadian juridical system, since the law is fragmented into multiple levels of competency, their differentiation by province, and their application to a sector or source of emissions. The Athabaskans denounce how making a legal claim in this system would be "unduly burdensome" even for those with wealth and resources, and materially impossible for Indigenous peoples.[25]

The difficulties for Indigenous peoples to access justice and redress mechanisms is a well-known problem of the national and international human rights system, as it reflects patterns of legal injustice. In the *Athabaskan Petition*, the Athabaskans cite *Hul'Qumi'Num Treaty Group v. Canada* (IACHR 2009), where it was found that there was no due process of law to protect the property rights of the Hul'-Qumi'Num Treaty Group[26] because the recourses available to claim right to their ancestral lands were not effective.[27] When accessing a court, Indigenous peoples must undertake a series of legal actions not typically applied to non-Indigenous peoples, such as the demonstration to be legally entitled to their lands. Courts may ask Indigenous peoples to prove their property rights and to document their customary land tenure in order to authenticate their continuous and traditional attachment to the territory in dispute (McNeil 1999). This process might result in a paradoxical narrative, where Indigenous peoples have to prove their continuous and traditional attachment to the land against the backdrop of colonization, and in court proceedings structured and informed by colonialism (Gilbert et al. 2018).

The *Athabaskan Petition* is an important witness to these patterns of climate injustice, and gives evidence to the inadequacy of the Canadian legal system to protect Indigenous peoples' rights and to address climate impacts in the Arctic. Whether their claims will succeed depend on their ability to link Canada's black carbon emissions to the widespread violations of Athabaskans' rights. However, the existence of such a claim constitutes an important opportunity to actively re-think how international human rights law should be re-shaped in light of a broad critical legal framework prescribing a paradigmatic shift and the decolonization of legal personhood.

Re-Thinking Human Rights-Based Approaches to Climate Litigation within the Critical Legal Framework

The *Athabaskan Petition* closely resembles another case brought before the IACHR, the *Petition to The Inter-American Commission on Human Rights Seeking Relief From Violations Resulting from Global Warming Caused By Acts and Omissions of the United States* (hereinafter *Inuit Petition*).[28] The two cases are characterized by similar geographical settings and comparable climate change impacts which indirectly caused violations of Indigenous peoples' human rights. The *Inuit Petition*, in its 167 pages, sought to hold the United States legally responsible for its acts and omissions that substantially contributed to global warming and, consequently, had severe impacts on the Arctic environment. These impacts, in turn, led to a violation of the human rights of the Inuit people. Thus, the *Inuit Petition* aimed to demonstrate the US' primary responsibility for global warming and consequent violation of Inuit's human rights, given its failure in lowering emissions' levels and its primacy in being the world's largest Green Houses Gases (GHGs) emitter.[29] The Inuit representatives argued that the permanent environmental changes in the Arctic are directly linked to the violation of their fundamental human rights, particularly the right to home, freedom of residence and movement, the right to property, and their means of subsistence.

However, in 2006 the IACHR rejected the *Inuit Petition* with a two-paragraph decision, affirming that:

> It will not be possible to process your petition at present because the information it contains does not satisfy the requirements set forth in those Rules [...] the information provided does not enable us to determine whether the alleged facts would tend to characterize a violation of rights protected by the American Declaration.[30]

No further explanation was given. Yet, one might consider that the *Inuit Petition* failed to expressly link US emissions to the specific effects of global warming in the Arctic. In other words, the Petitioners were not able to establish specific causation between the US failure to lower emissions and the violations of Inuit human rights, and this was cause for the IACHR to dismiss the *Inuit Petition* without making a determination on the alleged violation of human rights.

The inherent difficulties in establishing climate liabilities in association with human rights violations are a key issue of present climate litigation trends. When a court is faced with the verification of a State's or corporation's liability over climate change responsibility, this requires a thorough scientific demonstration of the correlation between emissions and a determined effect provoking a human right violation (Peel et al. 2015). The role of scientists and the progress of the so-called 'event attribution' science, which ultimately helps in linking emissions and violations of human rights, is becoming paramount in court proceedings. This GHG liability exists, or can potentially be established, in public and private law and can be partitioned in international, administrative, and criminal liability (public law), and tort liability (private law) (Faure et al. 2019). In the *Athabaskan Petition*, Canada's black carbon emissions are linked to specific consequences on the environment which, in turn, led to violations of fundamental human rights.[31] This approach, based on scientific affirmation linking black carbon to certain specific impacts, constitutes a novelty if compared to the claims of the *Inuit Petition*, and perhaps will lead to the success of the *Athabaskan Petition*.

Yet, uncertainties about the suitability of the current international human rights legal system in dealing with Indigenous climate justice claims remain. As demonstrated above, the *Athabaskan Petition* fully demonstrates the legal struggles Indigenous peoples face when attempting to access justice. Such difficulties arise because Indigenous people must actively translate their system of culture, beliefs, and knowledge into a viable means of understanding in Westernized human rights law, such as how they attempt to construct a right to subsistence in the *Athabaskan Petition*. The objective of such redress mechanisms, even if successful and because of its very structure, cannot go beyond establishing responsibility and commanding the State to modify its wrongful conduct (Wewerinke-Singh 2019). But this solution would be framed within the same context of the Western ontologies of State-centered, corporate-based solutions that cannot solve the dilemma of the dominant constitutions of liberal legal subjectivity. Thus, the objective of a critical legal interpretation of current developments in international human rights law is to unmask and reveal the flawed underpinnings of the dominant structure of the existing legal system that aims to establish liability without modifying its deep hierarchical construction.

Climate injustice as a theoretical framework is the foundational basis of an alternative paradigmatic order to current legal remedies provided by international law. It requires ethical consideration of law, politics, and governance, unravelling the deep interconnectedness between colonial and neo-colonial practices and the construction of the liberal legal subject. It interprets climate change as a form of systemic violence of the Westernized, capitalistic societies toward the most vulnerable who are disproportionately bearing the negative effects of the indiscriminate use of our planet's resources (Skillington 2017). The use of such a framework allows for a reconsideration of participatory structures that traditionally have excluded Indigenous views and cosmologies in the dominant legal paradigm, evidencing the need to challenge that context and restructure the system. Whilst the participatory paradigm of Indigenous peoples in international governance is progressively expanding through the institution of international fora, such the United Nations Permanent Forum on Indigenous Issues (UNPFII), the overall ratio of this system remains state-centered, whereas Indigenous peoples are "invited at the table" of the decision makers and have to conform their views to Westernized standards of operationalization (Lindroth 2006).

While international human rights law is perhaps one of the best tools at present to seek redress for climate injustices, this chapter has demonstrated how Indigenous peoples face unfair and unequal burdens than their non-Indigenous counterparts in filing cases before international human rights courts. As evidenced above, they need to translate their own systems of knowledge and claims into Westernized legal concepts in order to pursue recognition of a violation of their rights. These obstacles to justice directly relate to the rhetoric of representation of the legal subjects in human rights law, the translated image of a western cultural dominance connected to relations of power (Merry 2006). The paradox of human rights' *exclusive universalism* must still be overcome by a re-shaping of the Westernized, monolithic legal personhood that includes – and excludes – certain human embodiments. The paradoxical narrative of the attribution of legal personhood sees privileged groups and nations as the 'distributors of rights,' further advancing the colonial narrative, operating in a dialectical process of recognition that perpetrates exclusionary patterns (Hesford 2011). In this sense, the effectiveness of international human rights law as an instrument to seek redress in climate litigation will always be limited when applied by those people and individuals that do not correspond to the Westernized legal embodiment of the *Anthropos*.

Conclusion: Decolonizing Human Rights-Based Approaches to Climate Litigation

International human rights law is a powerful instrument by which climate litigation cases can be framed. However, this chapter has evidenced the structural limits of the international human rights legal system as it relates to the protection of Indigenous peoples' rights in the climate litigation realm. A process of decolonization of international human rights law is needed in order to recognize Indigenous history, cosmologies, and their holistic relationship with the natural world as an integral part of legal and human rights instruments.

Critical legal thought can work as a framework to re-think human rights-based approaches in climate litigation, and alternative views and epistemologies such as Indigenous cosmovisions and knowledge must be included into redress instruments to achieve climate justice. As demonstrated, Athabaskan cosmovisions entail a distinct cultural relationship with the environment that it is not substantially reflected in international human rights law. The legal instruments that the Athabaskan have used to try to protect their rights against Canada's omissions in cutting black carbon emissions demonstrate the inadequacy of the international human rights law system in addressing the complex issues around Indigenous people's rights and climate justice. Yet, the Athabaskans and other Indigenous peoples must rely on such flawed legal tools when appealing a national or international court, with little hope for success and redress for otherwise clear harms.

If the legal international community is sincere in their desire to address the rights of Indigenous peoples, human rights, and climate change harms, separately and together, a critical rethinking of international human rights law must address the oppressive hierarchies and the exclusive universality deeply embedded in their systems. The *Athabaskan Petition* has demonstrated how a shift in the legal epistemological perspectives and a decolonization of international human rights law is needed to achieve an inclusive, all-encompassing juridical order that considers Indigenous views as a legitimate source of law. The realization of such a shift should consider alternative paradigms in the conceptualization of the relationship between human beings and the environment, whereas the human dependence on ecological integrity should be recognized as a core value to be protected by legal means. Through this deconstruction of hierarchies, the decolonization of the juridical order, and the recognition of alternative views to the Westernized legal system, the peculiar relationship between Indigenous peoples' culture and the environment would be considered as a *tout court* right that does not necessitate an additional, creative function of courts in order to gain legitimate legal significance.

Notes

1 Dutch Supreme Court, *State of the Netherlands v. Urgenda Foundation,* 20 December 2019, at https://www.urgenda.nl/en/themas/climate-case/, last accessed February 2021.

2 Respectively, Treaty of Tordesillas and the Treaty of Saragossa.

3 Encomienda refers to a system used by the Spanish colonial government in the New World to reward conquerors. After the conquest of Mexico and Peru in 1500, conquerors were awarded the lands occupied by the Native Americans. The native inhabitants were defined as *encomendados* ("commended" or "entrusted") to the Spanish. They were expected to pay tribute to and work in the local fields or mines. Thus, the encomienda system was essentially a form of institutionally authorized slavery. See also: Riggs T. Gale *Encyclopedia of U.S. Economic History*, "Encomienda", Farmington Hills, MI, 2015.

4 See generally Doyle, C. M. *Indigenous Peoples, Title to Territory, Rights and Resources*, Routledge 2015, for a full account on the history of Indigenous peoples' consent requirement and how it was utilized, during colonization, as an instrument to obtain land and resources.

5 Green Facts & Intl Polar Foundation, Facts on Arctic Climate Change (2004), at http://www.greenfacts.org/en/arctic-climate-change/foldout-arctic-climate-change.pdf.

6 *Petition to the Inter-American Commission on Human Rights Seeking Relief from Violations of the Rights of Arctic Athabaskan Peoples Resulting from Rapid Arctic Warming and Melting Caused by Emissions of Black Carbon by Canada*, available at http://climatecasechart.com/non-us-case/petition-inter-american-commission-

human-rights-seeking-relief-violations-rights-arctic-athabaskan-peoples-resulting-rapid-arctic-warming-melting-caused-emissions/, last accessed February 2021.

7 The 1979 statute of the IACHR provides the Commission with the authority to uphold the rights protected in the American Declaration, following dispositions contained in Article 1: "2. For the purposes of the present Statute, human rights are understood to be:
   a The rights set forth in the American Convention on Human Rights, in relation to the States Parties thereto;
   b The rights set forth in the American Declaration of the Rights and Duties of Man, in relation to the other member states." See also: *Statute of the Inter-American Commission on Human Rights*, O.A.S. Res. 447 (IX-0/79), O.A.S. Off. Rec. OEA/Ser.P/IX.0.2/80, Vol. 1 at 88 (1979).

8 O.A.S. Secretary-General, *Annual Report of the Inter-American Commission on Human Rights*, OEA/Ser.L/V/J, doc. 5 rev. 1 (Mar. 7, 2011).

9 *Mary and Carrie Dann v. United States*, Case 11.140, Report No. 75/02, Inter-Am. C.H.R., Doc. 5 rev. 1 at 860 (2002).

10 *Mitchell v. Canada*, Petition 790/01, Inter-Am. C.H.R., Report No. 74/03, OEA/Ser.L/V/II.118, doc. 5 rev. 2 (2003).

11 However, a paradox in the use of the Declaration by Indigenous peoples can be seen here, as their views were not included in the drafting of the Declaration of the Rights and Duties of Man. Indigenous peoples indeed generally regard UNDRIP as the instrument that better protected their rights. Indeed, this instrument recognises multiple sources of Indigenous law and custom: Indigenous philosophies and views, treaties and agreements between Indigenous peoples and States, and historical claims and their remedies See also: Xanthaki A. *Indigenous Rights and United Nations Standards: Self-determination, Culture and Land*, Cambridge 2007.

12 *Supra* note 6.

13 *Athabaskan Petition*, Section IV. FACTS: RAPID ARCTIC WARMING CAUSED IN PART BY EMISSIONS OF BLACK CARBON, A SHORT-LIVED CLIMATE POLLUTANT, HARMS THE ARCTIC AND ARCTIC ATHABASKAN PEOPLES, *supra* note 6.

14 *Supra* note 6.

15 *Athabaskan Petition*, p. 68.

16 *Idem*, p. 69.

17 "Every person has the right to take part in the cultural life of the community, to enjoy the arts, and to participate in the benefits that result from intellectual progress, especially scientific discoveries", Declaration, Article XIII.

18 *Athabaskan Petition*, pp. 63–69.

19 *Idem*, p.30.

20 "The Commission acknowledged the fundamental nature of the right to property when it stated that "[v]arious international human rights instruments, both universal and regional in nature, have recognized the right to property as featuring among the fundamental rights of man. Such instruments include the Universal Declaration of Human Rights, the European Convention on Human Rights, and the African Charter on Human and Peoples' Rights", *Athabaskan Petition*, p. 64.

21 *Idem*, pp. 74–78.

22 See *The Mayagna (Sumo) Awas Tingni Community v Nicaragua,* [2001] Inter-Am Ct HR (Ser. C) No 79, 26; and *Kásek Indigenous Community v Paraguay*, [2010], Inter-Am Ct HR (Ser. C) No 214.

23 *Athabaskan Petition*, pp. 36–46.

24 *Idem*, pp. 81–4.

25 The *Athabaskan Petition* reads: "[…] Arctic Athabaskan peoples would have no reasonable chance of success in challenging Canada's insufficient regulation of black carbon emissions in Canadian courts because those courts have found that policy-related determinations concerning government regulation or the failure to regulate activities are non-justiciable," p. 81.

26 See Hul'qumi'num Treaty Group, at http://www.bctreaty.ca/treaties-and-agreements, last accessed May 2020.

27 "[…] it is not an effective mechanism to protect the right alleged by the alleged victims. Therefore, the first exception to the requirement of exhaustion of domestic remedies applies because there is no due process of law to protect the property rights of the HTG to its ancestral lands," IACHR. *Report No 105/09 on the admissibility of Petition 592-07, Hul'qumi'num Treaty Group*, Canada. October 30, 2009.

28 To access the case: http://climatecasechart.com/non-us-case/petition-to-the-inter-american-commission-on-human-rights-seeking-relief-from-violations-resulting-from-global-warming-caused-by-acts-and-omissions-of-the-united-states/, last accessed February 2021.

29 See Section D of the *Athabaskan Petition*, "BY ITS ACTS AND OMISSIONS, THE UNITED STATES VIOLATES THE HUMAN RIGHTS OF THE INUIT."

30 IACHR, Decision, 16/11/2006, available at http://blogs2.law.columbia.edu/climate-change-litigation/wp-content/uploads/sites/16/non-us-case-documents/2006/20061116_na_decision.pdf, last accessed May 2020.

31 *Athabaskan Petition*, pp. 11–16.

References

Acuña, R. M. (2015). "The Politics of Extractive Governance: Indigenous Peoples and Socio-Environmental Conflicts." *The Extractive Industries and Society* 2, no. 1, 85–92.

Anaya, S. J. (2004). *Indigenous Peoples in International Law*. Oxford: Oxford University Press.

Anisimov, O. A., Vaughan, D. G., Callaghan, T. V., Furgal, C., Marchant, H., Prowse, T. D., Vilhjalmsson, H. and Walsh, J. E. (2007). "Polar Regions (Arctic and Antarctic)." In *Climate Change 2007: Impacts, Adaptation and Vulnerability: Contribution of the Working Group II to the Fourth Assessment Report of the Intergovernmental Panel on Climate Change*, edited by L. M. Parry, O. F. Canziani, J. P. Palutikof, P. J. van der Linden and C. E. Hanson, 653–685. Cambridge: Cambridge University Press.

Antkowiak, T. (2013). "Rights, Resources, and Rhetoric: Indigenous Peoples and the Inter-American Court." *University of Pennsylvania Journal of International Economic Law* 35, no. 1, 113–187.

Baxi, U. (2006). *The Future of Human Rights*. Oxford: Oxford University Press.

CIEL. (2017). *Smoke and Fumes: The Legal and Evidentiary Basis for Holding Big Oil Accountable for the Climate Crisis*. Washington D.C.: Center for International Environmental Law.

Doyle, M. A. (2015). *Indigenous Peoples, Title to Territory, Rights and Resources*. Milton Park: Routledge.

Faure, M. and Peeters, M. (2019). "Liability and Climate Change." In *Oxford Research Encyclopaedia of Climate Change*, edited by H. von Storch. Oxford: Oxford University Press.

Federman, C. (2003). "Constructing Kind of Persons in 1886: Corporate and Criminal." *Law and Critique* 14: 167–189.

Foster, M. K. (1985). "Another look at the function of Wampum in Iroquois-White Councils." In *History and Culture of Iroquois Diplomacy: An Interdisciplinary Guide to the Treaties of the Six Nations and Their League* (pp. 99–114), edited by F. Jennings. Syracuse: Syracuse University Press.

Fuentes, A. (2017). "Protection of Indigenous Peoples' Traditional Lands and Exploitation of Natural Resources: The Inter-American Court of Human Rights' Safeguards." *International Journal on Minority and Group Rights* 24, no. 3, 229–253.

Gallas, A., Herr, H., Hoffer, F., and Scherrer, C. (2016). *Combating Inequality: The Global North and South*. Routledge.

Ganguli, G., Setzer, J., and Heyvaert, V. (2018). "If At First You Don't Succeed: Suing Corporations for Climate Change." *Oxford Journal of Legal Studies,* 38, no. 4, 841–868.

Gary, Kofinas, G. P., Chapin, III S. F., Shauna BurnSilver, S., Schmidt, J. I., Fresco, N. L., Kielland, K., Martin, S., Springsteen, A. and Rupp, T. S. (2010). *Resilience of Athabaskan Subsistence Systems to Interior Alaska's Changing Climate*. NRC Research Press.

Gilbert, Jéremie. (2016). "Land Grabbing, Investors, and Indigenous Peoples: New Legal Strategies for an Old Practice?" *Community Development Journal* 51(3), 350–366.

Gilbert, J., and Begbie-Clench, B. (2018). "Mapping for Rights: Indigenous Peoples, Litigation, and Legal Empowerment." *Erasmus Law Review* 11, 6.

Grear, A. (2015). "Deconstructing Anthropos: A Critical Legal Reflection on 'Anthropocentric' Law and Anthropocene "Humanity'." *Law Critique* 26, no. 3, 225–249.

Grear, A. (2011). "The Vulnerable Living Order: Human Rights and the Environment in a Critical and Philosophical Perspective." *Journal of Human Rights and the Environment* 2, no. 1, 23–44.

Guerrero, D. G. (2018). "The Limits of Capitalist Solutions to the Current Climate Crisis." In *The Climate Crisis: South African and Global Democratic Eco-Socialist Alternatives*, edited by Vishwas Satgar. Johannesburg: Wits University Press.

Heaton, J. W. (2012). "Athabaskans of the Interior: A New Perspective on Neglet in Alaska Historiography." *The Pacific Northwest Quarterly* 103, no. 3, 107–122.

Hesford, W. S. (2011). "Human Rights Rhetoric of Recognition." *Rhetoric Society Quarterly* 41, no. 3, 282–289.

Kásek Indigenous Community v Paraguay. 2010. (Ser. C) No 214. (Inter-Am Ct HR).

Langdon, S. J. (1992). *The Native People of Alaska: Traditional Living in a Northern Land*. Anchorage, Alaska: Greatland Graphics.

Lindroth, M. (2006). "Indigenous-State Relations in the UN: Establishing the Indigenous Forum." *Polar Record* 42, no. 3, 239–248.

Marks, G. C. (1992). "Indigenous Peoples and International Law: The Significance of Francisco de Vitoria and Bartolome de las Casas." *Australian Year Book of International Law* 1–52.

Mary and Carrie Dann v. United States. 2002. Case 11.140, Report No. 75/02, Doc. 5 rev. 1 (Inter-Am. C.H.R.).

McCrimmon, D. (2016). "The Athabaskan Petition to the Inter-American Human Rights Commission: Using Human Rights to Respond to Climate Change." *The Polar Journal* 6, no. 2, 398–416.

McNeil, K. (1999). "The Onus of Proof of Aboriginal Title (Canada)." *Osgoode Hall Law Journal* 37, no. 4, 775–803.

Merry, S. E. (2006). *Human Rights and Gender Violence: Translating International Law into Local Justice.* Chicago, IL: University of Chicago Press.

Mitchell v. Canada, Petition 790/01. 2003. Report No. 74/03, OEA/Ser.L/V/II.118, doc. 5 rev. 2 (Inter-Am. C.H.R.).

Natcher, D. C. (2004). "Implications for Fire Policy on Native Land Use in the Yukon Flats." *Human Ecology* 32, no. 4, 421–441.

Peel, J., and Osofsky, H. (2015). *Climate Change Litigation: Regulatory Pathways to Cleaner Energy.* Cambridge: Cambridge Studies in International and Comparative Law.

Petition to the Inter-American Commission on Human Rights Seeking Relief from Violations of the Rights of Arctic Athabaskan Peoples Resulting from Rapid Arctic Warming and Melting Caused by Emissions of Black Carbon by Canada. (2013). Interamerican Commission on Human and Peoples' Rights.

Setzer, J. and Byrnes, R. (2020). *Global Trends in Climate Change Litigation: 2020 Snapshot.* London: Grantham Research Institute on Climate Change and the Environment and Centre for Climate Change Economics and Policy, London School of Economics and Political Science.

Skillington, T. (2017). *Climate Justice and Human Rights.* New York: Palgrave Macmillan.

The Mayagna (Sumo) Awas Tingni Community v Nicaragua, Xákmok (Ser. C) No 79, 26. (2001). (Ser. C) No 79, 26 (Inter-Am Ct HR).

VanStone, J. W. (1974). *Athapaskan Adaptations: Hunters and Fishermen of the Subartic Forests.* Wheeling: Harlan Davidson.

Wewerinke-Singh, M. (2019). "Remedies for Human Rights Violations Caused by Climate Change." *Climate Law* 9: 224–243.

Williams, R. A. (1990). *The American Indian in Western Legal Thought: The Discourses of Conquest.* Oxford: Oxford University Press.

Xanthaki, Alexandra. (2007). *Indigenous Rights and United Nations Standards: Self-Determination, Culture and Land.* Cambridge: Cambridge University Press.

PART V

A Way Forward

Strategies to Expressly Consider Ethics in Climate Change Policy Formation

32

ETHICAL GUIDANCE FOR NEGOTIATING AND ARGUING ABOUT CLIMATE CHANGE

Hugh Breakey

Introduction

Argument is ethical. Through argument, it is possible to show respect, develop shared understandings, and craft mutually acceptable solutions to serious problems. Equally, it is possible to show disrespect and contempt through argument, and to mislead and manipulate. These ethical concerns become even more substantial when the stakes are high, and when much rests on the outcome of the argument – as is the case for debates surrounding climate change and potential policy responses to mitigate and adapt to it. Unfortunately, as I argue below, political discourse on climate change often involves the intentional use of misleading and manipulative rhetorical strategies, with damaging consequences.

The chapter proceeds as follows: The first section of this chapter overviews the ethics of argument. The second section shows why argument and deliberation are especially important in the context of climate change. The third section then explores the ethics of argument about climate change, noting problems that can occur, before the final section provides guidelines for helping civil society and policy-makers spot ethical concerns in arguments about climate change.

The Ethics of Argument

When we argue or deliberate with others, we engage in an activity that involves matters of consent, trust, influence, respect, and collective decision-making. For this reason, we can behave ethically in argument by showing respect and contributing constructively to collective outcomes. Equally though, we can behave in a morally wrong manner, failing to treat others appropriately.

There are two main ways this can occur: through interpersonal ethics (where we consider local issues that arise between those involved in the argument) and through political legitimacy (where we consider broader issues that arise in public discourse for polities and collective decision-making).

Focusing first on personal morality in argument, consider some familiar standards of argument:

- Equal standing: no speaker has any intrinsic authority;
- Good faith: speakers say what they believe;
- Open-mindedness: speakers genuinely listen and engage with opposing arguments; and
- Reasonableness: Speakers should aim at clear, persuasive logical argument that provides reasons and evidence that others can understand and accept.

DOI: 10.4324/9781003039860-38

These standards might seem sensible from a rational, philosophical point of view (Fogelin & Sinnott-Armstrong 2001; Eemeren & Grootendorst 2003). But they have ethical significance too (Ceva 2011; Correia 2012; Breakey 2019). Sometimes this is obvious. As a way of influencing people, staying focused on rational persuasion requires eschewing threats and coercion. So too, good faith ensures honesty, and aiming at logical, accessible argument requires foregoing wrongful actions like wanton abuse (as this involves the ad hominem fallacy). Observing the above standards also respects others' consent in entering argument with us: we obey the rules that allow the inherent goods of argument – knowledge, truth, understanding – to be mutually pursued.

More deeply though, sharing reasons with another person respects their rationality. It acknowledges they are a thinking being, equal to ourselves in important respects, willing and able to make decisions and form beliefs on an intelligent, reasoned basis. It also deepens our understanding of their views and helps correct biases and narrowness in our own minds. On a more fundamental level, it shows a commitment to the intrinsic importance of truth and knowledge, and to respecting our own and others' pursuit of this important good (Breakey 2012).

In addition, Regina Rini (2018) rightly stresses that argument specifically about *ethical* matters, discussing what we should do as individuals and societies, is a particularly important way of showing mutual respect. By engaging with our interlocutor on such issues, we recognize them as morally autonomous (able to understand shared reasons and come to their own principled standpoint) and also as part of our moral community (part of our shared society, with each of us as members who can give and receive reasons about how our society should work). Rini stresses that this requires genuinely listening to and engaging with the arguments and evidence others present to try and change our minds – even as they are required to be reciprocally open to us.

Despite the important relationship between rational discussion and respect, note that faulty and flawed argument is not immoral in itself. A person is not ethically bad merely because they hold a wrong position or give a fallacious argument. But the arguments above show that to do so deliberately or recklessly may well be immoral.

Let's now turn from interpersonal argument to consideration of public deliberation. The standards of ideal political discourse track closely the standards of interpersonal ethical argument described above, extolling "values of openness, equal chances to present alternatives, and full and impartial consideration of those alternatives" (Cohen 2002). This is unsurprising. The same methods of respecting others and pursuing goods like knowledge and understanding that work in the interpersonal case fill those same functions in larger political deliberation.

There are two important types of goods that can come from high standards political deliberation.

First, there are intrinsic goods that arise purely within the deliberation itself, irrespective of any ensuing changes in belief and knowledge, or informed policy change. These goods include showing respect for others and their political and moral views, and ensuring others feel listened to and respected, even if their views ultimately fail to persuade. In a political context, this respect is critical because it contributes to the legitimacy of the outcomes (such as policies and decisions). 'Legitimacy' refers to the status of an outcome or institution such that people have moral reason to respect and support it. Legitimacy can be created through many different means, but a critical part of political legitimacy involves deliberative justice (Breakey 2018).

In an ideal procedure of political deliberation

> participants regard one another as equals; they aim to defend and criticize institutions and programs in terms of considerations that others have reason to accept.... and they are prepared to cooperate in accordance with the results of such discussion, treating those results as authoritative.
>
> *(Cohen 2002)*

The idea that public reason – the capacity of the public to debate and together think through moral and political questions – can contribute to the legitimacy of political regimes and policies is a longstanding view of enlightenment philosophers like Hobbes and Locke, and recent thinkers like Rawls and Habermas (Ivison 1997). As well as intrinsically respecting the citizenry and their capacity for moral deliberation (as Rini highlighted above), legitimacy is important practically, because it promotes compliance and support.

Second, as well as these intrinsic goods, policies arising from an inclusive and high standards deliberation are more likely to be considered views, incorporating multiple perspectives, ideas, and sources of evidence, ultimately leading to better policy outcomes. Disagreement is not in itself a bad thing: when dealt with appropriately, it can lead to more informed outcomes and improved understanding (Leslie 2021).

Why Is Engagement and Discussion Important in This Context?

The twin goods of political deliberation noted above are desirable in any policy context. But global warming has features that make it more important than ever. One important feature in this respect is the nature of the *harms* being created by countries' emissions. One of the key ideas aminating public reason is that "the moral demands we make on others must be justifiable to those others by appeal to reasons they have, and not simply by appeal to the truth as we see it" (Quong 2018). This may seem confusing. Why shouldn't we simply act on the basis of the truth as we see it?

Consider an example. Imagine country X decided to do nuclear testing of its small nuclear weapons on their border with country Y, and by a river that runs to one of Y's major cities. Y raises immediate concerns about radioactivity and fallout. However, X has decided that these risks and their effects are overblown (and, some believe, perhaps just a conspiracy spread by nations like Y for their own purposes). Yet there is scant scientific evidence for X's position, and a great amount of evidence for Y's concerns, including by independent experts.

In such a case, it is obvious that X unilaterally coming to its own decision about harms and risks of harms, and then acting on that basis, is unacceptable. From Y's perspective, it has ample evidence that X is imposing grave risks of serious harm on Y, and X's blithe assertion that there are no significant risks is inadequate.

This is precisely the type of case that worried the political philosopher John Locke, and why he argued that settled law was so necessary for human societies. Unlike his more cynical forebear, Thomas Hobbes, Locke's central worry wasn't that people unrestrained by law would attack each other out of greed or fear (Locke 1690/1947). Rather, Locke realized that even morally good people would interpret their obligations and rights, and others' obligations and rights, differently. X would see harms where Y saw legitimate activities. Y would see unjustified assault where X saw righteous redress for past wrongs. As a result, X and Y would be driven into conflict, even as both genuinely saw themselves as rightly insisting upon moral law.

The case of the two countries and X's nuclear testing described above shows that these rule of law concerns do not arise only with respect to individual people or groups, but also to nation-states. Locke himself explicitly stressed this, seeing leaders of countries as exemplars of decision-makers in a state of nature, unrestrained by an overarching power, yet nevertheless needing to come to terms with others. Fortunately for our modern world, with the UN Charter and a commitment to the rules-based order, humanity has developed a system of international law that largely prevents the conflicts animating Locke from leading to international war and armed conflict (Pinker 2012).

But the narrowness of the UN Charter means that other types of harms, short of outright war and blockade, remain possible. We do not have (and possibly would not want, for a myriad of reasons) a world government capable of deciding such matters. But what this means is that we need other ways of resolving potential clashes over harms, and of engaging with other nations over potential harms.

In this respect, deliberation is crucial – especially deliberation between countries and deliberation that includes the concerns raised by countries on different sides of the dispute. In the story of X and Y above, this is obvious. Country X having an internal debate about whether there are any serious harms from their testing, and unilaterally proceeding on that basis, is plainly inadequate to maintaining proper respect towards, and appropriate relations with, Country Y.

Of course, climate change differs from the nuclear testing scenario. The harms are further away, and the causal relationship between X's actions and Y being harmed is weaker.

Nevertheless, the underlying point remains. If one country has serious and well-evidenced concerns about another country inflicting harms on them, then the process for resolving that issue should not be an internal debate within one country, coming to their own decision about the seriousness of those harms – much less whether it is within their economic interest to inflict those harms. This is precisely the type of unilateral decision-making that Locke saw as unjust and likely to have serious ramifications for relationships between aggrieved parties.

To avoid these outcomes, it is critical to have engagement with other parties that confer a wider legitimacy on nation' activities. Given that other legitimating devices (like overarching democratic law-making, for example) are not possible in international affairs, having fair and constructive deliberations stands as the most feasible option available (Breakey 2018).

The takeaway here is that, while unilateral decision-making (within a single country, and based wholly on that country's views about harms and benefits) is appropriate for many types of issues, climate change is not one of these issues (Breakey 2015a; Boran 2016). Deliberation is a key way forward, both internationally, to understand and respond to other views, and within nations to acknowledge that the task is not purely to work out what we as a nation think of the risks and harms of climate in/action, but to how we as a nation need to fairly respond to a case where other nations have different views about those risks and harms – and where the weight of current scientific opinion may be overwhelmingly on their side (Brown 2013). In sum, the idea that 'We have made up our minds, and that is enough' is not appropriate when others have substantial and interpersonally sharable reasons to believe they are being harmed.

A further distinct reason why deliberation is especially important in the context of climate change is because there is genuine complexity to many of the issues. This complexity does not mean that we can't come to clear positions about appropriate ways forward or rule out unconscionable options (Brown 2013). But it does mean that unearthing the truth of the matter may require a little digging, and a genuine back-and-forth exchange among parties.

For example, one party might reasonably insist that their country is making significant contributions to climate change, correctly citing a comparatively strong drop in emissions across recent years (see, e.g., Editorial 2021). Yet that same country might nevertheless still – even after those cuts – possess a much higher per-capita carbon footprint than comparable countries. Equally, the party might insist that the country is meeting its Paris commitments – meaning it is on track to meet its own individually determined contributions – but *not* that is making an equitable contribution to keeping global warming under 1.5°C, or at least under 2.0°C – the limits explicitly stated in the Paris Agreement (Editorial 2021). Yet the ethical and legal commitments under the Paris Agreement are those that deliver the promised limit on warming (Brown et al. 2018). In both cases, the party is airing evidence that is accurate and relevant, but in both cases, it is leaving out critical parts of the story that need to be acknowledged and incorporated, something that high standards public deliberation would promote and require.

As well as complexity in science and policy, there is also complexity in the way legitimacy works in global endeavors such as confronting climate change. There are many ways we can come to the view that a policy is ethically worthy of our support and compliance (Breakey 2015b, 2018). We may judge it on its *outcomes*: that it creates effective, fair results that solve problems and avoid disrupting settled expectations. We may judge it on its *processes*: that it was inclusive, informed, and democratic. We may judge it on our *relationship* to it: that we agreed to it or feel ownership over it.

One of the reasons it is useful to have these many strands of legitimacy is that it empowers collective problem-solving and dispute resolution. Even if we can't agree on what our *policy* should be, perhaps we can agree on the *process* (e.g., a vote) whereby we select that policy. This is critically important in climate change discussions for the reason noted above: there will be different views on what harms it is appropriate for one party to impose on another party. With this disagreement on substance, processes like deliberation and fair collective decision-making become critical.

However, the fact that we have many different types of legitimacy in play can also make it easier to criticize outcomes and processes. This is especially so for an issue like climate change, where effective response necessarily involves disrupting status quo expectations (e.g., of continued industrial activity) that in other cases, we might have good reason to respect (Brown 2017; Breakey 2022). It is also true that international governance institutions struggle to capture the types of strong legitimacy that governments (of liberal democratic countries) enjoy (consider, e.g., Buchanan & Keohane 2006). Decision-making at any of the structures of the United Nations, for example, simply is not democratic in the 'one person, one vote' manner of national elections.

What this means is that, with so many legitimacy factors in play, it can make it possible to talk around (rather than about) critical issues. It is reasonable to have concerns about the way international collective decision-making occurs, and the results of this for fairness and compliance. But at the same time, we need to be cognizant that concerns raised about these legitimacy dimensions do not distract from the most fundamental climate-related concerns: the fact that global warming is happening, and that if not responded to urgently, it will ultimately have disastrous outcomes for heavy emitters and vulnerable countries alike. Summing up, constructive argument about climate change is especially critical because carbon emissions involve inflicting harms and costs on other parties, and because there is genuine ethical complexity to the issue, and to the institutions we have for dealing with it.

The Ethics of Arguments about Climate Change

This section describes some of the typical ethical problems that occur in climate change discourse, and their potential harms and costs. There are several typical ethical problems that arise in debates over climate change. First, there is the misleading expression of somewhat true facts. As we saw above, some speakers will laud that they are 'meeting our Paris commitments,' when this suppresses the point that the commitments that they are referring to are entirely self-created (it is easy to fulfill commitments when one has chosen easily fulfillable commitments) and that they are failing to meet the collectively made commitments limiting warming to 1.5/2.0°C. The potential moral wrong here is one of manipulation: the speaker aims to lead the audience/interlocutor to their preferred conclusion, rather than frankly communicating and acknowledging relevant facts that empower them to come to their own conclusions.

Second, there can be an unwillingness to listen and respond to arguments leveled from one's interlocutors. Consider someone who argued that what *we* do as a country in restricting emissions will not ultimately matter, because so much of the problem is caused by other countries (any country could coherently mount this argument, even major emitters like the United States and China). As stated, there is nothing wrong with voicing this concern, as a 'first move' in considering the ethics of climate change. The ethical problem in argument arises when there is no attention given to the responses to these types of arguments, which are well-known and easily accessible. For example, this same type of problem occurs, often in a stronger form, in countless collective action situations – including voting in elections, a domain where no politician wants to point out that it doesn't matter what any individual voter does because the result will be determined by everyone else. But too often, these arguments are simply repeated as if the responses were never made. So rather than having a constructive, ethical dialogue, there is always a 'return to start' mechanism, and the argument cannot move forward. This is entirely different to defending a claim against an objection by dismantling or invalidating the objection. As Leslie (2021) argues, it can be advantageous to have debaters with rival commitments sticking to them. There isn't a moral demand to acquiesce to any given objection, only to acknowledge or respond.

Third, there can be a lack of awareness that this is not the type of problem that can be addressed unilaterally, and on the basis of national interest narrowly construed. Naturally, economic considerations are always an important factor in each nation's decision-making. But because of the costs involved in carbon emissions imposed on other nations, leading ultimately to grave outcomes for those nations, economic goals cannot be thought to simply trump larger discussions of ethics and fairness (Brown 2013). As the nuclear testing case above suggested, the nature and substance of an ethical deliberation must reflect more than what our *local* view of the stakes are, but whether other stakeholders have a different and better justified reading of those stakes.

Fourth, there is a tendency for conspiratorial thinking to occur in the context of climate change. Indeed, some research suggests that when those who doubt global warming are presented with conflicting evidence, their most common response is conspiratorial thinking (Lewandowsky & Cook 2020). Conspiratorial thinking doesn't automatically translate into unethical argument, but it is likely to do so. Such thinking is characterized by the willingness to hold contradictory views, significant imperviousness to evidence (e.g., by claiming sources of contrary evidence are either gulled by the conspiracy or complicit in it) and seeing themselves as privileged thinkers who have uncovered the 'truth' (Lewandowsky & Cook 2020). All these features can prompt conspiracy theorists to breach the norms of high standards deliberation. That said, high standards deliberation (with a focus on critical thinking, empathy, and respect) is almost certainly the best way of reaching conspiracy thinkers (see Lewandowsky & Cook 2020).

Two final issues require mention, which impacts on climate discussions even if they are not intrinsically part of them. The first is the use of corporate funding to develop and promote biased research outputs, which are then used as evidence in subsequent discussions and decision-making (Brown 2016; Farrell 2016). This action can be clearly immoral. For example, it may be done in bad faith, where the funding institution internally acknowledges that climate change is real but wishes to spread skepticism for their own purposes. Another type of dishonesty occurs when authorship and conflicts of interest are deliberately suppressed to mislead the citizenry about the source of the information.

The other final issue is the use of ulterior pressures to harass or coerce the other side. Sometimes this is done through bullying and threats from political partisans. But it can also happen in institutional ways. For example, it can be legitimate for public representatives to ask for further information or data on climate science, but any demands (much less coerced orders) must be reasonable, proportionate and apply to both sides of a debate equally (Biddle & Leuschner 2015).

Turning now to consider the harms and costs of poor argument about climate change, J. Anthony Blair (2011) describes how when the stakes rise, demands for truth-tracking and rational norms of argument become morally loaded. Just as we shouldn't *think* loosely and recklessly as individuals when people's lives are at stake, so too we shouldn't *argue* loosely and recklessly as interlocuters when people's lives are at stake. People should always be willing to entertain the question: what is the cost of this to others if I am wrong?

Care is required when attributing harm on the basis of our own conclusions. That is, X may be able to give an account of the harmfulness of Y's argument, based on the premise that X's view is correct. Intuitively, it will feel reasonable to X for her to calculate harms in this way, especially if there is considerable independent justification for her view. However, because everyone routinely believes that their views are well-justified, this policy leads to a situation where (at least in high stakes cases), everyone can hold that opposing arguments are not merely wrong but harmful or dangerous. For example, consider a religious devotee who believes in eternal life. She will think that atheistic arguments are not just wrong but harmful (as they cause people to forego the faith that would give them eternal life). But this assertion of harm clearly begs the question at hand: it is only a harm *if* her belief turns out to be true.

There are ethical worries with leveling accusations of harm on these bases – not least that they distract from exactly the type of high standards deliberation that such issues demand (Breakey 2021). Thus, the message isn't that we should shut down or manipulate discussions because people are dangerously wrong, but rather that because of the high stakes, we are additionally justified in insisting on high quality and constructive argument.

As well as the high stakes involved in climate policy, and the urgency that accompanies them, there are other costs to unethical argument. As we saw in Section "The Ethics of Argument", as well as quality argument contributing to better policy outcomes, it can also improve the legitimacy of policy responses by showing respect to others. Contrary-wise, low standards of argument increase public frustration, polarization, and anger, undermine faith in political institutions and reduce trust in fellow citizens. For example, if a liberal democracy cannot manage to have a mature, serious conversation about an issue as important and urgent as climate change, then people may begin to lose faith in liberal democracy, which carries its own costs.

Before we turn to practical responses to these points, note that ethical missteps can happen from either side of a debate. Simply because an arguer is on the 'right' side of an argument does not mean they will necessarily treat their interlocuter with respect and fairness, or that they will contribute to the legitimacy and constructiveness of collective decision-making. Indeed, firmness of conviction about being right can sometimes lead to impatience and frustration in the face of opposition, tempting an arguer to dispense with ethical considerations.

Practical Guidelines for Helping Civil Society and Policy-Makers Spot Ethical Concerns in Their Own and Others' Arguments about Climate Change

The ethics of argument provides important guidelines for how we should behave in argument, especially about urgent topics such as climate change. However, care must be taken when one feels that one is being mistreated in argument, as accusing an interlocuter of wrongdoing is usually unconstructive, and can escalate conflict (Glasl 1999; Breakey 2021). Often, the best response is to appeal to rational rather than moral norms, stressing that for the argument to achieve its goals of uncovering truth and promoting understanding, both interlocuters need to listen carefully and patiently to each other, and aim to respond logically. That said, sometimes 'calling out' wrongdoing is necessary, such as in clear cases of dishonesty or deception. While we shouldn't ever presume bad faith on the part of interlocuters, it must be acknowledged that dishonesty, manipulation, and conflicted interests do occur in the climate change context, and that evidence shouldn't always be taken at face value.

Three key guidelines are:

1 Insist on the importance of engaging with contrary positions, rather than holding firm on, and repeating, demonstrably flawed arguments. This involves being attentive to the 'return to start' mechanism that can occur when arguers return to arguments or positions that have previously been challenged and do so without acknowledging or responding to that challenge. This is especially important in the context of conspiratorial thinking. People are entitled to

consider the possibility of conspiracies (they do happen) and also whether problems of confirmation bias or groupthink could be influencing their opponents' beliefs. But neither of these devices is plausible in the context of climate change. In this context, we have countless decentralized scientists (across different disciplines) from myriad countries with different and often openly competing agendas and ideologies, all acknowledging and stressing the reality of climate change – over a period now of some decades. This is simply not how groupthink works, and it is not how viable conspiracies work. Both require a sufficiently close-knit community with a shared identity and common values.

2 Merely because a type of speech is legally protected does not make it morally permissible. Certain types of morally wrongful speech are made illegal – as captured by laws against hate speech, 'fighting words', defamation and so on. But for many types of action (and speech particularly), it is quite possible to obey the law and yet behave wrongfully. So, recognizing that a type of speech is legally permissible does not mean that it should not be given an ethical 'all clear.'

3 Ensure critical areas are not dropped out of the conversation. It is entirely appropriate that someone might be skeptical about international institutions and global governance. But such skepticism doesn't make the fact of climate change disappear, nor appropriately respond to the moral enormity of what unchecked global warming would look like.

In closing, one final reminder is not to lose sight of the fact that action – and proof of action – matters. It is not enough to talk the talk, even though talk is important in a world where action must be done collectively and in a coordinated, synergistic way. But walking the walk is even more critical, and we need to be alive at the risk of being enthused about ambitious declarations if they are not accompanied by detailed plans and observable strategies for implementation.

References

Biddle, J. B. & Leuschner, A. (2015) "Climate skepticism and the manufacture of doubt: Can dissent in science be epistemically detrimental?," *European Journal for Philosophy of Science* 5: 261–278.

Blair, J. A. (2011) "The moral normativity of argumentation," *Cogency* 3: 13–32.

Boran, I. (2016) "Principles of public reason in the UNFCCC: Rethinking the equity framework," *Science and Engineering Ethics* 23: 1253–1271.

Breakey, H. (2012) *Intellectual Liberty: Natural Rights and Intellectual Property,* London: Routledge.

Breakey, H. (2015a) "COP20's ethical fallout: The perils of principles without dialogue," *Ethics, Policy & Environment* 18: 156–169.

Breakey, H. (2015b) "Heating up climate change norms – Lessons from human rights," in H. Breakey, V. Popovski & R. Maguire (eds.) *Ethical Values and the Integrity of the Climate Change Regime*, Surrey: Ashgate.

Breakey, H. (2018) "It's right, it fits, we debated, we decided, I agree, it's ours, and it works: The gathering confluence of human rights legitimacy," *Law and Philosophy* 37: 1–28.

Breakey, H. (2023) "The ethics of arguing," *Inquiry* 66: 589–613.

Breakey, H. (2021) ""That's unhelpful, harmful and offensive!" Epistemic and ethical concerns with meta-argument allegations," *Argumentation* 35: 389–408.

Breakey, H. (2022). "When normal is normative: The ethical significance of conforming to reasonable expectations," *Philosophical Studies* 179: 2797–2821.

Brown, D. (2013) *Climate Change Ethics: Navigating the Perfect Moral Storm,* New York: Routledge.

Brown, D. (2016) *Why Exxon's and Other Fossil Fuel Companies' Funding of the Climate Change Disinformation Campaign Cannot be Excused As an Exercise in Free Speech but Must Be Understood as Morally Reprehensible Disinformation* [Online], Widener University. Available: https://ethicsandclimate.org/2016/07/29/why-the-climate-change-disinformation-campaign-cannot-be-excused-as-an-exercise-in-free-speech-but-should-be-understood-as-morally-reprehensible-disinformation/ [Accessed 29 April 2021].

Brown, D., Breakey, H., Burdon, P., Mackey, B. & Taylor, P. (2018) "A four-step process for formulating and evaluating legal commitments under the Paris agreement," *Carbon and Climate Law Review* 12: 1–12.

Buchanan, A. & Keohane, R. O. (2006) "The legitimacy of global governance institutions," *Ethics and International Affairs* 20: 405–437.

Ceva, E. (2011) "Just interactions in value conflicts: The adversary argumentation principle," *Politics, Philosophy & Economics* 11: 149–170.

Cohen, J. (2002) "Procedure and substance in deliberative democracy," in T. Christiano (ed.) *Philosophy and Democracy,* Oxford: Oxford University Press.

Correia, V. (2012 "The ethics of argumentation," *Informal Logic* 32: 222–241.

Editorial (2021) "Doing things our way on cutting greenhouse gases," *The Australian*, 26 April, 2021.

Eemeren, F. V. & Grootendorst, R. (2003) *A Systematic Theory of Argumentation,* Cambridge: Cambridge University Press.

Farrell, J. (2016) "Corporate funding and ideological polarization about climate change," *Proceedings of the National Academy of Sciences of the United States of America* 113: 92–97.

Fogelin, R. & Sinnott-Armstrong, W. (2001) *Understanding Arguments: An Introduction to Informal Logic,* Orlando: Harcourt College Publishers.

Glasl, F. (1999) *Confronting Conflict,* Gloucestershire: Hawthorn.

Ivison, D. (1997) "The secret history of public reason: Hobbes to Rawls," *History of Political Thought* 18: 125–147.

Leslie, I. (2021) *Conflicted: Why Arguments Are Tearing Us Apart and How They Can Bring us Together,* London: Faber.

Lewandowsky, S. & Cook, J. (2020) *The Conspiracy Theory Handbook*, Fairfax: Center for Climate Change Communication.

Locke, J. (1690/1947) *Two Treatises of Government,* New York: Hafner.

Pinker, S. (2012) *The Better Angels of Our Nature: Why Violence Has Declined,* London: Penguin Books.

Quong, J. (2018) "Public reason," in E. N. Zalta (ed.) *The Stanford Encyclopedia of Philosophy*, Stanford, CA: Stanford University.

Rini, R. (2018) "Abortion, ultrasound, and moral persuasion," *Philosophers' Imprint* 18: 1–20.

33

ETHICAL AND ECOLOGICAL EDUCATION

An Approach to Climate Change through the Lens of the Earth Charter

Mirian Vilela

Introduction

"There is no chocolate without cacao." This is a saying in Spanish that reminds us that without this most basic and fundamental ingredient, we cannot produce chocolate. Comparably, how do we expect to address the challenges of climate change without a society that is ecologically and ethically literate? In other words, how can we expect to have policy and decision-makers, engineers, lawyers, businesspeople, and government officials (such as ministers of economy or finance) conscious and sensitive to our human-nature connection, and to our responsibility toward the well-being of others, if this was not nurtured in their upbringing and process of education? The reality is that most professionals and the so-called well-educated people did not receive the kind of education needed to understand the causes and consequences of climate change, how to mitigate and adapt to this emergency, and, especially, what are our responsibilities to address it in a timely manner.

It seems that we live in a world where the drivers of the vehicles are blind, sometimes willfully so, but they continue to push the accelerator, taking the vehicles to and ofttimes over the cliff. The foundational aspect of our blindness has to do with our predominant worldview that does not make us see the profound human-nature connection, neither how our decisions affect the well-being of others or our obligations with future generations. Therefore, education is one of the most essential ingredients needed to address climate change given that this is what will make citizens everywhere see and understand our global-local challenges, the interdependence and complexities of our problems, the profound human-nature connection, and the eminent need to act and cooperate in an ethically responsible way. Education is what will clear our vision, and equip leaders, decision-makers, and everyone with the knowledge, values, and skills necessary to make the shift toward a more just and sustainable world. Without this key element, our society will continue to move with a blurred and blind vision, recklessly, without paying attention to what is essential in life.

Root Causes of Climate Change

One of the most important root causes of climate change is the predominant unsustainable patterns of production and consumption, which, as mentioned above, is happening mostly because of lack of consciousness, ethics, and appropriate education. Generally, people do not wake up in the

DOI: 10.4324/9781003039860-39

morning thinking about 'what are my consumption patterns?' and 'what do they have to do with the well-being of my neighbors,' let alone the well-being of humans far away, from other countries and cultures, and even less of future generations or non-human living species, not in their 'eye-sight.' People do not perceive it because we have been trained to see things with a fragmented and short-sighted approach and not to see the relationship of things. The same happens with those in the production systems, where people are largely not concerned with the social or environment impact of what is being produced or the food they choose to eat or if the shoes and clothes they decide to buy and wear are affecting the health of ecosystems, the Earth's regenerative capacity, the well-being of others, or the capacity of others to access their basic needs.

The result of this vacuum on ecological and ethical literacy is that people, especially leaders and decision-makers, make decisions that affect us all, as well as future generations, without seeing or caring about their impact. Imagine that this is like having many Mr. Magoos around the world who are not concerned nor responsible for what they do yet are driving our world to injustices and unsustainability.[1]

Ethical and Ecological Education

Despite the fact that the essential role of education in changing the course of things is recognized in many national and international policies and treaties, this is often neglected and undermined. Article 6 of the United Nations Framework Convention on Climate Change (UNFCCC) highlights the importance of "Education, Training and Public Awareness," Article 12 of the 2015 Paris Agreement on Climate Change reaffirms that, and the 2015 United Nations Sustainable Development Goal 13.3 also recognizes the important role of education and awareness-raising in the efforts of climate change mitigation and adaptation (UNFCCC 1992; Paris Agreement 2015; UNSDG 2015).

Education is often undermined, or disregarded, in the climate change discourse, for two reasons. First, education is often associated with something that is limited to primary and secondary schooling, meant only for children and youth. Therefore, naturally what comes to mind is that we cannot afford the time that will take until the next generation of leaders is raised to be in positions to address these urgent challenges. However, this is a limited perception of the scope of education. Education involves formal, non-formal, and informal processes. The formal realm also involves higher education; the non-formal realm involves executive education and the lifelong learning of policy makers and public and private sector officials; and informal education involves all spaces where information is shared, including media. It is time that we see the value in all of these dimensions of education to help address the challenge of climate change and build bridges between them. Second, because of our fragmented mind-set, solutions to the climate challenge seem to be limited to the realm of technology or economics.

It is important to emphasize that climate change, sustainability and environmental protection, and regeneration are not separate agendas, and to address them, necessary transformation of our education systems will be required. This, however, does not mean more education of the same kind, but a new kind of education: one that considers ethical and ecological literacy central to the learning process; one that helps learners connect these challenges of global warming to their wider, complex, inter-connected contexts and make it more meaningful to their reality. This will require a major transformation of our worldview and approach to education.

When it comes to ethical and ecological literacy and climate education, we need to consider that this is not just a matter of communicating what climate change is and that it requires urgent action, or the fact that the level of CO_2 and methane emissions are increasing in the atmosphere, and it is generating a huge negative impact on our planet (UNEP 2021). This does not mean much for most people because it does not help them see how this information is linked to their lifestyle choices and well-being. Who would visualize that the way they dispose the skin of the banana they eat or the kind of energy or electricity they use has to do with climate change? Indeed, not many people see the links of climate change with droughts, floods, and long-term food security. This requires education and awareness raising for all.

When it comes to bringing ethics to education, there is a perception by many that values education or ethical literacy is only limited to the field of philosophers and ethicists, or that it is a very abstract or theoretical field. But everybody, every day makes choices and decisions that have ethical implications. Ethical literacy is therefore necessary to help people consider the good and bad consequences of their decisions and actions and how they not only affect them, or those closest to them, but the collective, as well. From an Earth Charter perspective, it also means expanding that concern, and subsequently the responsibility for the consequences of their decisions, to the larger community of life; not only the whole human family, but the entire living and non-living world that allows life to exist and flourish. In this context, an ethically and ecologically literate person would feel concerned and responsible with the impact of what they are doing to that entire community of life in the short-, medium-, and long-term and act accordingly.

Our Worldviews

Many times, our worldviews blur our vision and perception of reality through narrow lens, preventing us from seeing the full picture or consequences of our choices, decisions, and actions, and how they impact others. We think and act according to the way we see the world, according to the stories we hear through our upbringing and through various formal, non-formal and [mostly] informal education, including the news and conversations we hear and the films and advertisement we watch. This means that our worldviews, fed by our cultural values, beliefs, and information we receive, shape the way we see, perceive, and understand the world and accordingly, how we act.

Therefore, from our current climate crises, we can say that the current predominant worldview of our societies is rooted in a myopic sense of human-nature disconnection, with short-sighted and fragmented thinking, and focused on self- or national-interest and individualism. Given this, it would be valuable to create opportunities, through education processes, to identify our worldviews, to question them, to be mindful of how they are formed, and to understand how they influence, sometimes with a lot of misperceptions, our lifestyles and choices. This would be an eye-opening process that would be of great help in our quest of moving our societies toward sustainability through an ecological civilization (Rockefeller 2008: 8; Hathaway 2009: 137).

The Earth Charter Worldview and the Ethical Dimension of Climate Change

The Earth Charter outlines a set of interrelated values and principles that can guide humanity toward a more just, sustainable, and peaceful world. It can be used to help us expand and deepen our consciousness with regards to how we ought to relate to other human beings and the larger living world. Over the past two decades, many groups and individuals have used it as an ethical reference for values clarification and orientation, a valuable instrument to deepen our understanding of the interrelated challenges humanity faces and on our shared responsibility to address them, as well as to nurture an ecological worldview.

The Earth Charter provides a clear and comprehensive vision of our interdependence and shared responsibility to address the current challenges and to move forward through cooperative action. It makes visible the underlying values and principles to consider if we are to take humanity down the path toward ecological civilization and move away from the current environment depletion and unsustainability track. Gus Speth, the founder and former president of the World

Resources Institute and former Administrator of the United Nations Development Programme (UNDP), refers to the Earth Charter "as the most sophisticated and participatory effort to frame values and principles for a sustainable future" (Speth 2004: 191–193).

From time to time, when it comes to conversations around climate change, some people make note that the Earth Charter does not explicitly mention climate change and therefore is not relevant in efforts to address it. However, anyone who looks closely into its text would see that it highlights not only the fundamental causes leading to climate change, but also core ideas and values necessary to address it. Therefore, the fact that it does not explicitly mention the words 'climate change' does not mean that it does not address it. We just need to clean our lenses a bit, to see that this idea is there, in the surface and depth of it.

Unpacking the Earth Charter Preamble

The Preamble of the Earth Charter is a great source of ideas and concepts to help us begin to think about the root causes of the current challenges humanity faces, including climate change. If we examine it carefully, it leads us to consider our ethical obligations and responsibility in addressing these challenges through our decisions and actions. In an inspirational and poetic way, the Preamble, with a tone of hope, despair, and emergency, lays out the global situation with its current challenges and possibilities and makes a call for action. It fulfills the purpose to spark our capacity to aspire and visualize a better world, while being conscious of the seriousness of the current situation of devastation.

For example, the second paragraph of the Preamble highlights the importance to preserve a healthy biosphere, essential for the community of life and the well-being of humanity and considers the protection of Earth's vitality a sacred trust. These understandings are clearly linked to the ethical dimension for addressing climate change. The third paragraph, in particular, makes a clear point on one of the most important root causes of the human contribution to climate change: the unsustainable patterns of production and consumption. It starts by highlighting the fact that "The dominant patterns of production and consumption are causing environmental devastation, the depletion of resources, and a massive extinction of species" (Earth Charter 2000). Toward the end of this paragraph, it highlights, with a sense of emergence, the sad fact that "The foundations of global security are threatened" (Earth Charter 2000).

The fourth paragraph then affirms that "fundamental changes are needed in our values, institutions, and ways of living" (Earth Charter 2000). This is an imperative if we are to address climate change because the predominant values and ways of living are taking us toward and over a cliff. By stating that 'the choice is ours,' it puts the decision on changing the course of things and where we are heading in our hands. Indeed, the Preamble opens by highlighting that we live in a time that we must *choose* our future. It is our responsibility to choose the way forward and to do so with urgency.

A core ethical idea that is articulated throughout the Earth Charter is the sense of universal and shared responsibility with the well-being of not only humans of present and future generations, but also the larger living world. This idea can be found in the fifth paragraph of the Preamble on Universal Responsibility, which starts with a call for action by stating that, "we must decide to live with a sense of universal responsibility, identifying ourselves with the whole Earth community as well as our local communities" (Earth Charter 2000). This exercise involves expanding our understanding and sensitivity to the fact that we are part of the community of life and that this belonging carries with it rights, responsibilities, and obligations. This idea is further clarified in Principle 2(b) which cultivates the consciousness that "with increased freedom, knowledge, and power comes increased responsibility" (Earth Charter 2000) to ensure the common good of all.

This sense of shared responsibility is a fundamental ethical principle for addressing climate change and one which can be found in the UNFCCC as "common but differentiated responsibilities," and note that it says "humankind" and not "only the citizens of your nation,"

> The Parties should protect the climate system for the benefit of present and future generations of humankind, on the basis of equity and in accordance with their common but differentiated responsibilities and respective capabilities.
>
> *(1992: Article 3.1.)*

It is also found in Principle 7 of the 1992 Rio Declaration on Environment and Development (Rio Declaration) when is affirms that,

> States shall cooperate in a spirit of global partnership to conserve, protect and restore the health and integrity of the Earth's ecosystem. In view of the different contributions to global environmental degradation, States have common but differentiated responsibilities. The developed countries acknowledge the responsibility that they bear in the international pursuit to sustainable development in view of the pressures their societies place on the global environment and of the technologies and financial resources they command.

From this perspective, what our educational experience needs to cultivate is the understanding that it would be unethical to generate problems for others, and particularly, without taking any responsibility to address it. Similarly, it would be unethical not to collaborate with others in efforts to adapt to and mitigate climate harms, particularly with those being affected and those unable to respond, if you have the power, knowledge, and capacity to do so. Ethical reflection such as this is what needs to happen. Closing our eyes, turning our backs, or being indifferent simply does not belong in the realm of ethics.

Earth Charter Principles Related to Climate Change

The first part of the Earth Charter, Respect and Care for the Community of Life, communicates the core ideas of the worldview that are articulated in the Charter, which lies in a life-centric and eco-centric approach. It intentionally orients us to think that our ethical obligation to respect and care is not limited to humans, but must be expanded to the entire community of life. This offers the foundations on which the rest of the principles build upon. Principle 2 and 4, and their supporting principles, offer special orientation to the climate change agenda. Principle 2 highlights our duties and responsibilities, such as the duty to prevent environmental harm and the responsibility for the common good, which again is expanded to the entire community of life. Principle 4(a) expands this sense of ethical responsibility with and to future generations as follows, "Principle 4. Secure Earth's bounty and beauty for present and future generations. (a) Recognize that the freedom of action of each generation is qualified by the needs of future generations" (Earth Charter 2000). The needs and rights of future generations require us to act with responsibility and resolve. Most people probably do not see, or realize, that their choices and decisions have an impact on the well-being of future generations, but it is an important ethical dimension that needs to be cultivated through education processes.

In its entirety, Pillar II on Ecological Integrity, with its four main principles and 20 supporting principles, are directly related to ways of addressing climate change or to prevent it. Namely, Principle 5, "Protect and restore the integrity of Earth's ecological systems, with special concern for biological diversity and the *natural processes that sustain life*" (Earth Charter 2000, emphasis added). Ecosystem conservation and restoration has a huge potential in mitigating climate change, and it contributes to the well-being of people. It also contributes to the achievement of several of the 2030 Sustainable Development Goals (SDG) targets including SDG 13 on Climate Action ("Take urgent action to combat climate change and its impacts") and SDG 15 on Life on Land ("Protect, restore and promote sustainable use of terrestrial ecosystems, sustainably manage forests, combat desertification, and halt and reverse land degradation and halt biodiversity loss") (UNSDG 2015). What seems to be more difficult for people to see is that this effort also contributes to SDG 1 on No Poverty ("End poverty in all its forms everywhere"), SDG 2 on Zero Hunger ("End hunger, achieve food security and improved nutrition and promote sustainable agriculture"), and SDG 3 on Good Health and Well-being ("Ensure healthy lives and promote well-being for all at all ages"), as well as to the economy.

Earth Charter Principles 5(e) and 5(f) speak specifically to climate adaptation and mitigation:

> 5(e) Manage the use of renewable resources such as water, soil, *forest products*, and marine life *in ways that do not exceed rates of regeneration and that protect the health of ecosystems*; and
> 5(f) Manage the extraction and use of non-renewable resources such as minerals and fossil fuels in ways that minimize depletion and cause no serious environmental damage.
>
> *(2000, emphasis added)*

We know that globally, electricity generation from burning coal, oil, and natural gas is the single largest source of global emissions increase and, therefore, is a major human influence on climate change. However, what most people do not see is that this has to do with our daily life, especially those who are part of the urban affluent society, given their greater dependency on, and use of, energy for nearly every product that is consumed, including food. Thus, our responsibility to see the global warming scenario is reflected in our dinner, and even more-so, on the dinner tables and in the homes of the wealthy. We just need to enhance our capacity to see this with the help of more ecological literacy.

Principle 7 of the Earth Charter and its supporting principles, underscore our ethical responsibility to look at our patterns of production and consumption and reorient it to move away from the path that is contributing to global warming and affecting the environment and the well-being of others. It states,

> "Principle 7. Adopt patterns of production, consumption, and reproduction that safeguard Earth's regenerative capacities, human rights, and community well-being.
>
> a Reduce, reuse, and recycle the materials used in production and consumption systems, and ensure that residual waste can be assimilated by ecological systems.
> b Act with restraint and efficiency when using energy and rely increasingly on renewable energy sources such as solar and wind.
> c Promote the development, adoption, and equitable transfer of environmentally sound technologies
> d Internalize the full environmental and social costs of goods and services in the selling price and enable consumers to identify products that meet the highest social and environmental standards.
> e Ensure universal access to health care that fosters reproductive health and responsible reproduction.
> f Adopt lifestyles that emphasize the quality of life and material sufficiency in a finite world" (2000).

Among others, this also relates to Principle 8 of the Rio Declaration,

> To achieve sustainable development and a higher quality of life for all people, States should reduce and eliminate unsustainable patterns of production and consumption and promote appropriate demographic policies.
>
> *(1992)*

And to SDG 12 on Responsible Consumption and Production ("Ensure sustainable consumption and production patterns"). As mentioned above, not many people realize that climate change action directly relates to their lifestyle choices, or they fail to see that the linkages between everything they consume (and produce) has an impact in the environment before and after its use.

The study on 'Natural Climate Solutions' highlights that global warming will likely be kept below 2°C if we avoid increases in fossil fuel emissions and if natural pathways such as conservation and restoration of land and ecosystems are implemented. But it particularly emphasizes the important contributions of NCS in the near future,

> Given the magnitude of fossil fuel emissions reductions required under any <2 °C scenario, and the risk of relying heavily on negative emissions technologies (NETs) that remain decades from maturity, immediate action on NCS should not delay action on fossil fuel emissions reductions or investments in NETs.
>
> *(Griscom 2017)*

Hence, the significance not only of Principle 5, 5(a), (b), (e), and (f) of the Earth Charter, but the entirety of Pillar II on Ecological Integrity is devoted to this undertaking (i.e., environmental conservation and rehabilitation, forest conservation and reforestation, and the protection of Earth's life support systems).

Principle 6 of the Earth Charter has a lot to do with the decisions we make at all levels, the importance to prevent harm, and the duty to act with precaution. It also articulates and advances the Precautionary Principle, while Principle 6(b) is related to the Polluter Pays Principles,

> "Principle 6. Prevent harm as the best method of environmental protection and, when knowledge is limited, apply a precautionary approach.
>
> a Take action to avoid the possibility of serious or irreversible environmental harm even when scientific knowledge is incomplete or inconclusive.
> b Place the burden of proof on those who argue that a proposed activity will not cause significant harm and make the responsible parties liable for environmental harm.
> c Ensure that decision making addresses the cumulative, long-term, indirect, long distance, and global consequences of human activities.
> d Prevent pollution of any part of the environment and allow no build-up of radioactive, toxic, or other hazardous substances.
> e Avoid military activities damaging to the environment" (Earth Charter 2000).

The Precautionary Principle can also be found in Article 3 of the UNFCCC,

> The Parties should take precautionary measures to anticipate, prevent or minimize the causes of climate change and mitigate its adverse effects. Where there are threats of serious or irreversible damage, lack of full scientific certainty should not be used as a reason for postponing such measures, taking into account that policies and measures to deal with climate change should be cost-effective so as to ensure global benefits at the lowest possible cost…
>
> *(1992)*

Principle 15 of the Rio Declaration,

> In order to protect the environment, the precautionary approach shall be widely applied by States according to their capabilities. Where there are threats of serious or irreversible damage, lack of full scientific certainty shall not be used as a reason for postponing cost-effective measures to prevent environmental degradation.
>
> *(1992)*

As well as its Principle 13,

> States shall develop national law regarding liability and compensation for the victims of pollution and other environmental damage.
>
> *(1992)*

The key point for our concern is that most people, especially decision makers in the public and private sector, are generally not aware of these principles and concepts, and their relevance to current times, let alone on how to translate them into action.

Earth Charter Principle 8 articulates the importance of knowledge exchange. Its supporting principle (a) affirms, "Support international scientific and technical cooperation on sustainability, with special attention to the needs of developing nations" (2000). Given the impact of global warming across borders, and the fact that no matter where we are, we will all be affected, (although in different ways and in varying levels of harm), information exchange, knowledge sharing, and scientific and technical cooperation and sharing is crucial. This ought to be regarded as an ethical obligation to the greatest extent possible.

This is also in line with Article 4 of the UNFCCC, where it states that all Parties shall,

> g Promote and cooperate in scientific, technological, technical, socio-economic and other research, systematic observation and development of data archives related to the climate system and intended to further the understanding and to reduce or eliminate the remaining uncertainties regarding the causes, effects, magnitude and timing of climate change and the economic and social consequences of various response strategies
>
> h Promote and cooperate in the full, open and prompt exchange of relevant scientific, technological, technical, socio-economic and legal information related to the climate system and climate change, and to the economic and social consequences of various response strategies.
>
> *(1992)*

Justice, democracy, and human rights (including the human right to a healthy environment) also demand that we make a concerted effort to take our ethical responsibility seriously and act collaboratively to address climate change. It is the responsibility of individuals, institutions, and governments to collaborate to ensure present and future generations will have access to a healthy environment; indeed, it is a prerequisite for a just and democratic governance system. Along these lines, the Earth Charter Principle 9(a) highlights as an ethical imperative the importance to "Guarantee the right to potable water, clean air, food security, uncontaminated soil, shelter, and safe sanitation, allocating the national and international resources required" (2000).

This inquiry into our ethical responsibility, guided by the Earth Charter and other key documents in global governance, helps us discern when we should or should not move ahead with a project or an action, as well as be mindful of the consequences and impact of our decisions and actions and our common but differentiated responsibilities. This notion needs to be brought to the processes of preparing future decision makers. The above-mentioned aspirations will more likely be realized and translated into practice once they are more strongly incorporate into education processes at all levels.

The Way Forward: Making the Invisible, Visible

The Earth Charter closes with The Way Forward,

> Life often involves tensions between important values. This can mean difficult choices. However, we must find ways to harmonize diversity with unity, the exercise of freedom with the common good, short-term objectives with long-term goals.
>
> *(2000)*

In order to address the kind of global and interdependent problems we face with the climate change emergency, good governance and collaboration at all levels is required. The "partnership of government, civil society, and business is essential for effective governance," and for that to happen, it is fundamental to count with shared vision and values on how to move forward (Earth Charter 2000).

The ethical dimensions are often ignored in climate change discourse because it is considered for many as not practical, besides most climate policy controversies have been articulated almost exclusively in scientific or economic language or because of the strong narrow-minded approach of national interest, which needs to be transcended. In the midst of turbulent waters and foggy contexts of the planetary challenges humanity face, decisions many times face ethical dilemmas. We need a beacon of light, reference points or a navigating instrument to guide decisions and help make the invisible visible. The Earth Charter can be used as that guide toward planetary consciousness and ethics.

Part of ethical inquiry is often to look at things from another angle, which can help us see things more clearly and better understand why:

- It is unethical to use renewable resources, such as water, soil, forests, and marine life, in a wasteful manner, or in ways that exceed their capacity to regenerate;
- It is unethical to use non-renewable resources, such as mineral and fossil fuels, with indifference to their depletion and consequences;
- It is unethical, nor logical, nor rational to consume or produce in an unsustainable manner.
- it is unethical to pollute a river, a watershed, or the ocean;
- it is unethical to act and live in a way that will take away the rights and opportunities of future generations to live in a healthy environment;
- It is unethical to make decisions disregarding their consequences in the short or long term; and
- It is unethical to make a decision that would only address one's own interest and negatively affect the collective.

According to a study on "What makes an effective leader?" to which 195 leaders in 15 countries, were asked to rate the most important competency they expect of a leader, 67 percent of the respondents selected "high ethical and moral standards" as the most important attribute of a leader (Giles 2016). Although this was the most rated attribute, educational institutions do not take the task of enhancing the ability of professionals and officials in the public or private sphere to be ethical leaders. This only reinforces the argument for the importance of ethical literacy, and in the case of global challenges such as climate change, an ethical and ecological literacy to nurture planetary consciousness and responsibility toward the well-being of the entire Earth community.

Although most of the principles articulated in the Earth Charter, or a variation of them, are already present in many international policies and laws, the reality is that there is still a significant lack of awareness about them and understanding of their meaning. Even more so in how to bring them to practice and decisions. Thus, although they can be found in texts, they are not reflected nor embedded in the dominant cultures, worldviews, and decisions. We cannot expect to have an ethical and ecologically conscious society without ethical and ecological literacy. Thus, to change the course of things, ethical and ecological literacy must be made central to education processes at all levels and contexts if we are serious about confronting our climate crises.

Note

1 Mr. Magoo is an old cartoon character who generates problems for others due to his extreme near-sightedness, but who always gets away with it.

References

Earth Charter Commission. (2000). *The Earth Charter.* Available at www.earthcharter.org

Earth Charter International. (2012). *Earth Charter and International Law: How Does the Earth Charter Correlate with International Law?* Available at https://earthcharter.org/the-earth-charter-and-international-law/

Giles, S. (2016) "The Most Important Leadership Competencies, According to Leaders Around the World." *Harvard Business Review.* March 15, 2016 Available at https://hbr.org/2016/03/the-most-important-leadership-competencies-according-to-leaders-around-the-world?utm_medium=social&utm_campaign=hbr&utm_source=facebook&tpcc=orgsocial_edit&fbclid=IwAR2LY_SjWhs0E3FMBVaDV5yLc5iMUcGRFC_h3WrjCDkzuiYEOO_AYdCau-M#

Griscom et al. (2017) *Natural Climate Solutions.* Published by Proceedings of the National Academy of Sciences of the United States of America (PNAS). Available at https://www.pnas.org/content/114/44/11645

Hathaway, M. and B. L. (2009). *The Tao of Liberation: Exploring the Ecology of Transformation.* New York: Orbis Books.

Rockefeller, S. (2008) "Crafting Principles for the Earth Charter." In Corcoran, P. B. & Wohlpart, A. J. (eds), *A Voice for Earth: American Writers Respond to the Earth Charter.* Athens: University of Georgia Press.

Speth, G. (2004). *Red Sky at Morning: America and the Crisis of the Global Environment.* New Haven: Yale University Press.

UNESCO. (2015) *Experts Meeting on Climate Change Education for Sustainable Development in Latin America and the Caribbean – Report and Recommendations* – May 2015.

United Nations Environment Programme and Climate and Clean Air Coalition – UNEP. (2021). *Global Methane Assessment: Benefits and Costs of Mitigating Methane Emissions.* Nairobi: United Nations Environment Programme.

United Nations Framework Convention on Climate Change (UNFCCC). (1992). Available at https://unfccc.int/process-and-meetings/the-convention/history-of-the-convention/convention-documents

United Nations Framework Convention on Climate Change (Paris Agreement) (2015). Available at unfccc.int/resource/docs/2015/cop21/eng/l09r01.pdf

United Nations Rio Declaration on Environment and Development (Rio Declaration). (1992). Available at https://www.un.org/en/development/desa/population/migration/generalassembly/docs/globalcompact/A_CONF.151_26_Vol.I_Declaration.pdf

United Nations Sustainable Development Goals (UNSDG). (2015). Available at https://documents-dds-ny.un.org/doc/UNDOC/GEN/N15/291/89/PDF/N1529189.pdf?OpenElement

34
FARMING KNOWLEDGE
Epistemic Injustice and Climate Information Services

Dina Lupin

Introduction

Public participation and access to climate information have long been recognized as important principles and practices in climate change policy and law. Article 6(a) of the United Nations Framework Convention on Climate Change (UNFCCC) calls for state parties to promote and facilitate "(ii) public access to information on climate change and its effects" and "(iii) public participation in addressing climate change and its effects and developing adequate responses."

Public participation plays important roles in climate change decision-making. It ensures decision-makers have access to relevant information and is often seen as crucial to the legitimacy and public acceptance of policy decisions (Dilling & Lemos 2011). In addition to the legal and political role of participation in decision-making processes, there is also an ethical dimension to participation. If a person is denied a say in matters that affects them, their agency and dignity are also denied.[1]

While the exclusion of affected parties from participation constitutes a legal and ethical injustice, certain kinds of inclusions in participatory processes can also amount to an injustice (Townsend & Townsend 2020). Specifically, certain engagement practices can expose participants to practices of *epistemic injustice*. Epistemic injustices, which arise when a person is wronged in their capacity as a knower (Fricker 2007), are often hard to detect, particularly when they happen in the context of an inclusive and participatory process. However, epistemic injustices may render a participation process unjust, undermining the legitimacy of decisions and denying people recognition of their agency.

This chapter examines practices of epistemic injustice in the context of climate change information services ('climate information services') provided to farming communities in Africa. Climate information services aim to ensure that people have access to the information they need in order to anticipate, respond, and adapt to the impacts of climate change (Furman et al. 2014). With their focus on climate knowledge in decision-making and accessible information, climate information services are only effective when they respond to the needs and contexts of their users. This chapter explores the sometimes-hidden ways in which farmers' inclusion in decision-making processes exposes them to practices of epistemic objectification and exploitation. This both denies the epistemic agency and authority of farmers and prevents their meaningful participation in decisions about the need, design, and content of climate information services programs that are meant to benefit them.

The chapter begins with a brief introduction to epistemic injustice and climate information services. It then turns to an assessment of participatory processes in climate information

DOI: 10.4324/9781003039860-40

services in the African context. Last, it considers some strategies that may be helpful in moving toward ethical participation practices that ensure farmers are heard, believed, and included in decision-making processes.

A Brief Introduction to Epistemic Injustice

Epistemic injustice captures the idea that certain practices of unfair discrimination mean that a person is wronged in their capacity as a knower. This occurs when someone is not recognized as having particular knowledge, or their reasonable beliefs are dismissed or ignored because of a prejudicial stereotype about their social identity. It also occurs when members of certain social groups are excluded from important epistemic communities or activities. The term, epistemic injustice, was originally coined by the feminist philosopher, Miranda Fricker (2007); however, the idea has been explored by a number of philosophers and theorists who have identified a wide array of epistemically unjust practices (Pohlhaus 2017).

Examples of epistemic injustice are sometimes found in practices of exclusion. For example, social prejudices may mean that certain individuals or groups are excluded from participating in various kinds of knowledge practices or are denied the opportunity to know what it is in their interests to know (Pohlhaus 2017). Beliefs about gendered roles or abilities might mean that girls are denied access to education or rural communities are denied information about planned mining activities (Goldman & O'Connor 2019). Patricia Collins has pointed out that the prejudicial belief that black women are only good for domestic labor means their views and insights have not been sought outside of that domain (Collins 2002).

While these kinds of exclusions mean that members of certain social groups do not get to participate and are denied a voice in certain matters, practices of epistemic injustice are also found when those facing unfair discrimination do get to speak. This may happen when a hearer fails to assign certain speakers the credibility that they deserve with the result that their speech is not believed or taken seriously.[2] For example, racial or gender bias may mean that police officers fail to believe and act on information provided by black and female witnesses or victims of crimes. In these cases, the speaker is not denied access to knowledge but rather they are wronged in their capacity as a knowledge-giver. Fricker argues,

> The capacity to give knowledge to others is one side of that many-sided capacity so significant in human beings: namely the capacity for reason. So wronging someone as a giver of knowledge … amounts to wronging that person as a knower, as a reasoner, and thus as a human being.
> *(Fricker 2007: 44)*

Paying attention to epistemic injustice in the context of climate information services (and in climate change decision-making more broadly) highlights the epistemological harms associated with the exclusion of certain groups from participating in knowledge production and decision-making processes. More importantly for purposes of this chapter, it also draws attention to the epistemological harms that can arise in participation processes when people are included. In the context of participation in climate information services, one finds practices of testimonial injustice, objectification, and exploitation.

A Brief Introduction to Climate Information Services

Farmers have always had to adapt to uncertainty, to shifting weather and changing conditions, to unforeseen threats and unpredictable risks (Vogel et al. 2019). For many farmers, however, climate change has meant unprecedented change and uncertainty (Coulibaly et al. 2015;

Tall et al. 2014). The start and end of seasons are shifting, rainfall patterns have become increasingly hard to predict, and the crops that have been grown for generations no longer thrive (Intergovernmental Panel on Climate Change [IPCC] 2014). As a result, many farmers are finding that generational knowledge and traditional methods of prediction are no longer sufficient. They are having to make decisions about what to do and what to grow in increasingly erratic conditions (Vogel et al. 2019).

Policy-makers have argued that adaptation in these circumstances is hindered by what they see as an information gap (Vincent et al. 2020). While scientists and forecasters are able to gather and analyze relevant meteorological data and generate more accurate predictions, this information is not always generated where it is needed and is not necessarily accessible to or useable by those who need it to make decisions about their farming practices (Georgeson et al. 2017). Many rural farming communities may lack either the means to access this information (through a lack of access to media or internet) or the information may not be accessible in a manner that allows for its application in daily decision-making (Hansen et al. 2019; Tall et al. 2014).

This idea that there is a gap between scientists (information generators) and farmers (among other decision-makers) has led to greater attention to and demand for climate information services. Climate information services are aimed at transforming climate and meteorological data and information, gathered through scientific research, into packaged information products (such as forecasts, trends and predictions) that enable different kinds of decision-makers to make sustainable and climate-responsive decisions (Leal Filho 2020). They include early warning systems, but also sustained weather information provision for long-term planning.

Climate Law, Policy, and Programs in Africa

Although climate information services have a long history, the last decade has seen a significant increase in interest and funding for these services, especially since the adoption of the UN Global Framework for Climate Services in 2009 (Hewitt et al. 2012; Tall et al. 2018). Climate information services are seen to be particularly important for countries in the Global South and have been a central component of several regional policies, institutions, and programs adopted by the African Union. These include the African Ministerial Conference on Meteorology, the African Adaptation Initiative, and the Integrated African Strategy on Meteorology.

A range of programs have been launched across the continent in the past decade that seek to either address the lack of infrastructure and expertise in relation to weather monitoring technology or focus on agricultural users of climate information and on what is needed to ensure their access to usable information. These include the Regional Climate Outlook Forum of the Consortium of International Agricultural Research Centers (CGIAR) funded by USAID, the United Nations Development Programme's project on Climate Information for Resilient Development in Africa (CIRDA), the World Meteorological Organization's Africa programs under the Global Framework for Climate Services, the UKAid-funded Weather and Climate Information Services for Africa (WISER), and the ClimDevAfrica program funded by the African Development Bank, among others.

Climate information services are targeted at a wide array of decision-makers, including law and policy-makers at the national and international level. These services can be part of a mitigation approach, informing decision-making on GHG-reduction plans, and they can be an adaptation measure, aimed at assisting those facing the effects of climate change to adapt to new conditions or avoid new threats. This chapter focuses on climate information services as an adaptation measure meant to better equip farmers with the information they need to anticipate and adapt to shifting weather patterns and coming events (Coulibaly et al. 2015).

Epistemic Injustice and Participation in Climate Information Services

The goal of climate information services is to generate useful and useable information for farmers and other decision-makers. Identifying and understanding the needs of those farmers and decision-makers is a necessary part of any successful climate information services program. This is by no means a simple exercise. Vincent et al have found:

> The process of identifying what information will be useful to the intended users is rarely simple or neutral. ...Heterogeneity amongst users and different needs of different people mean that wider inclusion is essential to avoid reinforcing power imbalances. ... In addition, needs are not static, but rather change over time. ... This process tends to require extensive engagement and iteration with potential users...
>
> *(Vincent et al. 2020: 2)*

Information services that do not take into account the contexts and conditions in which farmers live and work are unlikely to deliver usable services (Vogel et al. 2019). The recognition that there is no model 'user' has meant a growing focus on the inclusion and participation of targeted beneficiaries in the design of climate information services. While ensuring the engagement of affected parties in processes associated with climate information services is critical to avoiding the injustice of exclusion and non-participatory decision-making, efforts at inclusion can also expose participants to injustices, and specifically epistemic injustices.

Examining practices of engagement in climate information services suggests three ways in which inclusion can expose farming communities to epistemic injustices. First, farmers might be included in the planning of climate information services, but only as a consumer of the information and not as a producer. Second, farming communities might be included as producers (not only as users or consumers of information), but be objectified in the sense that they are treated as mere sources of information and not as informants or full epistemic agents (Craig 1990). Third, participants might be included as full epistemic agents, but without compensation or control over the manner and use of the knowledge they contribute.

Inclusion but as a User of Knowledge not as a Producer

The first way in which farmers are exposed to epistemically unjust practices is when they are treated merely as users, and not as producers, of knowledge. The primary goal of climate information services is often described as delivering "information that is salient, legitimate, and credible to end users" (Carr et al. 2016: 247). In delivering information and bridging the gap between scientists and farmers, however, climate information services all too often seem to envision a vertical path of information sharing, one in which knowledge authorities (scientists and information brokers) hand down information to information consumers or 'end users' (farmers and other decision-makers).

Farmers, particularly small-scale and subsistence farmers, are seen as highly vulnerable to climate change impacts and extreme weather events since their lives and livelihoods directly depend on the food that they grow and on bringing their produce to market (Vogel et al. 2019). The Foreword to the Integrated African Strategy on Meteorology states

> With agriculture being the mainstay of most of Africa's economy, of which over sixty percent of our citizens directly depend on for their day-to-day living; our farmers, who mainly rely on rainfall, are now confused about what and when to plant as climate change and variability impact upon the effectiveness of the traditional agricultural and crop-weather calendars.
>
> *(African Ministerial Conference on Meteorology [AMCOMET] 2012: 1)*

The idea of farmers as 'confused,' as lacking information and requiring the assistance of scientists and experts to continue farming in changing conditions, has resulted in an empty vessel approach to climate information services – farmers are seen as vessels to be filled with the information they need.

As Vogel et al. (2019) put it:

> The departure points for engagements are still tethered to the 'scientist' as expert who is 'communicating' and trying to embed climate information and knowledge into the daily realities of people. The brokerage of knowledge thus becomes the driving feature of the engagement process with a focus on communicating science that is then unfurled through value-chain type of approach to climate services.
>
> *(8)*

The focus is not on collaboration, partnership, or inclusion, but rather on the transmission of expertise from the knowers to the users.

Casting farmers into the role of passive recipients of information, rather than as information generators or as problem-solvers, means that they are only included in the process at an extremely late stage of decision-making. High-level decisions to pursue climate information services, to fund climate service programs, and to invest in the development of meteorological stations and scientific expertise rarely include the communities that are targeted as service users (Vincent et al. 2018). Users are only included once high-level decisions are already finalized and funded.

In addition, when seen only as users, farmers are often only provided with limited information. The efficient rolling out of climate information services then means only providing users with the information that a policy-maker, funder, or scientist (often far-removed from the farming community) has decided they need for the purposes of their climate-related decision-making (McOmber et al. 2013: 8). Information that is not seen as necessary to a farmer's decision-making is thought to hinder the ability of farmers to adapt to changes in the climate. Since information relating to, for example, international agreements or to adaptation funding arrangements are not directly necessary for making decisions about what to plant and when to harvest, this is information that is not made available to farming communities (Ciplet et al. 2012).

Vogel and O'Brien (2006) argue that the expectation that Africans will simply consume information because it has been given to them by experts abroad continues a long-standing practice of Africans being treated as the passive recipients of northern goods, aid, and information. Climate information services programs in African contexts (and in the global South more broadly) are often funded and supported by institutions located in the global North and are informed by a kind of epistemic paternalism in which policy-makers, scientists, program designers, and information brokers select and package information that they have determined is necessary for end users.

The act of only giving end users information that they are perceived to need to make better informed decisions about their farming activities suggests an assumption that farming communities have no *need* for information on, for example, the broader political, social, and scientific causes and responses to climate change. In addition, it suggests an assumption that there is no political, social, or scientific benefit in including farming communities in decision-making processes, discussions, and debates that extend beyond their own farming practices.

The view that these communities are merely end-users of information who do not need to be given information on or included in wider climate change issues and debated amounts to what Fricker refers to as a pre-emptive form of epistemic injustice. Fricker finds, "[t]he credibility of … a person on a given subject matter is already sufficiently in prejudicial deficit that their potential testimony is never solicited; so the speaker is silenced by the identity prejudice that undermines her credibility in advance" (Fricker 2007: 130). If policy-makers, funders, and scientists believed that providing farming communities with information so that they could engage in high-level debates would result in important and novel insights or options for responding to climate change, their insights would likely be sought.

Limiting access to information and participation in this way denies affected groups and individuals the information that is pertinent to their experiences and conditions and denies them the opportunity to consider and come up with other responses. Climate information services also appear to be a *fait accompli*, the only solution discussed or made available.

Inclusion as a Producer but as a Source of Information and Not as an Informant

A second way in which farmers might be exposed to practices of epistemic injustice is through their inclusion in participation processes, but merely as sources of information rather than as full agents or informants. In 2017, the World Food Programme and the Research Programme on Climate Change, Agriculture, and Food Security (CCAFS) published a report on a multi-year climate information services program carried out in Malawi and Tanzania. While there is no indication in the report that targeted beneficiary communities were included in the original decisions to conduct, fund, or design the program, they did participate in a base-line study. This suggests that these communities were not just seen as users of information, but in their participation in the study, also produced and contributed to the production of the project. The base-line study, however, was not aimed at assessing climate information services or national climate change policy, but rather at assessing the communities themselves. Participants were asked questions about their assets, their productivity, their sources of information, and how that information affected their decisions about their crops and livestock.

When farmers and other members of their communities are only engaged in a process in this way, they are not treated as full participants with ideas and knowledge to share. Rather, they are treated as a source of information about their living conditions. Fricker refers to this as "epistemic objectification" (Fricker 2007). Epistemic objectification happens when a person is treated as a mere source of information rather than as a "subject with a common purpose" and a full participant in an epistemic practice (Craig 1990, quoted in Fricker 1998: 56). When one is merely "objectified and observed," one is not treated as an active and credible participant (Code 2008). Rather, one is treated as a "passive state of affairs from which knowledge might be gleaned" (Fricker 2007: 132).

Of course, gathering information about a target community is critical to ensuring climate information services programs that are well-suited to the recipients (Gorman-Murray, McKinnon, & Dominey-Howes 2014). The simple fact that farmers might be involved in a base-line study does not, in and of itself, necessarily result in epistemic objectification. Involving farmers in a baseline study amounts to unjust treatment when this is the only or primary way in which they participate. In other words, when farmers are merely treated as sources of information and not also as agents and as active participants in climate decision-making, they are epistemically objectified (Tuvel 2015).

Inclusion as Informants but Without Compensation or Control over the Manner, Use or Uptake of Knowledge Production

The third practice of epistemic injustice that farmers may face, arises when farmers are included and recognized as informants, with knowledge to share, but are denied any control over the uptake of their knowledge and are not compensated for their contribution. In recent years, growing attention has been paid to the idea that climate information services should not be bestowed on farming communities by external actors, but rather that these services should be 'co-produced' (Bremer et al. 2019; Carter et al. 2019). Co-production has been defined as "bringing together different knowledge sources and experiences to jointly develop new and combined knowledge which is better able to support specific decision-making contexts" (Kniveton et al. 2016).[3] In the context of climate information services, co-production seems to offer a participatory alternative to top-down models of decision-making and planning that treats farmers as either mere consumers of information or as sources of information (Carter et al. 2019).

The idea of co-production has meant a shift in the approach of some climate service providers. For example, the *Manual for Co-production in African Weather and Climate Services (Manual)* (2019) reports

> In order to better understand the needs of the climate service users, approaches to climate services have been strengthening interaction with users, particularly in terms of incorporating users' expertise into the development and/or communication of climate information products in a user-producer dialogue process. This dialogue process often results in what is commonly referred to as 'co-production.'
>
> *(13)*

A co-production approach appears to address the concern of epistemic objectification of farmers in a very direct manner. It is an approach that seems to recognize the epistemic authority and agency of farmers and their communities and treats them as part of an epistemic community that collectively produces the services (Vincent et al. 2018).

While it has been argued that co-production "requires a different approach to knowledge production and reconfiguration of the hegemony of science" (Vincent et al. 2018: 49), the *Manual* seems to pre-empt the kinds of roles participants can play by maintaining a division between producers and users. The *Manual* states that citizens, "particularly those … directly impacted by climate-related risks, can provide context to help frame the approach as well as feedback on the suitability of possible solutions" and customary leaders and women's representatives "play key roles in developing trustworthy, contextualized and locally co-owned climate services that respect and blend with local and indigenous knowledge" (Carter et al. 2019: 21). In other words, citizens provide context to frame the approach, while leaders help ensure legitimacy. Neither of these are central roles concerned with the overall approach, conceptualization, or design of climate solutions or climate information services.

While farming communities might be engaged in a co-operative and inclusive co-production process, what they are co-producing (climate information services) is pre-determined. Farming communities in the global South often face a wide array of intersecting and interconnecting problems (including poverty, lack of access to basic services, social prejudices, and so on) that are exacerbated by climate change (Babugura et al. 2010; Furman et al. 2014). There are many possible problems and an array of political, social, technical, and financial responses to these problems. The decision to focus on the problem of access to climate and weather information, and on climate information services as the solution to that problem, represents a series of decisions already taken by policy-makers and funders. The fact that farmers are only brought in to co-produce a solution once the key decisions are made means that their participation is necessarily limited and geared toward serving the ends of someone else.

Climate information services are meant to benefit and empower farming communities so it might seem like their participation in co-production processes are in service to their own interests and not the interests of others. However, solicitations for farmer's knowledge of the co-production of climate information services need to be understood both in the broader process of decision-making and in the wider social and political context.

Climate information service programs are often targeted at farmers in the global South but are funded and organized by institutions in the global North. Gayatri Spivak (as cited in Tuvel 2015: 325) has argued that "efforts to gain third-world peoples' perspectives or to 'hear their voices' often serve the imperialist and globalizing projects of first-world nations." Solicitations of the knowledge of those living in the global South are an appropriation, "the retrieval of information from the South, not to encounter the Third World on its own terms, but for other, usually First World, purposes" (Kapoor 2004: 631).

While adaptation programs such as climate information services are often billed as being for the benefit of the world's most vulnerable and poor communities, they are in fact a solution to a problem that the world's (wealthiest) nations have created and are unable or unwilling to solve (Latin 2012; Tuvel 2015). The failure of the world's nations to take necessary mitigation action has been accompanied by growing attention to adaptation as a response to climate change.[4] Adaptation programs call on farming and other vulnerable communities to change the ways they live and work, to learn new techniques and skills, to move from their homes. The burden of responding to climate change has shifted onto those who are most vulnerable to its impacts.

Because adaptation programs like climate information services are seen to be for the good of the target farming communities, these communities are seen as beneficiaries of the services instead of experts and workers contributing to solving a global problem. Therefore, their time and expertise are often taken for granted in the co-production of those services and it is assumed they will be provided for free. For example, the UN World Food Programme was recently awarded a multi-million dollar grant for a project on adaptive capacity and climate information services for communities in the Congo River Basin (Adaptation Fund 2019). The proposal includes co-production with target communities "to allow for the identification of opportunities for blending scientific with local, indigenous knowledge" (Adaptation Fund 2019: 43). However, the detailed budget makes no provision for compensating those communities for their participation in co-production processes, for their time, or for their knowledge.

The solicitation of farmers' knowledge to further the ends of others without compensation constitutes a practice of epistemic exploitation. Epistemic exploitation "occurs when epistemic labor is coercively extracted from epistemic agents in the service of others" (Pohlhaus Jr 2017: 21). Nora Berenstain defines epistemic exploitation as "a variety of epistemic oppression marked by unrecognized, uncompensated, emotionally taxing, coerced epistemic labor" (2016: 570). Participation by farming communities in the design of climate information services is essential and should be done in a manner that recognizes the agency and knowledge of those communities. However, co-production processes that only involve farming communities in a limited capacity, for a predetermined outcome, and that fail to compensate participants for their time and knowledge, expose those communities to epistemic objectification and exploitation rather than epistemic empowerment.

Conclusion: Strategies for Creating a Just Space for Community Knowledge in Climate Information Services

While legal obligations to ensure participation might be satisfied through standard practices of consultation, an ethical problem arises when participation processes expose participants to practices of epistemic injustice. Attending to this ethical problem demands new strategies in both policy and practice. If participation, and even co-production, provides no guarantee that farming communities will be included in ways that recognize and value their epistemic agency, how can inclusion in decision-making processes happen in an epistemically just manner? This chapter now outlines some possible strategies for avoiding epistemic injustice and recognizing the agency and authority of farmers.

First, the classification of farmers as end users in climate information services risks building practices of epistemic objectification into the design and implementation of those services.

As Vogel et al. (2019) have argued, there is a need to move away from "extractive engagement techniques" that tend to be geared toward meeting funder objectives or pre-determined policy goals rather than ensuring farmers have a meaningful say in the adoption of climate solutions that suit them. Farmers should not be pigeon-holed into the role of user, but instead should occupy different roles, as both producer and user, as enquirer and informant (Medina 2013: 93). From a policy perspective, this means rethinking the ways people are classed and described but it also demands that participants are included from the earliest stages of decision-making and are equipped with the information they need to engage in key climate debates.

Practically speaking, a wide range of decisions in the context of climate policy-making at the national and domestic level have already been made and, while these processes are unlikely to be reopened to allow for the participation of farming communities, historic exclusions need to be addressed. The fact that farming communities have historically been excluded from high-level decision-making does not mean that these communities can never be engaged in an epistemically just way within the policy programs already adopted.

All participants should be made aware of the context of their participation, of the point at which they are being included, and of the processes and decisions that have come before. They should be provided with information about decisions already taken and what those decisions mean for farmers. In placing their participation in its historical context in this way, climate information services are not treated as something bestowed from above or as a *fait accompli,* but rather as the product of political and economic negotiations and trade-offs that can (and sometimes should be) questioned or resisted.

In addition, farming communities should be advised on how they can opt out of participation, and out of climate information services, without giving up the possibility of any climate adaptation assistance and funding. Despite the expectation of policy-makers or funders, communities may not be best served through the provision of climate information services. Researchers on climate information services in South Africa, for example, found that some participants did not want to talk about climate change, demanding instead that attention be given to the problem of poverty (Babugura et al. 2010: 59). Similarly, research in Senegal found that women farmers rely on men to sell their produce at the market (Carr et al. 2016). A lack of information about trading prices and economic fluctuations means that these women often did not know what their profits would be or how much money the men should return with from the market. As a result, women farmers, who have little control over what and when they farm, need economic and trading information services and not climate information services.

In both these cases, the farming communities can refuse climate information services, but in doing so they are likely refusing the only climate program targeted at their community (along with the information, training, and equipment that might have come with those services). The decision to provide climate information services – and not some other program – has already been made and, as a result, the funding has been earmarked and the relevant scientists and specialist consultants have been appointed. They are unlikely to be in a position to change direction, addressing poverty or offering economic services instead.

Policy and programs should be designed with sufficient flexibility to allow communities a way to refuse climate information services without giving up on any kind of climate assistance. Farming communities should be able to opt to work with funders and program administrators to address their needs and wants beyond the delivery of a pre-identified project.

Programs need to be especially flexible and adaptive when participatory processes reveal practices of exploitation or abuse within the targeted communities. Research into the accessibility of climate information services, for example, revealed practices of abuse and exploitation of women in some communities (Carr et al. 2016). As climate change results in more erratic weather patterns and reduced access to water, women's unpaid labor and demands on their resources will likely increase (McOmber et al. 2013).

There is a clear sense, however, in which climate information services are concerned with maintaining the status quo – they are designed to enable farmers to find ways to continue farming even in the face of rapidly changing conditions. They are not designed to provide women and other oppressed groups with support to escape abuse, nor are they geared toward providing information about alternative livelihoods or educational opportunities (McOmber et al. 2013). While outside interference into community dynamics and traditions raises additional ethical questions and concerns, participatory processes in climate information services should not be designed to ignore (or even reinforce) discriminatory practices.

The last strategy for avoiding epistemic injustice and exploitation in climate information services is, perhaps, the most important in that it is concerned with valuing the knowledge of farmers. Hügel and Davies (2020: 13) argue that there needs to be a "redistribution of expertise in decision-making structures." Farmers need to be recognized as experts on their own terms as well as privileged knowers in regard to their lives and experiences. In addition, solicitations of community knowledge must be done in a manner that is informed, inclusive, and compensated. This means recognizing the broader social service that farmers provide through their participation, as well as the particular benefits they provide to the funders, civil society organizations, and states who initiate climate information services programs. Critically, farmers must be compensated for their expertise and time. While all participants may benefit from efficient climate information services designed for the communities who use them, this benefit is not a substitute for payment for the time, labor, and expertise of farmers who co-produce those services.

Notes

1 For the purposes of this chapter, the link between dignity (or being recognized in one's humanness) and participation in public life is assumed. There are, however, a number of very different theoretical explanations and approaches that explore the connections between dignity, personhood, power, public decision-making, and rights to participation. For two very different accounts (see Freire 2018; Waldron 2012).

2 Fricker discusses this as a particular type of epistemic injustice, namely testimonial injustice (Fricker 2007).

3 It is important to clarify that the idea of co-production, as it is often used in the context of climate information services, refers to an applied method for involving users in the production side of information generation. This information is co-produced in the limited sense that scientists and users work together to produce knowledge and information products. This very applied understanding of co-production is to be distinguished from the theoretical concept as it is deployed by scholars such as Sheila Jasanoff (2004). See a discussion of the distinction between these uses of the concept in Nerlich (2015).

4 The IPCC's First Assessment Report in 1991 did not mention adaptation at all. In 2019, the World Bank announced it would be 'ramping up' its climate adaptation finance to match its investment in a low carbon future (Anon 2019).

References

Adaptation Fund. (2019) *Building Adaptive Capacity to Climate Change in Vulnerable Communities Living in the Congo River Basin*. Retrieved September 30, 2020 https://www.adaptation-fund.org/project/building-adaptive-capacity-to-climate-change-in-vulnerable-communities-living-in-the-congo-river-basin/.

African Ministerial Conference on Meteorology (AMCOMET). (2012) *Integrated African Strategy on Meteorology (Weather and Climate Services)*. Available at https://amcomet.wmo.int/sites/default/files/field/doc/pages/amcomet-integrated-african-strategy-meteorology-13677_en.pdf.

Anon. (2019) "World Bank Group Announces $50 Billion over Five Years for Climate Adaptation and Resilience." *World Bank*. Retrieved September 30, 2020 https://www.worldbank.org/en/news/press-release/2019/01/15/world-bank-group-announces-50-billion-over-five-years-for-climate-adaptation-and-resilience.

Babugura, A, N. Mtshali, and M. Mtshali. (2010) *Gender and Climate Change: South Africa Case Study*. Cape Town: Heinrich Böll Stiftung Southern.

Berenstain, N. (2016) "Epistemic Exploitation." *Ergo, an Open Access Journal of Philosophy* 3: 569–590. https://doi.org/10.3998/ergo.12405314.0003.022.

Bremer, S., A. Wardekker, S. Dessai, S. Sobolowski, R. Slaattelid, and J. van der Sluijs. (2019) "Toward a Multi-Faceted Conception of Co-Production of Climate Services." *Climate Services* 13: 42–50.

Carr, E. R., G. Fleming, and T. Kalala. (2016) "Understanding Women's Needs for Weather and Climate Information in Agrarian Settings: The Case of Ngetou Maleck, Senegal." *Weather, Climate, and Society* 8(3): 247–264.

Carter, S., A. Steynor, K. Vincent, E. Visman, and K. L. Waagsaether. (2019) "Co-Production in African Weather and Climate Services: Manual." *WISER and FCFA.*

Ciplet, D., J. Timmons Roberts, and M. Khan. (2012) "The Politics of International Climate Adaptation Funding: Justice and Divisions in the Greenhouse." *Global Environmental Politics* 13(1): 49–68.

Collins, P. (2002) *Black Feminist Thought: Knowledge, Consciousness, and the Politics of Empowerment.* London: Routledge.

Coulibaly, J. Y., J. Mango, M. Swamila, A. Tall, H. Kaur, and J. Hansen. (2015) *What Climate Services Do Farmers and Pastoralists Need in Tanzania? Baseline Study for the GFCS Adaptation Program in Africa.* Working Paper. CGIAR Research Program on Climate Change, Agriculture and Food Security (CCAFS).

Craig, E. (1990) *Knowledge and the State of Nature: An Essay in Conceptual Synthesis.* Oxford: Oxford University Press.

Dilling, L., and M. Carmen Lemos. (2011) "Creating Usable Science: Opportunities and Constraints for Climate Knowledge Use and Their Implications for Science Policy." *Global Environmental Change* 21(2): 680–689.

Freire, P. (2018) *Pedagogy of the Oppressed: 50th Anniversary Edition.* New York: Bloomsbury Publishing USA.

Fricker, M. (1998) "Rational Authority and Social Power: Towards a Truly Social Epistemology." In *Proceedings of the Aristotelian Society* 98(2): 159–177.

Fricker, M. (2007) *Epistemic Injustice: Power and the Ethics of Knowing.* Oxford: Oxford University Press.

Furman, C., C. Roncoli, W. Bartels, M. Boudreau, H. Crockett, H. Gray, and G. Hoogenboom. (2014) "Social Justice in Climate Services: Engaging African American Farmers in the American South." *Climate Risk Management* 2: 11–25.

Georgeson, L., M. Maslin, and M. Poessinouw. (2017) "Global Disparity in the Supply of Commercial Weather and Climate Information Services." *Science Advances* 3(5): e1602632.

Goldman, A., and C. O'Connor. (2019) "Social Epistemology." In *The Stanford Encyclopedia of Philosophy,* edited by E. N. Zalta. Metaphysics Research Lab, Stanford University.

Gorman-Murray, A., S. McKinnon, and D. Dominey-Howes. (2014) "Queer Domicide: LGBT Displacement and Home Loss in Natural Disaster Impact, Response, and Recovery." *Home Cultures* 11(2): 237–261.

Hansen, J. W., C. Vaughan, D. M. Kagabo, T. Dinku, E. R. Carr, J. Körner, and R. B. Zougmoré. (2019) "Climate Services Can Support African Farmers' Context-Specific Adaptation Needs at Scale." *Frontiers in Sustainable Food Systems* 3(21): 1–16.

Hewitt, C., S. Mason, and D. Walland. (2012) "The Global Framework for Climate Services." *Nature Climate Change* 2(12): 831–832.

Hügel, S., & Davies, A. R. (2020) "Public Participation, Engagement, and Climate Change Adaptation: A Review of the Research Literature." *WIREs Climate Change* 11(4), e645.

Intergovernmental Panel on Climate Change. (2014) *Fifth Assessment Report - Impacts, Adaptation and Vulnerability.*

Jasanoff, Sheila. (2004) *States of Knowledge: The Co-Production of Science and the Social Order.* Milton Park: Routledge.

Kapoor, I. (2004) "Hyper-Self-Reflexive Development? Spivak on Representing the Third World 'Other.'" *Third World Quarterly* 25(4): 627–647.

Kniveton, D., Visman, E., Daron, J., Mead, N., Venton, R. and Leathes, B. (2016) *A Practical Guide on How Weather and Climate Information Can Support Livelihood and Local Government Decision-Making: An Example from the Adaptation Consortium in Kenya.* Working draft. Exeter: Met Office.

Latin, H. A. (2012) *Climate Change Policy Failures: Why Conventional Mitigation Approaches Cannot Succeed.* Singapore: World Scientific.

Leal Filho, W. (2020) "Introducing Climate Services and Their Applications." In Walter Leal Filho and Daniela Jacob (Eds.) *Handbook of Climate Services.* Springer, pp. 3–9.

McOmber, C., A. Panikowski, S. McKune, W. Bartels, and S. Russo. (2013) *Investigating Climate Information Services through a Gendered Lens.* Working Paper. CGIAR Research Program on Climate Change, Agriculture and Food Security (CCAFS).

Medina, J. (2013) *The Epistemology of Resistance: Gender and Racial Oppression, Epistemic Injustice, and the Social Imagination*. Oxford: Oxford University Press.

Nerlich, Brigitte. (2015) "The Co-Production Confusion." *UoN Blogs*. Retrieved January 2, 2021 https://blogs.nottingham.ac.uk/makingsciencepublic/2015/03/20/the-co-production-confusion/.

Pohlhaus, G. (2017) "Varieties of Epistemic Injustice." In Ian James Kidd, José Medina, and Gaile Pohlhaus (Eds.) *The Routledge Handbook of Epistemic Injustice*. London: Routledge, pp. 13–26.

Tall, A., J. Y. Coulibaly, and M. Diop. (2018) "Do Climate Services Make a Difference? A Review of Evaluation Methodologies and Practices to Assess the Value of Climate Information Services for Farmers: Implications for Africa." *Climate Services* 11: 1–12.

Tall, A., P. M. Kristjanson, M. Chaudhury, S. McKune, and R. B. Zougmoré. (2014) *Who Gets the Information? Gender, Power and Equity Considerations in the Design of Climate Services for Farmers*. Working Paper.

Townsend, L., and D. L. Townsend. (2020) "Consultation, Consent, and the Silencing of Indigenous Communities." *Journal of Applied Philosophy* 7: 781–798.

Tuvel, R. (2015) "Sourcing Women's Ecological Knowledge: The Worry of Epistemic Objectification." *Hypatia* 30(2): 319–336.

Vincent, K., D. Conway, A. J. Dougill, J. Pardoe, E. Archer, A. Gajanan Bhave, R. Henriksson, N. Mittal, D. Mkwambisi, E. Rouhaud, and D. Tembo-Nhlema. (2020) "Re-Balancing Climate Services to Inform Climate-Resilient Planning – A Conceptual Framework and Illustrations from Sub-Saharan Africa." *Climate Risk Management* 29: 100242.

Vincent, K., M. Daly, C. Scannell, and B. Leathes. (2018) "What Can Climate Services Learn from Theory and Practice of Co-Production?" *Climate Services* 12: 48–58.

Vogel, C., and K. O'Brien. (2006) "Who Can Eat Information? Examining the Effectiveness of Seasonal Climate Forecasts and Regional Climate-Risk Management Strategies." *Climate Research* 33(1):111–122.

Vogel, C., A. Steynor, and A. Manyuchi. (2019) "Climate Services in Africa: Re-Imagining an Inclusive, Robust and Sustainable Service." *Climate Services* 15: 100107.

Waldron, J. (2012) "How Law Protects Dignity." *The Cambridge Law Journal* 71(1): 200–222.

35

STRATEGIES TO INCREASE MEDIA COVERAGE OF ETHICAL ISSUES THAT ARISE IN CLIMATE CHANGE POLICY FORMATION

Peter D. Burdon

Introduction

There is a systemic failure in the mainstream media to discuss and analyze the ethical dimensions of climate change (Brown 2020). This has been explained in various ways including the influence of the fossil fuel sector on media (Westervelt 2019) and attempts from the media to appear neutral and balanced in reporting (Muller 2020).

This chapter will extend the literature on this argument in two ways. First, it will provide a case study of the problem by examining *The Australian*, which is the only national broadsheet in Australia. The paper is also owned by the Murdoch media empire, which has vast holdings throughout the Western World. Thus, some of the pathologies that can be observed in *The Australian* will be familiar to anyone who has watched Fox News or read *The Times* in the United Kingdom. The analysis of *The Australian* will focus on the ways ethical analysis is displaced by climate skepticism and degraded forms of morality, such as the argument that Australia has a moral obligation to dig for coal. Following this, the chapter will provide a broader analysis of the reasons why the mainstream media has failed to engage in sustained ethical analysis. While noting the influence of the fossil fuel sector and the way some media organizations have become activist organizations that have no pretense at objectivity, structural arguments are prioritized in the analysis. More specifically, Edward S. Herman and Noam Chomsky's propaganda model (1988) is applied to argue that corporate advertisers play an active role in shaping the terrain of acceptable debate. Picking up on Donald Brown's (2017) concern with instrumental rationality, the chapter also examines the ways neoliberalism has shaped the way institutions think and act in increasingly economized ways.

Finally, the chapter concludes with strategies for increasing the coverage of the ethical dimensions of climate change. Because the diagnosis of the problem focuses on structures, the strategies respond in kind. Implicit in this approach is the belief that the problem needs to be addressed at its root, and simply providing more information or evidence to media organizations is insufficient. With this in mind, the role of independent (non-corporate) media should be promoted, as well as strategies for breaking up large media corporations and constructing narratives about climate change around the idea of 'public things.' This is a combination of personal activities that everyone can engage in immediately and larger political projects that will require more time and resources.

DOI: 10.4324/9781003039860-41

Case Study: *The Australian*

Western societies have an impoverished public conversation about the ethics of climate change. National studies have demonstrated that governments routinely fail to reference ethical arguments when setting national emission reduction commitments (Brown & Taylor 2015). In a similar way, major media companies routinely fail to provide editorial comment on ethical questions such as 'what responsibility do governments owe vulnerable communities?' or 'what constitutes a fair share of the global carbon budget?' The vacuum of ethical analysis is so stark that Donald Brown (2020) lampoons US media for their "total failure to cover numerous climate change issues that citizens need to understand to effectively critically evaluate both their government's response to climate change and the argument's most frequently made by opponents of proposed government climate change policies."

Brown's argument corresponds with analysis of the Australian media. Australia has one of the most concentrated rates of press ownership in the world (Bacon 2013; Muller 2016). This has a significant impact on how climate change is covered and the range of perspectives that news consumers are exposed to. Ethical issues are not only ignored, they are sometimes displaced by climate skepticism. Asking moral questions in this environment is like trying to grow a tree in a desert. The last comprehensive study of Australian media found that "11% of articles clearly rejected the notion of anthropogenic climate change and a further 21% were interpreted as suggesting doubt about it" (Bacon 2013: 77). In other words, a third of articles published in the Australian press did not accept the scientific consensus on anthropogenic climate change. In addition to this, articles that accepted the scientific consensus were almost never given a prominent position. For example, in print media, approximately 70% of articles on climate change appeared after page eight (Bacon 2013: 64). Finally, the knowledge claims made in Australian media only rarely draw on peer-reviewed science. To be more specific: "[o]nly 11% of all words in articles about climate science were dedicated to articles that explicitly referenced peer-reviewed climate science" (Bacon 2013: 14).[1]

Climate skepticism is most pronounced in media owned by the Murdoch Corporation (Hamilton 2007). This includes the only national broadsheet newspaper – *The Australian*. Examples are thick on the ground – on 12 January 2006, the paper claimed that "climate change may be a mirage." Two days later the paper described concerned citizens as "prophets of gloom" and argued that "it is an article of the green faith that the world's climate is changing for the worse, because coal-fired power plants pump greenhouse gases into the atmosphere… while environmental activists say science shows fossil fuels are responsible for global warming crisis, which may be right, they could just as easily be wrong." In 2009, *The Australian* published an article that described climate change as a "fraud" advanced by "warmaholics, drunk on government handouts and quasi-religious adulation from left-wing environmental organizations" (Jenkins 2009). And when the email account of climate scientists was hacked in the lead up to the Copenhagen Climate Conference, the paper reported: "In the past couple of weeks we have had a glimpse of the zealotry of the believers – and the gaps in their data – thanks to the exposure of the emails from the University of East Anglia" (quoted in Manne 2013: 42). The zealotry was, of course, their own and there was no editorial comment after successive inquires cleared the scientists of wrongdoing (Manne 2013: 42).

There are examples of misinformation right up to the present day. But like all forms of denialism, they are repetitive and lack creativity.[2] On several occasions the paper has issued public statements committing to action on climate change. In 2010, then editor Chris Mitchell argued that the paper had provided "longstanding support for a global response to limit greenhouse gas emissions." More recently, following international criticism on how the paper covered bushfires in Australia (Cave 2020), the paper published an editorial (Editorial 2020) that repeated a version of this statement: "The Australian's journalists report facts about how to tackle bushfires and about how to deal with the impact of climate change." It was not hard to find abundant evidence to the contrary. For example, the paper's coverage of the fires exaggerated the role of arson and failed to draw a connection with climate change (Abbott 2020). While the fire was still burning *The Australian* also ran an opinion piece from Ian Plimer (2019) which argued, "It has never been shown that human emissions of carbon dioxide drive global warming."[3]

Outside of promoting climate denial, *The Australian* also displaces serious ethical analysis by promoting a degraded form of morality. For example, the paper has never shown any concern with the fact that Australians are the highest per capita emissions of carbon. From an editorial perspective, there is no such thing as individual responsibility. All that matters is our overall contribution to carbon emissions. Robert Manne summaries the explicit message: "We contributed only 1.4% of global emissions. Nothing we could do would make any real difference" (Manne 2013: 45). This is consistent with the justifications given by successive Australian governments for their lack of ambition in setting emission reduction targets (Burdon 2015).

More pernicious perhaps is the paper's consistent argument that Australia has a moral responsibility to extract coal and export it to developing nations. In this alternative version of reality, coal companies are cast as benevolent aid organizations who are simultaneously tackling climate change by preventing 'dirty coal' from entering the market. This argument is repeated *ad nauseum*, but one recent example is from the paper's foreign editor, Greg Sheridan,

> UN projections are that Africa's population will increase to 4.5 billion by the end of this century. If any of them want to live above subsistence level, they will need cheap energy. Coal is sure to play a big part.
>
> *(2020)*

These are bad faith arguments that are made to disguise the paper's unconditional support for the fossil fuel sector. In service of these interests, the paper will sometimes argue that the risks of climate change might be mitigated by investment in technology such as 'clean coal' (Chambers 2020) and nuclear power (Lloyd 2019). It has also argued that the greatest contribution Australians could make to reducing emissions would be to export greater volumes of uranium (Manne 2013: 46). These policies can be promoted because they are consistent with the policy direction of the fossil fuel industry and the quarry vision of Australian politics (Pearse 2009). This also explains why, despite their potential for economic growth (Garnaut 2019), the paper has continually criticized renewable energy for their supposed failure to provide baseload electricity (Williams 2020). The antipathy to renewables is so high that their economics editor, Judith Sloan (2020), regularly advises the Australian public to "junk the obsession with renewable energy."

The Political Economy of the Mass Media

How do we account for the complicity of major media in failing to cover the ethical issues that arise out of climate change? So far, this chapter has highlighted how one newspaper shares common interests with the fossil fuel industry. However, nothing has been said about the nature of this relationship or how influence works more generally. To approach this topic, it is necessary to think about the larger political economy of mass media. This will be helpful for understanding not only why *The Australian* fails to engage the ethics of climate change, but other media organizations as well.

For the past thirty years, political analysis of the media has been written in the shadow of Edward S. Herman and Noam Chomsky's 1988 book, *Manufacturing Consent*. The key starting point of their argument is the claim that all systems of mass representation require some method for keeping the populous under control. Totalitarian regimes have the bludgeon and so it does not matter very much what the people think. However, democracies require more subtle forms of control and so develop tools of indoctrination or propaganda to "manufacture consent" or create "necessary illusions' in the minds of the people" (Chomsky quoted in Achbar 1994: 43).[4]

The media have a vital role to play in this process. Most of the work is done by a small number of agenda setting media. These are major corporations, and they use filters to set the basic intellectual framework within which smaller media providers operate (Chomsky 2002: 13–14).[5] Thus, for example, conversations about climate change take place within a framework with pre-determined goal posts. A conservative outlet like *The Australian* will cast doubt on the science of climate change and support the political agenda of the fossil fuel industry. By contrast, liberal outlets like The Guardian or The New York Times might make the case for government intervention and market mechanisms such as a carbon tax or trading mechanisms. Once this basic framework is set, it represents the markers of acceptable discourse. Arguments that countries ought to look beyond their immediate national interest or limit themselves to what would amount to a 'fair share' of the global carbon budget are reduced as radical or idealistic. Moreover, arguments about degrowth, a green new deal, or ending the extraction of fossil fuels can be cast as radical and extreme.

The reason state and corporate interest align in this example is not because the state or companies have direct control over the media. Hermann and Chomsky's argument does not rely on direct pressure or stand-over tactics. The reason that state, corporate, and media interests often align is because they share a common interest – maximizing profit. Chomsky notes,

> [T]he agenda-setting institutions are big corporations; in fact, they're mega-corporations, which are highly profitable and for the most part they're also linked into even bigger conglomerates. And they, like other corporations, have a product to sell and a market they want to sell it to: the product is their audiences, and the market is advertisers. So, the economic structure of a newspaper is that it sells readers to other businesses.
>
> *(2002: 14)*

In this mode of analysis, the worldview of advertisers is the limit of what media organizations will promote. Thus, even The New York Times and The Guardian have not committed themselves to a thorough-going analysis of the ethics of climate change. If Hermann and Chomsky are correct, one would also expect conservative media to wait until the last possible moment before abandoning the fossil fuel industry and their ideological agenda. In each instance, the media company is operating within the bounds that economic backers will tolerate.[6] As Chomsky (2002: 14) notes, papers like The Washington Post are "selling very privileged, elite audience to other business" and that the perspectives that emerge from this economic relationship "satisfy the needs and the interests and the perspectives of the buyers, the sellers and the market…it would be pretty surprising if that weren't the case." Ethical analysis may come in at the edges but so long as advertisers are either owned or connected to extractive industries, it will never become part of the dominant agenda. This is what Herman and Chomsky call the Propaganda Model and it is best seen as an empirical observation rather than a theory.

In the thirty years since this argument was first articulated, many of its basic elements have filtered into the mainstream. In the era of surveillance capitalism (Zuboff 2017) we have grown accustomed to the multiple ways our 'likes' and 'clicks' generate data that is sold to other companies for targeted advertising. News media are also open about their partisanship, and in some cases, have developed an entire business model that is predicated on unabashed propaganda. In these cases, what we have seen is the power of economic rationality to erode values such as truth or the role of the media in informing a democratic citizenry.[7] To the extent that these were ever solid commitments, they have been shown to be hollow and readily jettisoned for economic gain and political advantage.

Alongside this analysis, scholars have identified more insidious explanations for the lack of moral and ethical engagement on climate change in the media. For example, Donald Brown has highlighted the way instrumental rationality has led to values and ethics to be jettisoned from public discourse (2017, 2020). What is unique about this argument is that he is positioning instrumental rationality as a dominant form of reasoning that has come to pervade all of society. Commenting specifically on climate change, Brown argues:

> Most debates about environmental policies, at least in the United States, are classic examples of instrumental reasoning which ignore ethical reflection. Sociologists attribute this phenomenon to the power of opponents of environmental policies to frame public debates so that controversies only consider scientific and economic facts and the propensities of technically trained experts to admit that the intellectual tools on which their expertise is grounded are limited in resolving policy conflicts.
>
> *(2020)*

Wendy Brown has given the fullest account of this argument in her study of neoliberalism.[8] For her, neoliberalism is not just an ideology or set of economic policies. Rather, neoliberalism is also a dominant logic or way of thinking that plays a role in shaping the subjectivity of individuals. In Brown's (2015: 17) words, neoliberalism is a mode of reason that economizes everything and "configures all aspects of existence in economic terms." According to this perspective human beings are shaped as market actors and every field of activity is construed on the model of the firm (Brown 2015: 35). Brown's argument is not merely neoliberalism is corrupting institutions such as the media. Rather, she argues that *economization* or instrumental rationality has become "common sense, a reality principle, which remakes institutions and human beings wherever it settles." The reason I have defined economization with reference to instrumental rationality is to make the point that this process of at work even when money is not the issue (Brown 2015: 31).

The process of economization is vividly illustrated in recent campaigns to protect the Great Barrier Reef from climate change. Here campaigners rarely talk about the intrinsic value or the importance of the ecosystem for marine life. They are incentivized to talk in terms of tourism jobs or the loss that a dead Reef will bring to the Queensland economy (Smee 2018). Of course, these are valid concerns, but they are not the only or even the most important thing that will be lost if the reef collapses entirely. Wendy Brown offers another example in her reading of Barack Obama's 2013 State of the Union address in 2013,

> [E]very progressive value – from decreasing domestic violence to slowing climate change – Obama represented not merely as reconcilable with economic growth, but as driving it. Clean energy would keep us competitive – as long as countries like China keep going all-in on clean energy, so must we.
>
> *(Brown 2015: 25)*

Non-economic forms of economization are also easy to locate. Brown, for example, gives the example of dating services that speak to their clientele in terms of a "return on investment" (2015: 31). Those working inside of a university will also be familiar with encouragements to self-promote (act as entrepreneurs) to compete with colleagues for grants or the reduction of our work into measurable units. Even in the classroom it is not uncommon for students to reduce the knowledge process into a single sentence: "is this going to be in the exam."

The human condition today is one in which daily life has become economized. And if that is true, on what basis could anyone expect even the most liberally-minded media to behave any differently? Put another way – if economization and instrumental rationality have become our dominant logic, why would we expect the corporate media to be a beacon of ethics or value-reflective rationality?

Strategies to Increase Media Coverage of Ethical Issues

The prospects for sustained and ethical discourse look bleak but the situation is not hopeless, and neoliberalism is not such a totalizing discourse that we cannot think outside of it.[9] As Donald Brown (2020) has commented, the media's recent engagement with the science and ethical dimensions of COVID-19 ought to act as a model for how things might be done differently.

Independent Media

The oldest and most important response to the vagaries of the corporate media is the slogan:

'Don't hate the media, be the media.' One of the greatest success stories is the Independent Media Center (IMC) which was created in 1999 after the World Trade Organization protests in Seattle and has spread around the world.[10] The IMC was created to fill a vacuum in the way issues like globalization and protest were being covered by the agenda-setting media. The success of independent media has[11] rested on the fact that it provides an open publishing platform and journalists can share their content directly without going through an intermediary who might block or suppress content.[12] In addition to the immediate work of publishing journalism, independent media also acts as a training space where people learn skills related to writing, editing, and publishing (Democracy Now 2020).

Independent media is one of the few spaces where the ethical dimensions of climate change are routinely discussed. This is because they are free from corporate sponsorship and rely entirely on donations and subscriptions from readers. There is an obvious trade-off here – journalists have greater freedom, but the institution has less financial security. However, while there are examples of bankruptcy (Medea 2020), most independent media organizations have maintained or grown their audience (Wingerei 2019; Wingerei 2020). There are several explanations for this including the fact that most people receive their news online or directly from social media (Shearer 2018; Wingerei 2020). Moreover, during moments of crisis (like the COVID-19 pandemic) people have been turning to more local and reliable sources of information. For example, in Australia the readership of *The Australian* fell dramatically during 2020. By contrast, independent news such as The Saturday Paper increased its readership by 38 percent (Wingerei 2020). All of this shows that there is an audience for alternative, reflective, and ethical analysis. But those spaces are vulnerable and if we want to maintain or grow independent media than those with means must support them.

Break Up Large Media Corporations

A second strategy for promoting ethical analysis of climate change is to directly tackle the propaganda model described by Hermann and Chomsky and break up large media corporations. The argument here is not that this would *automatically* lead to more ethical analysis. Rather, the idea is to dilute the concentration of interests that own the media and which privilege certain kinds of instrumental reporting. Given the political power that media companies and big tech have this is a very difficult task (The Business Leader 2020). But jurisdictions have taken steps in this direction.[13]

For example, in October 2020, former Australian Prime Minister Kevin Rudd launched a petition that argued, "Australia's print media is overwhelmingly controlled by News Corporation, founded by Fox News billionaire Rupert Murdoch, with around two thirds of daily newspaper readership."[14] The petition goes on to accuse News Corp of using its power to "attack opponents in business and politics" and intimidating opponents into silence. To remedy this, the petition calls for a Royal Commission[15] into media diversity and for recommendations to be made about "policies to ensure optimal diversity across all platforms to help guarantee our nation's democratic future."

Illustrating the discontent many Australians feel about News Corp, the petition attracted more signatures than any other and it crashed the parliamentary website on opening weekend (Meade 2020). The petition was also promoted in the international press and by celebrities such as Hugh Grant and Bette Midler. When the petition closed, 501,876 people had signed, including a former conservative Prime Minister, Malcolm Turnbull. Despite such support, a Royal Commission has not been established. There is no threshold of signatures required for a petition to be debated in parliament and it would require the support of one of the major parties which so far has not occurred. However, one immediate consequence of the petition is that it has led to a significant public conversation about the power of News Corp and the way it has stood in the way of progress on issues like climate change. This culminated in an exchange where Malcolm Turnbull said the following to Paul Kelly, editor at large for *The Australian*,[16]

> The campaign on climate denial is just staggering and has done enormous damage to the world, to the global need to address global warming...I mean, it is so horrifically biased and such propaganda that Rupert's own son James can't stomach it...How offensive, how biased, how destructive does it have to be, Paul, before you will say – one of our greatest writers and journalists – 'It's enough, I'm out of it?'
>
> *(Cox 2020)*

Time will tell if this campaign is effective. But if News Corp is broken up and the concentration of media in Australia is diversified, then it may open spaces for alternative discourses and ethical reflection on issues like climate change.

Promote Stories in Defense of Public Things

The previous section argued that breaking up media corporations would diversify the range of voices contributing to the public conversation on climate change. For this to occur, journalists will need strategies for resisting neoliberal rationality and promoting value-reflective thinking. A lot could be said here, but the remaining space will highlight Bonnie Honig's argument that ethical consideration tends to coalesce around the defense of public things – like land, health care or public education. These have the potential to put cracks in the dominant logic and encourage alternative ways of valuing nature and protecting the dignity of all aspects of the Earth community. Moreover, Honig argues that they might form the "basis around which to organize, context, mobilize, defend or reimagine various modes of collective being together in a democracy" (2017: 24).

To illustrate this argument, Honig (2017: 3) highlights several examples including struggles to nationalize power in Quebec, and peasant resistance to the Monsanto empire. Alongside these examples, Honig also describes protests against the TransCanada pipeline which is being constructed over first nations land in British Columbia. In response to the law, Freda Huson, a member of the Unis'tot'en nation, argues: "They say they have developed C-51 for the purpose of protecting Canada's critical infrastructure projects, and what we're doing here is we're protecting our critical infrastructure" which includes "the land, itself and the medicine, berries, fish and wildlife hunting it provides" (Honig 2017: 22). For Huson, this is an act of reclaiming what is critical and has empowered her and other land defenders to communicate about non-instrumental values.

Honig (2017: 23) also provides an example from a short film called *Holding Their Ground* which shows how those at the Unis'tot'en camp have resisted the neoliberal rationality. In July 2015, representatives from Chevron came to the camp and said

> We're here today to talk to you about doing your work on your land and are requesting access onto your territory so that Wet'suwet'en people can work and to see benefits from the project. Will you allow...us to access the territory here today?

In response Huson replied:

> We've already said no to these projects. And that no pipelines will come on our territory and irregardless if you've got other Wet'suwet'en members working for you does not gain you access to our territory...You're trying to convince us to give up our way of life so that you can reap the benefits for government and investors and...while you're throwing crumbs to some of the partners that signed on...
>
> *(Honig 2017: 23)*

It is possible to read many things into this exchange. Certainly, there is a clash of cultures and values. However, Honig (2017: 23) also notes that the scene illustrates resistance to "neoliberal rationality" and the attempt to protect "vivid and cherished" aspects of culture against the logic of economization. Resistance here is "vivified in the public things of tribal life" and illustrates that for the Unis'tot'en nation the "sovereignty of the tribe and its relationship to public things are constitutively and inextricably intertwined." And importantly for the present argument, these alternative ways of valuing country have filtered their way into the mainstream media. For example, the Canadian Broadcasting Corporation have reported on the deep spiritual connection that First Nations people have with the land (Nicholson 2017); The Globe and Mail have written on First Nations guardianship obligations (Kane 2019); and The Toronto Star has given space to the importance of First Nations resistance (Issawi & Cruickshank 2019).

Something similar can be noted in cases where a private company seeks to control public land (Willingham 2019) or the emerging issue of companies polluting the night sky with satellites (Hall 2019). While not a performative 'speech act' there is something about naming these sites as "public things" that encourages reflection on ethics and morality. Of course, this will not always be the case but there are sufficient examples to illustrate that while neoliberalism might be ascendant, it is not totalizing. And to help us think outside of neoliberalism we might strategically construct and promote narratives about climate change that are grounded in the language of the commons and public goods.[17]

Conclusion

The conceit of this chapter is that the mainstream media does not adequately engage with the ethics of climate change. The Australian case study provided support for this claim. This paper was chosen because it is the only national broadsheet of its namesake and because it is owned by News Corp which has significant media holdings in the United States and the United Kingdom. While the case study alone is insufficient to prove the argument, it is put forward as an indicative example of the problem. Following this, several structural explanations were unpacked for the lack of ethical discourse amongst the agenda-setting media. This discussion focused on the way advertisers influence the scope of acceptable reporting and the role of neoliberalism in spreading instrumental rationality. Finally, the last section of the chapter discussed three strategies for increasing coverage of ethical issues. These included contributing to independent media, breaking up large media corporations and strategically promoting narratives about public things such as the global commons. These suggestions combine immediate things that we can do today and larger political projects that are designed to strike at the heart of the structural problems identified in the chapter. Taken together, they suggest that corporate coercion and instrumental rationality are problems we can overcome to promote ethical discourse.

Notes

1 These issues have only intensified since the 2013 study. For updated figures on media concentration, see Seccombe (2020).
2 For a full analysis see Manne (2013: 37–54).
3 For further commentary on how The Australian covered the bushfires, see Readfearn (2020).
4 An explicit statement of this perspective can be noted in Bernays (1928: 9): "The conscious and intelligent manipulation of the organized habits and opinions of the masses is an important element in democratic society."
5 The corporate structure of News Corp (which owns the Australian) can be seen here: https://corporatewatch.org/news-corp-company-profile/.
6 See further Chomsky (2002: 15).
7 This is the Jeffersonian ideal that the media is a "counterbalance to government power" (Rosenbaum 1989: 9).
8 Parts of this discussion were previously published in Burdon (2020).
9 Brown (2015: 222) also argues that we can think outside of neoliberalism. In fact, she expressly leaves open the possibility of identifying and countering neoliberalism and suggests that such work is necessary to "afford the slightest hope for a just, sustainable and habitable future."
10 Independent media is a global phenomenon. For an incomplete list see: https://indymedia.org. My definition of independent media is from here: https://theindependents.org.au/independent-media-portal/.
11 The IMC received 1.5 million hits in its first week and it remains a strong source of news 20 years after it was first established (Democracy Now 2019).
12 Independent media is not 'open access' and each site has specific rules that journalists must adhere to. For an example, see https://bristol.indymedia.org/editorial-guidelines/.
13 I am excluding from this discussion examples where protesters have ceased direct control of radio stations and other media outlets. For an example, see Lacey (2006).
14 The petition is online: https://www.aph.gov.au/petition_list?id=EN1938.
15 A Royal Commission is the highest form of public inquiry in Australia.
16 The exchange is worth watching in full: https://www.theguardian.com/global/video/2020/nov/10/qa-malcolm-turnbull-attacks-shocking-legacy-of-murdoch-and-news-corp-on-climate-crisis-video
17 For an example of this approach, see Taylor (1998).

References

Abbott, T. (2020) "Lives Matter, Not Political Points", *The Australian*, January 23.

Achbar, A. (1994) *Manufacturing Consent: Noam Chomsky and the Media*, London: Black Rose Books.

Bacon, W. (2013) "Skeptical Climate Part 2: Climate Science in Australian Newspapers", *Australian Centre for Independent Journalism*, October 30th, https://www.uts.edu.au/sites/default/files/Sceptical-Climate-Part-2-Climate-Science-in-Australian-Newspapers.pdf.

Bernays, E. (1928) *Propaganda*, New York: Horace Liveright.

Democracy Now. (2019) "Don't Hate the Media, Be the Media: Reflections on 20 Years of Indymedia, a Radical Media Movement", *Democracy Now*, November 27, https://www.democracynow.org/2019/11/27/indymedia_independent_media_seattle_wto_1999.

Brown, D. (2017) "Why Overcoming Instrumental Rationality In Climate Change Policy Controversies Is a First Order Problem Preventing Ethical Principles From Getting Traction to Guide Climate Change Policy Formation", *Ethics and Climate*, December 28, https://ethicsandclimate.org/2017/12/28/why-overcoming-instrumental-rationality-in-climate-change-policy-controversies-is-a-first-order-problem-preventing-ethical-principles-from-getting-traction-to-guide-climate-change-policy-formation/.

Brown, D. (2020) "What the Media Should Learn from Its Intense Coverage of COVID-19 about How to Cure Its Failed Reporting On the More Devastating Threat of Climate Change", *Ethics and Climate*, April 27 https://ethicsandclimate.org/2020/04/27/what-the-media-should-learn-from-its-intense-coverage-of-covid-19-about-how-to-cure-its-failed-reporting-on-the-more-devastating-threat-of-climate-change/.

Brown, W. (2015) *Undoing the Demos: Neoliberalism's Stealth Revolution*, Brooklyn: Zone Books.

Burdon, P. (2015) "A Change in the Weather? Australia and Climate Change Politics 2007–2014" in D. Brown & P. Taylor (eds.) *Ethics and Climate Change: A Study of National Commitments* (pp. 1–14), Gland: IUCN.

Burdon, P. (2020) "Neoliberalism in Legal Education Research", Chapter 3 in B. Golder, M. Nehme, A. Steel & P. Vines (eds.) *Imperatives for Legal Education Research Then, Now and Tomorrow*, Milton Park: Routledge.

Business Leader. (2020) "Yes, the Tech Giants Are Big - In Truth, Probably Too Big to Break Up", *The Guardian*, August 2, https://www.theguardian.com/business/2020/aug/02/tech-giants-too-big-break-up-panic-markets-facebook.

Cave, D. (2020) "How Rupert Murdoch Is Influencing Australia's Bushfire Debate", *The New York Times*, January 8, https://www.nytimes.com/2020/01/08/world/australia/fires-murdoch-disinformation.html.

Chambers, G. (2020) "Coalition Supports Coal Plant Upgrade", *The Australian*, October 8.

Chomsky, N. (2002) *Understanding Power: The Essential Chomsky*, New York: The New Press.

Cox, L. (2020) "Q+A: Malcolm Turnbull Clashes with News Corp's Paul Kelly Over Climate Coverage", *The Guardian*, November 10, https://www.theguardian.com/media/2020/nov/10/qa-malcolm-turnbull-clashes-with-news-corps-paul-kelly-over-climate-coverage.

Editorial. (2020) "Cool Heads Needed to Craft Response to Bushfire Crisis", *The Australian*, January 10, https://www.theaustralian.com.au/commentary/editorials/cool-heads-needed-to-craft-response-to-bushfire-crisis/news-story/40aa0a6d351e8e5578f700dd76717548.

Garnaut, R. (2019) *Superpower: Australia's Low-Carbon Opportunity*, Melbourne: Black Inc Press.

Hamilton, C. (2007) *Scorcher: The Dirty Politics of Climate Change*, Melbourne: Black Inc Press.

Herman, E. and Chomsky, N. (1988) *Manufacturing Consent*, New York: Pantheon Books.

Honig, B. (2017) *Public Things: Democracy in Disrepair*, New York: Fordham University Press.

Issawi, H. and Cruickshank, A. (2019) "Faces of the Resistance: Six Pipeline Opponents on Why They Oppose the Trans Mountain Expansion and What They're Prepared to do about it', *The Star*, August 15, https://www.thestar.com/vancouver/2019/08/15/faces-of-the-resistance-six-pipeline-protesters-on-why-they-oppose-the-trans-mountain-expansion-and-what-theyre-prepared-to-do-about-it.html.

Jenkins, J. (2009) "The Warmaholics' Fantasy", *The Australian*, January 6, http://web.archive.org/web/20090116082215/http://www.theaustralian.news.com.au/story/0,25197,24876451-7583,00.html.

Kane, L. (2019) "Indigenous Guardians Raise the Alarm on Impact of Climate Change in Canada", *The Globe and Mail*, March 14, https://www.theglobeandmail.com/canada/article-indigenous-guardians-raise-the-alarm-on-impact-of-climate-change-in/.

Lacey, M. (2006) "Mexican Protesters Keep Their Message Alive, and on the Air", *The New York Times*, October 31, https://www.nytimes.com/2006/10/31/world/americas/31mexico.html.

Lloyd, G. (2019) "The Nuclear Energy Option," *The Australian*, July 3, https://www.theaustralian.com.au/inquirer/report-argues-for-the-nuclear-energy-option/news-story/f3faf4befb3d68cedc01c6735d467976.

Manne, R. (2013) *Bad News: Murdoch's Australian and the Shaping of the Nation*, Melbourne: Quarterly Essay.

Meade, A. (2020) "More Than 150 Australian Newsrooms Shut Since January 2019 as Covid-19 Deepens Media Crisis", *The Guardian*, May 18, https://www.theguardian.com/media/2020/may/18/more-than-150-australian-newsrooms-shut-since-january-2019-as-covid-19-deepens-media-crisis.

Muller, D. (2016) "Fact Check: Is Australia's Level of Media Ownership Concentration One of the Highest in the World?", *The Conversation*, December 12, https://theconversation.com/factcheck-is-australias-level-of-media-ownership-concentration-one-of-the-highest-in-the-world-68437.

Muller, D. (2020) "Media 'Impartiality' on Climate Change Is Ethically Misguided and Downright Dangerous", *The Conversation*, January 31, https://theconversation.com/media-impartiality-on-climate-change-is-ethically-misguided-and-downright-dangerous-130778.

Nicholson, B. (2017) "Keystone XL Pipeline Faces Opposition From 'Historic Union' of Canada, U.S. Indigenous Tribes", *CBC*, May 16, https://www.cbc.ca/news/canada/calgary/keystone-xl-pipeline-indigenous-opposition-1.4117445.

Pearse, G. (2009) *Quarry Vision Coal: Climate Change and the End of the Resource Boom*, Melbourne: Quarterly Essay.

Plimer, I. (2019) "Let's Not Pollute Minds with Carbon Fears" *The Australian*, November 22.

Readfearn, G. (2020) "The Australian Says it Accepts Climate Science, so Why Does it Give a Platform to 'Outright Falsehoods'?" *The Guardian*, January 15, https://www.theguardian.com/media/2020/jan/15/the-australian-says-it-accepts-climate-science-so-why-does-it-give-a-platform-to-outright-falsehoods.

Rosenbaum, R. (1989) "Staying the Course in the Go-Go Years", *New York Times Book Review*, April 9.

Seccombe, M. (2020) "Murdoch Grab: The Other Story behind AAP's Sale" *The Saturday Paper*, September 12–18, https://www.thesaturdaypaper.com.au/news/media/2020/09/12/murdoch-grab-the-other-story-behind-aaps-sale/159983280010404.

Shearer, E. (2018) "Social Media Outpaces Print Newspapers in the U.S. as a News Source", *Pew Research*, December 10, https://www.pewresearch.org/fact-tank/2018/12/10/social-media-outpaces-print-newspapers-in-the-u-s-as-a-news-source/.

Sheridan, G. (2020) "Coal: Australia's Great and Costly Retreat", *The Australian*, September 24.

Sloan, J. (2020) "Only Reliable Electricity Can Give the Economic Jolt We'll Need," *The Australian*, May 17.

Taylor, P. (1998) *An Ecological Approach to International Law Responding to the Challenges of Climate Change*, London: Routledge.

Westervelt, A. (2019) "How the Fossil Fuel Industry Got the Media to Think Climate Change Was Debatable", *The Washington Post*, October 1, https://www.washingtonpost.com/outlook/2019/01/10/how-fossil-fuel-industry-got-media-think-climate-change-was-debatable/.

Williams, P. (2020) "Grid 'Unreliable' as Risks Mount", *The Australian*, February 24.

Willingham, R. (2019) "Federation Square Apple Store Doomed After Permit Refused by Heritage Authorities", *ABC News*, April 5, https://www.abc.net.au/news/2019-04-05/apple-store-federation-square-melbourne-plans-permit-refused/10975938.

Wingerei, K. (2019) "Independent Media Continues to Grow", *Pearls and Irritations Public Policy Journal*, June 13, https://johnmenadue.com/kim-wingerei-independent-media-continues-to-grow/.

Wingerei, K. (2020) "Independent Media Winning the Battle of the Online Audience", *Pearls and Irritations Public Policy Journal*, November 25, https://johnmenadue.com/independent-media-winning-the-battle-of-the-online-audience/.

Zuboff, S. (2017) *The Age of Surveillance Capitalism: The Fight for a Human Future at the New Frontier of Power*, New York: Public Affairs.

36

WHY WE NEED 'CULTURAL INTELLIGENCE' IN CLIMATE CHANGE COMMUNICATION, AND HOW INTEGRAL THEORY CAN HELP

David Storey

The Hard and the Soft

The lion's share of attention in climate discourse is devoted to the 'hard' side of the problem – (1) scientific evidence for climate change and projections about its future effects, (2) technological solutions, and (3) legislative and regulatory efforts to invest in and implement solutions – and rightfully so. Yet by now, it has become clear that solving the hard side is not sufficient and not possible with hard tools. Clear scientific understanding of and consensus about the reality and dangers of climate change, viable technological means to combat it, and an extant suite of legal and regulatory instruments to address it have not issued in adequate action. In short, we could be making more progress than we are given the tools we have on hand. Beyond this, to develop the hard tools, we will need not just to make progress in the medium term, but to solve the problem in the long term, we cannot rely on the hard tools we have. Put crudely: to increase scientific research and technological innovation in renewable energy systems, we need money; to get money, we need policy; and to get policy – robust, durable policy – we need political will among voters, donors, and elected officials. While political will is not a 'hard' problem, in the United States it is proving to be the hardest problem of all.

This essay focuses on the ethics of crafting and communicating climate policy to stakeholders and populations with different values and worldviews. When we think about climate ethics, we often focus on 'first-order' questions of global or intergenerational justice – e.g., 'What are the duties of the global North to the global South?' and 'What are our duties to future generations?' – and the best policies with which to pursue them. As important as these questions are, mustering the political will for action at scale requires us to ask different questions because the moral arguments and the policies they entail inevitably crash against the shoals of political and economic feasibility.

But this doesn't mean we should check ethics at the door and leave the policy-making to technocrats; it just means we aren't asking the right questions. 'Second-order' questions take us from ethics and policy to psychology and culture. They ask how climate change and the various policy proposals aimed at addressing it are ethically *perceived and interpreted* by people with different worldviews, moral psychologies, and value systems, and whether there are discernible patterns in this psychological and cultural landscape.

DOI: 10.4324/9781003039860-42

While philosophers doing climate change ethics do not pay enough attention to the social sciences, incorporating the social sciences into climate change communication is a necessary but insufficient step. It is necessary because however cogent and convincing moral arguments might be in theory, in practice we now know, based on work in fields like social psychology, evolutionary psychology, and behavioral economics, that humans are not the rational actors that modern intellectuals, scientists, and policy makers often take them to be. The 'information deficit model,' the notion that people would automatically change their beliefs, values, and actions if they just had correct and adequate scientific knowledge about climate change, is widely regarded as inadequate. All of this, made possible by social scientific research, is a sign of progress in our understanding of the 'soft' side of the problem. However, it is not enough to draw on the social sciences because doing so raises questions beyond their scope: questions that are unavoidably ethical and philosophical.

The social scientific turn is insufficient for two reasons. First, it purports to be descriptive rather than normative – to be 'value-neutral.' But as Max Weber pointed out, values bracket the process of knowledge production; they powerfully shape what is studied, why and how it is studied, and what is done with the results of study. What we do with the results of social scientific research, and what the conceptual blind spots of such research might be, cannot be determined solely with social scientific methods. Second, much climate change communication is laden with the 'values deficit model.' This recognizes the problem of the information deficit model and tries to correct it by aiming messaging not at scientific data, but at people's values. While this approach is an improvement, the presumption is often that one group – in the United States, generally conservatives – has the wrong values, and if we just reframe climate change using different words, they will come to see things our way and align with the right values – *our* values. The problem is not simply that this approach can be condescending and manipulative, but that it is unconscious of – or unconscious of the deficiencies of – its own worldview and value system. 'Third-order' questions, then, return to ethics and policy, asking how the knowledge gained from answering second-order questions can help us forge and frame climate policies that achieve *congruence* with multiple worldviews and value systems in a way that is not just empirically informed, but conceptually coherent and morally satisfying.

Rather than focusing on first-order ethical and policy questions, this article explores second- and third-order questions by examining different methods for solving the soft side of the climate problem. A popular approach will be examined, Jonathan Haidt's moral foundations theory, and rejected in favor of a developmental model recently proposed by philosopher Steve McIntosh. McIntosh's model, informed by Ken Wilber's integral theory, can help us better understand the nature of and relationships between the major worldviews and value systems in American culture – social conservative (traditionalists), libertarians, fiscal conservatives, and centrist Democrats (modernists), and progressives (postmodernists) – and how and why they see climate change in such different terms. Part of the argument is that policy is downstream from politics, politics is downstream from culture, and that the ethical values encoded in policy are ultimately rooted in culture. Hence, we require more of what McIntosh calls 'cultural intelligence' to build political will around climate change. It will conclude with reflections on how this approach can help us think about the Green New Deal.

Changing Climate, Changing Minds

By now, there is a vast academic literature on and even relatively broad public awareness of the problems around climate change communication. From the Yale Project on Climate Change Communication's 'Six Americas' to Robert Giffords' 'dragons of inaction,' from identity politics to ideological capture, from Daniel Kahneman's cognitive biases to Dan Kahan's cultural cognition, from Stephen Gardiner's 'perfect moral storm' to Oreskes and Conway's 'merchants of doubt' – the psychological obstacles to a critical mass of people recognizing and responding to climate change are well-documented, legion, and formidable. The aim here is not to survey and critique this literature, but to focus on one popular example that demonstrates a blind spot in how we talk about how to talk about climate change. Just as climate scientists insist that we should look at solutions on the hard side not in terms of a 'silver bullet' but as 'buckshot,' we should not assume that one communications strategy or psychological framework will solve the soft side. But aiming carefully, and being mindful of our blind spots, still matters.

In his 2012 book, *The Righteous Mind: Why Good People Are Divided by Politics and Religion*, social psychologist Jonathan Haidt applied Moral Foundations Theory (hereafter 'MFT') to the problem of political disagreement. MFT rests on three principles. First, people's moral responses are primarily moved by their intuitions (their 'elephant') and not strategic reasoning (the 'rider'). Second, there is more to morality than harm and fairness. Third, morality 'binds and blinds.'

According to MFT, the human mind is governed by six different 'moral foundations,' each of which represents a morally significant value and its corresponding violation: authority/subversion, loyalty/betrayal, sanctity/degradation, liberty/oppression, fairness/cheating, and care/harm (brief descriptions of the foundations can be found at moralfoundations.org). Haidt likens these foundations to taste buds. Just as people have different palates, they are sensitive to different moral foundations. The major finding in Haidt's research had political implications: while liberals tend to be high on fairness and care and low on the other foundations, conservatives have a more balanced moral palate. From this, Haidt drew two conclusions.

First, liberals and conservatives are caught in different 'moral matrices.' A moral matrix *binds* the in-group together, but it also *blinds* it to the humanity and virtues of the out-group. This disconnect fuels tribalism, culture war, negative partisanship, and political polarization more broadly. Thus, social psychology can help us see through the matrix and promote 'viewpoint diversity.'

The second conclusion is that because liberals have a narrower moral palate, the onus is on them to be more tolerant and understanding toward conservatives. While conservatives value all of the foundations, liberals – broadly construed as individuals who favor high government intervention in the economy to redress the social and environmental impacts of capitalism and are tolerant around social issues like abortion and same-sex marriage – score highly on only two foundations: care/harm and fairness/cheating. For Haidt, the regnant question is not 'what's the matter with Kansas?' but 'what's the matter with coastal elites?'

An important distinction here is that some foundations are 'individualizing,' while some are 'binding.' As Annukka Vainio and Jaana-Piia Mäkiniemi explain,

> The care/harm and fairness/cheating foundations are referred to as individualizing foundations, with their emphasis on the rights and welfare of individuals... The three other foundations – loyalty/betrayal, authority/subversion and sanctity/degradation – are referred to as binding foundations and have an emphasis on group-binding loyalty, duty and self-control.
> *(Vainio & Mäkiniemi 2016: 268)*

What happens when we look at climate change through the lens of MFT? There has been some research done on this in the United States and Finland, and the findings can give us some indication of how we might use this knowledge to think strategically about framing climate justice. Researchers found evidence of "the potential importance of moral foundations as drivers of intentions with respect to climate change action," which "suggests that compassion, fairness and to a lesser extent, purity, are potential moral pathways for personal action on climate change in the USA" (Dickinson et al. 2016: 1). This last finding with respect to purity is important because it is a potential pathway to engaging conservatives on climate change and because there is other research suggesting that the use of language that activates the sanctity/degradation foundation is associated with increased pro-environmental attitudes among conservatives.

These findings are also generally consistent with studies indicating that "those who endorse general binding foundations do not perceive climate change as a moral issue" (Vainio & Mäkiniemi 2016: 273). And as the authors of another study note, "where climate change is concerned, ethical discussions have focused mainly on non-harming and fairness, assuming that the other moral axes (those most prized by conservatives) are not relevant" (Dickinson et al. 2016: 3). The lion's share of the literature in climate ethics adopts this approach, potentially alienating potential allies. How, then, might we activate the binding foundations to encourage conservatives to see the moral relevance of climate change, and be motivated to take action?

Hanno Sauer describes the logic of such an approach by distinguishing between 'strategic' and 'genuine' appreciation of moral foundations:

> ...the strategic appreciation of conservative moral foundations occurs when liberals deliberately frame their concerns in terms which will resonate with the conservative, but which they would not prefer otherwise. One instructive example for this is climate change, where it can be shown that even though political conservatives are less likely to endorse measures against climate change, their position can be shifted in the direction of such measures if their importance is described in conservative terms....
>
> *(Sauer 2015: 157)*

It turns out that the study described in "How are moral foundations associated with climate-friendly consumption?" did just that. The study had subjects respond to questions about climate change that used language invoking each foundation. For instance, with respect to the authority/subversion foundation, they asked subjects to respond to the following statements: "Climate change increases wars and conflicts among nations," "Whether climate change mitigation is the duty of every citizen or not," and "The climate experts' recommendations ought to be followed." The following statements invoked the loyalty/betrayal foundation: "Climate change mitigation is fighting for the fatherland," "Whether climate change threatens the lives of [one's compatriots] or not," "Whether climate change is a threat to my family or not." In contrast, the following statements invoked the sanctity/degradation foundation: "Whether CC [climate change] is unnatural or not," "Whether CC will increase the number of infectious diseases or not," and 'CC will change [the] natural order created by God." The results of the study showed that both groups – those with individualizing foundations and those with binding foundations – saw climate change as a moral issue (Vainio & Mäkiniemi 2016: 276). While research using MFT is in its early stages, such results are promising. They indicate that framing climate change discussions using different moral categories can lead people who might otherwise not care about the issue to shift their attitude and take action.

However, there are reasons for caution. Since the release of his book, Haidt has become an enormously influential public intellectual. Political psychology in general, and Haidt's framework in particular, has widely been embraced as the skeleton key for making sense of, and hopefully defusing, the poisonous polarization that plagues our politics. Much like George Lakoff's 'framing' was in vogue during the 2000s and the heyday of the 'netroots,' MFT has penetrated popular consciousness and is regularly employed to make sense of political disagreement. Indeed, former George W. Bush staffer, evangelical, and 'Never Trumper' Peter Wehner recently profiled Haidt in *The Atlantic*, turning to him to help shed some light during a time in which pandemic and polarization fuel each other. Wehner writes:

> Haidt told me he considers himself to be a centrist, engaging with views from multiple sides in order to understand issues.... Haidt laments the state of contemporary American politics, believing that on both the right and the left we're seeing populism that responds to real problems but in illiberal ways.... But on the left, Haidt said, 'there's been a movement...that

> is focused instead on oppression and victimhood and victimization. And once you get into a framework of seeing your fellow citizens as good versus evil based on their group, it's kind of a mirror image of the authoritarian populism on the right.
>
> *(Wehner 2020)*

Haidt's opinion arguably does not do justice to progressives. Ironically, the tolerance, appreciation for diverse perspectives, non-judgmentalism, and aversion to ranking one morality higher than another – all hallmarks of Haidt's calls for 'viewpoint diversity' – is one of the key features of progressive thinking. Haidt's writings are peppered with personal anecdotes about the liberal scales dropping from his eyes when he began reading conservative writers like Burke and Hume, leading him to see the ways that liberals are often intolerant and, well, illiberal. In the face of such aggressive progressives, a number of public intellectuals like Haidt have gravitated toward 'classical liberalism,' which comforts itself by saying, in effect, 'There are bad people on both sides.' The primary problem in politics today, on this view, is dogmatic, self-righteous progressives. Haidt is partly correct: there is indeed an unhealthy, illiberal element found in some quarters of today's left. When it comes to environmental politics, this element often frames climate change in ways that trigger both conservatives and moderates. As Haidt put it during a presentation to the Citizens' Climate Lobby in 2019, the Green New Deal is "perfectly crafted to turn off as many conservatives, republicans and libertarians as humanly possible." As this article will discuss in the final section, Haidt is correct to point to the rhetorical problems in the Green New Deal, but his approach has the unintended effect of gaslighting progressives by making them think that their perspective in morality are on par with, if not deficient with respect to, conservatives, encouraging them to move toward 'centrism.'

Another problem with MFT is the prescriptive conclusion Haidt draws from his findings. Haidt has argued that since conservatives have access to a wider set of moral foundations than liberals, since political disagreements are due to people's different moral foundations, and since such disagreements cannot be resolved through rational discourse, the burden is on liberals to try and find common ground. This is the wrong conclusion to draw from the research on MFT because it does not leave room for any notion of moral development tied to the effects of education and rational reflection. This could be due to a social scientific preoccupation with description that leads Haidt to regard different moral psychologies as merely, well, different. The frame of 'tribalism' has powerfully shaped the way we think and talk about political affiliation over the last decade, and it seems to entail that political positions are akin to irrational religious convictions and/or 'hard-wired' evolutionary dispositions that cannot be changed. This frame resists the notion that some 'tribes' beliefs might be *truer* and their values *better* than others; while all tribes are blind to some extent, some may be more blind, and blind to more of reality, than others. Indeed, as Sauer perceptively argues, if we examine research done on MFT, "we see that a better education and sound debunking arguments leave 'liberal' judgments about harmful acts or rights-violations unaffected, undermining only 'conservative' moral beliefs, and that moral arguments based on liberal considerations often override those based on conservative ones" (Sauer 2015: 162). That is, there is evidence that conservatives can be led to revise their judgements based on rational testing, *but not the reverse*. This idea – that there is a certain developmental trajectory to types of values and worldviews – finds expression in the integral approach that is examined next.

The Need for an Integral Approach

Integral theory is predominantly associated with the American philosopher Ken Wilber, but over the last several years, researchers have begun applying it to environmental issues in general and climate change in particular. A comprehensive, or even cursory, sketch of integral theory is beyond the scope of this essay. Instead, it attempts to isolate an aspect of the integral approach that speaks directly to the 'soft side' of the climate problem, something that is generally neglected in extant approaches: namely, the *developmental* nature of consciousness and culture. Drawing on and distilling the work of developmental psychologists from Jean Piaget to Robert Kegan, integral theory posits that human beings – both individually and collectively – move through various developmental stages. Gail Hochachka explains:

> ...adult developmental theories provide a sequence of specific stages or levels, grounded in empirical science, that can help us understand key differences across a range of worldview stages. Without a sense of these developmental stages, climate change scholarship is missing that particular lens on how people perceive, relate to and act on climate change as they mature, grow and change through life.
>
> *(Hochachka 2020: 193)*

The stages most relevant to this discussion are the traditional, modern, and postmodern stages because (1) they represent the main worldviews and value systems in American culture; (2) the differences and tensions between them are responsible for the culture wars; and (3) the culture wars are the essential context for understanding why we disagree about climate change.

In *Developmental Politics: How America Can Grow into a Better Version of Itself*, Steve McIntosh presents a theory of cultural evolution aimed at defusing the hyperpolarization in American culture that obstructs action on climate change. Though drawing on the work of Haidt and other social scientists, he chiefly uses the resources of philosophy:

> While useful for understanding the problem of hyperpolarization, social science's value-neutral stance precludes it from providing a solution. To remedy this debilitating problem, we need political philosophy. Like the subject of ethics, the subject of values – which are the basis of culture – can only be adequately understood from a philosophical perspective.
>
> *(McIntosh 2020b)*

He describes his position as 'post-progressive.' FDR famously cast his New Deal as an attempt to 'save capitalism from the capitalists.' Likewise, McIntosh's approach could be cast as an attempt to save progressivism from progressives.

By 'post-progressive,' McIntosh does not mean 'anti-progressive.' A number of the intellectual circles with which Haidt is associated tend to define themselves in opposition to progressivism. McIntosh concedes that progressives often go against, or selectively follow, their stated views on tolerance and inclusion:

> Progressives are keen to demonstrate cultural sensitivity toward Islamic traditionalists, for instance, despite Islamic homophobia and contempt for liberal freedoms. If progressives showed a similar degree of cultural sensitivity toward Christian traditionalists, even while opposing them politically, this would help ameliorate the hyper-polarization that led to the rise of Trumpism.
>
> *(McIntosh 2020b)*

For McIntosh, however, anti-progressives are only seeing half of the truth, and they end up rejecting the key insights of healthy progressivism with the excesses of its illiberal iterations. Instead, we ought to recognize that progressivism, which emerged as a cultural force in the 1960s, is not some fashion or fad that is going away. It represents a new worldview and set of values – including environmentalism, multiculturalism, egalitarianism, and a heightened sensitivity to the suffering of others, particularly historically marginalized groups – that is good, that is here to stay, and that emerged precisely to correct for the injustices of the modernist culture it sought to counter. Postmodernism, McIntosh insists, should be seen "not merely as a critical branch of academia, but as a third major worldview in its own right" (McIntosh 2020a: 16). And it is a worldview, he argues, embraced by roughly 20 percent of the US population, with around 50 percent at modernism and 30 percent at traditionalism. McIntosh draws here on the World Values Survey data that, he writes, "clearly points to the existence of traditional religious values, secular-rational values, and what Inglehart terms 'post-material' values" (McIntosh 2020a: 17). Rather than aim to eradicate progressivism, then, we ought to temper its illiberal excesses and incorporate its insights.

This same method – *transcend* the biases and limitations of a worldview (what 'blinds,' in Haidt's terms), but *include* its valuable contributions – is what McIntosh calls 'cultural intelligence.' Here is where McIntosh's approach departs from, and exposes the limitations of, Haidt's, and from the conventional wisdom around what drives polarization. Instead of seeing political reality in *two* opposing camps (in '2-D'), it sees *three* distinct worldviews or cultures (in '3-D').

The first assumption that hamstrings the conventional approach is political dualism. This is the tendency to cut political reality into two opposed and mutually exclusive camps. It shows up not just in the juxtaposition of Democrats vs. Republicans, Left vs. Right or Blue vs. Red, but in the categories with which social scientists in recent years have divvied up the electorate: 'Open vs. Closed,' 'Anywheres vs. Somewheres,' 'Fixed vs. Fluid,' 'Prius vs. Pickup,' etc. These distinctions are useful and certainly track part of political reality, but they also blind us to deeper ones and, ironically, can end up reinforcing the tensions they aim to undo.

Instead of two camps, there are actually three: traditionalist, modernist, and postmodernist. Until the Trump era, the modern GOP more or less successfully combined the first two, while the Democratic party has struggled – and continues to struggle – to yoke the latter two. McIntosh points out how Haidt's moral foundations are correlated with these different worldviews: traditionalists (social conservatives) prioritize authority, sanctity, and loyalty; modernists (libertarians, classical liberals, fiscal conservatives) prioritize liberty; progressives prioritize fairness and care. And he follows Haidt in thinking that we must become more open and receptive to what each moral matrix has to offer.

> A post-progressive political perspective can see how all three major American worldviews... each include enduring values that our society needs, as well as accompanying pathologies that we must work to overcome. While this emerging perspective affirms progressive values such as environmentalism and greater inclusion, it transcends progressivism by affirming the legitimacy and ongoing necessity of modernist values and traditional values as well.
>
> *(McIntosh: 2020b)*

Just as progressivism emerged to address the problems created by modernism, such as economic inequality and environmental degradation, post-progressivism is emerging to address the problems created by progressivism itself: including the hyper-polarization we see in the form of a culture war of all against all. The great mistake is to think that one culture is 'right,' that the others are 'wrong,' and that one culture can or should 'win' and send its enemies into the dustbin of history. To paraphrase John Adams, values are stubborn things.

To this end, McIntosh offers a method called 'values integration':

> Developmental politics does not seek to change the values people hold dear. Rather than trying to convert or replace existing values, this new approach seeks to expand the scope of what people are able to value. Simply put, we don't want to erase or subtract people's values, we just want to add to them.
>
> *(McIntosh 2020a: 46)*

David Roberts, arguing that the only way to make progress on climate change is to "make [people] more liberal," provides an example of the mentality McIntosh criticizes:

> ...the drive to address climate change is ultimately a liberal project: It's about drawing together in cosmopolitan global unity as a species, thinking in long-term, non-zero-sum terms, sacrificing for and helping one another, and having the confidence and curiosity to embrace change, to experiment and learn and adapt on the fly... Those are all features of the climate project that draw on liberal personality traits. If climate hawks ever want to change the maddening, gridlocked political status quo in the US, they need to start thinking about how to bring those traits to the surface.
>
> *(Roberts 2019)*

Karen O'Brien points out the folly of this approach:

> ...it is essential to release the need for everyone to 'be on the same page' or 'at the same stage,' as most development research suggests that populations are spread across three or four major worldviews (i.e., traditional/mythic, modern/universalistic, or postmodern/pluralistic, with some emerging integral worldviews). Some environmentalists cling to a utopian ideal, where everyone is 'pluralistic' and 'worldcentric.' However, given the urgency of responding to climate change, this is an unrealistic scenario that may hinder rather than promote adaptive responses.
>
> *(O'Brien 2010)*

Instead, values integration can be pursued, McIntosh suggests, by "helping each worldview to see the others more clearly and more sympathetically." Rather than see the three worldviews as mutually exclusive, we must see them as interdependent. This involves first understanding what each worldview's 'core competencies' are:

> Modernism is focused on overcoming poverty and ignorance, and on liberating each individual to achieve their full potential. Postmodernism is concerned with righting the wrongs of modernity by building solidarity for the protection of the environment and the achievement of social justice. And traditionalism seeks to preserve our heritage and conserve the best of what has come before. Because each of these discrete problem sets continue to afflict our society, the problem-solving focus of these worldviews is still very much needed. Each form of culture thus has a necessary and ongoing job to do.
>
> *(McIntosh 2020a: 48)*

But it also entails seeing how each worldview supports, and is supported by, the others. For instance, traditionalism provides the foundation for modernity:

> ...modernism is not a complete or free-standing value system in its own right. In order to effectively use modernist values to achieve the freedom and prosperity that are the fruits of modernity, people need to have been previously socialized by traditionalism to obey the rule of law, to be honest and reasonably fair, and to continue to acknowledge the needs of the collective to at least some degree. Without an underlying foundation of functional traditional morality, modernism is unsustainable and indeed dangerous.
>
> *(McIntosh 2020a: 22–23)*

Modernism without the discipline of traditional morality gets us Gordon Gecko – 'greed is good' – which then leads progressives to reject capitalism itself, rather than a sick version of it. Conversely, modernism provides traditionalists with the freedom and prosperity to live their lives as they see fit. Meanwhile, postmodernism depends on modernism:

> ...in order to function properly, postmodern values are similarly dependent on the moral and physical capital of previous worldviews... Without the liberal protections of individual freedoms provided by modernist values, and the foundational norms of fair play and honesty provided by traditional values, postmodern culture would soon regress to a 'pretraditional' level of warring factions.
>
> *(McIntosh 2020a: 42)*

And a pre-traditional *tribalism*, rather than a postmodern pluralism, is precisely where we find ourselves: each of the three worldviews has temporarily regressed to what integral theory calls the 'warrior' level of consciousness, which is exactly what Thomas Hobbes was describing in his 'state of nature.' Indeed, in a chart laying out the healthy and sick forms of each worldview, McIntosh points out that the one thing all the sick forms have in common is "cultural bias and tribalism." Everyone feels attacked, oppressed, and victimized. The culture wars have flared so much in the last four years because everyone's 'warrior' consciousness has been on a constant simmer.

A major engine of the conflict is confusion about the very nature of culture, namely, that culture doesn't just *change*, but actually *develops*. Ideally, each new worldview transcends and includes the prior ones. This is perfectly captured in the cover of McIntosh's book, the painting 'Three Flags' by the artist Jasper Johns, which features three American flags on top of each other in a Russian doll arrangement. In practice, however, development is messy. New worldviews tend to not just *transcend* but *negate* their priors. Old worldviews tend to see a new worldview as the threat they were designed to neutralize: "each worldview tends to confuse its historical successor for its historical predecessor." Put another way, the dominant culture always sees what is 'above' as though it were 'below,' because what is above and what is below are *different* from it. For instance, many modernists, haunted by the history of the Cold War, mistakenly see the Green New Deal – a progressive policy vision through and through – as a regression to state socialism, and traditionalists might see it as an attack on the family and local communities subsuming all in the state.

The developmental dimension of McIntosh's approach, which adds depth or height, is the starkest point of divergence from Haidt's, which is confined to the two dimensions of left and right. Haidt's failure or refusal to explain why certain moral cultures are better than others is, ironically, a very progressive thing to do: tolerant, nonjudgmental, egalitarian. Recall Haidt has argued that since conservatives have access to a wider set of moral foundations than liberals, since political disagreements are due to people's different moral foundations, and since such disagreements cannot be resolved through rational discourse, the burden is on liberals to try and find common ground. This is the wrong conclusion to draw because it does not leave room for any notion of moral development tied to the effects of education and rational reflection. This is due to a social scientific preoccupation with description that leads Haidt to regard different moral psychologies as merely, well, different. Indeed, this is one reason McIntosh adopts a philosophical approach: "Philosophy is needed because supposedly value-neutral social science cannot establish a normative dimension of progress or otherwise clarify the direction of the good" (McIntosh 2020a: 76).

This asymmetry is critical. At a minimal level, we need to be able to say that a worldview that regards all people as equal regardless of their creed, color, or country – as modernism does – is indeed better than one that does not, as is sometimes the case with traditionalism, which can be ethnocentric. Likewise, a worldview that also recognizes the moral value of nonhuman animals and the environment – as postmodernism does – is better than those that do not. How, then, might this approach help us think about the most ambitious and controversial approach to climate change in recent years – the Green New Deal?

A New Green New Deal

While the Green New Deal (GND) is a resolution rather than a set of concrete legislative proposals, it immediately met with fierce resistance on both the right and the center for tying climate action to other progressive goals such as a federal jobs guarantee, universal healthcare, universal affordable housing, the elimination of nuclear power, as well as an estimated cost in the tens of trillions of dollars. Below will explain why the framing of the GND on the left is misguided, and why the Biden administration's approach up until now is prudent in part because it appears to reflect the concerns of Haidt and McIntosh.

In his 2012 book, *The New New Deal*, Michael Grunwald tells the story of how the Obama administration folded the most significant federal investment in clean energy in history into the American Recovery and Reinvestment Act by providing massive subsidies and loan guarantees to a number of energy companies. While the right-wing media fixated upon one of the investments that failed – Solyndra, which lost around $500 million – and criticized the administration for interfering in the 'free market,' by 'picking winners,' the program succeeded in both reviving renewable energy industry in the United States and recouping taxpayers' money.

Proponents of the GND should take heed. Author and activist Naomi Klein, one of its most strident defenders, embodies what both McIntosh and Haidt would regard as the wrong approach to climate politics. The subtitle of one of her books, *Capitalism vs. the Climate*, encapsulates the mistake that progressives often make in trying to achieve their political goals. The problem is not capitalism, but a particular form of it, often referred to as 'neoliberalism.' The language often used by the left – 'revolution,' 'radical,' and calls to 'dismantle capitalism,' as well as the ideas that liberalism and capitalism are inherently premised on 'white supremacy' – not only have the rhetorical effect of alienating traditionalists and modernists, but they also are neither feasible nor desirable. Without martialing the power of finance, business, and private industry, there is no hope for decarbonizing the global energy economy; the only thing that can run the industrial revolution in reverse, as it were, is another industrial revolution. Moreover, the association of the GND with the 'democratic socialism' of one of its co-sponsors, Rep. Alexandria Ocasio-Cortes, needlessly frames climate change as anti-capitalist.

Of course, it is true that *democratic* socialism is very different from the *state* socialism that the right associates with the Soviet Union; but that distinction is lost in the fog of political war. Though progressives do hold a more complex cognitive and moral perspective – the systems-thinking to understand the problem of climate change and the moral sensitivity to the suffering of distant, future, and marginalized people and nonhuman animals – they are a cultural minority, and often fail to appreciate the virtues of traditionalism and modernism. Roberts is right to be skeptical of attempts to convert conservatives to climate action, and it is important to stress that the GND is far from the Bolshevik plot its detractors make it out to be.

Nevertheless, the Biden administration has wisely distanced itself from the language of the GND while advancing a number of its core proposals and marrying action on climate to social and economic renewal under the brilliant banner 'Build Back Better.' As of this writing, Biden has already taken a page out of the Obama administration playbook by including a swath of climate provisions in its proposed pandemic relief legislation. 'Build' resonates with the modernist ethic of hard work and self-reliance, as well as the traditionalist value of patriotism and national pride. By treating climate change as an opportunity for national and economic renewal. 'Back' resonates with the traditionalist value of appreciating the past, and the idea that American heritage is a good thing worth recovering and preserving. And 'Better' resonates with the modernist emphasis on growth, improvement, and material progress, as well as the postmodern concern for sustainability and racial and environmental justice. From an integral perspective, the pathological modernism of which neoliberalism is the most obvious expression is not only responsible for the climate crisis, but for the psychological, cultural, and economic carnage that globalization has wrought on the rural communities that are the source of the right-wing populist movement that denies its existence. The prudent response is to frame the climate crisis in relation to this bigger picture, and craft and communicate solutions that will both appeal to, grow, and strengthen the connections between the healthy forms of traditional, modern, and postmodern worldviews and values.

The spirit of the GND is on point: for the last 40 years of the Reagan era, the balance between the public and the private sector has tilted too far toward the latter and is now poised to shift in the other direction, with a focus on industrial policy and long-term planning. But that is less a revolution, and more a return to the spirit of the New Deal era, albeit with the addition of addressing racial and environmental injustices. It does not require abandoning capitalism but making it work better, making it more free, more equal, more prosperous, and more supportive of thriving communities.

The integral perspective suggests a new approach to channeling the spirit of the GND into a suite of policy proposals while avoiding the political backlash inspired by its initial iteration, some of which the Obama administration was already doing, and the Biden administration appears to be pursuing. First, policy makers should decouple climate action from progressive social policies that are not obviously climate-specific, such as universal housing and healthcare, as well as a federal jobs guarantee. These policies trigger modernist concerns about undeserved entitlements, the welfare state, and 'big government.' At the same time, stress the public health benefits and healthcare savings of reducing carbon pollution, and the job creation, energy cost savings, and community renewal fostered by investments in green building infrastructure.

Second, on the legislative front, (1) fold as much climate policy into ostensibly non-climate bills as possible, such as a transportation bill to raise the gas tax and change how funds are allocated, and (2) pursue carbon pricing through a fee and dividend with a border-adjustment tax. Economists across the political spectrum insist that a carbon tax is the most efficient market-based policy tool to drive emissions reductions. The border-adjustment tax protects American firms and consumers (which traditionalists value), the fee is rooted in personal responsibility, incentivizes efficiency and innovation, and reduces the needs for regulations (which modernists value), and the dividend ensures that the poor do not shoulder the burden of higher energy prices (which progressives value).

Third, within the executive branch, the Biden administration should double-down on the Obama administration's interagency buckshot strategy, folding climate action into the missions of State, Defense, Transportation, Housing and Urban Development, Energy, Interior, and Agriculture. As Ben Adler writes in *Every Cabinet job is about climate change now*, "The climate cabinet is much larger than just the nominal climate team, because almost every cabinet job is actually a climate job. Most departments – from Interior to Transportation – can help reduce US greenhouse gas emissions and improve climate resilience" (Adler 2021). One objection here is that without major climate legislation, anything Biden does could be reversed by a Republican president; for an effective climate policy to stick, it has to be enshrined in law. Legislative options should indeed be exhausted, but more importantly, the context from the Obama years has changed. Though Trump rolled back many of Obama's executive actions on climate, future Republican presidents will face diminishing returns on such executive rollbacks, mainly because state and local governments and the private sector picked up the slack. The Biden administration's theory of the case is that a muscular push of industrial policy from the executive branch will create sufficient market pull that even absent climate legislation, no future administration will be able to reverse the tide.

Finally, on the messaging front, activists, politicians, and policy-makers should consider the following ideas. First, stop regarding climate as an 'environmental' issue, i.e., as a progressive issue. Second, frame climate less as a global issue – which summons the idea of our obligations to the global South – and more as a national, regional, and local issue, all of which resonate more with traditionalists. Third, expand environmental justice to include not just poor urban and minority communities, but poor rural and predominantly white communities, which are bastions of traditionalism. Fourth, emphasize that to the extent that there is a 'war on coal,' it is being waged by market forces, not government regulation; the real way to protect workers and communities dependent on coal production is to invest in infrastructure and low-carbon and renewable industries to create jobs *in* transitioning communities. Fifth, rather than talking about 'equity' – a concept that modernists mistakenly read as a call for equality of outcome that infringes on freedom and throttles the entrepreneurial spirit – emphasize that investment in renewable energy technologies and infrastructure is about creating more broad-based equality of opportunity and upward mobility. And finally, at the level of grand strategy, narrative, and vision, the president should frame the country's response to the climate crisis as a 'Place Race.' Much like the space race became both a real and symbolic proxy for the conflict between capitalism in the US and communism in the USSR, the goal to build a low-carbon energy economy, seen in the context of what many in the foreign policy community see as an emerging cold war with China, can become a long-term national project that binds the country to a higher purpose above tribal warfare. It can rally allied nations to unite in order to show that democratic capitalism, not authoritarian capitalism, is a superior model not just morally, but economically and ecologically. It constructively channels the ethnocentrism and patriotism of traditionalism, kindles the drive for innovation and love for freedom in modernists, and, obviously, addresses the environmental concerns of postmodernists.

These tools can help activists, politicians, policy makers, educators, community organizers, business executives, and church leaders to help size up the values and worldviews of the stakeholders and communities they work with, and to better educate and motivate them to work for climate justice – *however* they define it.

References

Adler, B. (2021) 'Every Cabinet job is about climate change now,' *Washington Post*, January 21st. Available at: https://www.washingtonpost.com/outlook/biden-cabinet-climate-change/2021/01/21/a288a12e-5b5d-11eb-b8bd-ee36b1cd18bf_story.html (Accessed January 25th, 2021)

Dickinson, J. et al. (2016) 'Which moral foundations predict willingness to make lifestyle changes to avert climate change in the USA?' *PLoS ONE*, 11(10), e0163852. Available at: DOI:10.1371/journal.pone.0163852 (Accessed January 9th, 2021)

Haidt, J. (2012) *The Righteous Mind: Why Good People Are Divided By Politics and Religion*, New York: Vintage.

Hochachka, G. (2020) 'The scenic route: A Developmental approach emphasizes the importance of human interiority in transformative approaches to climate change,' *Integral Review*, 16(1), pp. 187–214.

McIntosh, S. (2020a) *Developmental Politics: How American Can Grow Into a Better Version of Itself*, St. Paul: Paragon House.

——— (2020b) 'Polarization is more of a cultural problem than a political one,' *The Fulcrum*, February 15th [online]. Available at: https://thefulcrum.us/big-picture/polarization-is-more-of-a-cultural-problem-than-a-political-one (Accessed January 20th, 2021).

O'Brien, K. and Hochachka, G. (2010) 'Integral adaptation to climate change,' *Journal of Integral Theory and Practice*, 5(1), pp. 89–102.

Roberts, D. (2019) 'Don't bother waiting for conservatives to come around on climate change,' *Vox*, April 26th [online]. Available at: https://www.vox.com/energy-and-environment/2019/4/26/18512213/climate-change-republicans-conservatives (Accessed on January 15th, 2021)

Sauer, Hanno. (2015), 'Can't we all disagree more constructively? Moral foundations, moral reasoning, and political disagreement', *Neuroethics*, 8, pp. 153–159.

Vainio, A. and Mäkiniemi, J. (2016) 'How are moral foundations associated with climate-friendly consumption?' *Journal of Agricultural and Environmental Ethics*, 29, pp. 265–283.

Wehner, P. (2020) 'Jonathan Haidt is trying to heal America's divisions,' *The Atlantic*, May 24th [online]. Available at: https://www.theatlantic.com/ideas/archive/2020/05/jonathan-haidt-pandemic-and-americas-polarization/612025/ (Accessed January 13th, 2021).

Further Reading

Esbjörn-Hargens, S. and Zimmerman, M. (2008) *Integral Ecology: Uniting Multiple Perspective on the Natural World,* Boston: Integral Books. The definitive application of integral theory to ecology and environmental issues.

Hulme, Mike. (2009) *Why We Disagree About Climate Change: Understanding Controversy, Inaction, Opportunity,* New York: Cambridge University Press. An unusually wide-ranging interdisciplinary account of the roots of disagreement over climate drawing on science, economics, psychology, and other fields.

Inglehart, Ronald. (2108) *Cultural Evolution: People's Motivations Are Changing and Reshaping the World,* New York: Cambridge University Press. An overview of the status and implications of the World Values Survey.

Marshall, G. (2014) *Don't Even Think About It: Why Our Brains Are Wired to Ignore Climate Change,* New York: Bloomsbury. Survey of research on the psychology of climate change.

Watkins, A. and Wilber, K. (2015) *Wicked and Wise: How to Solve the World's Toughest Problems,* Great Britain: Urbane Publications. Analysis of climate change as a wicked problem from the perspective of integral theory.

37

ON THE EFFECTIVENESS AND LEGITIMACY OF 'SHAMING' AS A STRATEGY FOR COMBATTING CLIMATE CHANGE[1]

Behnam Taebi and Azar Safari

Introduction: Whose Actions, Whose Obligations?

The Paris Agreement at the Conference of Parties (COP21) to the United Nations Framework Convention on Climate Change (UNFCCC) marks a historic turn in climate change policy. Among other changes, states have agreed to substantially reduce their greenhouse gases (GHGs) in order to limit global warming to 2°C above pre-industrial levels. This agreement is based on accepting comprehensive national climate plans, or Intended Nationally Determined Contributions (INDCs), as the Paris agreement calls them. The Glasgow Climate Pact in 2021 built on the same principle and aimed to "reduce the gap between existing emission reduction plans and what is required to reduce emissions."[2] In achieving this level of GHG cuts, not only national states but also non-state actors such as municipalities and large corporations will play a crucial role. This could be problematic, because according to customary international law,[3] states may only be held responsible for the actions and omissions of their official organs. So states, as the main parties under obligation, are making promises that will be more or less effective depending on the active participation of many non-state actors. This chapter will focus on the role of one particular group of non-state actors, namely corporations – particularly large Multinational Corporations (MNCs), since they will play an essential role in making the global GHG cuts and making good on the national pledges.

One might argue that it is an ideal situation to introduce legally enforceable obligations for corporations. Assuming that "[a]voiding severe global catastrophe is a moral and legal imperative," a group of legal scholars and practitioners has published the Oslo Principles, in which they argue that "all States and corporations have an immediate moral and legal duty to prevent the deleterious effects of climate change" (Spier et al. 2015: 1). Climate negotiations during the last two decades have dealt with the extent of the state's legal duties. Corporations' moral and legal duties are a much less discussed subject. The authors of the Oslo Principles have explicitly included corporations as the second entity (in addition to states) because of their *ability* to accomplish the needed GHG reductions to avert climate change.[4] It is particularly relevant to consider the duties of these two most *capable* parties – namely states and corporations – in conjunction, because there is a relevant interaction between the two, i.e., states could play an important role in imposing certain restriction on corporations (legally binding or otherwise). Some scholars argue that corporations' (legal) obligations could best be conceived as a direct corporate responsibility under the auspices of international law (e.g., Adeyeye 2007) but this is not the chosen model for reducing GHGs internationally, given that the Paris Agreement is an international accord with nation states as the

 DOI: 10.4324/9781003039860-43

only parties. Alternatively, states could impose legal restrictions on corporations in accordance with international agreements. This is to an extent the current practice, but as we will discuss in Section "The Inadequacy of the Current Approach: States' Unwillingness and Inability," it has some serious shortcomings, i.e., states might be unwilling and sometimes even unable to take action with respect to MNCs; the phenomenon of the 'race to the bottom' is discussed in the second section, as well.

Short of effective legally enforceable rules, we need to explore alternatives such as incentivizing voluntary commitments among corporations along with (non-compulsory) compliance mechanisms. In this respect, it is worthwhile considering "naming and shaming" as one of the key strategies for penalizing noncompliance in voluntary approaches. In a sense, naming and shaming is implicitly the strategy for ensuring compliance among states as well. During the negotiations prior to the Paris conference, it became clear that a legally binding agreement would not have much chance of political success. The Obama administration, for instance, decided to forego a binding treaty, since such a treaty could not count on the required two-thirds majority in the US Senate; the administration's negotiators therefore explicitly promoted a strategy to 'name and shame' states into cutting their emissions (Davenport 2014).[5] So the Paris Agreement is based on accepting pledges by the parties (i.e. nation states) and regularly reviewing those pledges with "negative reputational consequences" for countries that fail to meet their targets (Jacquet & Jamieson 2016: 645). Acknowledging that reputational effects would have an impact beyond individuals,[6] Jacquet and Jamieson (2016: 643) suggest that the Paris Agreement could only succeed if "pledge and review" will be carried out beyond the nation states with "the power of shaming laggards." In this chapter, we investigate the effectiveness and the moral legitimacy of *shaming* as a strategy for incentivizing emission cuts among a very important group of non-state actors, namely large multinational corporations.

Indeed, there are already various mechanisms in place to incentivize and ensure non-state actors' contributions to cutting GHGs. Many non-state actors, including a considerable number of large corporations, have already committed themselves to several emission cuts. These cuts are, however, voluntary commitments without any external verification or compliance assurance. For such voluntary approaches to be effective, there are at least two requirements that need to be met. First, there should be an incentive mechanism to prompt more corporations to accept such voluntary emission cuts, because the success of the agreement very much relies on the participation of a large number of MNCs. Such an incentive mechanism should have high gains – such as public acknowledgment of a corporation's role in combatting climate change – and high losses, such as clear reporting on noncompliance and public exposure (naming), along with some kind of condemnation of such behavior (shaming) (building on Friman 2015b). Second, since reporting on voluntary cuts is often based on self-assessments, there needs to be a verification mechanism in order to ensure compliance with the agreed-upon cuts. The third section focuses on specifying the non-state actors and their roles in combatting climate change. The fourth section presents several *quasi-judicial*[7] (inter)governmental and non-governmental experiences with similar issues. By drawing comparisons with Corporate Social Responsibility (CSR) and with agreements to voluntarily commit to environmental programs supported by intergovernmental and non-state actors, we investigate the extent to which 'naming and shaming' could be a viable strategy for securing the contributions of MNCs for combatting climate change.

The fifth section deals with the issue of moral legitimacy of such commitment based voluntary approaches. One might argue that approaches based on voluntary commitments (and shaming the one who would not comply) might blur the fundamental moral obligations and are, therefore, ethically illegitimate. In this section, we will investigate the nature of moral obligations of corporations and the legitimacy of replacing such obligations with voluntary based approaches such as the one presented in this chapter. We argue that the effectiveness and legitimacy questions are closely tied together; provided that such voluntary based (quasi-judicial) approaches are the only alternative to legally imposed duties, they are most morally defensible if they would indeed be most effective in reducing the harmful GHGs. The fifth section further discusses the ethical limitations and pitfalls of voluntary approaches. The last section presents the conclusions.

The Inadequacy of the Current Approach: States' Unwillingness and Inability

As mentioned in the introduction, imposing legal restriction by nation states on MNCs might be rather problematic. One of the most significant challenges is the so-called 'race to the bottom.' That is, when host states look for more investment-driven development, they are willing to barter their power of regulation in exchange for short-term economic gains; "[i]n order to attract investment, many nations, particularly developing ones, will acquiesce to a corporation's needs … [by establishing] a corporate-friendly legal environment" (Macek 2002:104). To be more precise, states seem to be reluctant to impose limits on MNCs, thereby sacrificing their national interests in order to comply with international obligations. This race to the bottom often compels these states to lower their human rights standards, especially where labor rights are concerned, and to deregulate environmental and tax laws (Deva 2003, 2004; Milanovic 2009; Holland 2010).

In addition to being unwilling, some states rely on the revenues from MNCs to such an extent that they cannot afford to lose those revenues by imposing restrictive regulations. Hence, when a state's revenues are consolidated heavily in one entity, that state might simply be *unable* to impose restrictions on MNCs; a race to the bottom is then inevitable (Revak 2012). In a study performed in 2000, Anderson and Cavanagh showed that among the 100 biggest 'economies' in the world, 51 are corporations and 49 are countries (Anderson & Cavanagh 2000).[8] Moreover, "[t]he annual revenues of General Motors are greater than the GDP of more than 148 countries; while Wal-Mart's revenues exceed the combined GDP of sub-Saharan Africa, excluding South Africa and Nigeria" (Stiglitz 2007: 476). Indeed, that this is the case that these are large and powerful companies does not necessarily imply that they are unwilling to take responsibility for their environmental and climate related impact. Wal-Mart was one of nine massive companies that announced prior to the Paris negotiations that they would switch to 100 percent renewable energy; this was part of the RE100 Campaign as it will be discussed in the third section. Many corporations seem willing to contribute to global GHG cuts, because they acknowledge the importance of avoiding catastrophic climate change and also because they see long-term returns on their low carbon investments. But it should be mentioned that these nine corporations also epitomize the problem at hand; they have received extensive media exposure because they are on the list of Fortune 500 companies, meaning that only nine out of these 500 have pledged substantial action. Moreover, making pledges and commitments is one thing, but acting upon these promises and having the results externally audited is another. This issue is discussed at length in the following sections.

The Role of Non-State Actors in Global Emission Cuts

Since the Copenhagen Conference of Parties in 2009, there has been a new discourse surrounding climate change, focusing on limiting the increase of the global temperature above certain levels. The 2°C increase compared to pre-industrial levels was a limit agreed upon in Copenhagen; the Paris Agreement reiterated this target and made this goal more ambitious by presenting the more stringent target of 1.5°C above pre-industrial levels, "recognizing that this would significantly reduce the risks and impacts of climate change" (UNFCCC 2015: Article 2). In line with the

Copenhagen Agreement, recent policies regarding the target emission are, therefore, expressed in terms of an *emission gap* between business-as-usual and the desirable emission cuts that could help to meet the 2°C target. This has proven to be a helpful approach, since it shows that continuing the business-as-usual scenario would be utterly insufficient for meeting the target, as demonstrated by the United Nations Environment Program (UNEP) 'Bridging the Emission Gap' annual reports. The UNEP report from 2011 stressed that there is a need to cut 12 gigatons of CO_2 equivalent (Gt CO_2e)[9] globally, while national government pledges back then indicated a willingness to cut only half that amount (UNEP 2011). The Paris Agreement – if it proves successful – aims at substantially narrowing this gap in order to ensure that the global temperature would not rise above the indicated 2°C.[10]

Prior to this new thinking about emission cuts, scenarios seemed plausible in which "sovereign national governments agree under the UNFCCC on emission reductions; they subsequently introduce in their jurisdiction the right incentives for emission reductions; and finally, companies, municipalities, other organizations and individual citizens take measures to reduce their greenhouse-gas emissions" (Blok et al. 2012: 471). This top-down approach was, however, deemed insufficient. A new type of climate policy was needed, based on a combination of top-down and bottom-up approaches, in which non-state actors would also play a central role (Blok et al. 2012; Hsu et al. 2015; Jacquet & Jamieson 2016). This need was acknowledged by the introduction of the Non-state Actor Zone for Climate Action (NAZCA) at the Lima conference in 2014.[11] Likewise, the Paris Agreement welcomed "the efforts of non-Party stakeholders to scale up their climate actions, and *encourages* the registration of those actions in the Non-State Actor Zone for Climate Action platform" (UNFCCC 2015: 17, emphasis in original).

The term 'non-state actor' requires some explanation here. Strictly speaking, non-state actors are those actors that are not parties to the UNFCCC. Where climate change is concerned, there are various subnational and supranational non-state actors, including municipalities, regions, national and international non-governmental organizations (such as environmental organizations), and corporations. This chapter focuses on large corporations that have a crucial role to play in cutting GHGs. There are already a large number of International Cooperative Initiatives (ICIs) that report on non-state actors' efforts to reduce climate change. The 'Bridging the Emission Gap' publications report on these and other non-state actors' initiatives for global mitigation efforts.[12] Thirty such initiatives are currently listed on the Climate Initiatives Platform, "a new online portal for collecting, sharing and tracking information about International Climate Initiatives" (UNEP 2015).[13] Many large corporations have thus already committed themselves to voluntary GHG cuts in the coming years. Prominent initiatives include the Business Environmental Leadership Council (BELC), which is "the largest US-based group of corporations"; the Caring for Climate Initiative, which aims to advance the role of businesses in climate change; and RE100, which wants "at least 100 companies to make a global 100 percent renewable commitment" within a reasonable time frame (UNEP 2015: 41).

The current approach to corporations with respect to climate change has two main problems. First, a large number of the currently operable corporations that could have serious impact on GHG cuts are essentially MNCs, operating in different nation-states. As discussed in Section "The Inadequacy of the Current Approach: States' Unwillingness and Inability," imposing legal restrictions by individual states might be rather problematic and sometimes simply impossible. Second, all the aforementioned initiatives are based on the voluntary commitments of corporations or groups of corporations; reporting is mostly based on self-assessment and is often without third-party verification. The fourth section elaborates on this shortcoming of the existing mechanisms, by comparing it to similar situations in other areas.

Effectiveness of Shaming as a Strategy: Governmental and Non-Governmental Approaches

As mentioned earlier, countries have agreed to certain emission cuts – or the INDCs – in the Paris Agreement and the Glasgow Pact. While the detailed implementation of these INDCs remains unclear, it seems likely that their measuring, reporting, and verification (MRV) will not be in the hands of state actors alone. Earlier international negotiations (most notably in Bali and Cancun) have established mechanisms for MRV procedures. The international verification of national reports should take place through international review, "which is a process to increase the transparency of mitigation actions and their effects, and support needed" (UNFCCC 2014: 16). It seems clear that transparency and trust are the key issues in international reporting and verification, as has been emphasized in Article 13 of the Paris Agreement: parties need to build "mutual trust and confidence and to promote effective implementation, an enhanced transparency framework for action and support" (UNFCCC 2015: 28). Transparency and trust are also key to the domestic implementation of the Agreement within the states, and for the reporting and verification of the reductions as promised by non-state actors, including MNCs.[14] In this section, we first review some quasi-judicial experiences with similar issues, such as CSR, environmental management and sustainability. We then discuss the requirements for an effective naming and shaming strategy, also considering empirical evidence from the literature (Friman 2015a).

In the absence of international or domestic legally-binding responsibilities (also known as hard law) for MNCs, several internationally-recognized intergovernmental and non-governmental organizations have been formed within the realm of soft law. Soft law is a quasi-legal instrument that aims to institutionalize a social norm without exerting legally-binding force. Most efforts in this area have been devoted to addressing transparency and disclosure; they aim to create incentives for MNCs to behave ethically sound, specifically where human rights and CSR are concerned (Sutton 2003; Backer 2008). Two important instances of state-involved soft-law innovations are the initiatives of the Organization for Economic Co-operation and Development (OECD) and the United Nations Global Compact (UNGC). The OECD has developed principles that build on the notion of transparency, including the OECD Principles of Corporate Governance (OECD 2004) and the Guidelines for Multinational Corporations (OECD 2008). These provide "the contours for a system of monitoring and reporting that have a potentially significant application to issues of environmental transparency [...] without invoking the formal structures of the domestic legal orders of participating states" (Backer 2012: 110). The UNGC is perhaps one of the most recognizable initiatives for the collection and disclosure of CSR-related information (Akhtarkhavari 2010); it is organized around ten conduct-oriented principles covering subjects such as "human rights, labor, environmental and anticorruption values," all of which aim to create a framework for corporate accountability (Backer 2012: 115). Though the UNGC requires an Annual Communication of Progress (CoP), this communication is based on self-reporting. Nonetheless, failing to comply could change an organization's status to 'non-communicating' or 'inactive.'

Among the non-governmental initiatives, the Global Reporting Initiative (GRI), the International Organization of Standardization (ISO) and the Carbon Disclosure Project (CDP) are examples of organizations with huge international credit and publicity power. The GRI is a nonprofit that helps corporations and other organizations "understand and communicate the impact of business on critical sustainability issues such as climate change, human rights, corruption and many others."[15] Compared to the UNGC, the GRI demands a greater level of detail in its requests for disclosure and has much less media appeal; all the same, it has produced a successful and often used disclosure mechanism (e.g., Hedberg & von Malmborg 2003; Brown et al. 2009; Backer 2012). The ISO is mainly involved in developing standards for environmental transparency and for communication about environmental management; among other things, ISO provides for the most widely adopted voluntary environmental programs (Matthews 2003; Prakash & Potoski 2006; Backer 2012). Finally, the CDP is also a non-profit organization that provides reporting mechanisms. CDP's website explains that one of the benefits of becoming a signatory is "public recognition of your commitment to engaging with companies on issues of climate change."[16] Companies self-report to CDP by means of a questionnaire, indicating whether their report and assessment have been verified or whether such verification is still underway (UNEP 2015: 40). However, like many other voluntary initiatives, third-party verification does not seem to be standard practice.

What all the aforementioned intergovernmental and non-governmental mechanisms share is that they work based on voluntary commitments. Going back to the main focus of this chapter, we argue that an approach based on the voluntary engagement of MNCs in substantially reducing GHG emissions will only be successful if (1) there are incentive mechanisms for motivating more corporations to accept such voluntary emission cuts and (2) there is an independent monitoring and verification system. Let us elaborate on these two requirements.

An incentive mechanism should have high gains, such as public acknowledgment of the responsibility a corporation assumes for combatting climate change, such as the extensive media exposure that Wal-Mart and other eight Fortune 500 companies were receiving because they are moving toward to 100 percent renewable energy consumption. More broadly, accepting voluntary emission cuts, for instance by becoming a member of influential and credible international organizations such as the UNGC and CDP, has mostly reputational benefits for corporations. But there should also be high losses, such as clear reporting on noncompliance and public exposure (naming), as well as condemnation of such behavior (shaming). The question is whether such naming and shaming could be a potent deterrence strategy and if so, under what conditions. Let us elaborate on some key element of a system of naming and shaming as a policy instrument that could potentially maximize the impact on corporations' compliance. Building on Friman (2015a: 203), we distinguish between the three key steps of naming or "the public identification of noncompliance," shaming or "public condemnation of such behavior" and some kind of "material sanction" that could strengthen shaming.[17] Let us review these three steps in more detail.

An effective naming strategy hinges on the *credibility* of the organization that could name noncompliance and the *reliability* of the information that substantiates the naming. If states would be the parties that would *name* the noncommitting corporations, the issue of credibility does not seem to be a problem here.[18] Things become more complicated when it comes to the reliability of the information, since in a voluntary approach the information is based on self-assessments of corporations; a robust verification mechanism based one external audits seems to be indispensable here. There are two potential problems here. First, if we want shaming – as the next step that must follow naming – to be an effective strategy, the results of the external audits must become public. This could, in turn, make it more difficult for corporations to commit to such external audits. The relevant literature shows that corporations submit more readily to external audits if they know that the reports are not going to be public (e.g. Potoski & Prakash 2004). This gives rise to a dilemma between making shaming more effective on the one hand and increasing corporations' participation on the other. Second, there is the issue of *selective disclosure*, as defined by Marquis et al. (2016) as "a symbolic strategy whereby firms seek to gain or maintain legitimacy by disproportionately revealing beneficial or relatively benign performance indicators to obscure their less impressive overall performance." This problem emphasizes the need for clear and unified measurement, monitoring and verification practices, such that the final outcome could rule out selective disclosure. In sum, for naming to be possible, a robust measurement, monitoring, and verification mechanism is needed to ensure the reliability and quality of the information provided in corporations' self-assessments.

After the disclosure of the information about noncompliance, the next step is to move to shaming or condemnation of such an act. At first glance, being labeled noncompliant might seem to have a negative impact on corporations. Media can shape public opinion by providing people with evidence of such noncompliance. In most CSR cases known in the literature, after shaming MNCs have tried to showcase their collaboration with the international community by accepting public criticism and employing redressing measures to become compliant. At times, they have attempted to compensate for such violations by enhancing the support for consumers and workforces. Two interesting examples are worth mentioning here. The first is Nike that was involved in labor violations and other exploitative employment practices in the late 1990s. The company was then subjected to massive public disapproval, which resulted in loss of profit and reputation. Nike responded to this by accepting the critique and implementing a fairly strict supply chain (in accordance with CSR) (Waller & Conaway 2011; Shavers 2012).[19] The second example of successful shaming is when Shell proposed the disposal of the Brent Spar oil platform in the deep sea in North Sea; the proposal involved letting the platform sink rather than to dismantle and dispose of it onshore. Greenpeace started a powerful campaign resulting in, among other things, a widespread boycott of Shell Gas Stations in Northern Europe, which in turn let Shell abandon the plan (Gunningham 2009; Rosen-Zvi 2011).[20]

However, most of the known examples of when shaming has proven to be effective concern corporations that are in direct contact with their consumers; damage to their image could therefore directly undermine their business with individual consumers. Instead of such a Business-to-Consumer (B2C) relationship, many large corporations are engaged in Business-to-Business (B2B) or Business-to-Government (B2G) relationships, in which individual consumers' pressure is less and sometimes non-existing. Indeed, in a B2B and B2G settings shaming could still happen as a result of one corporation shaming the other corporation (as in B2B) or a government shaming a corporation (as in B2G). In a B2G situation, the impact of shaming seems evident. It seems, however, unlikely that governments will shame corporations (in a B2G relation) because, first, the 'race-to-the-bottom' problem could play a role in state-corporations interaction and, second, it neglects the de facto inability of economically less powerful states to impose restrictions on MNCs or to shame those powerful MNCs for not complying with commitments. The most problematic situation is the B2B relations in which neither the reputation damage nor a direct relation with the state would be a deterrent.

In sum, we could say that shaming could be an effective strategy for incentivizing corporations to cut GHG emissions if: (i) there would be high benefits associated with accepting those commitment (such as public acknowledgment of the responsibility a corporation) and high costs associated with non-compliance; and (ii) there would be an international independent measurement, monitoring and verification mechanism in place. Moreover, the literature suggests that shaming would be potentially most powerful in B2C relationships. In B2G relationships shaming could lose power (depending on the willingness and ability of the states); in B2B relationships shaming is most problematic as a strategy because, first, there is little public information about those interactions and, second, reputation damage is not a potential deterrent anymore.

Moral Legitimacy of Shaming as a Strategy

The previous section discussed shaming as a strategy for incentivizing emission cuts among large MNCs. In this section, we focus on the moral legitimacy of such approaches. One might argue that approaches based on voluntary commitments (and shaming the one who would not comply) might blur fundamental moral obligations and are, therefore, ethically illegitimate. In this section, we address the question of moral legitimacy by, first, focusing on the nature of moral obligations of corporations and, second, the legitimacy of replacing such obligations by voluntary based approaches such as the one presented in this chapter.[21]

Many people acknowledge that corporations have special obligations when it comes to protecting the environment, but there is no consensus about the nature and the extent of such obligations. Defending market solutions, Bowie (2013), for instance, argues that corporations have an obligation to abide by law and environmental legislation.[22] In his account, the moral responsibility for the environment lies mostly with consumers that could increase the demand by buying environmentally friendly products; the higher the demand, the more likely it becomes that corporations would act accordingly, as the argument goes. Against such minimalist account of corporations' obligations, Arnold and Bustos (2005, 2013) defend a historical account of obligations, i.e. corporations that have contributed to the accumulation of the harmful GHGs have a moral obligation to deal with the consequences.[23] So, they argue that large corporations (with massive energy consumption) have contributed to the existence and the perpetuation of global climate change. Bowies' minimalist account of obligations further fails because it is based on a hidden (and mostly unsubstantiated) assumption, namely that all corporations are in B2C relations, that is, consumers are always in a direct relationship with corporations and, by that in the position to steer corporations' policies by their purchases (or boycotts). This assumption is problematic for three reasons. First, as we have argued in Section "Effectiveness of Shaming as a Strategy: Governmental and Non-Governmental Approaches," many corporations (especially bigger ones that are most responsible for GHGs) have a strong relation with governments (B2G) or with other corporations (B2B). So, there is not always a B2C relationship that could put consumers in the right position to exert their influence. Second, even when corporations are in B2C relations, the consumer does not always have full disclosure about the supply chain and the subcontractors of the main corporation. So, even if the consumer would be willing and able to steer a corporation's behavior, she does not always have access to information regarding the whole supply chain. Third, such emphasis on consumers presupposes that all consumers are in the luxurious position to be able to afford to choose between environmentally friendly and unfriendly products. This neglects the issue of poverty, not only in developing countries and emerging economies (with lower standards of well-being) but also in industrialized countries. In sum, shifting environmental responsibilities from the corporations to consumers is an unhelpful and problematic approach.

Another way of approaching (moral) obligations of corporations is based on their *ability*, i.e., one could also argue that corporations and nation states are the two most *capable* parties that could contribute to the mitigation of climate change by substantially reducing the GHGs. The Oslo Principles take this stance in defending moral and legal obligations for corporations (Spier et al. 2015). Assigning obligations based on both historical accountability and the capability of the parties has been intensely debated when discussing distributions of responsibilities among nation states; the United Nations Framework Convention on Climate Change (of which the 2015 Conference of Parties meeting was held in Paris, leading to the Paris Agreement) is based on the notion of 'common but differentiated responsibility.' Many discussions in the field of climate ethics aim at spelling out the extent of these obligations for different states. It is not our intention to reiterate those arguments here,[24] but it is essentially the same rationale that could be expanded to include the role of corporations that have contributed to the existence of the problem on the one hand, and could contribute to its solution, on the other.

Let us now turn to the question of the moral legitimacy of replacing moral obligations with voluntary based approaches. If we agree that there are certain moral obligations for corporations, the most straightforward action would be to turn those obligations into legal duties. As mentioned in the first section, the duties of corporations can be seen either as a direct corporate responsibility under the auspices of international law or as duties that national states must impose on corporations. The international approach is not the chosen model in the Paris Agreement, and it is probably going to be difficult to include an international enforcement mechanism. Imposing legal duties by nation states might be rather problematic too because states might be unwilling or sometime unable to impose such restriction (as discussed in the second section). So, an alternative approach is a mechanism of (voluntary) pledge and review. The Paris Agreement presents such an approach for dealing with states, i.e., countries commit to voluntary emission cuts and report on their progress; the laggards will then be shamed for being late or non-compliant. This is called by Jacquet and Jamieson (2016) the "soft but significant power" of the Paris Agreement. In this chapter, we have extended the same rationale to include corporations in the analysis. Since climate change is probably the most complex problem, the world is facing and humanity has not been successful is averting it so far, all actors must get motivated to get involved (Jamieson 2014). The top-down approaches in the pre-Paris era have not been successful. These voluntary bottom-up approaches seem to be the only feasible alternative to comprehensive legally binding duties. In addressing the question of moral legitimacy, we argue that the effectiveness and legitimacy are tied together, that is, assuming that such voluntary based (quasi-judicial) approaches are the only alternative to legally imposed duties, such approaches are most morally defensible if they would indeed be the most effective in reducing the harmful GHGs.

Discussions and Conclusions: Under Certain Conditions, Shaming Could Work

Implementing the ambitious goals of the Paris Agreement and the Glasgow Pact depends on the contributions of both states and non-state actors, most notably large Multinational Corporations (MNCs). In the absence of relevant legally imposed regulations, national states might instead incentivize MNCs to engage in voluntary cooperation to cut their emission gases. An important challenge of such voluntary approaches is how to ensure compliance with the agreed upon commitments. In this chapter, we have investigated the effectiveness and the moral legitimacy of *shaming* as an approach for penalizing noncompliance. Let us first start by addressing the question of moral legitimacy. One might argue that having legally enforceable rules under the auspices of the international law would be the most defensible option, but unfortunately this is not the chosen practice in the Paris Agreement and the Glasgow Pact, and it seems unlikely that such legally enforceable rules will be introduced any time soon. Assuming that voluntary based approaches are the only alternative to legally imposed duties, we have argued that such approaches are a morally defensible solution if they would indeed be most effective in reducing the harmful GHGs. In our argument, effectiveness and legitimacy are strongly interlinked. This brings us to the question of effectiveness.

While eliciting voluntary cooperation might sound too informal and noncommittal, and thus ineffective, there are good reasons to believe that such incentives would work. In Section "Effectiveness of Shaming as a Strategy: Governmental and Non-Governmental Approaches," we have discussed several governmental and non-governmental initiatives that are quite successful in promoting CSR and incentivizing environmental management issues, despite a lack of legal enforcement mechanisms. A good example appears in one of the standards of the International Organization of Standardization (ISO) for environmental management: ISO14001 requires corporations to spend a good deal of money to create environmental management systems and yet, this is among the most widely applied standards (Prakash & Potoski 2006). In general, it seems fair to argue that voluntary based (non-binding) approaches could be effective too. As Rosen-Zvi (2011) correctly argues, the question is not *whether* voluntary based approaches are effective but *under which criteria* they could be effective. Likewise, we argue that under certain criteria, shaming could be an effective strategy for incentivizing emission cuts by large corporations. Those conditions could best be formulated by discussing the ethical pitfalls of voluntary approaches. We will recapitulate several pitfalls as discussed in this chapter, while discussing how they could best be responded to.

The first ethical pitfall is that voluntary commitments might shift the attention from fundamental moral obligations that corporations have in combatting climate change (as discussed in Section "Moral Legitimacy of Shaming as a Strategy") to "charity-based" commitments that corporations could make; this leads some corporations to choose to comply only selectively (Deva 2003). To be more precise, MNCs are free to choose among various regulations, to follow their desired implementation methods and to release only the information they are willing to disseminate. This is also referred to as *greenwashing* or selective disclosure, that is when corporations decide to reveal only beneficial or benign performance results (Marquis et al. 2016). We have argued that certain institutional arrangements need to be put in place in order to reduce the risk of greenwashing. More specifically, we think that after corporations have agreed to certain emission cuts, there must be an independent measurement, monitoring and verification mechanism in order to ascertain that the volunteered cuts are real and not just empty commitments.

As mentioned earlier, the literature suggests that 'naming and shaming' could contribute to a behavior change among corporations, but "the degree to which shame functions to change behaviour varies widely across firms and sectors" (Haufler 2015: 199). Likewise, we acknowledge that there is no one-size-fits-all for all corporations. More specifically, we zoom in on the different types of corporations' relationships, being Business-to-Consumer (B2C), Business-to-Government (B2G) and Business-to-Business (B2B). Before discussing the pitfalls of each type of relationship, two remarks seem to be in order. First, we don't claim that all corporations fall in one of the three categories; business corporations could be in a varying degree in business with direct consumers, the government, or other corporations. Second, our analysis did not have the ambition to be comparative; we have merely focused on how shaming could serve as a strategy to incentivize good behavior and to penalize non-compliance in each business relationship type and discussed the potential and the pitfalls of shaming as a strategy in each type of business relationship. Future empirical research needs to explore the effectiveness of shaming strategies in different business relationships and in different (industrialized and developing) states.

The most prominent examples in the literature, in which shaming has influenced corporations' behavior are in issues associated with CSR and environmental management. In these examples shaming has proven to be effective because those corporations were in direct contact with their consumer, or in a so-called B2C relationship. So, if shaming would be adopted as a strategy when corporations are voluntarily agreeing to emission cuts, those corporations that are in direct relationship with their individual consumers would be most likely sensitive to such a strategy. An important condition that needs to be met here is that there must be a reliable source of information about the corporations' behavior. So not only should there be an independent measurement and monitoring mechanism in place (pertaining to the first greenwashing pitfall) but also this information must be made public before shaming could be effective. This need for transparency, however, might cause some corporations to be less eager in accepting voluntary emission cuts in the first place, as discussed in the previous section.

When corporations have a relationship with states (in a B2G setting), the shaming mechanism works differently. States are in principle in a position to shame a corporation for lack of proper behavior (or lack of compliance with the agreed upon cuts) but there are at least three types of pitfalls. First, states might be inclined to wrongly blame corporations, for instance to cover for their own incompetence or lack of compliance with the international agreements. Second, in economically less powerful states, there might be a serious reluctance to impose any restrictions on corporations or to shame corporations for non-compliance (as discussed in the second section). This problem could be resolved by involving the (stronger) state in which parent companies of the MNCs are active or by involving international organizations. Indeed, this is presupposing that economically more powerful states in which the parent companies of MNCs are active would be willing to cooperate. The latter might prove to be an unsubstantiated presupposition, as we will discuss later in this section. This brings us to the third pitfall of shaming in a B2G setting, that is, the race to the bottom. A possible solution to this problem (both in economically powerful and less powerful states) is that when corporations voluntarily commit to any emission cuts, these commitments will be publicly disclosed so that independent international organizations[25] could be involved in the process of shaming. There are already several examples of such powerful international organizations that are involved in collecting independent and reliable data (as discussed in the previous section).

The third business type relationship is when corporations are in direct relation with each other. For shaming to be an effective strategy, we need to assume that (i) corporations would be informed about each other's voluntary commitments and (ii) willing to shame each other if one party does not comply. However, a lot of what is happening in B2B relationships is likely to stay outside of the public eye and public scrutiny. For shaming to be effective here, there needs to be a third overseeing party (either an international organization or a state) that could engage in shaming whenever needed. Of course, this is again presupposing that there is independent and reliable information available about the performances of each corporation.

To be sure, we do not claim to have found the silver bullet to ensure corporations' contributions to combatting climate change. We fully acknowledge the problems with this approach mostly for incentivizing MNC action in the developing countries, but also in some industrialized nations where the governments do not fully support the Paris Agreement and the Glasgow Pact. For instance, some politicians in the United States have consistently denied the human induced impact of climate change and the findings of the IPCC; the recently elected President Donald Trump seems to strengthen these voices. If the United States decides to not comply with its INDC or to withdraw from the Agreement altogether, it could seriously weaken the Paris Agreement, i.e. the world powers, most notably the biggest polluters, seem to have kept each other in a prisoners' dilemmatic balance in this agreement. Our argument rests on the assumption that the Paris Agreement will stay in place and that both states and non-state actors such as corporations will be engaged in making it a success.

Notes

1 This chapter is a slightly adjusted version of a paper published by the same authors in Science and Engineering Ethics (Taebi & Safari 2017).

2 https://unfccc.int/process-and-meetings/the-paris-agreement/the-glasgow-climate-pact-key-outcomes-from-cop26

3 This has also been emphasized in the Draft Articles on Responsibility of States for Internationally Wrongful Acts, also known as the DARSIWA 2001 (ILC 2001).

4 While the Principles document does not elaborate on this issue, the commentary that accompanies this document explicitly discusses this issue; see page 64 in the following commentary document: http://www.osloprinciples.org/macmillan/globaljustice/Oslo%20Principles%20Commentary.pdf.

5 In the international relations literature, naming and shaming has been thought to shape or change states' behavior. It is a strategy frequently used by human rights organizations; see, for instance, various contributions in (Friman 2015c). This chapter, however, asks how shaming could incentivize corporations to comply with the targets set forth in the Paris Agreement.

6 It has been empirically shown that the threat of disapproval (i.e., shaming) could lead to more cooperation at the individual level; see (Jacquet et al. 2011; Jacquet 2015).

7 A quasi-judicial mechanism is essentially judicial in character but it is not within the officially defined judicial power as for instance defined in constitutions. An important feature of a quasi-judicial mechanism is that there is no need to have a *de facto* judge in the procedure, meaning that arbitrators or even scientific experts may perform as deciding authorities. Another important difference with judicial mechanism is that quasi-judiciary mechanisms often lack an enforcement mechanism; naming and shaming or

other punitive actions such as suspension or termination of non-compliant parties' conventional rights seem to be the surrogate for actual legal enforcement. In Section "Effectiveness of Shaming as a Strategy: Governmental and Non-Governmental Approaches," we discuss several examples of quasi-judicial mechanism in discussions on Corporate Social Responsibility and environmental management.

8 The comparison is based on countries' GDPs and corporations' sales.

9 This is a unit for expressing the amount of greenhouse gases. Different GHGs have different impacts on global warming – methane, for instance, is a much more potent GHG than CO_2 – but we need to be able to express everything in a single unit. A CO_2 equivalent expresses all entities in an amount comparable to CO_2.

10 According to Stavins (2015), 96 percent of global emissions is accounted for in the Paris Agreement, since "186 of the 195 members of the UNFCCC submitted INDCs by the end of the Paris talks."

11 See the *Lima Call for Climate Action* Decision-/CP20 (UNFCCC 2014); http://unfccc.int/files/meetings/lima_dec_2014/application/pdf/auv_cop20_lima_call_for_climate_action.pdf.

12 See, for instance, Chapter 5 in (UNEP 2015).

13 Citation from the website of this initiative; see http://climateinitiativesplatform.org/index.php/Welcome. Consulted on 3 March 2016.

14 In more technical legal terms, in the international verification process the Biennial Update Reports (BURs) of each country will be reviewed in a procedure for International Consultation and Analysis (ICA). National verifications need to be organized domestically, and states need to report on their domestic verification in their BURs (UNFCCC 2014, 16).

15 Cited from the webpage of the GRI: https://www.globalreporting.org/information/about-gri/Pages/default.aspx. Consulted on 15 March 2016.

16 Adopted from the CDP website: https://www.cdp.net/en-US/Programmes/Pages/becoming-a-signatory.aspx. Consulted on 15 March 2016.

17 This is based on the synthesis analysis in the concluding chapter of the volume "The Politics of Leverage in International Relations: Name, Shame and Sanction" (Friman 2015c). While most contributions in this book are about the role of naming and shaming in international relations, the basic rationale of the arguments applies to the type of naming and shaming as discussed in this paper too.

18 We are assuming that states will not wrongfully 'name' a corporation, for instance in order to cover their own failure or incompetence. While this possibility exists, it seems unlikely that states would often undertake any 'naming and shaming' activity at all because of their unwillingness (race to the bottom) and inability (in case of less powerful states). So, the issue of the credibility of the organization (other than states) that does the 'naming' remains a concern.

19 For a detailed discussion of this case, see (Sutton 2003, 1170).

20 The Brent Spar case is a special case in that Shell seemed to have sufficient scientific evidence that their original proposal was the one that should have been preferred from an environmental point of view. Greenpeace suffered some reputation damage afterwards because their scientific evidence seemed to be flawed. We did include this example nonetheless because this case study seems to mark an important moment in which "major reputation sensitive companies" realized the power of shaming that could be "unleashed upon them by NGOs" (Gunningham 2009, 196).

21 Indeed, when fulfilling ethical obligations would become voluntary this would not lose their ethical stringency. However, when we agree that there are moral obligations but instead of turning those into legal obligations we leave the fulfilment of these moral obligations to voluntary approaches, it does require a reflection on the legitimacy of such voluntary approaches. This section deals with that question of legitimacy.

22 Environmental concerns and concerns for climate change are in principle distinct and only partly overlapping. Many authors, however, refer to them interchangeably, especially when it comes to questions of corporations' responsibilities for dealing with these concerns.

23 This argument is based on the assumption that corporation would have been aware of such consequences. The authors argue that this is a reasonable assumption, certainly after the wide scientific acknowledgement of anthropogenic climate change in the beginning of this century by the IPCC.

24 Interested readers could consult two leading anthologies on these and other related climate ethics subjects with regard to national and international responsibilities (Gardiner et al. 2010; Arnold 2011).

25 Indeed, the same problems and issues might again play at the level of such organization. Questions such as how the organizations should be financed, to whom does it report, how does it handle matters that are revealed to it "in confidence" etc. are relevant questions that need to be addressed prior to initiating any international organization for this purpose.

References

Adeyeye, A. 2007. "Corporate Responsibility in International Law: Which Way to Go?" *Singapore Yearbook of International Law* 11: 141–161.

Akhtarkhavari, A. 2010. *Global Governance of the Environment: Environmental Principles and Change in International Law and Politics*. Cheltenham: Edward Elgar Publishing.

Anderson, S., and J. Cavanagh. 2000. *The Rise of Corporate Global Power.* Washington, DC: Institute for Policy Studies.

Arnold, D. G. 2011. *The Ethics of Global Climate Change*. Cambridge: Cambridge University Press.

Arnold, D. G., and K. Bustos. 2005. "Business, Ethics, and Global Climate Change." *Business & Professional Ethics Journal* 24 (1/2): 103–130.

———. 2013. "Business Ethics and Global Climate Change." In *Ethical Theory and Business*, edited by D. G. Arnold, T. L. Beauchamp and N. Bowie, 9th ed., 462–475. Boston, MA: Pearson.

Backer, L. C. 2008. "From Moral Obligation to International Law: Disclosure Systems, Markets and the Regulation of Multinational Corporations." *Georgetown Journal of International Law* 39: 591–630.

———. 2012. "Transparency and Business In International Law - Governance Between Norm and Technique." In *Transparency In International Law*, edited by A. Bianchi and A. Peters, 477–501. Cambridge: Cambridge University Press.

Blok, K., N. Höhne, K. van der Leun, and N. Harrison. 2012. "Bridging the Greenhouse-Gas Emissions Gap." *Nature Climate Change* 2 (7): 471–474.

Bowie, N. 2013. "Morality, Money and Motor Cars." In *Ethical Theory and Business*, edited by D. G. Arnold, T. L. Beauchamp and N. Bowie, 9th ed., 456–462. Boston, MA: Pearson.

Brown, H. S., M. de Jong, and T. Lessidrenska. 2009. "The Rise of the Global Reporting Initiative: A Case of Institutional Entrepreneurship." *Environmental Politics* 18 (2): 182–200. https://doi.org/10.1080/09644010802682551.

Davenport, C. 2014. "Obama Pursuing Climate Accord in Lieu of Treaty." *The New York Times*, August 26, 2014.

Deva, S. 2003. "Human Rights Violations by Multinational Corporations and International Law: Where from Here?" *Connecticut Journal of International Law* 19: 1–57.

———. 2004. "Acting Extraterritorially to Tame Multinational Corporations for Human Rights Violations: Who Should 'Bell the Cat'?" *Melbourne Journal of International Law* 5: 37–65.

Friman, H. R. 2015a. "Conclusion: Exploring the Politics of Leverage." In *The Politics of Leverage in International Relations. Name, Shame, and Sanctions*, edited by H. R. Friman, 201–218. New York: Palgrave MacMillan.

———. 2015b. "Introduction: Unpacking the Mobilization of Shame." In *The Politics of Leverage in International Relations. Name, Shame, and Sanctions*, edited by H. R. Friman, 1–9. New York: Palgrave MacMillan.

———, ed. 2015c. *The Politics of Leverage in International Relations. Name, Shame, and Sanctions.* New York: Palgrave MacMillan.

Gardiner, S. M., S. Caney, D. Jamieson, and H. Shue, eds. 2010. *Climate Ethics: Essential Readings*. New York: Oxford University Press.

Gunningham, N. 2009. "Environment Law, Regulation and Governance: Shifting Architectures." *Journal of Environmental Law* 21 (2): 179–212.

Haufler, V. 2015. "Shaming the Shameless? Campaigning Against Corporations." In *The Politics of Leverage in International Relations. Name, Shame, and Sanctions*, edited by H. R. Friman, 185–200. New York: Palgrave MacMillan.

Hedberg, C. J., and F. von Malmborg. 2003. "The Global Reporting Initiative and Corporate Sustainability Reporting in Swedish Companies." *Corporate Social Responsibility and Environmental Management* 10 (3): 153–164. https://doi.org/10.1002/csr.38.

Holland, C. 2010. "Multinational Corporations and Global Governance: Is Corporate Social Responsibility Enough? A Study of Chevron's Yadana Gas Pipeline Operations in Burma." Covalense Intern Analyst Papers. Geneva: Covalence.

Hsu, A., A. S. Moffat, A. J. Weinfurter, and J. D. Schwartz. 2015. "Towards a New Climate Diplomacy." *Nature Climate Change* 5 (6): 501–503. https://doi.org/10.1038/nclimate2594.

ILC. 2001. "Draft Articles on Responsibility of States for Internationally Wrongful Acts." *International Law Commission (ILC).* http://untreaty.un.org/ilc/texts/instruments/english/draft /20articles/9-6-2001.pdf.

Jacquet, J. 2015. *Is Shame Necessary?: New Uses for an Old Tool.* New York: Pantheon.

Jacquet, J., C. Hauert, A. Traulsen, and M. Milinski. 2011. "Shame and Honour Drive Cooperation." *Biology Letters* 7: 899–901. https://doi.org/10.1098/rsbl.2011.0367.

Jacquet, J., and D. Jamieson. 2016. "Soft but Significant Power in the Paris Agreement." *Nature Climate Change* 6 (7): 643–646. https://doi.org/10.1038/nclimate3006.

Jamieson, D. 2014. *Reason in a Dark Time: Why the Struggle Against Climate Change Failed–and What It Means for Our Future*. Oxford: Oxford University Press. https://books.google.nl/books?hl=nl&lr=&id=e-vDIAgAAQBAJ&oi=fnd&pg=PP1&dq=reasons+in+a+dark+time&ots=fV4GwfbUla&sig=PxCh_cLABR5As9Tps8mvBl91J-0.

Macek, E. E. 2002. "Scratching the Corporate Back: Why Corporations Have No Incentive to Define Human Rights." *Minnesota Journal of Global Trade* 11: 101–124.

Marquis, C., M. W. Toffel, and Y. Zhou. 2016. "Scrutiny, Norms, and Selective Disclosure: A Global Study of Greenwashing." *Organization Science* 27(2): 483–504. http://papers.ssrn.com/abstract=1836472.

Matthews, D. H. 2003. "Environmental Management Systems for Internal Corporate Environmental Benchmarking." *Benchmarking: An International Journal* 10 (2): 95–106. https://doi.org/10.1108/14635770310469635.

Milanovic, M. 2009. "Norm Conflict in International Law: Whither Human Rights." *Duke Journal of Comparative & International Law* 20: 69–131.

OECD. 2004. *OECD Principles of Corporate Governance*. Paris: Organisation for Economic Co-opeartion and Development (OECD).

———. 2008. *OECD Guidelines for Multinational Enterprises*. Paris: Organisation for Economic Co-opeartion and Development (OECD).

Potoski, M., and A. Prakash. 2004. "The Regulation Dilemma: Cooperation and Conflict in Environmental Governance." *Public Administration Review* 64 (2): 152–163. https://doi.org/10.1111/j.1540-6210.2004.00357.x.

Prakash, A., and M. Potoski. 2006. *The Voluntary Environmentalists: Green Clubs, ISO 14001, and Voluntary Environmental Regulations*. Cambridge: Cambridge University Press.

Revak, H. 2012. "Corporate Codes of Conduct: Binding Contract or Ideal Publicity." *Hastings Law Journal* 63 (6): 1645–1670.

Rosen-Zvi, I. 2011. "You Are Too Soft: What Can Corporate Social Responsibility Do for Climate Change." *Minnesota Journal of Law, Science & Technology* 12: 527–570.

Shavers, A. W. 2012. "Human Trafficking, the Rule of Law, and Corporate Social Responsibility." *South Carolina Journal of International Law and Business* 9: 39–89.

Spier, J., A. Benjamin, M. Gerrard, T. Huydecoper, M. Kirby, M. C. Mehta, T. Pogge, et al. 2015. *Oslo Principles on Global Climate Change Obligations*. London: Global Justice Program at the Whitney and Betty MacMillan Center for International and Area Studies at Yale.

Stavins, R. 2015. *Paris Agreement—A Good Foundation for Meaningful Progress*. December 12, 2015. http://www.robertstavinsblog.org/2015/12/12/paris-agreement-a-good-foundation-for-meaningful-progress/.

Stiglitz, J. E. 2007. "Regulating Multinational Corporations: Towards Principles of Cross-Border Legal Framework in a Globalized World Balancing Rights with Responsibilities." *American University International Law Reviewnt'l Law Review* 23 (3): 451–558.

Sutton, M. 2003. "Between a Rock and a Judicial Hard Place: Corporate Social Responsibility Reporting and Potential Legal Liability under Kasky v. Nike." *UMKC Law Review* 72: 1159–1185.

Taebi, B., and A. Safari. 2017. "On Effectiveness and Legitimacy of 'Shaming' as a Strategy for Combatting Climate Change." *Science and Engineering Ethics* 23 (5): 1289–1306.

UNEP. 2011. *Bridging the Emissions Gap. A UNEP Synthesis Report*. Nairobi: United Nations Environment Programme (UNEP).

———. 2015. *The Emissions Gap Report 2015. A UNEP Synthesis Report*. Nairobi: The United Nations Environment Programme (UNEP).

UNFCCC. 2014. *Handbook on Measurement, Reporting and Verification for Developing Country Parties*. Bonn, Germany: United Nations Framework Convention on Climate Change (UNFCCC).

———. 2015. *Adoption of the Paris Agreement. Draft Decision -/CP.21*. Paris: United Nations Framework Convention on Climate Change (UNFCCC).

Waller, R. L., and R. N. Conaway. 2011. "Framing and Counterframing the Issue of Corporate Social Responsibility: The Communication Strategies of Nikebiz.Com." *The Journal of Business Communication (1973)* 48 (1): 83–106. https://doi.org/10.1177/0021943610389752.

38

THE ART OF DISOBEDIENCE

Climate Justice Activism

Benjamin J. Richardson

Disobeying Climate Inaction: The New Protest Movements

Environmentalists increasingly accept they must disobey in order to be heard and foster change. To disobey, explains French philosopher Frédéric Gros (2020), is to reject the docile culture of 'surplus obedience' that pervades modern life and replace it with the ethical and political responsibility to act for a just and compassionate future – even if this entails disobeying the law. It is well known to legal philosophers that the law is not necessarily aligned with justice and ethics, amply demonstrated today by the failures of governments to mitigate climate change with the urgency and ambitiousness needed to avoid catastrophic harm and suffering. Appalled, a new generation of activists has mobilized for action on interconnected grievances of climate breakdown, economic inequality, and social injustice. 'Climate justice' is the confluence of these intersecting grievances and speaks to the reality that global warming not only endangers the biosphere, it also threatens humankind itself (Schlosberg & Collins 2014).

This disobedience is being spearheaded by Extinction Rebellion (XR), a radical grassroots movement launched in the United Kingdom (UK) in April 2018, and since establishing branches in numerous countries, with a strategy of massive civil disobedience to compel authorities to take urgent climate action (Richardson 2020a). Another strand of the new activism is the youth-led climate strikes associated with Greta Thunberg, in which school students skip Friday classes to participate in protests against climate inaction (Thunberg 2019). Localized protests targeting specific grievances also arise, such as Indigenous First Nations blockading oil pipelines or coal mines imposed on their traditional homelands (Jolley & Rickards 2020; Lindeman 2020). Not all protesters, however, share the same views about how to address climate change. France's 'Yellow Vests' movement (*Mouvement des gilets jaunes*, for instance, is stirred by the perceived economic injustice of higher fuel taxes that fall disproportionately on the working class and farmers (Grossman 2019).

Civil disobedience is not new, of course. Its recent history includes the Occupy sit-ins (responding to the 2008 Global Financial Crisis), the anti-globalization campaigns (opposing neoliberal economic policies) and, of earlier vintage, the anti-nuclear movement, and they in-turn built on the Black civil rights movement and the Suffragettes (Giugni 2004). The recent climate protests share some characteristics with these antecedents but also some notable differences: the language of 'emergency'; the prominent involvement of children and youth as participants and organizers; and the hyper-aesthetic character of many protests, as evident in mass 'die-ins,' artistic paraphernalia, social-media platforms, and catchy figurative language. Some of these characteristics were

DOI: 10.4324/9781003039860-44

Figure 38.1 Anti-nuclear march, February 1961, London; CC BY-SA 3.0

not unknown to earlier protest movements – for instance, the anti-nuclear campaigners spoke of the urgency to stop the Doomsday Clock and they deployed media-grabbing colorful posters, banners, and slogans (Rucht 1990) – but these characteristics are far more pronounced and pivotal to the climate justice movement (Figure 38.1).

The protests against climate injustice respond to the underlying failures of environmental law. Its modern form emerged from the late 1960s and brought some gains, including substantive improvements in some environmental conditions, as well as enhanced opportunities for public participation in decision-making (Richardson & Razzaque 2006). Lawmakers had hoped that better due process for affected communities, including rights to consultation and access to the courts, would channel any grievances off the streets and into forums for orderly deliberation and adjudication. Whilst such participatory rights generally continue in many jurisdictions, their value is diminishing as governments, such as in Australia and the United States, introduce anti-protest laws that thwart environmental activism, including through broad police powers, vague offences, and harsh penalties (Kaufman 2020; Richardson 2020b). Such hostility, coupled with inaction on climate change, exemplifies what Naomi Klein (2009) calls "the shock doctrine": the collusion of business and political elites to force through controversial changes to the law under the pretext of a national emergency. The COVID-19 pandemic has given additional cover for such unscrupulous tactics, such as the implementation of lockdowns to protect public health, but that collaterally serve to restrict environmental protest (Sukdeo & Richardson 2020). Protesting against climate inaction thus involves disobeying intransigent or hostile regimes unwilling to take serious action against global warming and also unwilling to give protesters the space to be heard.

Whilst the paucity of legal action on climate breakdown in the face of abundant scientific evidence of looming catastrophe reflects the vested political and economic interests in maintaining the status quo, the difficulties in mobilizing social pressure for change also reflect the aesthetic characteristics of the climate emergency. Climate breakdown defies perception as an emergency in the conventionally understood sense, such as one fleeing a flood or fire (Frantz & Mayer 2009). Instead, global warming touches most people only vicariously, often via media news and scientific studies. Certainly, some people encounter the horrors directly, such as the mega forest fires that scorched southeast Australia and the western United States during 2020. Yet many other impacts either gestate slowly over several decades, such as rising sea levels or the acidification of the oceans, and/or they occur in spaces remote from human communities, such as starving polar bears deprived of their ice-laden habitat and food sources. Sensory deficits of these temporally and spatially displaced impacts in turn may diminish the sense of urgency to act on climate change (Richardson 2017a). As the proverb says, 'out of sight, out of mind.'

Humans' sensory deficits of the upheavals of climate change not only affect the political will of governments to act, but they also pose a challenge for environmental activists that seek to win the hearts and minds of the masses. It is insufficient to simply publicize climate scientists' advice though data, charts, and graphs to generate the support needed to carry green parties to electoral success or provoke mass public demonstrations. It has thus become vital for climate activists to harness the arts and use aesthetic means to engage the public at more emotional or visceral levels to garner media attention and political influence (Miles 2010).

This chapter thus aims to identify the key characteristics of the aestheticized activism of climate protest and assess their significance for promoting climate justice. It begins with an examination of the role of aesthetics in the methods and goals of climate activism. It then turns to case studies of the XR movement and youth climate strikes, respectively, followed by a discussion on the impact and influence of aesthetic activism in leveraging changes in climate politics and law. The chapter concludes with brief reflections on the future of aesthetic activism for climate justice.

The Aesthetic and Artistic Dimensions of Climate Change

A growing body of environmental aesthetics research suggests that the aesthetic domain matters to how people relate to, experience, and govern their interactions with the natural world (Berleant 1997; Carlson 2000). Legal, policy, and social responses to climate change, and other environmental problems, may ostensibly reflect scientific evidence and economic calculations, but the natural world is also experienced as an animated, sensory realm. It acquires aesthetic meaning through visual, acoustic, and other sensory stimuli, and includes the aesthetic consequences of harmful, anthropogenic behavior, such as unsightly and degraded landscapes from pollution and deforestation (Richardson 2019). Adjectivally, 'aesthetics' refers to how people value or characterize sensual experiences, such as when encountering scenery perceived as beautiful. As a noun, aesthetics means the principles or philosophy behind aesthetic judgements (Steckler 2010). Art and aesthetics are overlapping domains: art can be created and admired for its beauty, yet we can also appreciate the aesthetic of non-artistic objects and the value of art for non-aesthetic reasons (e.g., its political messages or social commentary). The meaning of art and aesthetics has altered and broadened significantly over the past century due to innovations in conceptual and performance art, as well as the attribution of aesthetics to wider contexts beyond fine art and natural beauty, such as through various expressive activities in the media, business, and consumer culture (Saito 2008). Recognition of the aestheticized character of recent climate activism thus feeds off this evolving understanding of aesthetics and art, affirming how environmental protest responds both to the altered sensory experiences of climate change, as well as articulating its own artistic characteristics that can inform that sensory experience and political and legal responses.

The arts have a long history of engagement with the natural world beyond the traditions of decorative landscape painting and other familiar fine art genres (Richardson 2019). The national

Figure 38.2 The "Blue Marble," December 1972, NASA

parks movement that arose in the late 19th century benefited from artists who helped lawmakers and the general public to appreciate wilderness areas they had never visited (Strochlic 2016). In our era, new technologies have enabled people to visualize the environment in unprecedented ways, such as with the famous NASA photographs of Earth, including the "Blue Marble," captured in December 1972 by the Apollo 17 crew (see Figure 38.2). This image has been identified as influential in spurring the global environmental movement by revealing a new aesthetic of the planet's beauty and vulnerability in the great void of the universe (Kelsey 2011). In turn, the aesthetic qualities of a more valued natural world have become an object of legal protection; for example, the UK's Countryside and Rights of Way Act 2000 and the United States' Wild and Scenic Rivers Act 1968 protect aesthetic values, in addition to ecosystems and wildlife. This is also seen in international treaties, such as the Convention of International Trade in Endangered Species of Flora and Fauna of 1973. Artists have also depicted environmental degradation in their work, such as the mining and industrial blight dramatized in the photography of Edward Burtynsky (2009), and many have worked directly with environmental groups, whether by providing captivating imagery of sublime landscapes that they wish to protect, or provocative symbols and slogans used for protesting.

Climate change politics has potent aesthetic qualities. One dimension is politically engaged artists who explore environmental upheaval and reject traditional artistic conventions where "nature ends up objectified as an ontology divorced from social, political and technological processes," as explained by art historian Demos (2016: 44). Climate breakdown has become a major preoccupation of artists whose work depicts coral bleaching, polar melting, climate refugees, and wildlife extinctions, as well as the politics of climate governance and inaction (Taplin 2014). To a lesser extent, some musicians also do so; for example, the song 'Kyoto Now,' released in 2002 by the rock group Bad Religion, became an anthem for championing better international climate action. Australian musician Paul Kelly in 2020 also released a song 'Sleep, Australia, Sleep,' which attacks the incompetence or insouciance of his government in addressing the climate crisis.

Climate activist art has also been bolstered by the backing of organizations and networks. The Australian-based CLIMARTE was founded with the mission of "harnessing the creative power of the arts to inform, engage, and inspire action on climate change." CLIMARTE brings together artists, gallery owners, patrons, and scholars for the exchange of ideas and public engagement for climate action. Its co-director, Guy Abrahams (2016), believes in the "creative capacity of art to raise awareness and engage people in both local and global efforts to tackle climate change." A similar synergistic role is played by Cape Farewell, launched in England, and the Canary Project, in the United States.

Extinction Rebellion (XR)

The most aesthetically expressive stance in climate politics is in the grassroots, activist movement known as XR. It was founded in England in May 2018 by a cohort of activists and academics, led by Roger Hallam, an organic farmer, and Gail Bradbrook, a molecular biophysicist, who perceived a need for a new style of activism (Berglund & Schmidt 2020). Going beyond earlier protest movements' modes of campaigning, XR's climate justice activism has centered on developing a stronger, more coherent, and readily identifiable aesthetic. XR rejects 'reformist' green politics in favor of mass mobilizations to cause civil disruption that compels governments to meet its three core demands: tell the truth about the climate emergency; drastically cut carbon emissions; and convene citizens' assemblies to develop new policies for social and environmental well-being (Reed & Burgess 2020: 171–172).

The visual arts, music, and theatre permeate XR's zeitgeist. Its aesthetic has been carefully cultivated to distinguish the movement from other climate and environmental activist groups, such as Greenpeace, with its less confronting, "hippy," and warmer vibe (Weyler 2004). XR's street protests are awash with placards, posters, banners, flags, leaflets, costumes, and chants in a sea of color and sound (see Figure 38.3).

Figure 38.3 XR's 'Dead Sea March,' October 3, 2020, Hobart, Australia; photograph by Benjamin J. Richardson

Figure 38.4 Extinction Rebellion's global symbol

XR's distinctive aesthetic serves to boost media publicity and recruit members. Its retro font logo, depicting a black-and-white stylized hourglass within a circle, represents that time is running out to save the planet (see Figure 38.4). Taking a cue from the corporate world where trademarks and logos are recognized as crucial to brand recognition, the XR symbol helps to "transcend language differences and distill a lot of meaning into a simple form," explain its proponents (Morgan 2019). Another distinctive part of the XR repertoire is its 'Red Rebels,' the silent choir of red-robed witnesses whose street performances convey the gruesome tragedy of the looming climate catastrophe. Whilst XR is decentralized, operating through organically established chapters in different cities and regions, the movement is united globally through its shared philosophy and connectivity via digital platforms (e.g., Signal & MatterMost) and social media that allow its performers, musicians, crafters, and writers to develop and swap creative outputs. The movement accommodates a degree of aesthetic variety and experimentation, as local XR chapters typically have their own art groups to design costumes, banners, and performances customized to convey local issues and concerns. For instance, in the XR Tasmania chapter, to which the author belongs, the group has developed bespoke banners and posters that speak to local controversies, such as major forest fires and marine pollution.

The catchy and simple messaging of XR is another important strategy. Figurative language is an aesthetic medium, most evident in specific literary devices such as alliteration, metonyms, and metaphors. Important metaphors for the modern environmental movement include 'Mother Earth' (nature as a giant organism, rather than a machine), the 'web of life' (the ecological interconnections that bind humans and other species together), and 'think global, act local' (that humans' environmental behaviors have ramifications at multiple scales) (Verhagen 2008). Common XR slogans include 'there is no Planet B,' 'this is a climate emergency,' and 'extinction is forever.' During the COVID-19 pandemic, additional messages appeared, such as the banners shown in Hobart, Australia, inscribed with 'Covid Bad; Climate Worse; Act Now.' Pithy and provocative protest slogans help to garner media attention while also conveying a more complex story to stimulate public curiosity and debate.

XR's distinctive aesthetic also informs its emphasis on non-violent dissent. One approach is staging mass 'die-ins,' where protesters simultaneously feign being dead (see Figure 38.5). The act

Figure 38.5 Extinction Rebellion 'die in,' October 3, 2020, Hobart, Australia; photograph by Benjamin J. Richardson

of playing dead evokes the demise of humankind and other species if we continue with business-as-usual on climate change. The 'deceased' may occupy an area only briefly or stay indefinitely, such as to block street traffic until forcibly removed by police. Some die-ins are augmented by polemical artistic touches, such as wearing fake, blood-stained attire, and simulating writhing death throes. The policing of such protests is itself a theatrical event; for example, one of the practices of protesters is to 'go floppy' when arrested, which then requires multiple officers to intervene and adds to the visual mayhem.

The aesthetic vibrancy of XR's tactics also reflects the movement's desire to make protesting fun. This stance is most evident in the music, chanting, and flamboyant costumes that accompany XR protests. The protagonists' aim is not merely to attract more supporters but also to reduce the stress for themselves when put in tension-filled arrestable situations. The XR organizers also hope that the humorous antics may also calm the police and observers, thereby helping to create a more congenial and less dangerous atmosphere for all.

XR is a movement sustained by volunteers and the resources and time necessary to prepare its artistic paraphernalia and protest actions are considerable. One protest, for example, can take months of staging and planning. It will be crucial to see how XR is able to continue to expand its membership and innovate its work, while also managing heavy administrative burdens and burn-out. While bringing art into protest may make it fun, the laborious preparation can be much less so.

Youth Climate Strikes

Paralleling the rise of XR is youth-led climate activism, coalesced in recent years around school strikes for climate, sometimes referred to as 'Fridays for Future' or 'Youth Strike for Climate.' Reflective of the movement's growth, in 2019 the *Collins Dictionary* picked 'climate strike' as the phrase of the year, defined as "a form of protest in which people absent themselves from education or work in order to join demonstrations demanding action to counter climate change" (Manister 2019). The movement took off after Swedish teenager Greta Thunberg staged a protest in August 2018 outside the Swedish parliament building with a sign reading *Skolstrejk för klimatet* ("school strike for climate"), vowing to continue until Sweden met the carbon emissions target under the Paris Agreement of 2015.

Thunberg's raw and impassionate approach quickly captured media interest, and soon several thousand climate strikes were staged in some 150 countries during March, May, and September 2019. One of these events, on 20 September, attracted some 4 million participants – probably the largest climate protest in history (Laville & Watts 2019). Their concerns are best summed up in the words of Thunberg and her climate co-strikers, in an open letter to the *Guardian* newspaper (Global Coordination Group of the Youth-led Climate Strike 2019):

> *We, the young, are deeply concerned about our future. [...] We are the voiceless future of humanity. We will no longer accept this injustice. [...] We finally need to treat the climate crisis as a crisis. It is the biggest threat in human history, and we will not accept the world's decision-makers' inaction that threatens our entire civilization. [...] We, the young, have started to move. We are going to change the fate of humanity, whether you like it or not. [...}. We demand the world's decision-makers take responsibility and solve this crisis. You have failed us in the past. If you continue failing us in the future, we, the young people, will make change happen by ourselves. The youth of this world has started to move, and we will not rest again.*

Although organized by youth, climate strikes have attracted and included a broad demographic, including many adults and parents accompanying their young children. A climate strike in Hobart, Australia in September 2019 attracted an estimated 22,000 protesters – representing about 10 percent of the city's population – and included people of all ages and backgrounds (see Figure 38.6).

The involvement of defiant young people is significant. In her analysis of children activists, Nicole Rogers argues that the climate strikes constitute a significant reframing in the status of children from passive, naïve, or gullible bystanders to smart, disruptive agitators responding to the climate emergency (Rogers 2020: 113–115). She remarks,

> [d]enied the vote [...] and in the absence of any official role in modern democracies, children are nevertheless making their voices heard. [...] As some of the placards carried by youth climate strikers read, 'You'll die of old age, I'll die of climate change.'
>
> *(Rogers 2020: 121)*

Despite this impressive display of civic engagement, Australian Prime Minister Scott Morrison told children to go back to school, "what we want is more learning and less activism in schools" (Australian Associated Press 2019). Yet not even a year later, COVID-19 led to the rapid government closure of schools. The irony of when public health did or did not matter was not lost on the climate activists. Whilst the involvement of children in the climate strikes is historically unprecedented, it is less so in the Global South, where "young people have historically been prominent among environmentalists." The reason is two-fold: poorer countries have much younger demographics and children are among the most affected by environmental degradation (Bowman 2020).

Like with XR, the youth climate strikers draw on a clever arsenal of aesthetic tools to rally supporters and garner politicians' attention. According to the 'Global Climate Strike' website, "[a]rt is a powerful way we can continue to build connections, engage people, tell our stories, and demand change." The strikes commonly involve street marches accompanied by rich

Figure 38.6 Youth climate strike, September 20, 2019, Hobart, Australia; photograph by Benjamin J. Richardson

artistic paraphernalia, including colorful posters, banners, and flags, many inscribed with messages about the urgency of climate action. Organizers have even prepared 'Climate Strike Art Kits,' made available online to help with designing banners and other creations. Youth strikers also rely extensively on digital activism, which has only increased during the social restrictions imposed during the Covid-19 pandemic. The sheer presence of thousands of young people marching is itself a powerful visual statement, and when diverse and colorful art is added, the media is able to capture and offer the world extensive, potent images of a new generation demanding climate action.

Whilst the climate strikes generally do not involve unlawful, civil disobedience, it is not unknown. Most strikes involve rallies approved by police or other local authorities along specified routes, typically in major city centers. Some youth, however, are involved in XR groups where they are far more likely to engage in unauthorized protests. Yet, the climate strikes are defiant in a significant sense in that they constitute an abandonment of mandatory school time in favor of public activism. Some schools have developed a policy of tacit acceptance that some students will skip classes, and though they may not be severely punished, they may suffer by reduced marks for missed tests or class attendance (Shine 2019) (Figure 38.7).

Relevantly to their disobedience, a number of climate strikes have involved students convening alternate 'climate classrooms,' strategically placed around the lawns of parliaments or government offices. These classrooms involve serious discussion and learning about climate change, but also represent a form of performance art, mimicking a regular school day with a series of lessons. As one organizer in Tasmania explained, "[i]n the past, we've been criticized for taking students out of the classroom to protest climate change, so this week we're bringing the classroom with us" (ABC 2020).

Figure 38.7 A youth climate striker's explanation for skipping classes; September 20, 2019, Hobart, Australia; photograph by Benjamin J. Richardson

Influence of Aesthetic Activism

Whilst innovative and novel, the aesthetics of climate protest and activism sit within a broader trend in contemporary political life in which public debate and campaigning are increasingly mediated through artistic, musical, and theatrical forms. The advent of 'image politics,' as DeLuca (2005) calls it, feeds off the hyper-aesthetic culture saturated with mass television and internet communications. In *Picture Theory*, William Mitchell (1995) argues that Western culture has taken a "pictorial turn" in which images have become the key means of social and political communication. Likewise, Mark Taylor and Esa Saarinen (1994: 15) believe that "in our era, we must philosophize with images rather than concepts." Image-based politics, however, are problematic to the extent that they overly simplify or distort important messages or skew attention away from a serious debate about the issues at stake, such as how to address climate change. Relevantly here, Jean Baudrillard (1994) critiques the growing culture of spectacle in Western societies, in which the public assumes the role of enfeebled spectator and consumer of a commodified 'hyper-reality.' In this milieu, critical public debate is susceptible to being replaced by seductive corporate marketing and government propaganda. Thus, while the aesthetic turn in climate politics is welcomed by many activists as a positive strategy, the strategy is also available to their opponents – consider, for instance, the inflammatory theatre in the political rallies of former US President Donald Trump.

It remains indeterminable at this stage whether XR's creative prowess and aesthetic embellishments can galvanize mass civil disobedience to leverage fundamental changes to political and economic systems (Gunningham 2020). Some commentators have already identified a discernible impact, such as XR's ability to challenge the meanings people habitually ascribe to their social spaces, disrupting the obsequious consumer culture with new narratives that question the environmental and social sustainability of the status quo (Axon 2019). Mass die-ins and street blockades may be particularly effective in questioning the docile, 'surplus obedience' of the masses (Gros 2020)

The XR vision is premised on a theory of change that when reformist politics have failed, rebellion can succeed if the critical threshold of 3.5 percent of the population participates in disobedience. That number comes from the conclusions of social movement researchers Erica Chenoweth and Maria Stephan (2011), whose analysis of past social uprisings found that when at least 3.5 percent of the population mobilized non-violently, the campaign's demands would be met approximately 50 percent of the time, and more often successful compared to violent dissent. Pioneering research by Gene Sharp (1973) on the power of nonviolent disobedience of the Gandhian tradition also supports the aspirations of XR strategists.

Some scholars and activists have criticized that belief, however, on two principal grounds. Firstly, the 3.5 percent number is derived from examples of political resistance to oppressive regimes, especially in the Global South, where the protesters themselves were directly affected (Ahmed 2019). By contrast, climate change does not yet directly and personally impinge on the lives of most people in the societies where XR is active, notably in Australia, the United Kingdom, and other Western liberal democracies. Secondly, XR offers little guidance to solve specific challenges in addressing climate change, such as choosing effective policy instruments (e.g., carbon taxes versus direct regulation) and allocating fairly the costs and benefits of decarbonizing economies. It is assumed that such details will be determined or guided by citizens' assemblies, yet there can be no assurance that such forums – even if governments agreed to convene them – would lead to acceptable climate action. In all likelihood, future generations would still not have a voice in those assemblies, and many in the current generation would continue to be self-serving and short-sighted [in the scale and scope of the crisis].

Thus far, the impact of XR and the climate strikes can be gauged in a variety of ways. Undoubtedly, and as academic researchers are starting to quantify, they have garnered massive media attention and attracted many thousands of supporters (Thackeray et al. 2020). Documentary films in mainstream cinemas are also expanding the range of audiences that encounter their messages, such as 'I Am Greta,' released in 2020, Some governments are also heeding their demands to declare a 'climate emergency,' such as in May 2019, when the British Parliament became the world's first national legislature to accept, without a formal vote, a motion to declare a climate emergency (Brown 2019). The Climate Emergency Declaration website tracks such initiatives worldwide and has identified hundreds of examples. In declaring a climate emergency, a government authority or legislature effectively concedes the urgency to address climate change and that existing measures are insufficient. The courts, too, are starting to recognize that climate protesters arrested for civil disobedience may have a lawful excuse. Courts have upheld the defense in two English cases but rejected it in two Queensland trials (Richardson 2020b). One success involved a London jury that acquitted XR activists Roger Hallam and David Durant of criminal charges that had resulted from spray chalking a university building with anti-fossil-fuel slogans. Whilst the Queensland courts, where the defense has been tested thus far in Australia, have rejected it, in one case taken to an appeal court the defendants succeeded in having their criminal convictions set aside, with the court accepting that the activists' climate concerns 'lessen[ed] their moral culpability' (*EH v QPS*; *GS v QPS* [2020] QDC 205, at para. 77). Given the difficulty of empirically measuring cause and effect, it is not possible to further quantify how climate activism is influencing the specific climate policies of governments, business corporations, or other actors, but the influence is growing.

Closing Reflections

In our media-inebriated cultures, the deployment of images, music, and other creative expression is an indispensable tool for social agitation, especially as traditional means of dissent such as street parades and petitions lose their potency. Disobedience can be communicated through art, on its own or with verbal and textual additions. In *The Art of Moral Protest*, James Jasper (1999: 13, 369) identifies artists on the "cutting edge of society's understanding of itself as it changes" by putting "into concrete form new ways of seeing and judging the world, new ways of thinking and feeling about it." Academic researchers (Curtis et al. 2014) have corroborated how art can convey information about environmental issues in novel and thought-provoking ways, and that art can be morally persuasive by creating empathy or compassion for its subject-matter. Furthermore, Rob Rosenthal and Richard Flacks (2010) identify how activist art and music can help in "reframing" key ideas and narratives for challenging prevailing assumptions. These are vital contributions in times when climate change skepticism or insouciance remains prevalent. Aesthetic activism can help distill esoteric discourses and ideas to be materially visible and knowable to a wider audience in society, injecting greater political salience into the debates. Furthermore, unlike spoken or written languages, many forms of visual arts and music are universally or widely cognizable across diverse cultures, thus enabling aesthetic activism to attain greater global reach.

Aesthetic activism, however, generally cannot replace conversation and consultation, because we need spaces for audiences to respond and debate rather than just be recipients of artistic statements. For instance, artists can dramatize the urgency of stopping climate change, and help people to emotionally engage with the challenges, but the hard choices about how to decarbonize economies and address the social and economic inequalities of capitalism must still be debated in global conferences, national legislatures, and other governance spaces. Apart from limitations in its capacity to inform climate policy-making, aesthetic activism needs additional allies to have political strength. Gunningham's (2020) analysis of XR's impact suggests it will more likely succeed through coalitions with sympathetic actors in civil society and business sectors, where efforts to accelerate climate action are intensifying, such as through the fossil fuels divestment movement (Richardson 2017b). These cross-sectoral collaborations can enrich opportunities to garner media attention, reach bigger audiences, and foster changes in public attitudes that ultimately may impact legal and policy change. The timing of aesthetic activism is also important to boost its impact and maximize the reach of activists' messages, such as staging protests to coincide with international climate treaty negotiations or the release of major scientific studies; for example, alongside the annual Conference of the Parties of the UNFCCC or the reports of the IPCC.

In closing, it is indisputable that aesthetic activism is vital for enriching public awareness and opening human imagination to novel and interesting ways for understanding and responding to the climate crisis. The artistic activism of XR and the youth climate strikes taps into the visceral emotional and cultural power of aesthetics, as reflected in enduring adages such as 'seeing is believing,' 'where words fail, music speaks,' and 'a picture tells a thousand words.' Yet, like the solidarity required for a public protest, the art of disobedience cannot be a solo effort. Other actors and processes must be involved to truly transform corporate and government behavior. This includes the people and policy-makers as seen in court action and climate protection legislation, but also the myriad actions required by individuals in their own lives in taking personal responsibility for our changing climate.

References

ABC (Australian Broadcasting Corporation) (2020) "Global School Strike for Climate Change Movement Resumes, With Protests Talking Place across Australia," *ABC News*, 25 September, <https://www.abc.net.au/news/2020-09-25/global-student-strike-for-climate-action/12702434>.

Abrahams, G. (2014) "Foreword," in G. Abrahams, K. Gellatly and B. Johnson (eds), *ART+CLIMATE=CHANGE THE BOOK*, Melbourne: Melbourne University Press.

Ahmed, N. (2019) "The Flawed Social Science behind Extinction Rebellion's Change Strategy," *Insurge Intelligence*, 29 October, <https://medium.com/insurge-intelligence/the-flawed-science-behind-extinction-rebellions-change-strategy-af077b9abb4d>.

Australian Associated Press. (2019) "Scott Morrison Tells Students Striking Over Climate Change to be 'Less Activist'," *The Guardian,* 26 November 2018, <https://www.theguardian.com/environment/2018/nov/26/scott-morrison-tells-students-striking-over-climate-change-to-be-less-activist>.

Axon, S. (2019) "Warning: Extinction Ahead! Theorizing the Spatial Disruption and Place Contestation of Climate Justice Activism," *Environment, Space, Place* 11(2): 1–26.

Baudrillard, J. (1994) *Simulacra and Simulation*, trans S. F. Glaser, Ann Arbor: University of Michigan Press.

Brown, L. (2019) "Climate Emergency: What Is a Climate Emergency?" *BBC News,* 3 May <https://www.bbc.com/news/newsbeat-47570654>.

Berglund, O. and Schmidt, D. (2020) *Extinction Rebellion and Climate Change Activism,* Berlin: Springer.

Berleant, A. (1997) *Living in the Landscape: Towards an Aesthetics of Environment,* Lawrence: University of Kansas Press.

Bowman, B. (2020) "'They Don't Quite Understand the Importance of What We're Doing Today': The Young People's Climate Strikes as Subaltern Activism," *Sustain Earth* 3(16): 1–13, <https://doi.org/10.1186/s42055-020-00038-x>.

Burtynsky, E. (2009) *Oil,* Gottingen: Steidl Verlag.

Carlson, A. (2000) *Aesthetics and the Environment: The Appreciation of Nature, Art and Architecture,* New York: Routledge.

Chenoweth, E. and Stephan, M. J. (2011) *Why Civil Resistance Works: The Strategic Logic of Nonviolent Conflict,* New York: Columbia University Press.

Climate Emergency Declaration. (undated) <https://climateemergencydeclaration.org/climate-emergency-declarations-cover-15-million-citizens>.

CLIMARTE. (undated) <https://climarte.org>.

Curtis, D., Reid, N. and Reeve, I. (2014) "Towards Ecological Sustainability: Observations on the Role of the Arts," S*.A.P.I.EN.S: Surveys and Perspectives Integrating Environment and Societ*y 7: 1.

DeLuca, K. (2005) *Image Politics: The New Rhetoric of Environmental Activism,* London: Routledge.

Demos, T. J. (2016) *Decolonizing Nature: Contemporary Art and the Politics of Ecology,* Berlin: Sternberg Press.

Frantz, C. M. and Mayer, F. S. (2009) "The Emergency of Climate Change: Why Are We Failing to Take Action?" *Analyses of Social Issues and Public Policy* 9(1): 205–222.

Giugni, M. (2004) *Social Protest and Policy Change: Ecology, Antinuclear, and Peace Movements in Comparative Perspective*, New York: Rowman and Littlefield.

Global Climate Strike. (undated) <https://globalclimatestrike.net/art>.

Global Coordination Group of the Youth-led Climate Strike (2019) "Climate Crisis and a Betrayed Generation: Letters," *The Guardian,* 1 March, <https://www.theguardian.com/environment/2019/mar/01/youth-climate-change-strikers-open-letter-to-world-leaders>.

Gros, F. (2020) *Disobey! The Philosophy of Resistance*, London: Bloomsbury.

Grossman, E. (2019) "France's Yellow Vests – Symptom of Chronic Disease," *Political Insight* 10(1): 30–34.

Gunningham, N. (2020) "Can Climate Activism Deliver Transformative Change? Extinction Rebellion, Business and People Power," *Journal of Human Rights and the Environment* 11(special issue): 10–31.

Jasper, J. (1999) *The Art of Moral Protest: Culture, Biography, and Creativity in Social Movements*, Chicago: University of Chicago Press.

Jolley, C. and Rickards, L. (2020) "Contesting Coal and Climate Change Using Scale: Emergent Typologies in the Adani Mine Controversy," *Geographical Research* 58(1): 6–23, <https://doi.org/10.1111/1745-5871.12376>.

Kaufman, A. C. (2020) "States Quietly Pass Laws Criminalizing Fossil Fuel Protests Amid Coronavirus Chaos," *Huffington Post,* 28 March, <www.huffingtonpost.com.au/entry/pipeline-protest-laws-coronavirus_n_5e7e7570c5b6256a7a2aab41?ri18n=true>.

Kelsey, R. (2011) "Reverse Shot: Earthrise and Blue Marble in the American Imagination," in E. H. Jazairy (ed), *Scales of the Earth,* Cambridge: Harvard University Press, 10–16.

Klein, K. (2009) *The Shock Doctrine: The Rise of Disaster Capitalism,* Toronto: Knopf Canada.

Laville, S. and Watts, J. (2019) "Across the Globe Millions Join Biggest Climate Protest Ever," *The Guardian*, 21 September, <https://www.theguardian.com/environment/2019/sep/21/across-the-globe-millions-join-biggest-climate-protest-ever>.

Lindeman, T. (2020) "'The Revolution Is Alive': Canada Protests Spawn Climate and Indigenous Rights Movement," *The Guardian,* 28 February, <https://www.theguardian.com/world/2020/feb/28/canada-pipeline-protests-climate-indigenous-rights>.

Manister, R. (2019) "'Climate Strike' Named Word of the Year by Collins Dictionary," *CBS News*, 7 November, <https://www.cbsnews.com/news/word-of-the-year-collins-dictionary-names-climate-strike-word-of-the-year-for-2019>.

Miles, M. (2010) "Representing Nature: Art and Climate Change," *Cultural Geographies* 17(1): 19–35.
Mitchell, W. J. T. (1995) *Picture Theory: Essays on Verbal and Visual Representation,* Chicago: University of Chicago Press.
Morgan, G. (2019) "Extinction Rebellion's Graphics Became the Symbol of Environmental Uprising" *NewStatesman,* 18 September, <https://www.newstatesman.com/politics/environment/2019/09/how-extinction-rebellions-graphics-became-symbol-environmental-uprising>.
Reed, R. and Burgess, C. (2020) "Extinction Rebellion and Environmental Activism – the XR Interviews' *Journal of Human Rights and the Environment* 11(special issue): 171–180.
Richardson, B. J. (2017a) *Time and Environmental Law: Telling Nature's Time,* Cambridge: Cambridge University Press.
Richardson, B. J. (2017b) "Divesting from Climate Change: The Road to Influence," *Law and Policy* 39(4): 325–348.
——— (2019) *The Art of Environmental Law: Governing with Aesthetics,* Oxford: Hart Publishing / Bloomsbury.
——— (ed.) (2020a) *From Student Strikes to Extinction Rebellion: New Protest Movements Shaping Our Future,* Cheltenham: Edward Elgar Publishing.
——— (2020b) "Protesting Against Climate Breakdown: Novel Legal Options," *Australian Environment Review,* 35(1): 21–25.
——— and Razzaque, J. (2006) "Public Participation in Environmental Decision-Making," in B. J. Richardson and S. Wood (eds), *Environmental Law for Sustainability,* Oxford: Hart Publishing, 165–194.
Rogers, N. (2020) "Victim, Litigant, Activist, Messiah: The Child in Time of Climate Change," *Journal of Human Rights and the Environment* 11(special issue): 103–121.
Rosenthal, R. and Flacks, R. (2010) *Playing for Change: Music and Musicians in the Service of Social Movements,* London: Routledge.
Rucht, D. (1990) "Campaign, Skirmishes and Battles: Anti-nuclear Movements in the USA, France and West Germany," *Industrial Crisis Quarterly* 4(3): 193–222.
Saito, Y. (2008) *Everyday Aesthetics,* New York: Oxford University Press.
Schlosberg, D. and Collins, L. B. (2014) "From Environmental to Climate Justice: Climate Change and the Discourse of Environmental Justice," WIREs *Climate Change* 5(3): 359–374.
Sharp, G. (1973) *The Politics of Nonviolent Action,* New York: Porter Sargent Publishers.
Shine, R. (2019) "National School Strike for Climate Sees Students across the Country Planning to Skip School," *ABC News,* 20 September, <https://www.abc.net.au/news/2019-09-20/national-climate-strike-students-across-australia-skip-school/11528100>.
Steckler, R. (2010) *Aesthetics and the Philosophy of Art,* New York: Rowman and Littlefield Publishers.
Strochlic, N. (2016) "We Have a Painter to Thank for Yellowstone," *National Geographic Magazine,* April, <www.nationalgeographic.com/magazine/2016/05>.
Sukdeo, V. and Richardson, B. J. (2020) "Collateral Damage: Relaxing Environmental and Labour Standards during the Pandemic," *Corona Journal,* 7 May, <https://crisis-diary.net/2020/05/07>.
Taplin, R. (2014) "Contemporary Climate Change Art as the Abstract Machine: Ethico-Aesthetics and Futures Orientation," *Leonardo* 47: 509–510.
Taylor, M. and Saarinen, E. (1994) *Imagologies: Media Philosophy,* New York: Routledge.
Thackeray, S., et al. (2020) "Civil Disobedience Movements such as School Strike for the Climate are Raising Public Awareness of the Climate Change Emergency," *Global Change Biology* 26(3): 1042–1044.
Thunberg, G. (2019) *No One Is Too Small to Make a Difference,* London: Penguin Books.
Verhagen, F. C. (2008) "Worldviews and Metaphors in the Human-Nature Relationship: An Ecolinguistic Exploration through the Ages," *Language and Ecology* 2(3): 1.
Weyler, R. (2004) *Greenpeace: How a Group of Journalists, Ecologists and Visionaries Changed the World,* Vancouver: Raincoast Books.

39

ENVIRONMENTAL PERSONHOOD

A Novel Principle of Law for Climate Justice?

Alessandro Pelizzon

The Emergence of an Ecological Jurisprudence

Over the past decade, Rights of Nature Constitutional, legislative and judicial initiatives have emerged in a growing number of jurisdictions, from Bolivia to India, from New Zealand to Colombia, marking the practical articulation of this novel legal theory as one of the fastest growing legal movements of the 21st century. Over such a relatively brief period of time, three phases can already be identified within the movement: after an initial focus on the acceptance of Nature as a right-holder, a shift occurred toward the identification of said right-holder as a legal person. The movement now appears to inhabit a third phase, one where the focus has become the articulation of a novel category of personhood, tentatively called 'environmental personhood.' Albeit still protean, the emergence of such a novel category nonetheless represents a fertile terrain for inter-jurisdictional and inter-normative dialogue around climate justice and the ethical obligations of global governance institutions, as well as the focus of this chapter. In a time of global climate-induced social uncertainty, this chapter ultimately asks, could 'environmental personhood' emerge as a novel principle of law with particular significance for climate and environmental action and ethics?

"Contemporary public concern for protecting nature's ecological equilibrium should lead to the conferral of standing upon environmental objects to sue for their own preservation," Justice Douglas of the US Supreme Court wrote in his dissenting opinion in *Sierra Club v Morton* in 1972. Such a powerful assertion was inspired by the publication, not long prior, of an article titled *Should Trees Have Standing?* by Professor Christopher Stone of the University of Southern California. Professor Stone had concluded one of his property law classes by asking a provocative question of his student cohort

> What would a radically different law-driven consciousness look like? ... One in which Nature had rights ... Yes, rivers, lakes, ... trees ... animals ... (I may have ventured 'rocks'; I am not certain.) How would such a posture *in law* affect a community's view of *itself*?
>
> *(Stone 2010: xi)*

The question, triggered by an interest in property issues, soon became one of the key pillars of a novel consideration of environmental legal issues, destined to blossom into a veritable new branch of legal theory. Stone's reflection led him to assert that "[i]t is not inevitable, nor is it wise, that natural objects should have no rights to seek redress in their own behalf" (Stone 1972: 457). It is

DOI: 10.4324/9781003039860-45

this radically novel position (that is, novel for a world used to considering the environment as the passive background against which the legal narratives of humanity unfolded) that informed the above-mentioned case of *Sierra Club v Morton*.

The case had been initiated by the environmental preservation organization Sierra Club, a long-time defender of the southern Sierra Mountains, in order to stop the development by the Walt Disney Company of a ski resort in the Mineral King valley, since such development would have caused the destruction of a significant portion of the local old growth forest within the Sequoia National Park. Albeit the case was purportedly concerned with the issue of standing under the *Administrative Procedure Act*, it became far better known for the ground-breaking possibilities entailed by Justice Douglas's dissenting opinion.

The issue of 'standing to sue' – that is, "[w]hether a party has a sufficient stake in an otherwise justiciable controversy to obtain judicial resolution of that controversy" – was resolved in the case against the Sierra Club, who had "failed to allege that it or its members would be affected in any of their activities or pastimes by the Disney development" (*Sierra Club v Morton*: 732), and thus lacked standing in the case. However, Stone's argument had already suggested that if standing was the necessary pre-requisite to prevent damages from occurring, and since damages did indeed occur to a large swathe of a pristine old growth forest, maybe standing ought to be attributed to the forest itself. This argument resonated powerfully in Justice Douglas's own words, who wrote, almost poetically, that

> [T]he voice of the inanimate object, therefore, should not be stilled … '[t]he critical question of "standing" would be simplified and also put neatly in focus if we fashioned a federal rule that allowed environmental issues to be litigated before federal agencies or federal courts in the name of the inanimate object about to be despoiled, defaced … Contemporary public concern for protecting nature's ecological equilibrium should lead to the conferral of standing upon environmental objects to sue for their own preservation. … Inanimate objects are sometimes parties in litigation. A ship has a legal personality, a fiction found useful for maritime purposes. The corporation sole -- a creature of ecclesiastical law -- is an acceptable adversary, and large fortunes ride on its cases. The ordinary corporation is a 'person' for purposes of the adjudicatory processes, whether it represents proprietary, spiritual, aesthetic, or charitable causes. So it should be as respects valleys, alpine meadows, rivers, lakes, estuaries, beaches, ridges, groves of trees, swampland, or even air that feels the destructive pressures of modern technology and modern life.
>
> *(Sierra Club v Morton: 741–743)*

Justice Douglas's dissenting opinion undoubtedly represents the first judicial articulation of the possibility of legal personhood being attributed to environmental entities in modern courts.

However, Professor Stone and Justice Douglas's invitation to re-imagine and enlarge the category of subjects inhabiting the Western legal world[1] remained virtually unheeded for the following three decades, until self-described 'geologian'[2] Father Thomas Berry introduced the world to what he and South African environmental lawyer Cormac Cullinan called an 'Earth Jurisprudence'. After authoring the seminal *Wild Law* in 2002, Cullinan described

> Earth Jurisprudence' as 'a philosophy of law and human governance … based on the idea that humans are only one part of a wider community of beings and that the welfare of each member of that community is dependent on the welfare of the Earth as a whole.

Cullinan continues,

> From this perspective, human societies will only be viable and flourish if they regulate themselves as part of this wider Earth community and do so in a way that is consistent with the fundamental laws or principles that govern how the Universe functions.
>
> *(Cullinan 2011: 13)*

Over the last decade, Earth Jurisprudence has often been described as a 'neo-natural law theory', even though "many of its practitioners incorporate significant elements of legal positivist thought" (Mylius 2018: 3). While the possibilities entailed by the emergence of an Earth Jurisprudence actually transcend the classical theoretical dichotomy, within Western legal theory, between natural law theory (at least in its more modern, post-Aquinian interpretation) and legal positivism, it is undoubtable that as a novel legal theory Earth Jurisprudence opened the theoretical space for a dialogue between law and ecology. Earth Jurisprudence, Ben Mylius writes, "seeks to ecologize legal philosophy." Furthermore,

> [W]here earlier natural law theories asked about the relationship between law and morality, Earth jurisprudence is interested in the relationship between law and ecology: and more specifically, in whether ecological norms, principles, or laws can and/or should have a bearing on the content, form, and validity of human laws.
>
> *(Mylius 2018: 5)*

In this sense, the emergence of Earth Jurisprudence represents the theoretical culmination of a broader trajectory within environmental law, one that can be observed unfolding over the past few decades. Such a trajectory shows a progressive shift in environmental legal discourse from a reactive response to a series of contingent – although ultimately interconnected – environmental predicaments, toward a homeostatic – and yet dynamic – balance between anthropogenic and non-anthropogenic forces in order to maintain existing environmental conditions, capable of preserving both human and non-human life in its present conditions. Such a trajectory can be evinced from the cruder quantification and regulation of climate-altering emissions to a far deeper and more nuanced interrogation of legal theory itself, by gesturing toward an epistemological normativity that transcends the traditional confines of Western legal theory. I have previously called this trajectory an *ecological jurisprudence* (Pelizzon 2014), and, while a full exploration of this trajectory is well beyond the intent of this chapter, it can nonetheless be symbolically captured, for the purpose of this writing, in Berry's and Cullinan's oeuvre. "Ecology is not a part of law." Berry aptly concludes; rather, "law is an extension of ecology" (Berry 1999).

From Rights to Personhood

If Berry's and Cullinan's theoretical forays into an Earth Jurisprudence symbolically represent a key threshold within Western legal theory, their practical implications are etched in the Ecuadorian Constitution of 2008. It is with this legal document that the pursuit of environmental justice was finally and fully informed, in a practical sense, by the theoretical framework discussed in the previous section. Foreshadowed by the inclusion of local ecosystems as legal subjects within municipal ordnances drafted with the assistance of the Community Environmental Legal Defense Fund in the United States from as early as 2006,[3] Articles 71–74 of the Ecuadorian Constitution assert that

> Nature, or Pacha Mama, where life is reproduced and occurs, has the right to integral respect for its existence and for the maintenance and regeneration of its life cycles, structure, functions and evolutionary processes. All persons, communities, peoples and nations can call upon public authorities to enforce the rights of nature.[4]

Known as the 'Montecristi' Constitution (from the town where the Assembly was first convened), this represents the first major legal document to recognize legally enforceable ecosystem rights, or 'Rights of Nature.' As such, this document is to the praxis of an ecological jurisprudence what Berry's and Cullinan's Earth Jurisprudence is to its theory.

The Constitutional innovation led by Ecuador was soon followed by a great number of Constitutional, legislative, judicial, and political initiatives in a great number of jurisdictions. In 2010, following the failure of the 2009 United Nations Climate Change Conference held in Copenhagen, Bolivia held the *World People's Conference on Climate Change and the Rights of Mother Earth* in the town of Cochabamba, leading to the release of the *Universal Declaration on the Rights of Mother Earth*, a document whose ecological aspirations mirror the more anthropocentric ones of the *United Nations Universal Declaration of Human Rights*. Not content with a merely aspirational document – albeit one submitted for consideration to the United Nations General Assembly – Bolivia followed by including ecological considerations in its Constitution first, and then by more explicitly articulating ecosystem rights in two subsequent legislative documents.

Ecuador and Bolivia's examples have been followed, within less than half a decade, by jurisdictions as diverse as Mexico (exemplified by the language of the Mexico City Constitution), Brazil (with the legislation passed by the Municipalities of Bonito, Paudalho, and Florianopolis), Costa Rica, and Uganda (with its *National Environmental Act* of 2019). Of particular significance have been Aotearoa New Zealand's *Te Urewera Act* which, in 2014, granted legal recognition to the *Te Urewera* forest in its own right, and *Te Awa Tupua (Whanganui Claims Settlement) Act*, which, in 2017, declared the Whanganui River a legal person, with 'all [attendant] rights, powers, duties, and liabilities' (*Te Awa Tupua Act* 2017: s14(1)). Many other jurisdictions – such as the European Union, Sweden, Nepal, France, the United Kingdom, Holland, Spain, and the Philippines – are presently contemplating the adoption of Constitutional, national, or local Rights of Nature provisions.

Equally, particularly in the United States, Tribal law has been strategically used to articulate a pluralist legal intersection articulated around an eco-jurisprudential axis, with, among others, the Ho-Chunk nation's vote for a Rights of Nature tribal constitutional amendment in 2016, the Ponca Nation's adoption of customary law explicitly recognizing Rights of Nature in 2017, and the adoption of the 'Rights of Manoomin' law by the White Earth band of the Chippewa Nation in 2018, granting legal rights to the *manoomin*, or wild rice.

In 2009, the international community echoed the national momentum, with the United Nations proclaiming that the 22nd of April is 'International Mother Earth Day' and establishing the Harmony with Nature Program within the General Assembly. At the same time, civil society has been mobilizing with gusto: the Global Alliance for the Rights of Nature was formed in 2010, and it sponsored the first Rights of Nature Tribunal in 2014.

Spearheaded by Constitutional and legislative initiatives worldwide, numerous cases have also been successfully litigated. Since the *Vilcabamba* case, the first successful Rights of Nature case in the world, Ecuador has seen scores of cases being found in favor of the Constitutional Rights of Nature provisions (Kauffman & Martin 2017), culminating in the recent Constitutional Court – the highest Ecuadorian court – hearing about the application of 'Rights of Nature' provisions to halt all mining concessions in the Los Cedros protected rainforest, a hearing whose *amicus curiae* brief has been undersigned by world-leading scientists such as Jane Goodall, E. O. Wilson, and more than 1200 colleagues from around the world. In 2017, the High Court of Uttarakhand, India, handed down two judgments granting legal personhood to the Ganges and Yamuna rivers and their tributaries, and to their glaciers, lakes, air, meadows, dales, jungles, forests, wetlands, grasslands, springs and waterfalls. In 2016, Colombia's Constitutional Court recognized the Atrato river as a legal entity and a subject of rights, to be represented by a 'commission of guardians of the Atrato river' (*comisión de guardianes del río Atrato*), composed of both a representative of the

Colombian government and a representative nominated by the local Indigenous and afro-Colombian communities. A suite of subsequent cases in different Colombian courts gave rise to a host of similar decisions, which recognized various parts of Colombia – such as the Colombian Amazon basin, the Páramo in Pisba (a high Andean ecosystem), and a number of rivers – as subjects of rights. In 2019, the High Court Division of the Supreme Court of Bangladesh delivered another landmark decision, giving all rivers in Bangladesh legal protection by granting them legal personhood.

It is undoubtable that the speed at which an ecological jurisprudence has been emerging over the past decade, as paradigmatically exemplified by the rapidly growing Rights of Nature movement, has very few parallels within the contemporary legal world. This presents those involved in the pursuit of climate and environmental justice with great hope, suggesting the possibility that legal change may be able to keep pace with (and maybe even outpace) the unprecedented pace of environmental changes that scientists have been describing over the course of the past few decades. While many challenges – both theoretical and practical – still remain, the historical legal evolution of the Rights of Nature movement has most certainly captured the imagination of legal practitioners and the general public alike at an exponential pace.

Writing almost a decade ago, before many of the above events could be even imagined, Anne Schillmoller and I wrote that 'at this stage in its development, the theory and practice of Earth jurisprudence occupies both unruly and creative territory, one which requires scholars and practitioners to negotiate a heterogeneous terrain' (Schillmoller and Pelizzon 2013: 32). We then noted that the very idea of nature – somewhat unquestioningly embraced then by many contemporary theoreticians within the movement – was inscribed within a contested terrain, one "built on power relations, discursively constituted through 'technologies of truth' which function as a dividing practice" (Schillmoller & Pelizzon 2013: 32). While it is true that many of the questions we raised then have not yet been fully answered at a theoretical level, in many of the more recent instances discussed above pragmatic answers to such questions emerged almost organically, embracing the concept of nature as an inherently "promiscuous" one (Soper 1995: 15), and almost effortlessly moving "beyond nature and culture" (Descola 2013).

The recognition of the Yarra River/Birrarung in Australia as "one living and integrated natural entity" (*Yarra River Act* 2017: s 1(a)), the description of the Whanganui River/Te Awa Tupua in New Zealand as a "spiritual and physical entity … that supports and sustains both the life and natural resources within the Whanganui River and the health and well-being of the iwi, hapū, and other communities of the River" (*Te Awa Tupua Act* 2017: s 13(a)), and the identification of the Ganges as *Ganga Mata* (the divine mother) and the Yamuna as *Yami* (the Lady of Life) in the Indian cases cited above, are all powerful examples of a profound ontological and epistemological shift within the movement toward a deeper appreciation of ontologically and ethically pluralist standpoints, which are essential to a radical (and necessary) reconstitution of global environmental governance.

The organic reimagination of the very idea of nature that is at the core of the metaphysical shift just described (and that, consequently, expands the existing ontological horizon of the environmental ethical discourse) has been due, in significant part, to the very active and leading role played by Indigenous activists and theorists. Indigenous peoples, Erin O'Donnell and others, "have demonstrated a nuanced strategic approach to using 'rights of Nature' as a way to support a collective approach to environmentally sustainable and culturally appropriate development by raising the profile of both natural entities and Indigenous peoples" (O'Donnell et al. 2020: 14). Moreover, the development of a legal theory that inherently transcends the contemporary Western dichotomy between natural law theory and legal positivism allows for a comparative platform whereby distinct ontological conceptions of the non-human other can be negotiated by reference to equally distinct normative frameworks and parameters.

Furthermore, while Stone's argument most certainly contained ecocentric undertones, Berry and Cullinan's work explicitly adopts a whole-of-system approach that does not distinguish between biotic and abiotic components of the system, thus allowing rivers to be conceived of as subjects of rights as much as individual non-human animals can. The shift toward an ecological jurisprudence has thus already been marked by a clear transition from *biocentrism* – whereby the emphasis is placed on a number of individual components of any given ecosystem, mostly selected from within the biosphere – to *ecocentrism* – whereby the system as a whole takes center stage, thus radically expanding the horizons of climate and environmental action.

This shift is best captured by the evolution of the language of legislative initiative and cases alike over the past decade. Both the Ecuadorian Constitutional provisions and the Vilcabamba case (the first successfully litigated Rights of Nature case), albeit profoundly innovative in terms of their content, display a relatively traditional and somewhat objective language, still gesturing toward a more materialistic tradition – at least descriptively, if not necessarily ontologically. The contrast with latter cases is, indeed, striking. To reflect the connection that the Whanganui *iwi* (the local Māori community) has with the Te Awa Tupua, the river, the third reading of the *Te Awa Tupua (Whanganui River Claims Settlement) Act 2017* (New Zealand) was prefaced by the singing of a *waiata*, a traditional and ceremonial Māori song ritually sung at particularly significant events. The High Court of Uttarakhand extensively quoted the work of Wangarī Maathai on the essential role of song, stories and cultures in connecting with (and thus preserving and protecting) Nature. Equally, the Yarra River, which flows through the Australian city of Melbourne, is defined in the *Yarra River Protection (Wilip-gin Birrarung murron) Act 2017* (Vic) as 'one living and integrated natural entity', a linguistic recognition of the *Birrarung* – the traditional *Wurundjeri* name of the river – as more than a flowing body of water, but rather as a metaphysical 'entity and pathway in which stories, songs, and law are emplaced, through the creative journeys of ancestral beings' (Clark et al. 2019: 823).

The conservativism with which some of the initial Rights of Nature advocates approached the movement at its inception (notwithstanding Berry's unquestioned challenge to a materialistic and dualistic metaphysical tradition), which led to an articulation of proposals still fully located within the confines of traditional Western legal theory, has thus been displaced by many actual examples of initiatives around the world that embrace the pluralist possibilities entailed by the emergence of an ecological jurisprudence. The most immediate outcome has been a growing recognition of Nature, both as a whole and in its multifaceted instantiations, as a *subject*. No longer a passive background against which the tapestry of human history unfolds, the above examples gesture toward Nature as a complex counterpart to the human *umwelt*, a network of entities and identities of equal metaphysical and moral standing as human beings, inscribed as an active participant within climate and environmental action.

This ultimately metaphysical shift, certainly strange for many traditional legal practitioners still located within a de-physicalized Western legal tradition (in which law is best exemplified as a body of abstract, almost mathematical, rules), is mirrored by a discernible pattern within the evolution of Rights of Nature initiatives described above. If the Montecristi Constitution is adopted as the starting threshold for the Rights of Nature movement, three phases can be clearly discerned within this movement over the past decade. While each phase has contributed directly to an expansion of environmental ethics, the trajectory itself is worthy of attention.

The first phase, explicitly focused on the articulation of the fundamental *rights* to be attributed to (or recognized for) Nature, is epitomized by the Ecuadorian Constitutional provisions and their related court cases, as well as the local ordinances drafted in the United States with the assistance of the Community Environmental Legal Defence Fund. While contemporaneous in its passing of parallel 'Mother Earth' legislation, Bolivia decided to focus its gaze not on a set of rights, but rather onto their Hohfeldian correlative, the collective duties and responsibilities owed toward the non-human world as well as other human beings, seen as inseparable. In addition to articulating a language ('Mother Earth') in direct opposition to the purportedly neutral and detached depiction of 'Nature' as a collection of relatively inanimate objects, the Bolivian experience is also actively focused on the development of a "socio-communitarian productive education model" devoted to the "integral management of life systems" (Zambrana-Flores 2019). However, notwithstanding the slightly different Bolivian path, the first phase of the movement was undoubtedly marked by a clear focus on the extension of rights to the non-human world, with all extant critiques commonly advanced within the broader rights discourse.

While many discussions on the nature, extent, enforceability (and even feasibility) of Rights of Nature are certainly still at the fore of the current discourse within the movement, a few years after the introduction of the Montecristi Constitution the focus markedly shifted from a focus on *rights* to one on *right-holder(s)*. Anticipated by the passing in New Zealand of the *Te Urewera Act* of 2014, which transformed a previous national park into a 'legal entity' managed by a newly established Te Urewera Board, the shift was fully realized with the enactment of the *Te Awa Tupua Act* in 2017. As mentioned above, the Act explicitly declared the Te Awa Tupua as a legal person, with "all [attendant] rights, powers, duties, and liabilities" (*Te Awa Tupua Act* 2017: s 14(1)). In 2017, thus, the attribution of legal personhood to (or recognition for) natural entities unquestionably took center stage and captured the legal world's imagination, as exemplified by the immediately subsequent – and already cited – examples in India, Colombia and Bangladesh (to cite the most prominent ones). Interestingly, rivers seemed – and still seem – to be the primary focus of this phase, perhaps because, as Mari Margil writes, they "are the arteries of the earth, and lifelines for humanity and millions of other animals and plants" (Margil 2017), or perhaps because, as natural entities, their riparian abode is clearly bounded – from source to estuary – and thus more easily tractable than less geographically individuated natural entities.

Whatever the reason for the rivers' centrality within this second phase of the movement, the legal categories employed gave rise to a host of critiques, both in the forms of judicial responses and complex commentary. These critiques form one of the two pillars of the third phase of the movement, one currently unfolding, and one in which the extension of existing legal categories to the natural world is interrogated, questioned, and ultimately criticized, thus opening the discourse to novel possibilities and creating a broader space for ethical engagement where to consider deeper and more complex ontological, ethical and normative questions.

Testing the Limits of Personhood

The reason for the shift from *rights* to *right-holders*, which is perhaps intuitively inevitable, is aptly captured by O'Donnell and others, who write that while "the creation (or recognition) of broad 'existence rights' for Nature, such as the right for species to exist, or the right for ecosystems to function," readily captured the legal imagination of legal and activists alike in a number of jurisdictions, "their implementation has been challenging, as it is not always clear when (or if) they give rise to a cause of action in law, or what kind of remedies they afford." It is thus unsurprising that these rights "have crystallized around recognition of natural entities as legal persons," since "[l]egal personality is articulated as the capacity to bear rights and duties in law" (O'Donnell et al. 2020: 6), and thus identifies a specific subject that can attempt to resolve the previous uncertainty. The attempt to more clearly identify the natural subject of novel environmental rights has direct implications for climate and environmental action, by expanding the community of environmental and climate actors and actants.

The recognition of legal personhood for natural entities, however, soon attracted a number of complex questions in and of itself, showing that the above uncertainty could not be fully resolved via the mere identification of a legal subject of Rights of Nature in the form of a legal person. Not long after Sharma J of the High Court of Uttarakhand variously described the Ganges and Yamuna rivers as "juristic entities," "juridical persons," "juristic persons," "legal persons/living persons," and, "juristic/legal persons/living entities having the status of a legal person with all corresponding rights, duties and liabilities of a living person" (*Ganges and Yamuna case*: 12–16), the Supreme Court of India heard an appeal against this ruling, and in so doing, overturned it (O'Donnell 2018). Of grave concern for the court was the extension of 'duties and liabilities' to the river, since the very possibility of enforcing such duties and liabilities would give rise to a confused and confusing plethora of questions. Would it ever be possible to sue a river for flooding? And even assuming such a possibility, what kind of reparation could ever be expected, and how would it ever be enforced? The negative implications for climate and environmental action are indeed readily apparent.

Another critique advanced against the attribution of personhood to natural features more confusedly presented the quest to establish Rights of Nature and the attribution of personhood to Nature as two competing pursuits, which are not necessarily – and inherently – interdependent. While such a critique can easily be dismissed by a quick reference both to the introductory paragraph to this section and to the intertwined legal categories of *things*, *persons* and *actions* discussed below, other authors present a more nuanced critique of the impact that the application of legal personhood has onto existing legal and political struggles.

Virginia Marshall, for example, recently addressed the unintended, and yet profound, impact that the attribution of legal personhood for Nature can have on Australian First Nations legal struggles (Marshall 2020). While Marshall presents the overall Rights of Nature as both monolithic and with already clearly defined theoretical artefacts – a picture that is indeed far from the case, as the movement is far more in flux and far more multifaceted (and internally contradictory) than her description accounts for – she correctly and astutely points out that

> [T]he western environmental discourse of legal personhood and legal personality are couched within the environmental and philosophical construction of nature as separate from humankind, whereby components of nature such as water, forests and minerals would be held in trust as public assets.
>
> *(Marshall 2020: 4)*

As a result, she highlights the inherent risk that the novel application of legal personhood would entail for the existing struggles that Australian Indigenous peoples have conducted over the past two centuries, and in particular since the mid-1900: "[t]he construct of nature as a legal entity invariably replaces accepted western concepts of property and also undermines the exercise of Indigenous self-determination and inherent governance" (Marshall 2020: 4). Indigenous Peoples' struggles (not exclusively in Australia) have been articulated, over the past few decades, against a precise set of colonial ontological assumptions, and thus, as a result, a complex set of strategies and adaptations have been deployed to engage with various colonial counterparts. The sudden proposed ontological shift entailed by the emergence of an ecological jurisprudence, Marshall aptly notes, imposes an additional burden on Indigenous advocates to suddenly adapt their strategies in order to pursue self-determination and a host of rights already enshrined (but yet to be realized) in the *United Nations Declaration on the Rights of Indigenous Peoples* within the new ontological parameters.

The risk of environmental colonialism, or, as Michael Cepek defines it, of 'environmentality' (Cepek 2011) – "the risk that environmentalist programs and movements operate as forms of governmentality in a Foucauldian sense" (Pelizzon 2014: 34) – has been present from the onset of the Rights of Nature movement, and Marshall's cautionary words are an apt reminder of its ongoing presence. However, contra Marshall's caution, James Morris and Jacinta Ruru had already hypothesized, in 2010, the use of legal personality as a strategic tool to empower Indigenous Peoples' relationship to water and waterways. Their argument was fully vindicated, seven years later, by the *Te Awa Tupua Act*, thus presenting a more nuanced and less dichotomic discourse surrounding the emergence of legal personhood in relation to natural features, involving complex exchanges between Indigenous and colonial participants, as mentioned in the previous section. As the example shows, it behooves the climate and environmental justice movement to be aware of the multifaceted and complex issues that surround the extension of personhood to the non-human world, as the risk of normative colonialism remains ever present.

Finally, the unclear and uncertain terrain occupied by the attribution of legal personhood to Nature and natural features is often confusingly cast against the backdrop of a historically situated, yet often ignored, evolution of the very concept of legal personhood, one that requires further clarification in order for the discourse to proceed further. What is, indeed, a 'person'? Ngaire Naffine asserts that

> [P]erhaps the greatest political act of law is the making of a legal person (simply put, he who can act in law) and, in the same move, the making of legal non-persons (those who cannot act in law and who are generally thought of as property).
>
> *(Naffine 2003: 347)*

This sharp divide between legal persons and non-persons is further complicated by the Lockean equation of the 'person' with the 'self',[5] as well as the Kantian definition of a person as "the subject whose actions are susceptible to imputation" (Kant 1959 trans.: 24), and it is for this reason that Edward Mussawir and Connall Parsley suggest that contemporary doctrine around legal personhood is "in thrall to a naturalized image of the human in whose service it curtails its own potential operations" (2017: 45). As a result, the disentanglement of the self and the person, as well as the person and its legal theoretical implications, is a necessary precondition for a complete appraisal of the current position occupied by the emergence of a discourse that attributes legal personhood to Nature.

The Birth of the Persona

A clearer understanding of the original idea of personhood is thus fundamental to correctly guide climate and environmental action. Roberto Esposito writes that in Roman law, where the concept is to be originally located, "no human being was a person by nature … since human beings arrived into life from the world of things, they could always be thrust back into it" (Esposito 2012a: 3). Rather than being an inherent attribute of an individual, Esposito shows, the concept of the person functions, in Foucauldian terms, as a *dispositif*, "a way of arranging the relation between the human and animal in contemporary subjectivity" (Esposito 2012b: 30). As a result, "the concept of *persona* … is necessary in law in order to separate the identity of a real living being from that of a purely artificial, fabricated role that is reserved and instituted at the level of juridical existence" (Mussawir & Parsley 2017: 47). The original concept of the person, therefore, is one whereby "the human being is not a person before the law because he [sic] is a human being, but because the law calls him or her 'person'" (Gaecker 2016: 295).

The origin of this concept is often articulated around the metaphor of the mask. Hannah Arendt identifies the Roman *dramatis persona* with the Greek's πρόσωπον, the mask ancient actors used to wear in a theatrical play. The mask, Arendt writes,

> …Obviously has two functions: it had to hide, or rather to replace, the actor's own face and countenance, but in a way that would make it possible for the voice to sound through. At any rate, it was in this twofold understanding of a mask through which a voice sounds that the word persona became a metaphor and was carried from the language of the theatre into legal terminology. The distinction between a private individual in Rome and a Roman citizen was

> that the latter had a persona, a legal personality, as we would say; it was as though the law had affixed to him the part he was expected to play on the public scene, with the provision, however, that his own voice would be able to sound through. The point was that 'it is not the natural Ego which enters a court of law. It is a right-and-duty-bearing person, created by the law, which appears before the law.' Without his persona, there would be an individual without rights and duties, perhaps a 'natural man' – that is, a human being or homo in the original meaning of the word, indicating someone outside the range of the law and the body politic of the citizens, as for instance a slave – but certainly a politically irrelevant being.
>
> *(Arendt 1963: 106–107)*

In this sense, as Jeanne Gaecker paraphrases, "legal personhood, while bestowed on the human person at birth in the Western legal systems as we know them, is itself a fiction precisely because it is a creation, i.e., a construction of law" (2016: 288). Without it, she adds, "we would 'merely' be individuals, or even 'unpersons' in the Orwellian sense, thought to have never existed in terms of the law and obviously without anything resembling a human right." Moreover, "Such individuals would be treated as objects, as was the case with slaves, women, and children in ancient Greece and Rome and even in less ancient times in Western civilization, and as still is the case with animals." The original meaning of personhood is thus inherently legal, and only subsequently, through Boetius's definition of persona as *rationalis naturae individua substantia* and later Thomas Aquinas's assertion that *vera persona est rei rationabilis individua substantia*, the concept was extended to encompass the underlying individual in his or her totality, as defined by Locke.

Furthermore, the codification of Roman law ordered by the Emperor Justinian in the 6th century CE, within the *Corpus Iuris Civilis*, provided the tripartite division of the books of private law in *personae*, *res*, and *actiones* (or persons, objects and actions).[6] Such a division, which we can paraphrase in modern terms between subjects and objects (of substantive law), and procedure, was incorporated by William Blackstone in his *Commentaries on the Laws of England*, and thus informs the totality of the Western legal tradition to the present day. In fact, in John Gray's words, "a 'person' is often used as meaning a human being, but the technical legal meaning of a 'person' is a subject of legal rights and duties" (Gray 1921: 27).

The extension of the legal category of the *persona* beyond the limits of the human began in earnest in the 11th century CE, coinciding with the institution of the *universitas*. Originally a corporate body akin to an ecclesiastical *collegium*, the *universitas*

> Could not be excommunicated or be found guilty of a civil or criminal offence and subsequently punished, on the view that the universitas did not have a corporal body and neither did it have a soul or a free will, being only a legal name.
>
> *(Gaecker 2016: 281)*

This changed when, in the late 11th century CE, Emperor Frederick Barbarossa awarded a special status to the schools of Bologna devoted to the study of law, and local authorities increasingly recognized the *universitas* as a separate entity, "an autonomous corporation with specific privileges for the students such as tax exemption" (Gaecker 2016: 292). The development of a concept of personhood applied to corporate entities

> Accelerated with the growth of commerce throughout Europe, as kings and emperors granted charters. From then on, corporations other than church institutions could also hold property, enter into contracts in their own name, govern themselves by boards, and, if necessary, sue and be sued.
>
> *(Gaecker 2016: 292)*

The emergence of corporate personality as an extended category of legal personhood gave rise to the current dichotomy that currently exhausts the possibilities of personhood accepted by the Western legal tradition: the *natural* person (i.e. any human) and the *artificial* person.[7] Such distinction, more clearly present within the Civil Law family than in the Common Law, was nonetheless aptly captured by Thomas Hobbes in the *Leviathan*

> A Person is he whose words or actions are considered, either as his own or as representing the words or actions of an other man, or of any other thing to whom they are attributed, whether Truly or by Fiction. When they are considered as his owne, then is he called a Naturall Person: And, when they are considered as representing the words and actions of an other, then is he a Feigned or Artificial person.
>
> *(Hobbes 1651: 83)*

This brief foray into the conceptual origin of the person has hopefully not distracted the reader from the original focus of this paper. Rather, it hopefully allows a return to the issue of legal personhood attributed to natural features with renewed clarity, thus leading to a better understanding of the third and current phase of a global (and often very loosely connected) Rights of Nature movement.

Towards an Environmental Person

The classical distinction between *natural* and *artificial* persons in law, both historically situated legal creations, still dominates the contemporary legal landscape. However, their exhaustiveness is currently in question, with challenges emerging from animal rights, environmentalism, artificial intelligence, and a renewed interest in corporate personhood. It is thus within this yet uncertain *milieu* that the legal personhood for Nature discourse is best located, and it is at this intersection that its relevance for climate and environmental justice becomes apparent. The limits of a traditional conception of legal personhood, when vested upon Nature and natural entities, are here made bare.

Firstly, Nature as a subject of rights does not allow for an easy invocation of such rights. Assuming, with Kelsen, that the distinction between 'natural' and 'unnatural' is inherently normative and cannot be evinced by empirical observation alone, Nature then becomes co-terminus with the whole cosmos. As a result, the identification of specific natural features as the subjects of the rights recognized to Nature by the many legal, judicial, and constitutional initiatives presented at the beginning of this paper constitutes an inevitable necessity. It follows that the pursuit of rights of (or for) Nature and the identification of the subjects of those rights are inextricably intertwined and interdependent. Furthermore, as some authors have noted, the use of apparently distinct terms – such as *subjects* and variously defined *persons* (juristic, legal, et cetera) – reveals a false distinction. Instead, these are all permutation of the same element of the Justinian tripartition discussed in the previous section. An example of such false distinction is the English translation of the Colombian definition of the Atrato as an *entidad sujeto de derechos* (literally an 'entity subject of rights') as potentially failing to fully articulate the river as a legal person. To the mind of the author of the judgment, however, the distinction did not exist, and was merely a result of a semantic difference within translation. Rights and legal persons are, therefore, two elements of a single and inseparable legal continuum.

Yet, the attribution of personhood to natural features highlights the uncomfortable space that Nature inhabits in relation to the classical Western dichotomy between *natural* and *artificial* legal persons. Natural features are not *natural* persons in a traditional legal sense, since such space is exclusively inhabited by (mostly living) humans. At the same time, natural features are not, nor could they ever be, purely *artificial* creations, since they do not exist in a purely abstract state. Rather, they exist in a physical, embodied form, while at the same time transcending the limits of what a *natural* person is often constituted of in the Western legal and philosophical tradition alike, as defined by the Lockean collation of personhood and 'self'. As a result, the two existing categories of legal personhood are inherently incapable of capturing the *persona* imagined by the many advocates of an emerging ecological jurisprudence. Admittedly, the broader personhood discourse currently attempts to extend beyond the natural/artificial divide in a number of distinct instances: animals, AI and robots, for example, all constitute potential legal subjects uncomfortably aligned with the traditional divide. In this sense, the extension of personhood to natural features is part of this larger trend. However, natural features are intrinsically unique in that they exist irrespective of human intervention (unlike robots and, at present, AI constructs) while at the same time extending their existence beyond the individual, toward the collective and the relational (unlike individual animals).

In noting the constraints of the existing categories of personhood when engaging with natural elements, Mari Margil proposed, in 2018, to constitute an entirely novel category of personhood, one that she provisionally and experimentally termed 'nature personhood.' The invitation echoes Christina Woods's suggestion of a *sui generis* trust construct in order to establish a global approach to nature (2014). While Margil's invitation fully addresses many of the problems raised by the discussions encountered thus far, the term 'nature' to refer to a novel category of personhood is certainly too linguistically proximate with the term 'natural' (person) traditionally attributed to humans alone. As a result, a tentative and alternative term proposed to capture the shift toward a novel category of personhood for natural elements could be that of an 'environmental person.'

The desire to transcend the traditional Western dichotomy, to gesture toward a novel category of 'environmental personhood' is evinced in the language of many of the cases considered throughout this paper. What these instances reveal is a yet unspecified theoretical space, and it is this space that the current phase of the Rights of Nature movement seems to be inhabiting. Far from being predetermined, what this novel 'environmental person' is constituted of is yet to be fully defined, and its emerging definition opens a 'pluriversal' dialogical space (Escobar 2020), one in which Marshall's cautionary words can be hopefully heeded. Jon Garthoff, for example, suggests that legal personhood be decomposed into at least two elements – standing and liability – and that legal systems reject the principle that an entity possesses one just in case it possesses the other (2019: 967). Sam Campbell and Laura Gurney argue that the Ganges judgment "presents three overlapping ontologies of the Ganges – the river as socio-economic resource, ecosystem, and spiritual being – and invokes them simultaneously as grounds for granting legal personality" (Campbell & Gurney 2020: 1). Indeed, the court writes that "[a]ll the Hindus have deep Astha in rivers Ganga and Yamuna and they collectively connect with these rivers," (*Ganges and Yamuna case*: 17) and, even more strikingly, "[r]ivers Ganga and Yamuna are breathing, living and sustaining the communities from mountains to sea." (*Ganges and Yamuna case*: 17). Similarly, in the more recent case of the Turag River in Bangladesh, the High Court of Bangladesh offered an innovative interpretation of a number of constitutional provisions to "express a further opinion on the interdependence of humanity and nature within Bangladesh" (Islam & O'Donnell 2020: 165). And after all, the idea that a natural entity, feature, or system was to be inherently more than a mere 'legal person' was already contained in the words of the first judgment that contemplated this possibility:

> So it should be as respects valleys, alpine meadows, rivers, lakes, estuaries, beaches, ridges, groves of trees, swampland, or even air that feels the destructive pressures of modern technology and modern life. The river, for example, is the living symbol of all the life it sustains or nourishes - fish, aquatic insects, water ouzels, otter, fisher, deer, elk, bear, and all other

animals, including man, who are dependent on it or who enjoy it for its sight, its sound, or its life. The river as plaintiff speaks for the ecological unit of life that is part of it. Those people who have a meaningful relation to that body of water - whether it be a fisherman, a canoeist, a zoologist, or a logger - must be able to speak for the values which the river represents and which are threatened with destruction.

(Sierra Club v Morton: 742)

The sense of respect entailed by the words of all these judgments reveal a clear desire to transcend the anthropocentric confines within which the conceptualization of Nature has been contained, within Western legal thought, over the past few centuries. In this sense, the emergence of an 'environmental person' represents the current frontier of the movement toward an ecological jurisprudence, one where an ethics of care is extended well beyond the boundaries of human concerns, one where the category of the 'other' is removed from the world of the merely inanimate and is rather cast as an interactive participant in a normative dialogue with the cosmos.

At the same time, the emergence of an 'environmental person' is replete with possibilities, to engage multiple legal orders in the collective definition of what such a novel category of personhood will ultimately look like. While transcending the limits of anthropocentrism, the novel category of 'environmental personhood' still needs to avoid the risk of being atomistic rather than relational, and, in so doing, to continue to perpetrate a form of ontological (if not necessarily legal) colonialism. With such normative humility in mind, however, the establishment of a novel category of 'environmental personhood' constitutes both a more advanced strategic tool within the toolset of the growing Rights of Nature movement and a deep epistemic bridge between distinct legal orders, one that can only profoundly enrich climate and environmental action and ethics alike.

Notes

1 I refer here to what most comparative lawyers term the Western legal tradition, comprised of the civil law, common law, and now post-soviet law family.
2 Although often described as an eco-theologian, Berry believed the term to be too reductive and preferred to call himself a 'geologian' or Earth scholar.
3 Since the first ordinance recognizing Nature as a subject of rights, drafted by the small community of Tamaquah Borough, in Schuylkill County, Pennsylvania in 2006 with the assistance of the Community Environmental Legal Defense Fund (CELDF). CELDF has assisted more than 30 communities to follow suit, the most prominent of which is likely the one passed unanimously by the City Council of Pittsburgh, Pennsylvania.
4 In the original, 'La naturaleza o Pacha Mama, donde se reproduce y realiza la vida, tiene derecho a que se respete integralmente su existencia y el mantenimiento y regeneración de sus ciclos vitales, estructura, funciones y procesos evolutivos. Toda persona, comunidad, pueblo o nacionalidad podrá exigir a la autoridad pública el cumplimiento de los derechos de la naturaleza.'
5 In this context, for Locke, a person is 'a thinking intelligent being, that has reason and reflection, and can consider itself as itself, the same thinking thing in different times and places'.
6 The division is often attributed to the Roman jurist Gaius, in 2nd century CE.
7 Though it is important to note that artificial does not mean fictitious. An artificial person is not imaginary, but certainly real.

References

Arendt, H. (1976, originally published 1963) *On Revolution*, New York: Penguin.
Berry, T. (1999) *The Great Work. Our Way Into the Future*, New York: Three Rivers Press.
Campbell, S., and Gurney, L. (2020) "Mapping and navigating ontologies in water governance: The case of the Ganges" *Water International* 45(7–8), 847–864.
Cepek, M. L. (2011) "Foucault in the forest: Questioning environmentality in Amazonia' *American Ethologist* 38(3), 501.

Clark, C., Emmanouil, O., Page, J., and Pelizzon, A. (2019) "Can you hear the rivers sing? Legal personhood, ontology and the nitty gritty of governance" *Ecology Law Quarterly,* 45, 787–823.

Constitución de la Republica del Ecuador 2008

Constitutición del Estado Plurinacional de Bolivia 2008

Corte Constitucional, Decision T-622 10/11/2016 (Colombia) (*Atrato case*)

Cullinan, C. (2002) *Wild Law: A Manifesto for Earth Justice,* Capetown: Siber Ink.

Cullinan, C. (2011) "A history of wild law," in Burdon, P. (ed.), *Exploring Wild Law: The Philosophy of Earth Jurisprudence* (pp. 12–23), Adelaide: Wakefield Press.

Descola, P. (2013) *Beyond Nature and Culture,* Chicago: The University of Chicago Press.

Escobar, E. (2020) *Pluriversal Politics. The Real and the Possible,* Durham: Duke University Press.

Esposito, R. (2012a) "The *Dispositif* of the Person", *Law, Culture and the Humanities,* 8(1), 17–30.

Esposito, R. (2012b) *The Third Person,* Cambridge: Polity.

Gaecker, J. (2016) "'Sua cuique persona?' A Note on the Fiction of Legal Personhood and a Reflection on Interdisciplinary Consequences" *Law & Literature,* 28(3), 287–317.

Garthoff, J. (2019) "Decomposing Legal Personhood", *Journal of Business Ethics,* 154, 967–974.

Global Alliance for the Rights of Nature <https://www.rightsofnaturetribunal.com>.

Global Alliance for the Rights of Nature. (2020) *Press Release Los Cedros Case – Amicus Curiae to Ecuador's Highest Court* <https://therightsofnature.org/press-release-los-cedros-case-amicus-curiae-to-ecuadors-highest-court/>.

Gray, J. C. (1921) *The Nature and Sources of the Law,* New York: Macmillan.

Hobbes, T. (1987, originally published 1651) *Leviathan*, London: Penguin Classics.

Human Rights and Peace for Bangladesh v Government of Bangladesh and others [2016] HCD (Writ Petition No. 13989 of 2016, judgment declared on February 3, 2019, translation from Bangla by MS Islam).

Kant, I. (1959, originally published 1785) *Foundations of the Metaphysics of Morals*, Indianapolis: Bobbs-Merrill.

Kauffman, C. M., and Martin, P. M. (2017) "Can Rights of Nature Make Development More Sustainable? Why Some Ecuadorian lawsuits Succeed and Others Fail" 9*World Development*, 92, 130–142.

Lalit Miglani v State of Uttarakhand & others (March 30, 2017) MCC 139/2017 (India) (*Glaciers case*).

Ley De Derechos De La Madre Tierra, Ley N 71, 21 de Diciembre de 2010

Ley marco De La Madre Tierra y Desarrollo Integral Para Vivir Bien, Ley N 300, 15 de Octubre de 2012.

Margil, M. et al. (2017) "Now rivers have the same legal status as people, we must uphold their rights" *The Guardian*, <https://www.theguardian.com/global-development-professionals-network/2017/apr/21/rivers-legal-human-rights-ganges-whanganui>.

Marshall, V. (2020) "Removing the veil from the 'rights of nature': The Dichotomy between first nations customary rights and environmental legal personhood", *Australian Feminist Law Journal,* 45(2), 233–248.

Morris, J. D. K., and Ruru, J. (2010) "Giving voice to rivers: Legal personality as a vehicle for recognising indigenous peoples' relationship to water?" *Australian Indigenous Law Review,* 14(2), 49–62.

Mussawir, E., and Parley, C. (2017) "The law of persons today: At the margins of jurisprudence", *Law and Humanities,* 11, 44–46.

Mylius, B. (2018) "Earth jurisprudence," *Encyclopedia of the Anthropocene,* 4, 271–276.

Naffine, N. (2003) "Who are law's persons? From Cheshire cats to responsible subjects", *Modern Law Review,* 66(3), 346–367.

O'Donnell, E. (2018) "At the intersection of the sacred and the legal: Rights for nature in Uttarakhand, India", *Journal of Environmental Law,* 30(1), 135–144.

O'Donnell, E., Poelina, A., Pelizzon, A., and Clark, C. (2020) "Stop burying the lede: The essential role of indigenous law(s) in creating rights of nature", *Transnational Environmental Law,* (online) 1–14.

Pelizzon, A. (2014) "Earth laws, rights of nature and legal pluralism", in Maloney, M., and Burdon, P. (eds.), *Wild Law – In Practice,* Sydney: Routledge.

Schillmoller, A., and Pelizzon, A. (2013) "Mapping the terrain of earth jurisprudence: Landscape, thresholds and horizons", *Barry Environmental and Earth Law Journal,* 3(1), 3–32.

Sierra Club v Morton 405 US 727 (1972).

Sohidul Islam, M. S., and O'Donnell, E. (2020) "Legal rights for the Turag: Rivers as living entities in Bangladesh", *Asia Pacific Journal of Environmental Law,* 23(2), 160–177.

Soper, K. (1995) *What Is Nature? Culture, Politics and the Non-Human*, New Jersey: John Wiley and Sons.

Stone, C. D. (1972) "Should trees have standing? Towards legal rights for natural objects", *Southern California Law Review*, 45, 450.

Stone, C. D. (2010) *Should Trees Have Standing? Law, Morality and the Environment,* Oxford: Oxford University Press.

Te Awa Tupua (Whanganui Claims Settlement) Act 2017 (N.Z.) s14(1).

Te Urewera Act 2014, New Zealand.

United Nations Harmony with Nature <http://www.harmonywithnatureun.org>.

Wheeler et al. v. Director de la Procuraduria General del Estado en Loja, Corte Provincial de Justicia de Loja, March 31, 2011, Judgment 11121-2011-0010 (*Vilcabamba case*).

Yarra River Protection (Wilip-gin Birrarung murron) Act 2017 (Vic) (Australia).

Zambrana-Flores, I. (2020) "Mother Earth approach in the implementation of Education and Climate Change", *UN Harmony with Nature Program*, <http://files.harmonywithnatureun.org/uploads/upload808.pdf>. Retrieved March 20, 2020

INDEX

Note: **Bold** page numbers refer to tables; *Italic* page numbers refer to figures and page numbers followed by "n" denote endnotes.

Taylor & Francis Group
an informa business

Taylor & Francis eBooks

www.taylorfrancis.com

A single destination for eBooks from Taylor & Francis with increased functionality and an improved user experience to meet the needs of our customers.

90,000+ eBooks of award-winning academic content in Humanities, Social Science, Science, Technology, Engineering, and Medical written by a global network of editors and authors.

TAYLOR & FRANCIS EBOOKS OFFERS:

- A streamlined experience for our library customers
- A single point of discovery for all of our eBook content
- Improved search and discovery of content at both book and chapter level

REQUEST A FREE TRIAL

support@taylorfrancis.com

Routledge
Taylor & Francis Group

CRC Press
Taylor & Francis Group